Plates and Shells for Smart Structures

Plates and Shells for Smart Structures

Classical and Advanced Theories for Modeling and Analysis

Erasmo Carrera, Salvatore Brischetto and Pietro Nali

Politecnico di Torino, Italy

WILEY

A John Wiley & Sons, Ltd., Publication

This edition first published 2011

Registered office
John Wiley & Sons Ltd, The Atrium, Southern Gate, Chichester, West Sussex, PO19 8SQ, United Kingdom

For details of our global editorial offices, for customer services and for information about how to apply for permission to reuse the copyright material in this book please see our website at www.wiley.com.

Library of Congress Cataloguing-in-Publication Data

Carrera, Erasmo.
Plates and shells for smart structures : classical and advanced theories for modeling and analysis / Erasmo Carrera, Salvatore Brischetto, Pietro Nali. – 1st ed.
p. cm.
Includes bibliographical references and index.
ISBN 978-0-470-97120-8 (hardback)
1. Shells (Engineering) 2. Plates (Engineering) 3. Smart structures.
I. Brischetto, Salvatore. II. Nali, Pietro. III. Title.
TA660.S5C276 2011
624.1′776–dc23

2011019535

A catalogue record for this book is available from the British Library.

Print ISBN: 9780470971208
ePDF ISBN: 9781119950011
oBook ISBN: 9781119950004
ePub ISBN: 9781119951124
Mobi ISBN: 9781119951131

FSC
www.fsc.org
MIX
Paper from responsible sources
FSC® C013604

Contents

About the Authors

Erasmo Carrera

After earning two degrees (Aeronautics, 1986, and Aerospace Engineering, 1988) at the Politecnico di Torino, Erasmo Carrera received his PhD in Aerospace Engineering in 1991, jointly at the Politecnico di Milano, Politecnico di Torino, and Università di Pisa. He began working as a Researcher in the Department of Aeronautics and Space Engineering at the Politecnico di Torino in 1992 where he held courses on Missiles and Aerospace Structures Design, Plates and Shells, and the Finite Element Method, and where he has been Professor of Aerospace Structures and Aeroelasticity since 2000. He has visited the Institut für Statik und Dynamik, Universität Stuttgart twice, the first time as a PhD student (6 months in 1991) and then as Visiting Scientist under a GKKS Grant (18 months from 1995). In the summer of 1996 he was Visiting Professor at the ESM Department of Virginia Tech. He was also Visiting Professor for two months at SUPMECA, Paris, in 2004, and at CRP H. Tudor, G.D. Luxembourg, in 2009. His main research topics are: composite materials, FEM, plates and shells, postbuckling and stability, smart structures, thermal stress, aeroelasticity, multibody dynamics, non-classical lifting systems and multifield problems. Professor Carrera has made significant contributions to these topics. In particular, he proposed the Carrera Unified Formulation to develop hierarchical beam/plate/shell theories and finite elements for multilayered structure analysis as well as the generalization of classical and advanced variational methods for multifield problems. He has been responsible for various research contracts with the EU and national and international agencies/industries. Presently, he is Full Professor and Deputy Director of his department. He is the author of more than 300 articles, many of which have been published in international journals. He serves as a referee for many journals, and as contributing editor for *Mechanics of Advanced Materials and Structures, Composite Structures and Journal of Thermal Stresses*. He has also served on the Editorial Boards of many international conferences.

Salvatore Brischetto

After earning his degree in Aerospace Engineering at the Politecnico di Torino in 2005, Salvatore Brischetto received his PhD in Aerospace Engineering (Politecnico di Torino) and in Mechanics (Université Paris Ouest–Nanterre La Défense) in 2009. He won the excellence prize for PhD students at the Politecnico di Torino in 2008. Dr Brischetto worked as a Research Assistant in the Department of Aeronautics and Space Engineering at the Politecnico di Torino from 2006 to 2010, and has been Assistant Professor in the same department since 2010. His main research topics are: smart structures, composite materials, multifield problems, functionally graded materials, thermal stress analysis, carbon nanotubes, inflatable structures, and plate and shell finite elements. He is the author of more than 50 articles on these topics, more than half of which have been published in international journals. He serves as a reviewer for some international journals, such as *Composite Structures*, *Journal of Mechanics of Materials and Structures*, *Applied Mathematical Modeling*, *Journal of Applied Mechanics*, *Journal of Composite Materials*, etc. He has also been Guest Editor for *Mechanics of Advanced Materials and Structures* for the Special Issues entitled "Modeling and analysis of functionally graded beams, plates and shells, Parts I and II." He has been Teaching Assistant at the Politecnico di Torino for courses on computational aeroelasticity, structures for aerospace vehicles and nonlinear analysis of aerospace structures since 2007.

Pietro Nali

After earning his degree in Aerospace Engineering in 2005 at the Politecnico di Torino, Pietro Nali held a traineeship at the European Space Agency/ESTEC, Structures and Thermal Division, from August 2005 to February 2006. He was the candidate from the Politecnico di Torino for an ESA NPI (Networking/Partnering Initiative) position in 2006. He received his PhD in Aerospace Engineering (Politecnico di Torino) and in Mechanics (Université Paris Ouest–Nanterre La Défense) within the framework of the NPI, in 2010, for the topic "Modeling and validation of multilayered structures for spacecraft, including multifield interactions." Since 2010, Dr Nali has worked as a Research Assistant in the Department of Aeronautics and Space Engineering at the Politecnico di Torino. His main research topics are: finite elements, multilayered plate modeling, smart structures, composite materials, failure criteria, multifield problems, thermal stress analysis, and structure nonlinearities.

Preface

Smart structures involve interactions between mechanical and electric fields. Classical models for beams, plates, and shells were originally developed to compute stress fields due to the application of mechanical loadings. These classical models have demonstrated certain difficulties and limitations in the analysis of smart structures. Electrical loadings are in fact "field loadings" which require the use of advanced structural models. Smart structures, in most applications, are layered structures with piezoelectric patches/layers. Layered structures have, by definition, several "interfaces." Interfaces lead to discontinuous distributions along the thickness of both the electrical and mechanical properties. This book presents a detailed analysis of classical and advanced structural models that are able to deal with mechanical and electric field loadings. Assumptions are made on displacements, transverse stresses, electric potential, and transverse electric displacements. Extensions of the principle of virtual displacements (PVD) and of the Reissner mixed variational theorem (RMVT) are used to derive governing equations and finite element matrices of laminated plate/shell structures embedding piezoelectric layers. Assumptions on the unknown variables are introduced through the application of the Carrera Unified Formulation, where the accuracy of the models can be enriched by preserving the form of governing equations and finite element matrices, which are written in terms of a few fundamental nuclei. A large variety of plate/shell models are built and compared. Classical theories, based on Kirchhoff–Love and Reissner–Mindlin assumptions, are obtained as particular cases. The classical and advanced structural models discussed in this book have been coded using the academic in-house software MUL2 (MULtifield problems for MULtilayered structures). MUL2 has been used in most of the quoted numerical calculations. An updated version of these codes is available to buyers of this book at http://www.mul2.com.

www.wiley.com/go/carrera

1

Introduction

In many national and international declarations, it has been stated that developments in advanced structures, in the automotive and shipbuilding industries, as well as in aeronautical and space sciences, are subordinate to the development of so-called *smart structures*.

The definition of smart structures has been extensively discussed since the late 1970s. A workshop was organized by the US Army Research Office in 1988 in order to propose a definition of smart systems/structures to be adopted by the scientific community (Ahmad 1988):

> A system or material which has built-in or intrinsic sensor(s), actuator(s) and control mechanism(s) whereby it is capable of sensing a stimulus, responding to it in a predeterminated manner and extent, in a short/appropriate time, and reverting to its original state as soon as the stimulus is removed.

According to design practices, smart structures are systems that are capable of sensing and reacting to their environment, through the integration of various elements, such as sensors and actuators. Smart structures can allow their shape to be varied to very high precision and without using classical mechanical actuators, alleviate vibrations and acoustic noise, and even monitor their own structural health.

Piezoelectric, piezomagnetic, electrostrictive, and magnetostrictive materials are of interest when designing smart structures. Shape memory alloys,

Plates and Shells for Smart Structures: Classical and Advanced Theories for Modeling and Analysis, First Edition.
Erasmo Carrera, Salvatore Brischetto and Pietro Nali.

electrorheological fluids, and fiber optics should also be mentioned. This book deals with smart structures, taking advantage of piezoelectric effects.

Nowadays, it is difficult to foresee whether smart structures will be employed to any great extent in the future. However, interest in a better understanding of the topic appears essential and could lead to many other uses related to other extensive domains of application.

1.1 Direct and inverse piezoelectric effects

Piezoelectricity was discovered by Jacques and Pierre Curie in 1880, when they realized that several kinds of crystals were able to generate positive or negative electric charges when subjected to mechanical pressure (Curie and Curie 1880, 1881). When dealing with piezoelectric materials, a charge is generated when molecular electrical dipoles are caused by a mechanical loading: that is, the *direct effect (sensor configuration)*. Conversely, when an electric charge is applied, a slight change occurs in the shape of the structure: that is, the *inverse effect (actuator configuration)*. It has been demonstrated that piezoelectric materials can be used at the same time as actuators and sensors, obtaining the so-called *self-sensing piezoelectric actuator* (Dosh *et al.* 1992).

Piezoelectricity is a feature of some natural crystals (such as quartz and tourmaline) or synthetic crystals (lithium sulfate), and several kinds of polymers and polarized ceramics. The most common piezoelectric materials are the piezoceramic *barium titanate* ($BaTiO_3$) and *piezo lead zirconate titanate* (PZT). The crystal lattice of piezoelectric materials is of the *face-centered cubic* (FCC) kind. Metallic atoms are located at the vertex of the cube, while oxygen atoms remain at the center of the cube's faces. A heavier atom is located at the center of the cube and it can shift slightly to positions with less energy, with a consequent deformation of the crystal lattice (metastable structure). If an electric field is applied to the structure, the central atom can exceed the potential energy threshold and move to a lower energy configuration. This is followed by a rupture of symmetry and the creation of an electric dipole (Figure 1.1). The previous phenomenon is possible only below the so-called *Curie temperature*. Above this temperature, the piezoelectric effect disappears due to high thermal agitation. Polarized piezoceramics are obtained by heating them above their Curie temperature and subjecting the material to an intense electric field during thermal cooling. In so doing, all the dipoles become oriented in the same direction and the material obtains a stable polarization. Moreover, apart from a residual polarization, the crystal lattice of the polarized piezoceramic will also undergo a residual deformation. After the polarization process, a very small electric potential will be sufficient to obtain a temporary deformation and vice versa.

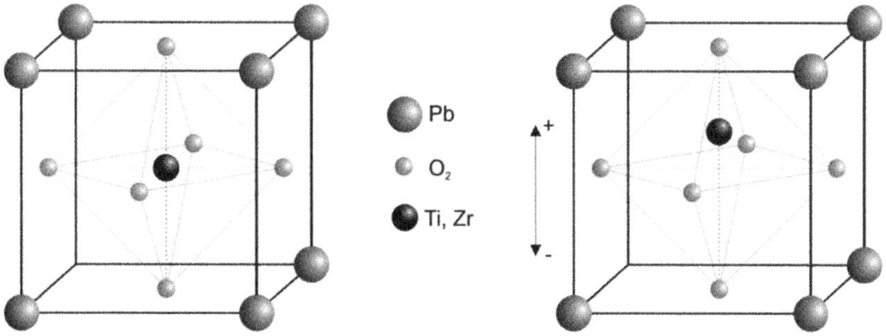

Figure 1.1 Piezoceramic cell before (left) and after (right) polarization.

Even if the electro mechanical coupling is a nonlinear phenomenon, piezoelectric problems are usually studied through linear analysis. This leads to the adoption of assumptions, which will be discussed in Chapter 2. Additional details on this topic can be found in the works by Cady (1964), Tiersten (1969), and Ikeda (1996).

1.2 Some known applications of smart structures

Smart structures have been used in sensing, actuating, diagnosing, and assessing the health of structures, depending on the external stimuli. Sensors and actuators should be integrated into the complete structures and this leads to unusual design solutions, compared to traditional structural design solutions (Srinivasan and McFarland 2001). In the most advanced design concepts, smart structures could have the ability to save and analyze information in order to perform a learning process.

Nowadays, smart structures are applied in many different domains, but they all share the common feature of having a highly cross-disciplinary design. Among other applications, the following current/potential ones can be mentioned.

Structural health monitoring The strain field of some critical locations of a generic structural system can be measured using embedded sensors in order to identify possible damage and retain structural safety and reliability. Damage is intended here as a variation of the material and/or geometric properties, which could affect the performances of the systems. *Self-diagnostic* ability plays a crucial role in the aeronautical and space industry, where sensing the strain field of some relevant structural subcomponents helps in the conduction of an appropriate maintenance program and in avoiding crack propagation. This topic appears of particular interest for composite materials, whose failure prediction is a challenging task. Composites are being progressively employed more and

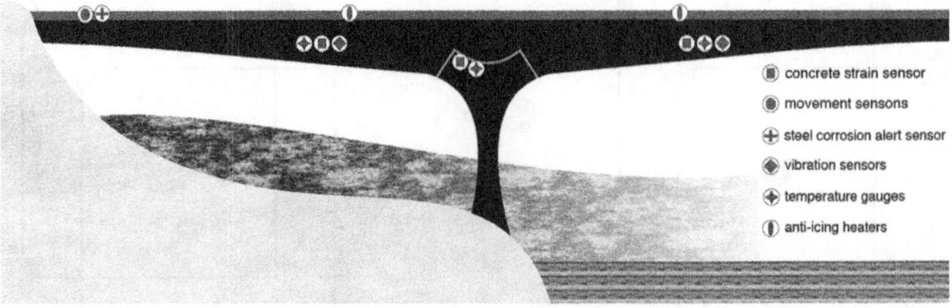

Figure 1.2 Smart system scheme of the Saint Anthony Falls Bridge for structural health monitoring.

more in aerospace engineering in order to replace metallic structures. As a consequence, structural health monitoring will become a very important task in the near future. In principle, crack propagation could be restrained by producing compressive stresses around the failure through a proper network of embedded actuators (Rogers, 1990). Rogers (1990) also mentioned the possibility of using skin-like tactile piezoelectric sensors to sense temperatures and pressures. Structural health monitoring is also applied extensively in civil engineering. The most well-known examples refer to the remote monitoring of bridge deflections, mode shapes, and the corresponding frequencies (Deix *et al.* 2009; Spuler *et al.* 2009). The scheme of the Saint Anthony Falls Bridge in Figure 1.2 represents an example of a smart system with embedded devices that offers optimal diagnostics (Foster 2009). Monitoring is usually performed by analyzing the dynamic response of a system through an array of properly located sensors. Periodic observations and comparisons to previous measurements and numerical simulations can indicate some local damage or structural/material degradation resulting from the operational environment.

Vibration control Due to their high strain sensitivity (Sirohi and Chopra 2000), piezoelectric sensors and actuators are easily employed for vibration damping/attenuation/suppression (Inman *et al.* 2001). Piezoceramics are used to reduce noise and improve the comfort of vehicles, such as cars, trucks, and helicopters, and to improve the performances of machine tools. The same technique is often employed in spacecraft carrying equipment in a pure operational dynamic environment. Active vibration control is usually applied in engineering practice in order to suppress dangerous vibrations over a certain range of frequencies, as in the case of helicopter blades (Chopra 2000). Piezoelectric materials are also effective in passive damping: a part of the mechanical energy introduced into the structural system is converted into electrical energy, according to the piezoelectric effect. Piezoelectric passive damping devices are commonly embedded in high-performance sports devices, such as tennis rackets, baseball bats, and skis (Gaudenzi 2009).

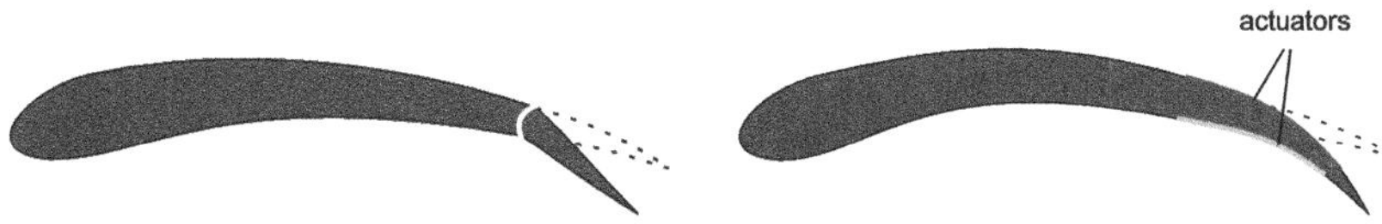

Figure 1.3 Wings with conventional flaps (left) and with smart flexible flaps (right).

Shape morphing Among the possible shape morphing industrial applications of structural components, focusing on the aeronautics field, it is worth mentioning the advantages of a wing with variable shape. Commercial aircraft have to respect increasing efficiency requirements and reduce emissions. One possible solution is to propose a variable shape wing that is able to optimize performances in all phases of the mission. The means that can be employed to vary the shape of the wing are quite challenging and can vary in complexity, depending on which properties have to be modified: sweep angle, profile, aspect ratio, etc. Swept wings (as a solution to reduce wave drag) were first used on jet fighter aircraft. Variable shape wings, in a broad sense, could play a significant role in future aircraft designs. The elementary wing shape changes for take-off/cruise/landing are currently obtained by means of rigid body motions of movable parts, e.g., flaps, slats, ailerons, and spoilers. It is understood that a smart flexible wing, without secondary parts, that would be able to perform proper shape changes, would lead to a remarkable reduction in drag, weight, and overall system complexity; see Figure 1.3 for an example of a hingeless flap that can be obtained from shape morphing.

Active optics Active optics, which are usually employed in large reflector telescopes and can be considered as a particular case of shape morphing, allow the shape of mirrors to be monitored and readjusted during operation. In this way, it is possible to avoid effects due to gravity or wind (in the case of an Earth-based telescope) or deformations due to thermo mechanical coupling or structural imperfections (in the case of space telescopes). The use of accurate actuators, together with an algorithm that is able to quantify the quality of images, allow a precision to be obtained that goes well beyond the possibilities of conventional reflector telescopes. Active optics are currently employed in 10 m class telescopes and are also going to be applied in the next generation of 40 m telescopes (Preumont *et al.* 2009).

Microelectromechanical systems (MEMS) MEMS consist of extremely small mechanical devices driven by electricity. A device's dimensions vary from 20 μm to 1 mm. MEMS devices can be used as multiple microsensors and microactuators (Varadan and Varadan 2002). MEMS are particularly promising in the medical field, where they can be employed as blood sugar sensors, insulin delivery pumps, micromotor capsules that unclog arteries, or filters that expand after insertion into a blood vessel in order to trap blood clots (Srinivasan and McFarland 2001).

Many potential benefits can be obtained due to the extensive use of smart structures in industrial applications. Reducing maintenance costs, in the case of self-diagnostic structural health monitoring, should be mentioned. In fact, maintenance time is a crucial point for airlines, which, according to the low-cost business philosophy, has greatly reduced profit margins. Another benefit consists of the possibility of producing new components, according to new design concepts, like shape morphing and the integration of MEMS in structures. It should also be emphasized that MEMS are currently enlarging medical perspectives, and opening up new scenarios for the future of health care programs.

Aim of this book This book aims to illustrate the classical techniques and some advanced models that are able to describe mechanical and electrical variables in plate/shell structures that have piezoelectric layers embedded in the lamination stacking sequence. Two-dimensional axiomatic models are considered through analytical and finite element approaches. Classical models (e.g., Kirchhoff, Mindlin, and equivalent single-layer kinematic descriptions) are compared to advanced theories (mixed, layer wise, and higher order descriptions) through several numerical examples. Most of the presented theories are derived on the basis of the Carrera Unified Formulation, which probably is one of the most modern and advanced tools for dealing with the theory of structures.

References

Ahmad I 1988 Smart structures and materials. In *Proceedings of US Army Research Office Workshop on Smart Materials, Structures and Mathemetacal Issues.*

Cady WG 1964 *Piezoelectricity*. Dover.

Chopra I 2000 Status of application of smart structures technology to rotorcraft systems. *J. Am. Helicopter Soc.* **45**, 228–252.

Curie J and Curie P 1880 Développement par compression de l'électricitè polaire dans les cristaux hémièdres a faces inclinées. *C. R. Acad. Sci. Paris* **91**, 294–295.

Curie J and Curie P 1881 Contractions et dilatations produites par des tensions électriques dans des cristaux hémièdres a faces inclinées. *C. R. Acad. Sci. Paris* **93**, 1137–1140.

Deix S, Ralbovsky M, Stuetz R, and Wittmann SM 2009 Structural health monitoring using wireless sensor networks. In *Proceedings of the IV ECCOMAS Thematic Conference on Smart Structures and Materials.*

Dosh JJ, Inman DJ, and Garcia E 1992 Self-sensing piezoelectric actuator for collocated control. *J. Intell. Mater. Syst. Struct.* **3**, 166–185.

Foster D 2009 The bridge to smart technology. In *Bloomberg Businessweek.*

Gaudenzi P 2009 *Smart Structures: Physical Behaviour, Mathematical Modelling and Applications*. John Wiley & Sons, Ltd, UK.

Ikeda T 1996 *Fundamentals of Piezoelectricity*. Oxford University Press.

Inman DJ, Ahmadihan, M and Claus RO 2001 Simultaneous active damping and health monitoring of aircraft panels. *J. Intell. Mater. Syst. Struct.* **12**, 775–783.

Preumont A, Bastaits R, and Rodrigues G 2009 Active optics for large segmented mirrors: scale effects. In *Proceedings of the IV ECCOMAS Thematic Conference on Smart Structures and Materials*.

Rogers CA 1990 Intelligent material systems and structures. In *Proceedings of US Japan Workshop on Smart/Intelligent Materials and Systems*.

Sirohi J and Chopra I 2000 Fundamental understanding of piezoelectric strain sensors. *J. Intell. Mater. Syst. Struct.* **11**, 246–257.

Spuler T, Moor G, and Berger R 2009 Modern remote structural health monitoring: an overview of available systems today. In *Proceedings of the IV ECCOMAS Thematic Conference on Smart Structures and Materials*.

Srinivasan AV and McFarland MD 2001 *Smart Structures: Analysis and Design*. Cambridge University Press.

Tiersten HF 1969 *Linear Piezoelectric Plate Vibrations*. Plenum Press.

Varadan VK and Varadan VV 2002 Microsensors, microelectromechanical systems (MEMS) and electronics for smart structures and systems. *Smart Mater. Struct.* **9**, 953–972.

2

Basics of piezoelectricity and related principles

The phenomenon of piezoelectricity is described by referring to the most common piezoelectric materials. Fundamental piezoelectricity equations are discussed, the meaning of the coupling coefficients is dealt with in detail, and some available data concerning electromechanical properties are given. The physical and variational principles of piezoelectricity are introduced. First, the principle of virtual displacements is extended to the electromechanical case by simply adding the internal electrical work. Then, three extensions of the Reissner mixed variational theorem, which permits one to consider a priori some transverse mechanical and electrical variables, are briefly discussed. A clear definition of the field variables is given. The constitutive equations of piezoelectricity are explained in detail for the different variational statements that are proposed.

2.1 Piezoelectric materials

The phenomenon of piezoelectricity is a particular feature of certain classes of crystalline materials. The piezoelectric effect is due to a linear energy conversion between the mechanical and electric fields. The linear conversion between the two fields is in both directions, and it thus defines a direct or converse piezoelectric effect. The *direct piezoelectric effect* generates an electric polarization

Plates and Shells for Smart Structures: Classical and Advanced Theories for Modeling and Analysis, First Edition.
Erasmo Carrera, Salvatore Brischetto and Pietro Nali.

by applying mechanical stresses. The *converse piezoelectric effect* instead induces mechanical stresses or strains by applying an electric field. These two effects represent the coupling between the mechanical and electric fields. The first applications were in the field of submarine detection during World War I. Interest increased after the introduction of *piezoceramic* PZT (Lead Zirconate Titanate) at the end of the first half of the twentieth century. These ceramic materials offered much higher performances and have thus broadened the possible field of applications. These applications, however, were still limited to sound and ultrasound devices. A description of the early piezoelectric materials can be found in Cady (1964). Kawai (1979), in the late 1970s, discovered another class of piezoelectric materials, the so-called *polyvinylidene fluoride* (PVDF), a semi-crystalline polymer with high sensor capability. In recent years, piezoelectricity has been the subject of renewed interest, as inactive intelligent structures with self-monitoring and self-adaptive capabilities. Interesting reviews on these topics can be found in Chopra (2002), Tani *et al.* (1998), and Rao and Sunar (1994).

The first applications of piezoelectric materials were in sound and ultrasound sensors and sources. These applications are still topical, but in recent years a new range of applications has evolved. The use of piezoelectric materials in the so-called *adaptive structures* or *smart structures* has opened up a new and interesting field over the last 20 years. A typical and very simple example of a smart structure is the plate indicated in Figure 2.1, where a network of sensors and actuators is embedded to control the deformations and apply corrections.

Different applications require different properties, such as high- or low-frequency actuation, high deformation, high sensory capabilities, and so on. To this end, different materials can have advantages in certain fields. Alternative

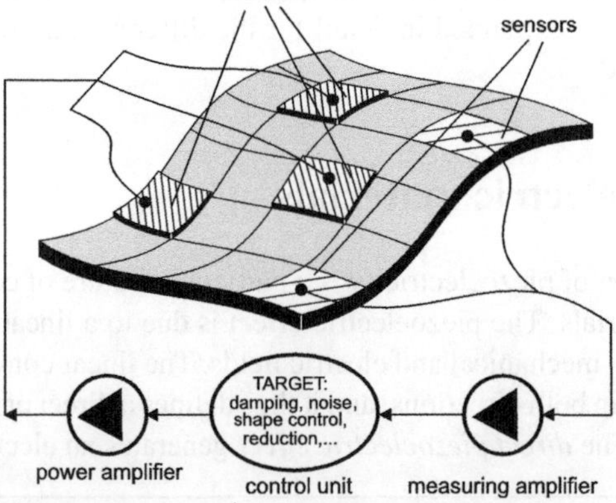

Figure 2.1 Example of a smart structure: the sensor–actuator network for a plate.

Table 2.1 Material properties of some piezoelectric ceramics (PZT) and polymers (PVDF).

Material:		PZT-5H	PZT-5A	PIC 151	PVDF
Producer:		Morgan	Morgan	PI Ceramic	Kynar
E_{11}	[GPa]	71	69	79	2
E_{33}	[GPa]	111	106	77	2
ν	[—]	0.31	—	—	—
ρ	[kg/m^3]	7450	7700	7800	1800
$\varepsilon_{11}/\varepsilon_0$	[—]	—	1700	1980	12
$\varepsilon_{33}/\varepsilon_0$	[—]	3400	1730	2100	12
d_{33}	[m/V] $\times 10^{-12}$	593	374	450	−33
d_{31}	[m/V] $\times 10^{-12}$	−274	−171	−210	23
d_{15}	[m/V] $\times 10^{-12}$	741	585	580	—
T_C	[°C]	195	365	250	—

smart materials to those treated in this work are shape memory alloys (SMAs), polymer gels (PGs), or electromagnetostrictive materials, as described in Carrera *et al.* (2009a,b).

In this book, the adaptive materials that have been considered for smart structures are crystalline materials that show piezoelectric properties, and piezoelectric polymers and semi-crystalline polymers with ferroelectric properties. The crystalline material group includes natural crystals (e.g., quartz (SiO_2), Rochelle salt ($KNa(C_4H_4O_6) \cdot 4H_2O$), tourmaline ($SiO_2 + B, Al$)) and manufactured ceramics (e.g., barium titanate ($BaTiO_3$), lead zirconate titanate (PZT)). The second group considers piezoelectric polymers and semi-crystalline polymers, such as polyvinylidene fluoride (PVDF). Table 2.1 shows the basic parameters of some typical piezoceramics (PZT) and typical piezoelectric polymers (PVDF) (see also Ikeda 1996; Rogacheva 1994). E_{11} and E_{33} are Young's moduli, ν is the Poisson ratio, and ρ is the mass density. T_C is the Curie temperature, and the relative permittivities $\varepsilon_{11}/\varepsilon_0$ and $\varepsilon_{33}/\varepsilon_0$ are expressed with respect to the reference permittivity $\varepsilon_0 = 8.85 \times 10^{12} As/V\,m$. The meaning of the piezoelectric coefficients d_{33}, d_{31}, and d_{15} is clarified later on.

Crystalline materials must be polarized to express a piezoelectric effect. Polarized domains exist at the microscopic level, but their directions are randomly distributed. An external polarization is necessary to activate the material. If a sufficiently high electric field, expressed by the potential Φ_P, is applied to the crystalline material, the domains reorder more or less in the same direction and macroscopic polarization is produced. After poling, the material has a remanent polarization and a remanent elongation, as can be seen from the hysteresis

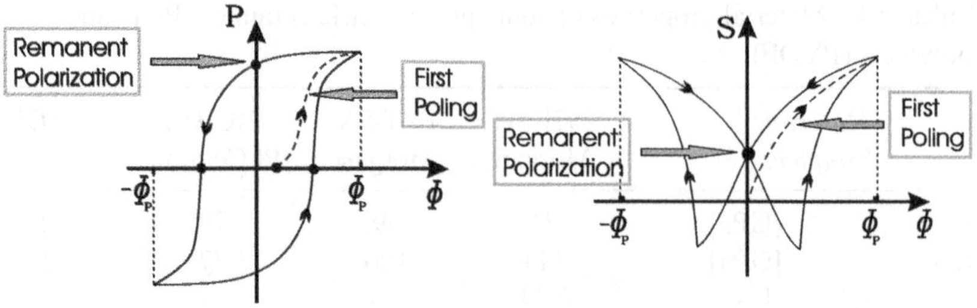

Figure 2.2 Poling of piezoelectric materials: hysteresis of polarization P (left), hysteresis of strain S (right).

curves in Figure 2.2. In this activated state, any applied lower potential than the polarization potential Φ_P leads to a temporary deformation and vice versa. The chosen coordinate system for the polarization is indicated in Figure 2.3. The definition of an appropriate reference system is fundamental: depending on the chosen polarization, piezoelectric materials have a different coupling between the electric field and the mechanical deformations or stresses. The materials considered in this work are polarized in direction 3, as indicated in Figure 2.3. For further details on this topic, readers can refer to Ikeda (1996), Rogacheva (1994), and Yang and Yu (1993).

The effect of the electric field on the elastic field (converse effect), and of the elastic field on the electric field (direct effect), is assumed to be linear. The coupling is therefore represented by linear factors: the piezoelectric coefficients. The mechanical system is represented by the stresses $\boldsymbol{\sigma}$ and the strains $\boldsymbol{\epsilon}$, while the electrical system is represented by the dielectric displacement $\boldsymbol{\mathcal{D}}$ and the electric field $\boldsymbol{\mathcal{E}}$. Four possible definitions of the coefficients for the coupling of the two systems are given in Ikeda (1996). These definitions are summarized in

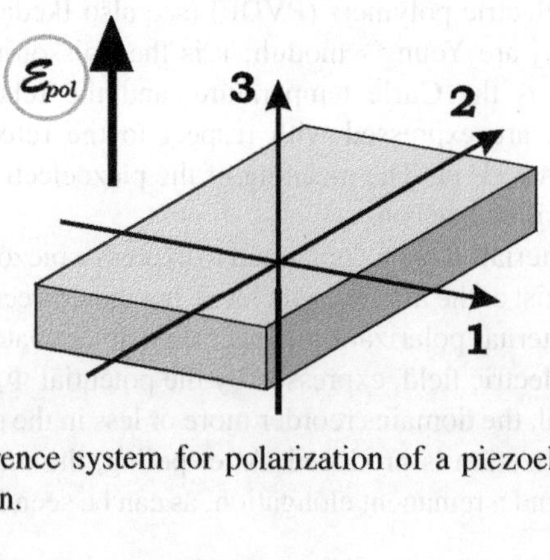

Figure 2.3 Reference system for polarization of a piezoelectric material in transverse direction.

Table 2.2 Piezoelectric coefficients. Coupling between mechanical and electrical fields.

Piezoelectric coefficient	Converse effect	Direct effect
$\boldsymbol{a}$	$\sigma = \boldsymbol{a}^T\, \mathcal{D}$	$\mathcal{E} = \boldsymbol{a}\, \epsilon$
$\boldsymbol{d}$	$\epsilon = \boldsymbol{d}^T\, \mathcal{E}$	$\mathcal{D} = \boldsymbol{d}\, \sigma$
$\boldsymbol{b}$	$\epsilon = \boldsymbol{b}^T\, \mathcal{D}$	$\mathcal{E} = \boldsymbol{b}\, \sigma$
$\boldsymbol{e}$	$\sigma = \boldsymbol{e}^T\, \mathcal{E}$	$\mathcal{D} = \boldsymbol{e}\, \epsilon$

Table 2.2 . For each type of coefficient there exist different components which relate one electric field component to one component of the mechanical field.

The typical notation is explained, as an example, for the converse effect in the case of $\boldsymbol{d}$. The components of the electric field $\boldsymbol{\mathcal{E}}$ are named $\mathcal{E}_1$, $\mathcal{E}_2$, and $\mathcal{E}_3$ in the 1, 2, and 3 directions, respectively. The components of the strain tensor in engineering notation are referred to as 1 to 6, and they represent the components 11, 22, 33, 23, 13, and 12. The different components of $\boldsymbol{d}$ are thus named d_{ij}, with i referring to the electric field direction and j to the stress or strain components. In the case of polarized polycrystalline ceramic materials with the crystal symmetry associated to the crystallographic class, five different coefficients d_{ij} exist: d_{15}, d_{24}, d_{31}, d_{32}, and d_{33}, of which the first and second pair have the same value. The array form of the piezoelectric coefficients states:

$$\boldsymbol{d} = \begin{bmatrix} 0 & 0 & 0 & 0 & d_{15} & 0 \\ 0 & 0 & 0 & d_{24} & 0 & 0 \\ d_{31} & d_{32} & d_{33} & 0 & 0 & 0 \end{bmatrix} \tag{2.1}$$

The meaning of coefficients d_{33}, d_{31}, and d_{15} is explained in Figure 2.4.

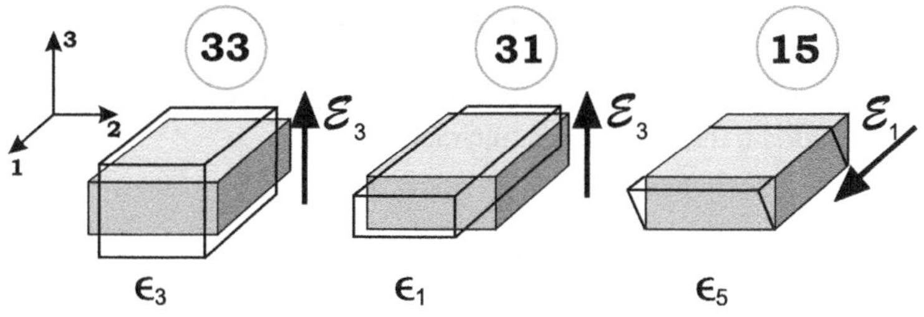

Figure 2.4 Meaning and effects of some piezoelectric coefficients.

2.2 Constitutive equations for piezoelectric problems

The individual material and its reaction to applied loads are characterized by constitutive equations. Their form, for the electromechanical problem, is obtained in this section according to the form reported in Carrera *et al.* (2007, 2008). The general coupling between the mechanical, electric, and thermal fields can be determined using thermodynamical principles and Maxwell's relations (Altay and Dökmeci 1996a; Tiersten 1969). To this end, it is necessary to define a *Gibbs free-energy function* G and a *thermopiezoelectric enthalpy density* H (Ikeda 1996; Nowinski 1978):

$$G(\epsilon_{ij}, \mathcal{E}_i, \theta) = \sigma_{ij}\epsilon_{ij} - \mathcal{E}_i \mathcal{D}_i - \eta\theta \tag{2.2}$$

$$H(\epsilon_{ij}, \mathcal{E}_i, \theta, \vartheta_i) = G(\epsilon_{ij}, \mathcal{E}_i, \theta) - F(\vartheta_i) \tag{2.3}$$

where σ_{ij} and ϵ_{ij} are the stress and strain components, $\mathcal{E}_i$ is the electric field vector, $\mathcal{D}_i$ is the electric displacement vector, η is the variation in entropy per unit of volume, and θ is the temperature considered with respect to the reference temperature T_0. The function $F(\vartheta_i)$ is the dissipation function which depends on the spatial temperature gradient ϑ_i:

$$F(\vartheta_i) = \frac{1}{2}\kappa_{ij}\vartheta_i\vartheta_j - \tau_0\dot{h}_i \tag{2.4}$$

where κ_{ij} is the symmetric, positive, semi-definite conductivity tensor. In the second term, τ_0 is a thermal relaxation parameter and $\dot{h}_i$ is the temporal derivative of the heat flux h_i. The thermal relaxation parameter is usually omitted in the proposed multifield problems. Further details about the dissipation function $F(\vartheta_i)$ can be found in Altay and Dökmeci (1996b), Yang *et al.* (2006), and Cannarozzi and Ubertini (2001), where interesting considerations are made about the inclusion or lack of inclusion of the dissipation function $F(\vartheta_i)$ (e.g., it must be considered in the thermo mechanical analysis of a structure with temperature imposed on the surfaces).

The thermopiezoelectric enthalpy density H can be expanded in order to obtain a quadratic form for a linear interaction:

$$\begin{aligned} H(\epsilon_{ij}, \mathcal{E}_i, \theta, \vartheta_i) = {} & \frac{1}{2}Q_{ijkl}\epsilon_{ij}\epsilon_{kl} - e_{ijk}\epsilon_{ij}\mathcal{E}_k - \lambda_{ij}\epsilon_{ij}\theta \\ & -\frac{1}{2}\varepsilon_{kl}\mathcal{E}_k\mathcal{E}_l - p_k\mathcal{E}_k\theta - \frac{1}{2}\chi\theta^2 - \frac{1}{2}\kappa_{ij}\vartheta_i\vartheta_j \end{aligned} \tag{2.5}$$

where Q_{ijkl} is the elastic coefficient tensor considered for an orthotropic material in the problem reference system (Reddy 2004). e_{ijk} are the piezoelectric coefficients and ε_{kl} are the permittivity coefficients (Rogacheva 1994). λ_{ij} are thermo mechanical coupling coefficients, p_k are the pyroelectric coefficients, and $\chi = \rho C_v / T_0$, where ρ is the material mass density, C_v is the specific heat per unit mass, and T_0 is the reference temperature (Ikeda 1996).

For the piezoelectricity problems proposed in this book, the thermal contributions are not considered and the piezoelectric enthalpy density H coincides with the Gibbs free-energy function G. Equation (2.5) can be rewritten as:

$$H(\epsilon_{ij}, \mathcal{E}_i) = \frac{1}{2} Q_{ijkl} \epsilon_{ij} \epsilon_{kl} - e_{ijk} \epsilon_{ij} \mathcal{E}_k - \frac{1}{2} \varepsilon_{kl} \mathcal{E}_k \mathcal{E}_l \tag{2.6}$$

The constitutive equations are obtained by considering the following relations:

$$\sigma_{ij} = \frac{\partial H}{\partial \epsilon_{ij}}, \qquad \mathcal{D}_k = -\frac{\partial H}{\partial \mathcal{E}_k} \tag{2.7}$$

The constitutive equations for the electromechanical problem are obtained by considering Equations (2.6) and (2.7):

$$\sigma_{ij} = Q_{ijkl} \epsilon_{kl} - e_{ijk} \mathcal{E}_k \tag{2.8}$$

$$\mathcal{D}_k = e_{ijk} \epsilon_{ij} + \varepsilon_{kl} \mathcal{E}_l \tag{2.9}$$

These equations can be written in single-subscript notation by using the indexes $m = q = 1, 2, 3, 4, 5, 6$ and $i = j = 1, 2, 3$:

$$\sigma_m = Q_{mq} \epsilon_q - e_{mi} \mathcal{E}_i \tag{2.10}$$

$$\mathcal{D}_i = e_{iq} \epsilon_q + \varepsilon_{ij} \mathcal{E}_j \tag{2.11}$$

From the equations written in single-subscript notation, it is very easy to write their matrix form; the matrices and vectors are indicated in bold type. Considering a generic multilayered structure, Equations (2.10) and (2.11) are written for a generic layer k in the problem reference system (x, y, z) as:

$$\boldsymbol{\sigma}^k = \boldsymbol{Q}^k \boldsymbol{\epsilon}^k - \boldsymbol{e}^{kT} \boldsymbol{\mathcal{E}}^k \tag{2.12}$$

$$\boldsymbol{\mathcal{D}}^k = \boldsymbol{e}^k \boldsymbol{\epsilon}^k + \boldsymbol{\varepsilon}^k \boldsymbol{\mathcal{E}}^k \tag{2.13}$$

The 6 × 1 stress and strain component vectors are:

$$\boldsymbol{\sigma}^k = \begin{Bmatrix} \sigma_{xx} \\ \sigma_{yy} \\ \sigma_{zz} \\ \sigma_{yz} \\ \sigma_{xz} \\ \sigma_{xy} \end{Bmatrix}^k, \quad \boldsymbol{\epsilon}^k = \begin{Bmatrix} \epsilon_{xx} \\ \epsilon_{yy} \\ \epsilon_{zz} \\ \gamma_{yz} \\ \gamma_{xz} \\ \gamma_{xy} \end{Bmatrix}^k \tag{2.14}$$

The 3 × 1 electric field $\boldsymbol{\mathcal{E}}^k$ and electrical displacement $\boldsymbol{\mathcal{D}}^k$ vectors are:

$$\boldsymbol{\mathcal{E}}^k = \begin{Bmatrix} \mathcal{E}_x \\ \mathcal{E}_y \\ \mathcal{E}_z \end{Bmatrix}^k, \quad \boldsymbol{\mathcal{D}}^k = \begin{Bmatrix} \mathcal{D}_x \\ \mathcal{D}_y \\ \mathcal{D}_z \end{Bmatrix}^k \tag{2.15}$$

The elastic coefficients matrix $\boldsymbol{Q}^k$ of Hooke's law in the problem reference system for an orthotropic material is:

$$\boldsymbol{Q}^k = \begin{bmatrix} Q_{11} & Q_{12} & Q_{13} & 0 & 0 & Q_{16} \\ Q_{12} & Q_{22} & Q_{23} & 0 & 0 & Q_{26} \\ Q_{13} & Q_{23} & Q_{33} & 0 & 0 & Q_{36} \\ 0 & 0 & 0 & Q_{44} & Q_{45} & 0 \\ 0 & 0 & 0 & Q_{45} & Q_{55} & 0 \\ Q_{16} & Q_{26} & Q_{36} & 0 & 0 & Q_{66} \end{bmatrix}^k \tag{2.16}$$

The matrix $\boldsymbol{\varepsilon}^k$ of the permittivity coefficients has 3 × 3 dimensions:

$$\boldsymbol{\varepsilon}^k = \begin{bmatrix} \varepsilon_{11} & \varepsilon_{12} & 0 \\ \varepsilon_{12} & \varepsilon_{22} & 0 \\ 0 & 0 & \varepsilon_{33} \end{bmatrix}^k \tag{2.17}$$

The piezoelectric coefficients matrix $\boldsymbol{e}^k$ has 3 × 6 dimensions:

$$\boldsymbol{e}^k = \begin{bmatrix} 0 & 0 & 0 & e_{14} & e_{15} & 0 \\ 0 & 0 & 0 & e_{24} & e_{25} & 0 \\ e_{31} & e_{32} & e_{33} & 0 & 0 & e_{36} \end{bmatrix}^k \tag{2.18}$$

In order to use the relations proposed in Equations (2.12) and (2.13) in the variational statements presented in the following sections, it is convenient to split them into in-plane components (subscript p) and out-of-plane components (subscript n). Another two new subscripts are introduced: the subscript C for those variables in the variational statements which need the substitution of the constitutive equations; the subscript G for those variables in the constitutive equations which need the substitution of the geometrical relations (the latter are introduced in the next section). The proposed constitutive equations are valid for both plate and shell geometries; for this reason, a curvilinear reference system (α, β, z) is introduced in place of the less general rectilinear one (x, y, z). Geometrical relations for shells are obtained in Section 2.3 and their degeneration into geometrical relations for plates is discussed. The split stress and strain component vectors are:

$$\boldsymbol{\sigma}_{pC}^k = \begin{Bmatrix} \sigma_{\alpha\alpha} \\ \sigma_{\beta\beta} \\ \sigma_{\alpha\beta} \end{Bmatrix}^k, \quad \boldsymbol{\sigma}_{nC}^k = \begin{Bmatrix} \sigma_{\alpha z} \\ \sigma_{\beta z} \\ \sigma_{zz} \end{Bmatrix}^k, \quad \boldsymbol{\epsilon}_{pG}^k = \begin{Bmatrix} \epsilon_{\alpha\alpha} \\ \epsilon_{\beta\beta} \\ \gamma_{\alpha\beta} \end{Bmatrix}^k, \quad \boldsymbol{\epsilon}_{nG}^k = \begin{Bmatrix} \gamma_{\alpha z} \\ \gamma_{\beta z} \\ \epsilon_{zz} \end{Bmatrix}^k \tag{2.19}$$

The 3×1 vectors of the electric field and electrical displacement, split into in-plane and out-of-plane components, are:

$$\boldsymbol{\mathcal{E}}_{pG}^k = \begin{Bmatrix} \mathcal{E}_\alpha \\ \mathcal{E}_\beta \end{Bmatrix}^k, \quad \boldsymbol{\mathcal{E}}_{nG}^k = \{\mathcal{E}_z\}^k, \quad \boldsymbol{\mathcal{D}}_{pC}^k = \begin{Bmatrix} \mathcal{D}_\alpha \\ \mathcal{D}_\beta \end{Bmatrix}^k, \quad \boldsymbol{\mathcal{D}}_{nC}^k = \{\mathcal{D}_z\}^k \tag{2.20}$$

The split form of Equations (2.12) and (2.13), considering Equations (2.19) and (2.20), is:

$$\boldsymbol{\sigma}_{pC}^k = \boldsymbol{Q}_{pp}^k \boldsymbol{\epsilon}_{pG}^k + \boldsymbol{Q}_{pn}^k \boldsymbol{\epsilon}_{nG}^k - \boldsymbol{e}_{pp}^{kT} \boldsymbol{\mathcal{E}}_{pG}^k - \boldsymbol{e}_{np}^{kT} \boldsymbol{\mathcal{E}}_{nG}^k \tag{2.21}$$

$$\boldsymbol{\sigma}_{nC}^k = \boldsymbol{Q}_{np}^k \boldsymbol{\epsilon}_{pG}^k + \boldsymbol{Q}_{nn}^k \boldsymbol{\epsilon}_{nG}^k - \boldsymbol{e}_{pn}^{kT} \boldsymbol{\mathcal{E}}_{pG}^k - \boldsymbol{e}_{nn}^{kT} \boldsymbol{\mathcal{E}}_{nG}^k \tag{2.22}$$

$$\boldsymbol{\mathcal{D}}_{pC}^k = \boldsymbol{e}_{pp}^k \boldsymbol{\epsilon}_{pG}^k + \boldsymbol{e}_{pn}^k \boldsymbol{\epsilon}_{nG}^k + \boldsymbol{\varepsilon}_{pp}^k \boldsymbol{\mathcal{E}}_{pG}^k + \boldsymbol{\varepsilon}_{pn}^k \boldsymbol{\mathcal{E}}_{nG}^k \tag{2.23}$$

$$\boldsymbol{\mathcal{D}}_{nC}^k = \boldsymbol{e}_{np}^k \boldsymbol{\epsilon}_{pG}^k + \boldsymbol{e}_{nn}^k \boldsymbol{\epsilon}_{nG}^k + \boldsymbol{\varepsilon}_{np}^k \boldsymbol{\mathcal{E}}_{pG}^k + \boldsymbol{\varepsilon}_{nn}^k \boldsymbol{\mathcal{E}}_{nG}^k \tag{2.24}$$

The explicit forms of the new matrices in Equations (2.21)–(2.24) are:

- elastic coefficient matrices

$$\boldsymbol{Q}_{pp}^k = \begin{bmatrix} Q_{11} & Q_{12} & Q_{16} \\ Q_{12} & Q_{22} & Q_{26} \\ Q_{16} & Q_{26} & Q_{66} \end{bmatrix}^k, \quad \boldsymbol{Q}_{pn}^k = \begin{bmatrix} 0 & 0 & Q_{13} \\ 0 & 0 & Q_{23} \\ 0 & 0 & Q_{36} \end{bmatrix}^k \tag{2.25}$$

$$\boldsymbol{Q}_{np}^k = \begin{bmatrix} 0 & 0 & 0 \\ 0 & 0 & 0 \\ Q_{13} & Q_{23} & Q_{36} \end{bmatrix}^k, \quad \boldsymbol{Q}_{nn}^k = \begin{bmatrix} Q_{55} & Q_{45} & 0 \\ Q_{45} & Q_{44} & 0 \\ 0 & 0 & Q_{33} \end{bmatrix}^k \tag{2.26}$$

- piezoelectric coefficients

$$\boldsymbol{e}_{pp}^k = \begin{bmatrix} 0 & 0 & 0 \\ 0 & 0 & 0 \end{bmatrix}^k, \quad \boldsymbol{e}_{pn}^k = \begin{bmatrix} e_{15} & e_{14} & 0 \\ e_{25} & e_{24} & 0 \end{bmatrix}^k \tag{2.27}$$

$$\boldsymbol{e}_{np}^k = \begin{bmatrix} e_{31} & e_{32} & e_{36} \end{bmatrix}^k, \quad \boldsymbol{e}_{nn}^k = \begin{bmatrix} 0 & 0 & e_{33} \end{bmatrix}^k \tag{2.28}$$

- permittivity coefficients

$$\boldsymbol{\varepsilon}_{pp}^k = \begin{bmatrix} \varepsilon_{11} & \varepsilon_{12} \\ \varepsilon_{12} & \varepsilon_{22} \end{bmatrix}^k, \quad \boldsymbol{\varepsilon}_{pn}^k = \begin{bmatrix} 0 \\ 0 \end{bmatrix}^k \tag{2.29}$$

$$\boldsymbol{\varepsilon}_{np}^k = \begin{bmatrix} 0 & 0 \end{bmatrix}^k, \quad \boldsymbol{\varepsilon}_{nn}^k = \begin{bmatrix} \varepsilon_{33} \end{bmatrix}^k \tag{2.30}$$

2.3 Geometrical relations for piezoelectric problems

We define a thin shell as a three-dimensional body bounded by two closely spaced curved surfaces, the distance between the two surfaces being small in comparison to the other dimensions. The middle surface of the shell is the locus of the points that lie midway between these surfaces. The distance between the surfaces measured along the normal to the middle surface is the *thickness* of the shell at that point (Leissa 1973). Shells may be seen as generalizations of a flat plate (Leissa 1969); conversely, a flat plate is a special case of a shell with no curvature (see Figure 2.5). Geometrical relations for plates are considered as a particular case of those for shells. The material is assumed to be linearly elastic and homogeneous, and displacements are assumed to be small, thereby yielding linear equations; shear deformation and rotary inertia effects are neglected, and the thickness is taken to be small.

In the case of shells with constant radii of curvature (see Figure 2.5), the in-plane (p) and out-of-plane (n) strain components are linked to the displacement vector by the equations:

$$\boldsymbol{\epsilon}_{pG}^k = [\epsilon_{\alpha\alpha}, \epsilon_{\beta\beta}, \gamma_{\alpha\beta}]^{kT} = \left(\boldsymbol{D}_p^k + \boldsymbol{A}_p^k\right) \boldsymbol{u}^k \tag{2.31}$$

$$\boldsymbol{\epsilon}_{nG}^k = [\gamma_{\alpha z}, \gamma_{\beta z}, \epsilon_{zz}]^{kT} = \left(\boldsymbol{D}_{np}^k + \boldsymbol{D}_{nz}^k - \boldsymbol{A}_n^k\right) \boldsymbol{u}^k \tag{2.32}$$

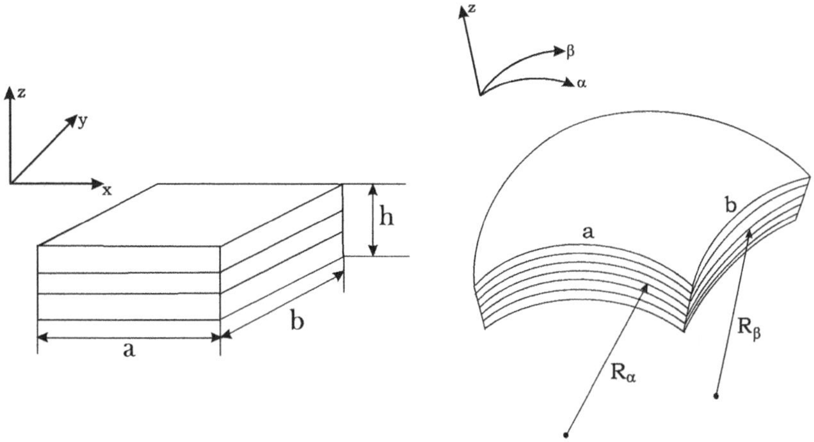

Figure 2.5 Geometry and notations for generic multilayered plates and shells.

where, for each layer k, the vector of displacement components is $\boldsymbol{u} = (u, v, w)$. The explicit form of the introduced arrays follows:

$$\boldsymbol{D}_p^k = \begin{bmatrix} \frac{\partial_\alpha}{H_\alpha^k} & 0 & 0 \\ 0 & \frac{\partial_\beta}{H_\beta^k} & 0 \\ \frac{\partial_\beta}{H_\beta^k} & \frac{\partial_\alpha}{H_\alpha^k} & 0 \end{bmatrix}, \quad \boldsymbol{D}_{np}^k = \begin{bmatrix} 0 & 0 & \frac{\partial_\alpha}{H_\alpha^k} \\ 0 & 0 & \frac{\partial_\beta}{H_\beta^k} \\ 0 & 0 & 0 \end{bmatrix}, \quad \boldsymbol{D}_{nz}^k = \begin{bmatrix} \partial_z & 0 & 0 \\ 0 & \partial_z & 0 \\ 0 & 0 & \partial_z \end{bmatrix} \tag{2.33}$$

$$\boldsymbol{A}_p^k = \begin{bmatrix} 0 & 0 & \frac{1}{H_\alpha^k R_\alpha^k} \\ 0 & 0 & \frac{1}{H_\beta^k R_\beta^k} \\ 0 & 0 & 0 \end{bmatrix}, \quad \boldsymbol{A}_n^k = \begin{bmatrix} \frac{1}{H_\alpha^k R_\alpha^k} & 0 & 0 \\ 0 & \frac{1}{H_\beta^k R_\beta^k} & 0 \\ 0 & 0 & 0 \end{bmatrix} \tag{2.34}$$

Details on Equations (2.31)–(2.34) are given in Carrera and Brischetto (2007a,b). The symbols ∂_α, ∂_β and ∂_z indicate the partial derivatives with respect to the α, β, and z coordinates, respectively. The parametric coefficients are $H_\alpha^k = (1 + z^k/R_\alpha^k)$ and $H_\beta^k = (1 + z^k/R_\beta^k)$, where R_α^k, and R_β^k are the radii of curvature in the α and β directions, respectively. The geometrical relations, which link the electric field to the electric potential Φ^k, are also given in Carrera and Brischetto (2007a,b):

$$\boldsymbol{\mathcal{E}}_{pG}^k = [\mathcal{E}_\alpha, \mathcal{E}_\beta]^{kT} = -\boldsymbol{D}_{ep}^k \, \Phi^k \tag{2.35}$$

$$\boldsymbol{\mathcal{E}}_{nG}^k = [\mathcal{E}_z]^k = -\boldsymbol{D}_{en}^k \, \Phi^k \tag{2.36}$$

where the meaning of the arrays is:

$$\boldsymbol{D}^k_{ep} = \begin{bmatrix} \dfrac{\partial_\alpha}{H^k_\alpha} \\ \dfrac{\partial_\beta}{H^k_\beta} \end{bmatrix}, \quad \boldsymbol{D}^k_{en} = \left[\, \partial_z \,\right] \tag{2.37}$$

The geometrical relations for the shells in Equations (2.31)–(2.37) degenerate into geometrical relations for plates when the radii of curvature R^k_α and R^k_β are infinite. The parameters H^k_α and H^k_β are therefore equal to one, and the orthogonal curvilinear coordinates (α, β, z) degenerate into rectilinear ones (x, y, z):

$$\boldsymbol{\epsilon}^k_{pG} = [\epsilon_{xx}, \epsilon_{yy}, \gamma_{xy}]^{kT} = \boldsymbol{D}_p\, \boldsymbol{u}^k \tag{2.38}$$

$$\boldsymbol{\epsilon}^k_{nG} = [\gamma_{xz}, \gamma_{yz}, \epsilon_{zz}]^{kT} = (\boldsymbol{D}_{np} + \boldsymbol{D}_{nz})\, \boldsymbol{u}^k \tag{2.39}$$

$$\boldsymbol{\mathcal{E}}^k_{pG} = [\mathcal{E}_x, \mathcal{E}_y]^{kT} = -\boldsymbol{D}_{ep}\, \Phi^k \tag{2.40}$$

$$\boldsymbol{\mathcal{E}}^k_{nG} = [\mathcal{E}_z]^{k} = -\boldsymbol{D}_{en}\, \Phi^k \tag{2.41}$$

The new differential operators do not depend on the k layer:

$$\boldsymbol{D}_p = \begin{bmatrix} \partial_x & 0 & 0 \\ 0 & \partial_y & 0 \\ \partial_y & \partial_x & 0 \end{bmatrix}, \quad \boldsymbol{D}_{np} = \begin{bmatrix} 0 & 0 & \partial_x \\ 0 & 0 & \partial_y \\ 0 & 0 & 0 \end{bmatrix}, \quad \boldsymbol{D}_{nz} = \begin{bmatrix} \partial_z & 0 & 0 \\ 0 & \partial_z & 0 \\ 0 & 0 & \partial_z \end{bmatrix}$$

$$\boldsymbol{D}_{ep} = \begin{bmatrix} \partial_x \\ \partial_y \end{bmatrix}, \quad \boldsymbol{D}_{en} = \left[\, \partial_z \,\right] \tag{2.42}$$

The symbols in the differential operators matrices are: $\partial_x = \partial/\partial x$, $\partial_y = \partial/\partial y$ and $\partial_z = \partial/\partial z$. Further details on the geometrical relations of plates can be found in Carrera *et al.* (2007) for electromechanical problems.

2.4 Principle of virtual displacements

Recently, several two-dimensional approaches have successfully been extended to multifield problems (Chopra 2002; Correira *et al.* 2000; Ossadzow-David and Touratier 2003). Refined models can be obtained via the extension of the principle of virtual displacements (PVD) to the electromechanical case.

The PVD of an electroelastic medium can be derived from the Hamilton principle, as indicated in Carrera *et al.* (2007, 2008):

$$\delta \int_{t_0}^{t} (E_c - E_p)dt = 0 \Rightarrow \delta \int_{t_0}^{t} E_c dt - \delta \int_{t_0}^{t} E_p dt = 0 \tag{2.43}$$

where E_c and E_p are the kinetic and potential energy, respectively. δ is the variational symbol, t_0 the initial time, and t a generic instant (Carrera *et al.* 2008). The total potential energy E_p includes the piezoelectric enthalpy density, H, as described in Equation (2.6), and the work done by surface tractions $\bar{t}_j$ and electric charge $\bar{Q}$ on the displacements u_j and electric potential Φ, respectively:

$$E_p = \int_V H dV - \int_\Gamma (\bar{t}_j u_j - \bar{Q}\Phi) d\Gamma \tag{2.44}$$

where V is the volume and Γ the boundary of the reference surface Ω.

The variation in the kinetic energy E_c is the well-known relation (Carrera *et al.* 2007, 2008):

$$\begin{aligned}\delta \int_{t_0}^{t} E_c dt &= \delta \int_{t_0}^{t} dt \int_V \left(\tfrac{1}{2}\rho \dot{u}_i \dot{u}_i\right) dV = \int_{t_0}^{t}\int_V \rho \dot{u}_i \delta \dot{u}_i dV dt \\ &= \int_V \rho \dot{u}_i \delta u_i dV \,|_{t_0}^{t_1} - \int_{t_0}^{t}\int_V \rho \ddot{u}_i \delta u_i dV dt\end{aligned} \tag{2.45}$$

and since δu_i vanishes at t_0 and t_1, the following expression can be obtained:

$$\delta \int_{t_0}^{t} E_c dt = -\int_{t_0}^{t}\int_V \rho \ddot{u}_i \delta u_i dV dt = -\int_{t_0}^{t} \delta L_{in} dt \tag{2.46}$$

where ρ is the mass density, $\dot{u}_i$ and $\ddot{u}_i$ are the first and second temporal derivatives of displacement u_i, respectively, and δL_{in} is the virtual variation of the work done by the inertial loads.

The variation in the potential energy E_p can be rewritten according to Equations (2.2)–(2.6):

$$\delta \int_{t_0}^{t} E_p dt = \delta \int_{t_0}^{t} \left[\int_V \left(G(\epsilon_{ij}, \mathcal{E}_i) \right) dV - \int_\Gamma (\bar{t}_j u_j - \bar{Q}\Phi) d\Gamma \right] dt \tag{2.47}$$

where G is the Gibbs free-energy function which is coincident with the piezoelectric enthalpy density H in the proposed case (Altay and Dökmeci 1996b; Yang *et al.* 2006; Cannarozzi and Ubertini 2001). The variables in the problem are the strain vector ϵ_{ij} and the electric field $\mathcal{E}_i$. The contribution given by the external loads is the virtual variation in the external work:

$$\delta L_e = \delta \int_{\Gamma} (\bar{t}_j u_j - \bar{Q}\Phi) d\Gamma \tag{2.48}$$

Using Equation (2.48), Equation (2.47) can be rewritten as:

$$\delta \int_{t_0}^{t} E_p dt = \delta \int_{t_0}^{t} \int_V \Big(H(\epsilon_{ij}, \mathcal{E}_i) \Big) dV dt - \int_{t_0}^{t} \delta L_e dt \tag{2.49}$$

Differentiating each term in Equation (2.49):

$$\delta \int_{t_0}^{t} E_p dt = \int_{t_0}^{t} \int_V \Big(\frac{\partial H}{\partial \epsilon_{ij}} \delta \epsilon_{ij} + \frac{\partial H}{\partial \mathcal{E}_i} \delta \mathcal{E}_i \Big) dV dt - \int_{t_0}^{t} \delta L_e dt \tag{2.50}$$

Considering the relations given in Equation (2.7), Equation (2.50) can be rewritten as:

$$\delta \int_{t_0}^{t} E_p dt = \int_{t_0}^{t} \int_V \Big(\sigma_{ij} \delta \epsilon_{ij} - \mathcal{D}_i \delta \mathcal{E}_i \Big) dV dt - \int_{t_0}^{t} \delta L_e dt \tag{2.51}$$

where σ_{ij} are the stress components and $\mathcal{D}_i$ is the vector containing the electric displacement components. Combining Equation (2.51) and Equation (2.46), then according to Equation (2.43), the final version of the PVD in the case of the electromechanical problems is:

$$\int_{t_0}^{t} \int_V \Big(\sigma_{ij} \delta \epsilon_{ij} - \mathcal{D}_i \delta \mathcal{E}_i \Big) dV dt = \int_{t_0}^{t} \delta L_e dt - \int_{t_0}^{t} \delta L_{in} dt \tag{2.52}$$

By discarding the dependence on the time t and introducing the vectorial form of the constitutive equations, split into in-plane (p) and out-of-plane (n) components (see Equations (2.21)–(2.24)), Equation (2.52) can be rewritten as:

$$\int_V \Big(\delta \boldsymbol{\epsilon}_{pG}^T \boldsymbol{\sigma}_{pC} + \delta \boldsymbol{\epsilon}_{nG}^T \boldsymbol{\sigma}_{nC} - \delta \boldsymbol{\mathcal{E}}_{pG}^T \boldsymbol{\mathcal{D}}_{pC} - \delta \boldsymbol{\mathcal{E}}_{nG}^T \boldsymbol{\mathcal{D}}_{nC} \Big) dV = \delta L_e - \delta L_{in} \tag{2.53}$$

The bold letters in Equation (2.53) denote vectors; T stands for the transpose of a vector. Subscripts C and G suggest the substitution of constitutive and geometrical relations, respectively. V is the total volume of the considered multilayered plate or shell.

The general form of the governing equations is:

$$\boldsymbol{K}_{uu} \boldsymbol{u} + \boldsymbol{K}_{u\Phi} \boldsymbol{\Phi} = \boldsymbol{p}_u - \boldsymbol{M}_{uu} \ddot{\boldsymbol{u}} \tag{2.54}$$

$$\boldsymbol{K}_{\Phi u} \boldsymbol{u} + \boldsymbol{K}_{\Phi\Phi} \boldsymbol{\Phi} = 0 \tag{2.55}$$

The matrices $\boldsymbol{K}$ are already considered to be assembled at a multilayer level and expanded for the chosen order in the thickness direction. The vectors contain the degrees of freedom for the displacement $\boldsymbol{u}$ and the electric potential Φ. $\boldsymbol{M}_{uu}$ is the inertial matrix and $\ddot{\boldsymbol{u}}$ is the second temporal derivative of the displacement vector. The mechanical load is $\boldsymbol{p}_u$.

2.4.1 PVD for the pure mechanical case

In the case of pure mechanical problems, the PVD has only the displacement $\boldsymbol{u}$ as the primary variable (Reddy 2004). The variational statement, the constitutive equations, and the governing equations can be considered as particular cases of the most general case illustrated in the previous section. The variational statement is simplified, on the basis of Equation (2.53), by discarding the internal electrical work:

$$\int_V \left(\delta\boldsymbol{\epsilon}_{pG}^T\boldsymbol{\sigma}_{pC} + \delta\boldsymbol{\epsilon}_{nG}^T\boldsymbol{\sigma}_{nC}\right)dV = \delta L_e - \delta L_{in} \tag{2.56}$$

The relative constitutive equations are the well-known Hooke's law (Reddy 2004), which can be considered as a particular case of the constitutive equations given in Equations (2.21)–(2.24):

$$\boldsymbol{\sigma}_{pC}^k = \boldsymbol{Q}_{pp}^k\boldsymbol{\epsilon}_{pG}^k + \boldsymbol{Q}_{pn}^k\boldsymbol{\epsilon}_{nG}^k \tag{2.57}$$

$$\boldsymbol{\sigma}_{nC}^k = \boldsymbol{Q}_{np}^k\boldsymbol{\epsilon}_{pG}^k + \boldsymbol{Q}_{nn}^k\boldsymbol{\epsilon}_{nG}^k \tag{2.58}$$

By substituting Equations (2.57) and (2.58) into the variational statement of Equation (2.56), the governing equation is obtained for the pure mechanical case:

$$\boldsymbol{K}_{uu}\boldsymbol{u} = \boldsymbol{p}_u - \boldsymbol{M}_{uu}\ddot{\boldsymbol{u}} \tag{2.59}$$

It is important to notice that Equation (2.59) can be obtained in a simpler way by deleting the second line and the second column in Equations (2.54) and (2.55); in fact, matrix $\boldsymbol{K}_{uu}$ is the same for both the pure mechanical and electromechanical cases.

2.5 Reissner mixed variational theorem

The Reissner mixed variational theorem (RMVT) (Reissner 1984) allows one to assume two independent sets of variables: a set of primary unknowns as in the PVD case, and a set of extensive variables which are a priori modeled in the thickness direction. The main advantage of using the RMVT is that a complete fulfillment of the C_z^0 requirements is obtained a priori for the modeled extensive variables (Carrera 2001). Different extensions of the RMVT are given, in the case of electromechanical problems, by starting from the electromechanical PVD: transverse shear/normal stresses $\boldsymbol{\sigma}_n$ as the extensive variables; transverse normal electric displacement $\boldsymbol{\mathcal{D}}_n$ as the extensive variables; both $\boldsymbol{\sigma}_n$ and $\boldsymbol{\mathcal{D}}_n$ as the extensive variables. These three cases can be obtained as follows: a Lagrange multiplier is added for $\boldsymbol{\sigma}_n$ in the former case, a different Lagrange multiplier is considered for $\boldsymbol{\mathcal{D}}_n$ in the second case, and two Lagrange multipliers are added in the last case. When a new Lagrange multiplier is added (Reissner 1984), the constitutive equations must be rearranged in order to explicitly model the variables. For this reason, each proposed extension of the RMVT should not be seen as a particular case of the other two.

2.5.1 RMVT($\boldsymbol{u}$, $\boldsymbol{\Phi}$, $\boldsymbol{\sigma}_n$)

RMVT($\boldsymbol{u}$, $\boldsymbol{\Phi}$, $\boldsymbol{\sigma}_n$) is obtained, considering the variational statement in Equation (2.53) for the electromechanical PVD, by a priori modeling the transverse shear/normal stresses $\boldsymbol{\sigma}_{nM}$ (the new subscript M is introduced to show that the transverse stresses are now modeled and not obtained via constitutive equations). The added Lagrange multiplier is $\delta\boldsymbol{\sigma}_{nM}^T(\boldsymbol{\epsilon}_{nG} - \boldsymbol{\epsilon}_{nC})$. The condition that is necessary to add this multiplier is that the transverse strains $\boldsymbol{\epsilon}_n$, calculated by means of geometrical relations (G) and using the constitutive equations (C), must be the same or almost the same. In this way, the balance of the internal work does not change or remains almost the same:

$$\int_V \Big(\delta\boldsymbol{\epsilon}_{pG}^T\boldsymbol{\sigma}_{pC} + \delta\boldsymbol{\epsilon}_{nG}^T\boldsymbol{\sigma}_{nM} + \delta\boldsymbol{\sigma}_{nM}^T(\boldsymbol{\epsilon}_{nG} - \boldsymbol{\epsilon}_{nC}) - \delta\boldsymbol{\mathcal{E}}_{pG}^T\boldsymbol{\mathcal{D}}_{pC}$$

$$-\delta\boldsymbol{\mathcal{E}}_{nG}^T\boldsymbol{\mathcal{D}}_{nC}\Big)dV = \delta L_e - \delta L_{in} \tag{2.60}$$

The relative constitutive equations are obtained from Equations (2.21)–(2.24) considering the transverse stresses $\boldsymbol{\sigma}_n$ as being modeled (M) and the transverse strains $\boldsymbol{\epsilon}_n$ as being obtained from constitutive equations (C):

$$\boldsymbol{\sigma}_{pC}^k = \hat{\boldsymbol{C}}_{\sigma_p\epsilon_p}^k\boldsymbol{\epsilon}_{pG}^k + \hat{\boldsymbol{C}}_{\sigma_p\sigma_n}^k\boldsymbol{\sigma}_{nM}^k + \hat{\boldsymbol{C}}_{\sigma_p\mathcal{E}_p}^k\boldsymbol{\mathcal{E}}_{pG}^k + \hat{\boldsymbol{C}}_{\sigma_p\mathcal{E}_n}^k\boldsymbol{\mathcal{E}}_{nG}^k \tag{2.61}$$

$$\boldsymbol{\epsilon}_{nC}^k = \hat{\boldsymbol{C}}_{\epsilon_n\epsilon_p}^k\boldsymbol{\epsilon}_{pG}^k + \hat{\boldsymbol{C}}_{\epsilon_n\sigma_n}^k\boldsymbol{\sigma}_{nM}^k + \hat{\boldsymbol{C}}_{\epsilon_n\mathcal{E}_p}^k\boldsymbol{\mathcal{E}}_{pG}^k + \hat{\boldsymbol{C}}_{\epsilon_n\mathcal{E}_n}^k\boldsymbol{\mathcal{E}}_{nG}^k \tag{2.62}$$

$$\boldsymbol{\mathcal{D}}_{pC}^k = \hat{\boldsymbol{C}}_{\mathcal{D}_p\epsilon_p}^k\boldsymbol{\epsilon}_{pG}^k + \hat{\boldsymbol{C}}_{\mathcal{D}_p\sigma_n}^k\boldsymbol{\sigma}_{nM}^k + \hat{\boldsymbol{C}}_{\mathcal{D}_p\mathcal{E}_p}^k\boldsymbol{\mathcal{E}}_{pG}^k + \hat{\boldsymbol{C}}_{\mathcal{D}_p\mathcal{E}_n}^k\boldsymbol{\mathcal{E}}_{nG}^k \tag{2.63}$$

$$\boldsymbol{\mathcal{D}}_{nC}^k = \hat{\boldsymbol{C}}_{\mathcal{D}_n\epsilon_p}^k\boldsymbol{\epsilon}_{pG}^k + \hat{\boldsymbol{C}}_{\mathcal{D}_n\sigma_n}^k\boldsymbol{\sigma}_{nM}^k + \hat{\boldsymbol{C}}_{\mathcal{D}_n\mathcal{E}_p}^k\boldsymbol{\mathcal{E}}_{pG}^k + \hat{\boldsymbol{C}}_{\mathcal{D}_n\mathcal{E}_n}^k\boldsymbol{\mathcal{E}}_{nG}^k \tag{2.64}$$

According to Carrera *et al.* (2007, 2008), the meaning of coefficients $\hat{C}$ in Equations (2.61)–(2.64) is:

$$
\begin{aligned}
&\hat{C}^k_{\sigma_p\epsilon_p} = Q^k_{pp} - Q^k_{pn} Q^{k\ -1}_{nn} Q^k_{np}, \quad \hat{C}^k_{\sigma_p\sigma_n} = Q^k_{pn} Q^{k\ -1}_{nn} \\
&\hat{C}^k_{\sigma_p\mathcal{E}_p} = Q^k_{pn} Q^{k\ -1}_{nn} e^{kT}_{pn} - e^{kT}_{pp}, \quad \hat{C}^k_{\sigma_p\mathcal{E}_n} = Q^k_{pn} Q^{k\ -1}_{nn} e^{kT}_{nn} - e^{kT}_{np} \\
&\hat{C}^k_{\epsilon_n\epsilon_p} = -Q^{k\ -1}_{nn} Q^k_{np}, \quad \hat{C}^k_{\epsilon_n\sigma_n} = Q^{k\ -1}_{nn} \\
&\hat{C}^k_{\epsilon_n\mathcal{E}_p} = Q^{k\ -1}_{nn} e^{kT}_{pn}, \quad \hat{C}^k_{\epsilon_n\mathcal{E}_n} = Q^{k\ -1}_{nn} e^{kT}_{nn} \\
&\hat{C}^k_{\mathcal{D}_p\epsilon_p} = e^k_{pp} - e^k_{pn} Q^{k\ -1}_{nn} Q^k_{np}, \quad \hat{C}^k_{\mathcal{D}_p\sigma_n} = e^k_{pn} Q^{k\ -1}_{nn} \\
&\hat{C}^k_{\mathcal{D}_p\mathcal{E}_p} = e^k_{pn} Q^{k\ -1}_{nn} e^{kT}_{pn} + \varepsilon^k_{pp}, \quad \hat{C}^k_{\mathcal{D}_p\mathcal{E}_n} = e^k_{pn} Q^{k\ -1}_{nn} e^{kT}_{nn} + \varepsilon^k_{pn} \\
&\hat{C}^k_{\mathcal{D}_n\epsilon_p} = e^k_{np} - e^k_{nn} Q^{k\ -1}_{nn} Q^k_{np}, \quad \hat{C}^k_{\mathcal{D}_n\sigma_n} = e^k_{nn} Q^{k\ -1}_{nn} \\
&\hat{C}^k_{\mathcal{D}_n\mathcal{E}_p} = e^k_{nn} Q^{k\ -1}_{nn} e^{kT}_{pn} + \varepsilon^k_{np}, \quad \hat{C}^k_{\mathcal{D}_n\mathcal{E}_n} = e^k_{nn} Q^{k\ -1}_{nn} e^{kT}_{nn} + \varepsilon^k_{nn}
\end{aligned}
\tag{2.65}
$$

The governing equations are obtained by using the variational statement in Equation (2.60), the constitutive relations in Equations (2.61)–(2.64), and an opportune two-dimensional plate/shell model. In symbolic form, these equations are:

$$K_{uu} u + K_{u\sigma} \sigma_n + K_{u\Phi} \Phi = p_u - M_{uu} \ddot{u} \tag{2.66}$$

$$K_{\sigma u} u + K_{\sigma\sigma} \sigma_n + K_{\sigma\Phi} \Phi = 0 \tag{2.67}$$

$$K_{\Phi u} u + K_{\Phi\sigma} \sigma_n + K_{\Phi\Phi} \Phi = 0 \tag{2.68}$$

The transverse shear/normal stresses σ_n in Equations (2.66)–(2.68) are primary variables of the problem, and are directly obtained from the governing equations; this permits one to have transverse stresses that are a priori and completely fulfill the C^0_z requirements. This RMVT form has three variables (u, σ_n, and Φ), while the PVD for the electromechanical case has two variables (u and Φ). It is important to notice that the matrices K for the PVD in Equations (2.54) and (2.55) are completely different from those in the RMVT in Equations (2.66)–(2.68): this is because a Lagrange multiplier has been added and the constitutive equations have been rewritten.

2.5.1.1 Pure mechanical case

The RMVT, with the transverse shear/normal stresses modeled a priori for the case of pure mechanical problems, has the displacements u and the transverse stresses σ_n as variables. The variational statement can be obtained as a particular case of RMVT($\boldsymbol{u}$, $\boldsymbol{\Phi}$, $\boldsymbol{\sigma}_n$) simply by discarding the internal electrical work, and it is therefore possible from Equation (2.60) to obtain (Carrera and Demasi 2002a,b):

$$\int_V \left(\delta\boldsymbol{\epsilon}_{pG}^T\boldsymbol{\sigma}_{pC} + \delta\boldsymbol{\epsilon}_{nG}^T\boldsymbol{\sigma}_{nM} + \delta\boldsymbol{\sigma}_{nM}^T(\boldsymbol{\epsilon}_{nG} - \boldsymbol{\epsilon}_{nC})\right)dV = \delta L_e - \delta L_{in} \tag{2.69}$$

No new Lagrange multipliers are added to RMVT($\boldsymbol{u}$, $\boldsymbol{\Phi}$, $\boldsymbol{\sigma}_n$), therefore the constitutive equations can be considered as a particular case of those in Equations (2.61)–(2.64):

$$\boldsymbol{\sigma}_{pC}^k = \hat{\boldsymbol{C}}_{\sigma_p\epsilon_p}^k\boldsymbol{\epsilon}_{pG}^k + \hat{\boldsymbol{C}}_{\sigma_p\sigma_n}^k\boldsymbol{\sigma}_{nM}^k \tag{2.70}$$

$$\boldsymbol{\epsilon}_{nC}^k = \hat{\boldsymbol{C}}_{\epsilon_n\epsilon_p}^k\boldsymbol{\epsilon}_{pG}^k + \hat{\boldsymbol{C}}_{\epsilon_n\sigma_n}^k\boldsymbol{\sigma}_{nM}^k \tag{2.71}$$

The governing equations can be obtained using Equation (2.69) and Equations (2.70) and (2.71) (Carrera 1999a,b):

$$\boldsymbol{K}_{uu}\boldsymbol{u} + \boldsymbol{K}_{u\sigma}\boldsymbol{\sigma}_n = \boldsymbol{p}_u - \boldsymbol{M}_{uu}\ddot{\boldsymbol{u}} \tag{2.72}$$

$$\boldsymbol{K}_{\sigma u}\boldsymbol{u} + \boldsymbol{K}_{\sigma\sigma}\boldsymbol{\sigma}_n = 0 \tag{2.73}$$

The governing equations in Equations (2.72) and (2.73) can be simply obtained by eliminating the third column and the third line in Equations (2.66)–(2.68). The remaining matrices $\boldsymbol{K}$ are the same as those in RMVT($\boldsymbol{u}$, $\boldsymbol{\Phi}$, $\boldsymbol{\sigma}_n$) and RMVT($\boldsymbol{u}$, $\boldsymbol{\sigma}_n$). The matrix $\boldsymbol{K}_{uu}$ is completely different from that in Equation (2.59), for the pure mechanical PVD, because of the introduction of a Lagrange multiplier and the consequent rearrangement of the constitutive equations.

2.5.2 RMVT($\boldsymbol{u}$, $\boldsymbol{\Phi}$, $\boldsymbol{\mathcal{D}}_n$)

RMVT($\boldsymbol{u}$, $\boldsymbol{\Phi}$, $\boldsymbol{\mathcal{D}}_n$) is obtained, considering the variational statement in Equation (2.53) for the electromechanical PVD($\boldsymbol{u}$, $\boldsymbol{\Phi}$), by a priori modeling the transverse normal electric displacement $\boldsymbol{\mathcal{D}}_{nM}$ (here, the new subscript M is introduced to show that the transverse normal electric displacement is now modeled and not obtained via the constitutive equations). The added Lagrange multiplier is $\delta\boldsymbol{\mathcal{D}}_{nM}^T(\boldsymbol{\mathcal{E}}_{nG} - \boldsymbol{\mathcal{E}}_{nC})$. The condition necessary to add this multiplier is that the transverse normal electric field $\boldsymbol{\mathcal{E}}_n$, calculated by means of geometrical relations (G) and using the constitutive equations (C), must be the same or almost the same. In this way the internal work does not change or remains almost the same:

$$\int_V \Big(\delta\boldsymbol{\epsilon}_{pG}^T\boldsymbol{\sigma}_{pC} + \delta\boldsymbol{\epsilon}_{nG}^T\boldsymbol{\sigma}_{nC} - \delta\boldsymbol{\mathcal{E}}_{pG}^T\boldsymbol{\mathcal{D}}_{pC} - \delta\boldsymbol{\mathcal{E}}_{nG}^T\boldsymbol{\mathcal{D}}_{nM}$$
$$-\delta\boldsymbol{\mathcal{D}}_{nM}^T(\boldsymbol{\mathcal{E}}_{nG} - \boldsymbol{\mathcal{E}}_{nC})\Big)dV = \delta L_e - \delta L_{in} \tag{2.74}$$

The relative constitutive equations are obtained from Equations (2.21)–(2.24) considering the transverse normal electric displacement $\mathcal{D}_n$ as being modeled (M) and the transverse normal electric field $\mathcal{E}_n$ as being obtained from constitutive equations (C):

$$\boldsymbol{\sigma}^k_{pC} = \bar{\boldsymbol{C}}^k_{\sigma_p\epsilon_p}\boldsymbol{\epsilon}^k_{pG} + \bar{\boldsymbol{C}}^k_{\sigma_p\epsilon_n}\boldsymbol{\epsilon}^k_{nG} + \bar{\boldsymbol{C}}^k_{\sigma_p\mathcal{E}_p}\boldsymbol{\mathcal{E}}^k_{pG} + \bar{\boldsymbol{C}}^k_{\sigma_p\mathcal{D}_n}\boldsymbol{\mathcal{D}}^k_{nM} \tag{2.75}$$

$$\boldsymbol{\sigma}^k_{nC} = \bar{\boldsymbol{C}}^k_{\sigma_n\epsilon_p}\boldsymbol{\epsilon}^k_{pG} + \bar{\boldsymbol{C}}^k_{\sigma_n\epsilon_n}\boldsymbol{\epsilon}^k_{nG} + \bar{\boldsymbol{C}}^k_{\sigma_n\mathcal{E}_p}\boldsymbol{\mathcal{E}}^k_{pG} + \bar{\boldsymbol{C}}^k_{\sigma_n\mathcal{D}_n}\boldsymbol{\mathcal{D}}^k_{nM} \tag{2.76}$$

$$\boldsymbol{\mathcal{D}}^k_{pC} = \bar{\boldsymbol{C}}^k_{\mathcal{D}_p\epsilon_p}\boldsymbol{\epsilon}^k_{pG} + \bar{\boldsymbol{C}}^k_{\mathcal{D}_p\epsilon_n}\boldsymbol{\epsilon}^k_{nG} + \bar{\boldsymbol{C}}^k_{\mathcal{D}_p\mathcal{E}_p}\boldsymbol{\mathcal{E}}^k_{pG} + \bar{\boldsymbol{C}}^k_{\mathcal{D}_p\mathcal{D}_n}\boldsymbol{\mathcal{D}}^k_{nM} \tag{2.77}$$

$$\boldsymbol{\mathcal{E}}^k_{nC} = \bar{\boldsymbol{C}}^k_{\mathcal{E}_n\epsilon_p}\boldsymbol{\epsilon}^k_{pG} + \bar{\boldsymbol{C}}^k_{\mathcal{E}_n\epsilon_n}\boldsymbol{\epsilon}^k_{nG} + \bar{\boldsymbol{C}}^k_{\mathcal{E}_n\mathcal{E}_p}\boldsymbol{\mathcal{E}}^k_{pG} + \bar{\boldsymbol{C}}^k_{\mathcal{E}_n\mathcal{D}_n}\boldsymbol{\mathcal{D}}^k_{nM} \tag{2.78}$$

According to Carrera *et al.* (2008), the meaning of coefficients $\bar{\boldsymbol{C}}$ in Equations (2.75)–(2.78) is:

$$\begin{aligned}
\bar{\boldsymbol{C}}^k_{\sigma_p\epsilon_p} &= \boldsymbol{Q}^k_{pp} + \boldsymbol{e}^{kT}_{np}\boldsymbol{\varepsilon}^{k\ -1}_{nn}\boldsymbol{e}^k_{np}, \quad \bar{\boldsymbol{C}}^k_{\sigma_p\epsilon_n} = \boldsymbol{Q}^k_{pn} + \boldsymbol{e}^{kT}_{np}\boldsymbol{\varepsilon}^{k\ -1}_{nn}\boldsymbol{e}^k_{nn}\\
\bar{\boldsymbol{C}}^k_{\sigma_p\mathcal{E}_p} &= \boldsymbol{e}^{kT}_{np}\boldsymbol{\varepsilon}^{k\ -1}_{nn}\boldsymbol{\varepsilon}^k_{np} - \boldsymbol{e}^{kT}_{pp}, \quad \bar{\boldsymbol{C}}^k_{\sigma_p\mathcal{D}_n} = -\boldsymbol{e}^{kT}_{np}\boldsymbol{\varepsilon}^{k\ -1}_{nn}\\
\bar{\boldsymbol{C}}^k_{\sigma_n\epsilon_p} &= \boldsymbol{Q}^k_{np} + \boldsymbol{e}^{kT}_{nn}\boldsymbol{\varepsilon}^{k\ -1}_{nn}\boldsymbol{e}^k_{np}, \quad \bar{\boldsymbol{C}}^k_{\sigma_n\epsilon_n} = \boldsymbol{Q}^k_{nn} + \boldsymbol{e}^{kT}_{nn}\boldsymbol{\varepsilon}^{k\ -1}_{nn}\boldsymbol{e}^k_{nn}\\
\bar{\boldsymbol{C}}^k_{\sigma_n\mathcal{E}_p} &= \boldsymbol{e}^{kT}_{nn}\boldsymbol{\varepsilon}^{k\ -1}_{nn}\boldsymbol{\varepsilon}^k_{np} - \boldsymbol{e}^{kT}_{pn}, \quad \bar{\boldsymbol{C}}^k_{\sigma_n\mathcal{D}_n} = -\boldsymbol{e}^{kT}_{nn}\boldsymbol{\varepsilon}^{k\ -1}_{nn}\\
\bar{\boldsymbol{C}}^k_{\mathcal{D}_p\epsilon_p} &= \boldsymbol{e}^k_{pp} - \boldsymbol{\varepsilon}^k_{pn}\boldsymbol{\varepsilon}^{k\ -1}_{nn}\boldsymbol{e}^k_{np}, \quad \bar{\boldsymbol{C}}^k_{\mathcal{D}_p\epsilon_n} = \boldsymbol{e}^k_{pn} - \boldsymbol{\varepsilon}^k_{pn}\boldsymbol{\varepsilon}^{k\ -1}_{nn}\boldsymbol{e}^k_{nn}\\
\bar{\boldsymbol{C}}^k_{\mathcal{D}_p\mathcal{E}_p} &= \boldsymbol{\varepsilon}^k_{pp} - \boldsymbol{\varepsilon}^k_{pn}\boldsymbol{\varepsilon}^{k\ -1}_{nn}\boldsymbol{\varepsilon}^k_{np}, \quad \bar{\boldsymbol{C}}^k_{\mathcal{D}_p\mathcal{D}_n} = \boldsymbol{\varepsilon}^k_{pn}\boldsymbol{\varepsilon}^{k\ -1}_{nn}\\
\bar{\boldsymbol{C}}^k_{\mathcal{E}_n\epsilon_p} &= -\boldsymbol{\varepsilon}^{k\ -1}_{nn}\boldsymbol{e}^k_{np}, \quad \bar{\boldsymbol{C}}^k_{\mathcal{E}_n\epsilon_n} = -\boldsymbol{\varepsilon}^{k\ -1}_{nn}\boldsymbol{e}^k_{nn}\\
\bar{\boldsymbol{C}}^k_{\mathcal{E}_n\mathcal{E}_p} &= -\boldsymbol{\varepsilon}^{k\ -1}_{nn}\boldsymbol{\varepsilon}^k_{np}, \quad \bar{\boldsymbol{C}}^k_{\mathcal{E}_n\mathcal{D}_n} = \boldsymbol{\varepsilon}^{k\ -1}_{nn}
\end{aligned} \tag{2.79}$$

It is possible to obtain the governing equations using the variational statement in Equation (2.74), the constitutive relations in Equations (2.75)–(2.78), and an opportune two-dimensional plate/shell model. In symbolic form, these are:

$$\boldsymbol{K}_{uu}\boldsymbol{u} + \boldsymbol{K}_{u\Phi}\boldsymbol{\Phi} + \boldsymbol{K}_{u\mathcal{D}}\boldsymbol{\mathcal{D}}_n = \boldsymbol{p}_u - \boldsymbol{M}_{uu}\ddot{\boldsymbol{u}} \tag{2.80}$$

$$\boldsymbol{K}_{\Phi u}\boldsymbol{u} + \boldsymbol{K}_{\Phi\Phi}\boldsymbol{\Phi} + \boldsymbol{K}_{\Phi\mathcal{D}}\boldsymbol{\mathcal{D}}_n = 0 \tag{2.81}$$

$$\boldsymbol{K}_{\mathcal{D}u}\boldsymbol{u} + \boldsymbol{K}_{\mathcal{D}\Phi}\boldsymbol{\Phi} + \boldsymbol{K}_{\mathcal{D}\mathcal{D}}\boldsymbol{\mathcal{D}}_n = 0 \tag{2.82}$$

The transverse normal electric displacement $\mathcal{D}_n$ in Equations (2.80)–(2.82) is a primary variable of the problem, and it is directly obtained from the governing equations; this permits one to have the transverse normal electric displacement that a priori and completely fulfills the C_z^0 requirements. This RMVT has three variables ($\boldsymbol{u}$, $\boldsymbol{\Phi}$, and $\mathcal{D}_n$), while the PVD for the electromechanical cases has two variables ($\boldsymbol{u}$ and $\boldsymbol{\Phi}$). It is important to notice that the matrices $\boldsymbol{K}$ in Equations (2.54) and (2.55) are completely different from those in the RMVT in Equations (2.80)–(2.82).

2.5.3 RMVT(u, Φ, σ_n, $\mathcal{D}_n$)

The starting point is the variational statement in Equation (2.53) for the electromechanical PVD($\boldsymbol{u}$, $\boldsymbol{\Phi}$). RMVT($\boldsymbol{u}$, $\boldsymbol{\Phi}$, $\boldsymbol{\sigma}_n$, $\mathcal{D}_n$) is obtained by a priori modeling both the transverse shear/normal stresses $\boldsymbol{\sigma}_{nM}$ and the transverse normal electric displacement $\mathcal{D}_{nM}$ (here, the new subscript M is introduced to show that the transverse stresses and normal electric displacements are now modeled and not obtained via the constitutive equations). The added Lagrange multipliers are $\delta\boldsymbol{\sigma}_{nM}^T(\boldsymbol{\epsilon}_{nG} - \boldsymbol{\epsilon}_{nC})$ and $\delta\mathcal{D}_{nM}^T(\mathcal{E}_{nG} - \mathcal{E}_{nC})$. The condition necessary to include these multipliers is that the transverse strains $\boldsymbol{\epsilon}_n$ and the normal electric field $\mathcal{E}_n$, calculated by means of geometrical relations (G) and using constitutive equations (C), must be the same or almost the same. In this way the internal work does not change or remains almost the same (Carrera and Brischetto 2007a,b):

$$\int_V \Big(\delta\boldsymbol{\epsilon}_{pG}^T\boldsymbol{\sigma}_{pC} + \delta\boldsymbol{\epsilon}_{nG}^T\boldsymbol{\sigma}_{nM} + \delta\boldsymbol{\sigma}_{nM}^T(\boldsymbol{\epsilon}_{nG} - \boldsymbol{\epsilon}_{nC}) - \delta\mathcal{E}_{pG}^T\mathcal{D}_{pC} - \delta\mathcal{E}_{nG}^T\mathcal{D}_{nM}$$
$$-\delta\mathcal{D}_{nM}^T(\mathcal{E}_{nG} - \mathcal{E}_{nC})\Big)dV = \delta L_e - \delta L_{in} \tag{2.83}$$

The relative constitutive equations are obtained from Equations (2.21)–(2.24), by considering the transverse stresses $\boldsymbol{\sigma}_n$ and the normal electric displacement $\mathcal{D}_n$ as being modeled (M) and the transverse strains $\boldsymbol{\epsilon}_n$ and the normal electric field $\mathcal{E}_n$ as being obtained from constitutive equations (C):

$$\boldsymbol{\sigma}_{pC}^k = \tilde{\boldsymbol{C}}_{\sigma_p\epsilon_p}^k\boldsymbol{\epsilon}_{pG}^k + \tilde{\boldsymbol{C}}_{\sigma_p\sigma_n}^k\boldsymbol{\sigma}_{nM}^k + \tilde{\boldsymbol{C}}_{\sigma_p\mathcal{E}_p}^k\mathcal{E}_{pG}^k + \tilde{\boldsymbol{C}}_{\sigma_p\mathcal{D}_n}^k\mathcal{D}_{nM}^k \tag{2.84}$$

$$\boldsymbol{\epsilon}_{nC}^k = \tilde{\boldsymbol{C}}_{\epsilon_n\epsilon_p}^k\boldsymbol{\epsilon}_{pG}^k + \tilde{\boldsymbol{C}}_{\epsilon_n\sigma_n}^k\boldsymbol{\sigma}_{nM}^k + \tilde{\boldsymbol{C}}_{\epsilon_n\mathcal{E}_p}^k\mathcal{E}_{pG}^k + \tilde{\boldsymbol{C}}_{\epsilon_n\mathcal{D}_n}^k\mathcal{D}_{nM}^k \tag{2.85}$$

$$\mathcal{D}_{pC}^k = \tilde{\boldsymbol{C}}_{\mathcal{D}_p\epsilon_p}^k\boldsymbol{\epsilon}_{pG}^k + \tilde{\boldsymbol{C}}_{\mathcal{D}_p\sigma_n}^k\boldsymbol{\sigma}_{nM}^k + \tilde{\boldsymbol{C}}_{\mathcal{D}_p\mathcal{E}_p}^k\mathcal{E}_{pG}^k + \tilde{\boldsymbol{C}}_{\mathcal{D}_p\mathcal{D}_n}^k\mathcal{D}_{nM}^k \tag{2.86}$$

$$\mathcal{E}_{nC}^k = \tilde{\boldsymbol{C}}_{\mathcal{E}_n\epsilon_p}^k\boldsymbol{\epsilon}_{pG}^k + \tilde{\boldsymbol{C}}_{\mathcal{E}_n\sigma_n}^k\boldsymbol{\sigma}_{nM}^k + \tilde{\boldsymbol{C}}_{\mathcal{E}_n\mathcal{E}_p}^k\mathcal{E}_{pG}^k + \tilde{\boldsymbol{C}}_{\mathcal{E}_n\mathcal{D}_n}^k\mathcal{D}_{nM}^k \tag{2.87}$$

According to Carrera and Brischetto (2007a,b), the meaning of coefficients $\tilde{\boldsymbol{C}}$ in Equations (2.84)–(2.87) is:

$$\begin{aligned}
\tilde{\boldsymbol{C}}^k_{\sigma_p\epsilon_p} &= \boldsymbol{Q}^k_{pp} - \boldsymbol{Q}^k_{pn}\boldsymbol{Q}^{k\,-1}_{nn}\boldsymbol{Q}^k_{np} - \left(\boldsymbol{Q}^k_{pn}\boldsymbol{Q}^{k\,-1}_{nn}\boldsymbol{e}^{kT}_{nn} - \boldsymbol{e}^{kT}_{np}\right)\left(\boldsymbol{e}^k_{nn}\boldsymbol{Q}^{k\,-1}_{nn}\boldsymbol{e}^{kT}_{nn} + \boldsymbol{\varepsilon}^k_{nn}\right)^{-1} \\
&\quad \times \left(\boldsymbol{e}^k_{np} - \boldsymbol{e}^k_{nn}\boldsymbol{Q}^{k\,-1}_{nn}\boldsymbol{Q}^k_{np}\right) \\
\tilde{\boldsymbol{C}}^k_{\sigma_p\sigma_n} &= \boldsymbol{Q}^k_{pn}\boldsymbol{Q}^{k\,-1}_{nn} - \left(\boldsymbol{Q}^k_{pn}\boldsymbol{Q}^{k\,-1}_{nn}\boldsymbol{e}^{kT}_{nn} - \boldsymbol{e}^{kT}_{np}\right)\left(\boldsymbol{e}^k_{nn}\boldsymbol{Q}^{k\,-1}_{nn}\boldsymbol{e}^{kT}_{nn} + \boldsymbol{\varepsilon}^k_{nn}\right)^{-1}\left(\boldsymbol{e}^k_{nn}\boldsymbol{Q}^{k\,-1}_{nn}\right) \\
\tilde{\boldsymbol{C}}^k_{\sigma_p\mathcal{E}_p} &= \boldsymbol{Q}^k_{pn}\boldsymbol{Q}^{k\,-1}_{nn}\boldsymbol{e}^{kT}_{pn} - \boldsymbol{e}^{kT}_{pp} - \left(\boldsymbol{Q}^k_{pn}\boldsymbol{Q}^{k\,-1}_{nn}\boldsymbol{e}^{kT}_{nn} - \boldsymbol{e}^{kT}_{np}\right)\left(\boldsymbol{e}^k_{nn}\boldsymbol{Q}^{k\,-1}_{nn}\boldsymbol{e}^{kT}_{nn} + \boldsymbol{\varepsilon}^k_{nn}\right)^{-1} \\
&\quad \times \left(\boldsymbol{e}^k_{nn}\boldsymbol{Q}^{k\,-1}_{nn}\boldsymbol{e}^{kT}_{pn} + \boldsymbol{\varepsilon}^k_{np}\right) \\
\tilde{\boldsymbol{C}}^k_{\sigma_p\mathcal{D}_n} &= \left(\boldsymbol{Q}^k_{pn}\boldsymbol{Q}^{k\,-1}_{nn}\boldsymbol{e}^{kT}_{nn} - \boldsymbol{e}^{kT}_{np}\right)\left(\boldsymbol{e}^k_{nn}\boldsymbol{Q}^{k\,-1}_{nn}\boldsymbol{e}^{kT}_{nn} + \boldsymbol{\varepsilon}^k_{nn}\right)^{-1} \\
\tilde{\boldsymbol{C}}^k_{\epsilon_n\epsilon_p} &= -\boldsymbol{Q}^{k\,-1}_{nn}\boldsymbol{Q}^k_{np} - \boldsymbol{Q}^{k\,-1}_{nn}\boldsymbol{e}^{kT}_{nn}\left(\boldsymbol{e}^k_{nn}\boldsymbol{Q}^{k\,-1}_{nn}\boldsymbol{e}^{kT}_{nn} + \boldsymbol{\varepsilon}^k_{nn}\right)^{-1}\left(\boldsymbol{e}^k_{nn}\boldsymbol{Q}^{k\,-1}_{nn}\boldsymbol{Q}^k_{np} - \boldsymbol{e}^k_{np}\right) \\
\tilde{\boldsymbol{C}}^k_{\epsilon_n\sigma_n} &= \boldsymbol{Q}^{k\,-1}_{nn} - \boldsymbol{Q}^{k\,-1}_{nn}\boldsymbol{e}^{kT}_{nn}\left(\boldsymbol{e}^k_{nn}\boldsymbol{Q}^{k\,-1}_{nn}\boldsymbol{e}^{kT}_{nn} + \boldsymbol{\varepsilon}^k_{nn}\right)^{-1}\boldsymbol{e}^k_{nn}\boldsymbol{Q}^{k\,-1}_{nn} \\
\tilde{\boldsymbol{C}}^k_{\epsilon_n\mathcal{E}_p} &= \boldsymbol{Q}^{k\,-1}_{nn}\boldsymbol{e}^{kT}_{pn} - \boldsymbol{Q}^{k\,-1}_{nn}\boldsymbol{e}^{kT}_{nn}\left(\boldsymbol{e}^k_{nn}\boldsymbol{Q}^{k\,-1}_{nn}\boldsymbol{e}^{kT}_{nn} + \boldsymbol{\varepsilon}^k_{nn}\right)^{-1}\left(\boldsymbol{e}^k_{nn}\boldsymbol{Q}^{k\,-1}_{nn}\boldsymbol{e}^{kT}_{pn} + \boldsymbol{\varepsilon}^k_{np}\right) \\
\tilde{\boldsymbol{C}}^k_{\epsilon_n\mathcal{D}_n} &= \boldsymbol{Q}^{k\,-1}_{nn}\boldsymbol{e}^{kT}_{nn}\left(\boldsymbol{e}^k_{nn}\boldsymbol{Q}^{k\,-1}_{nn}\boldsymbol{e}^{kT}_{nn} + \boldsymbol{\varepsilon}^k_{nn}\right)^{-1} \qquad (2.88) \\
\tilde{\boldsymbol{C}}^k_{\mathcal{D}_p\epsilon_p} &= \boldsymbol{e}^k_{pp} - \boldsymbol{e}^k_{pn}\boldsymbol{Q}^{k\,-1}_{nn}\boldsymbol{Q}^k_{np} - \boldsymbol{e}^k_{pn}\boldsymbol{Q}^{k\,-1}_{nn}\boldsymbol{e}^{kT}_{nn}\left(\boldsymbol{e}^k_{nn}\boldsymbol{Q}^{k\,-1}_{nn}\boldsymbol{e}^{kT}_{nn} + \boldsymbol{\varepsilon}^k_{nn}\right)^{-1}\left(\boldsymbol{e}^k_{np} - \boldsymbol{e}^k_{nn}\right. \\
&\quad \left.\boldsymbol{Q}^{k\,-1}_{nn}\boldsymbol{Q}^k_{np}\right) - \boldsymbol{\varepsilon}^k_{pn}\left(\boldsymbol{e}^k_{nn}\boldsymbol{Q}^{k\,-1}_{nn}\boldsymbol{e}^{kT}_{nn} + \boldsymbol{\varepsilon}^k_{nn}\right)^{-1}\left(\boldsymbol{e}^k_{np} - \boldsymbol{e}^k_{nn}\boldsymbol{Q}^{k\,-1}_{nn}\boldsymbol{Q}^k_{np}\right) \\
\tilde{\boldsymbol{C}}^k_{\mathcal{D}_p\sigma_n} &= \boldsymbol{e}^k_{pn}\boldsymbol{Q}^{k\,-1}_{nn} - \boldsymbol{e}^k_{pn}\boldsymbol{Q}^{k\,-1}_{nn}\boldsymbol{e}^{kT}_{nn}\left(\boldsymbol{e}^k_{nn}\boldsymbol{Q}^{k\,-1}_{nn}\boldsymbol{e}^{kT}_{nn} + \boldsymbol{\varepsilon}^k_{nn}\right)^{-1}\boldsymbol{e}^k_{nn}\boldsymbol{Q}^{k\,-1}_{nn} \\
&\quad - \boldsymbol{\varepsilon}^k_{pn}\left(\boldsymbol{e}^k_{nn}\boldsymbol{Q}^{k\,-1}_{nn}\boldsymbol{e}^{kT}_{nn} + \boldsymbol{\varepsilon}^k_{nn}\right)^{-1}\boldsymbol{e}^k_{nn}\boldsymbol{Q}^{k\,-1}_{nn} \\
\tilde{\boldsymbol{C}}^k_{\mathcal{D}_p\mathcal{E}_p} &= \boldsymbol{\varepsilon}^k_{pp} + \boldsymbol{e}^k_{pn}\boldsymbol{Q}^{k\,-1}_{nn}\boldsymbol{e}^{kT}_{pn} - \boldsymbol{e}^k_{pn}\boldsymbol{Q}^{k\,-1}_{nn}\boldsymbol{e}^{kT}_{nn}\left(\boldsymbol{e}^k_{nn}\boldsymbol{Q}^{k\,-1}_{nn}\boldsymbol{e}^{kT}_{nn} + \boldsymbol{\varepsilon}^k_{nn}\right)^{-1}\left(\boldsymbol{e}^k_{nn}\boldsymbol{Q}^{k\,-1}_{nn}\right. \\
&\quad \left.\times\,\boldsymbol{e}^{kT}_{pn} + \boldsymbol{\varepsilon}^k_{np}\right) - \boldsymbol{\varepsilon}^k_{np}\left(\boldsymbol{e}^k_{nn}\boldsymbol{Q}^{k\,-1}_{nn}\boldsymbol{e}^{kT}_{nn} + \boldsymbol{\varepsilon}^k_{nn}\right)^{-1}\left(\boldsymbol{e}^k_{nn}\boldsymbol{Q}^{k\,-1}_{nn}\boldsymbol{e}^{kT}_{pn} + \boldsymbol{\varepsilon}^k_{np}\right) \\
\tilde{\boldsymbol{C}}^k_{\mathcal{D}_p\mathcal{D}_n} &= \boldsymbol{e}^k_{pn}\boldsymbol{Q}^{k\,-1}_{nn}\boldsymbol{e}^{kT}_{nn}\left(\boldsymbol{e}^k_{nn}\boldsymbol{Q}^{k\,-1}_{nn}\boldsymbol{e}^{kT}_{nn} + \boldsymbol{\varepsilon}^k_{nn}\right)^{-1} + \boldsymbol{\varepsilon}^k_{pn}\left(\boldsymbol{e}^k_{nn}\boldsymbol{Q}^{k\,-1}_{nn}\boldsymbol{e}^{kT}_{nn} + \boldsymbol{\varepsilon}^k_{nn}\right)^{-1} \\
\tilde{\boldsymbol{C}}^k_{\mathcal{E}_n\epsilon_p} &= -\left(\boldsymbol{e}^k_{nn}\boldsymbol{Q}^{k\,-1}_{nn}\boldsymbol{e}^{kT}_{nn} + \boldsymbol{\varepsilon}^k_{nn}\right)^{-1}\left(\boldsymbol{e}^k_{np} - \boldsymbol{e}^k_{nn}\boldsymbol{Q}^{k\,-1}_{nn}\boldsymbol{Q}^k_{np}\right)
\end{aligned}$$

$$\tilde{\boldsymbol{C}}^k_{\mathcal{E}_n\sigma_n} = -\left(\boldsymbol{e}^k_{nn}\boldsymbol{Q}^{k\ -1}_{nn}\boldsymbol{e}^{kT}_{nn} + \boldsymbol{\varepsilon}^k_{nn}\right)^{-1}\boldsymbol{e}^k_{nn}\boldsymbol{Q}^{k\ -1}_{nn}$$

$$\tilde{\boldsymbol{C}}^k_{\mathcal{E}_n\mathcal{E}_p} = -\left(\boldsymbol{e}^k_{nn}\boldsymbol{Q}^{k\ -1}_{nn}\boldsymbol{e}^{kT}_{nn} + \boldsymbol{\varepsilon}^k_{nn}\right)^{-1}\left(\boldsymbol{e}^k_{nn}\boldsymbol{Q}^{k\ -1}_{nn}\boldsymbol{e}^{kT}_{pn} + \boldsymbol{\varepsilon}^k_{np}\right)$$

$$\tilde{\boldsymbol{C}}^k_{\mathcal{E}_n\mathcal{D}_n} = \left(\boldsymbol{e}^k_{nn}\boldsymbol{Q}^{k\ -1}_{nn}\boldsymbol{e}^{kT}_{nn} + \boldsymbol{\varepsilon}^k_{nn}\right)^{-1}$$

The governing equations are obtained from the variational statement in Equation (2.83), the constitutive relations in Equations (2.84)–(2.87), and using an opportune two-dimensional plate/shell model. In symbolic form, these equations are:

$$\boldsymbol{K}_{uu}\boldsymbol{u} + \boldsymbol{K}_{u\sigma}\boldsymbol{\sigma}_n + \boldsymbol{K}_{u\Phi}\boldsymbol{\Phi} + \boldsymbol{K}_{u\mathcal{D}}\boldsymbol{\mathcal{D}}_n = \boldsymbol{p}_u - \boldsymbol{M}_{uu}\ddot{\boldsymbol{u}} \tag{2.89}$$

$$\boldsymbol{K}_{\sigma u}\boldsymbol{u} + \boldsymbol{K}_{\sigma\sigma}\boldsymbol{\sigma}_n + \boldsymbol{K}_{\sigma\Phi}\boldsymbol{\Phi} + \boldsymbol{K}_{\sigma\mathcal{D}}\boldsymbol{\mathcal{D}}_n = 0 \tag{2.90}$$

$$\boldsymbol{K}_{\Phi u}\boldsymbol{u} + \boldsymbol{K}_{\Phi\sigma}\boldsymbol{\sigma}_n + \boldsymbol{K}_{\Phi\Phi}\boldsymbol{\Phi} + \boldsymbol{K}_{\Phi\mathcal{D}}\boldsymbol{\mathcal{D}}_n = 0 \tag{2.91}$$

$$\boldsymbol{K}_{\mathcal{D}u}\boldsymbol{u} + \boldsymbol{K}_{\mathcal{D}\sigma}\boldsymbol{\sigma}_n + \boldsymbol{K}_{\mathcal{D}\Phi}\boldsymbol{\Phi} + \boldsymbol{K}_{\mathcal{D}\mathcal{D}}\boldsymbol{\mathcal{D}}_n = 0 \tag{2.92}$$

The transverse shear/normal stresses $\boldsymbol{\sigma}_n$ and the normal electric displacement $\boldsymbol{\mathcal{D}}_n$ in Equations (2.89)–(2.92) are primary variables of the problem, and they can be obtained directly from the governing equations; this fact permits their C^0_z requirements to be fulfilled a priori and completely. The proposed RMVT has four primary variables ($\boldsymbol{u}$, $\boldsymbol{\sigma}_n$, $\boldsymbol{\Phi}$, and $\boldsymbol{\mathcal{D}}_n$). It is important to notice that the matrices $\boldsymbol{K}$ in Equations (2.89)–(2.92) are completely different from those of the PVD and the other two extensions of the RMVT: two different Lagrange multipliers have been added.

References

Altay GA and Dökmeci MC 1996a Fundamental variational equations of discontinuous thermopiezoelectric fields. *Int. J. Eng. Sci.* **34**, 769–782.

Altay GA and Dökmeci MC 1996b Some variational principles for linear coupled thermoelasticity. *Int. J. Solids Struct.* **33**, 3937–3948.

Cady WG 1964 *Piezoelectricity*. Dover.

Cannarozzi AA and Ubertini F 2001 A mixed variational method for linear coupled thermoelastic analyis. *Int. J. Solids Struct.* **38**, 717–739.

Carrera E 1999a Multilayered shell theories accounting for layerwise mixed description. Part I: governing equations. *AIAA J.* **37**, 1107–1116.

Carrera E 1999b Multilayered shell theories accounting for layerwise mixed description. Part II: numerical evaluations. *AIAA J.* **37**, 1117–1124.

Carrera E 2001 Developments, ideas and evaluation based upon Reissner's mixed variational theorem in the modelling of multilayered plates and shells. *Appl. Mech. Rev.* **54**, 301–329.

Carrera E and Demasi L 2002a Classical and advanced multilayered plate elements based upon PVD and RMVT. Part 1: derivation and finite element matrices. *Int. J. Numer. Methods Eng.* **55**, 191–231.

Carrera E and Demasi L 2002b Classical and advanced multilayered plate elements based upon PVD and RMVT. Part 2: numerical implementation. *Int. J. Numer. Methods Eng.* **55**, 253–291.

Carrera E and Brischetto S 2007a Piezoelectric shell theories with a priori continuous transverse electromechanical variables. *J. Mech. Mater Struct.* **2**, 377–399.

Carrera E and Brischetto S 2007b Reissner mixed theorem applied to static analysis of piezoelectric shells. *J. Intell. Mater. Syst. Struct.* **18**, 1083–1107.

Carrera E, Boscolo M, and Robaldo A 2007 Hierarchic multilayered plate elements for coupled multifield problems of piezoelectric adaptive structures: formulation and numerical assessment. *Arch. Comput. Methods Eng.* **14**, 383–430.

Carrera E, Brischetto S, and Nali P 2008 Variational statements and computational models for multifield problems and multilayered structures. *Mech. Adv. Mater. Struct.* **15**, 182–198.

Carrera E, Brischetto S, Fagiano C, and Nali P 2009a Mixed multilayered plate elements for coupled magneto-electroelastic analysis. *Multidiscip Modeling Mater. Struct.* **6**, 251–256.

Carrera E, Di Gifico M, Nali P, and Brischetto S 2009b Refined multilayered plate elements for coupled magneto-electroelastic analysis. *Multidiscip Modeling Mater. Struct.* **5**, 119–138.

Chopra I 2002 Review of state of art of smart structures and integrated systems. *AIAA J.* **40**, 2145–2187.

Correira VMF, Gomes MAA, Suleman A, Soares CMM, and Soares CAM 2000 Modeling and design of adaptive composite structures. *Comput. Methods Appl. Mech. Eng.* **185**, 326–346.

Ikeda T 1996 *Fundamentals of Piezoelectricity*. Oxford University Press.

Kawai H 1979 The piezoelectricity of polyrinydene fluoride. *Jpn. J. Appl. Phys.* **8**, 975–976.

Leissa AW 1969 *Vibration of Plates*. NASA SP-160.

Leissa AW 1973 *Vibration of Shells*. NASA SP-288.

Nowinski JL 1978 *Theory of Thermoelasticity with Applications*. Sijthoff & Noordhoff.

Ossadzow-David C and Touratier M 2003 Multilayered piezoelectric refined plate theory. *AIAA J.* **41**, 90–99.

Rao SS and Sunar M 1994 Piezoelectricity and its use in disturbance sensing and control of flexible structures: a survey. *Appl. Mech. Rev.* **47**, 113–123.

Reddy JN 2004 *Mechanics of Laminated Composite Plates and Shells; Theory and Analysis*. CRC Press.

Reissner E 1984 On a certain mixed variational theory and a proposed application. *Int. J. Numer. Methods Eng*. **20**, 1366–1368.

Rogacheva NN 1994 *The Theory of Piezoelectric Shells and Plates*. CRC Press.

Tani J, Takagi T, and Qiu J 1998 Intelligent materials systems: application of functional materials. *Appl. Mech. Rev.* **51**, 505–521.

Tiersten HF 1969 *Linear Piezoelectric Plate Vibrations*. Plenum Press.

Yang Q, Stainer L, and Ortiz M 2006 A variational formulation of the coupled thermo-mechanical boundary-value problem for general dissipative solids. *J. Mech. Phys. Solids* **54**, 401–424.

Yang JS and Yu JD 1993 Equations for laminated piezoelectric plate. *Arch. Mech*. **45**, 653–664.

3

Classical plate/shell theories

Two-dimensional plate/shell models are introduced in the axiomatic framework. First, classical theories for shells/plates, such as the classical lamination theory and the first-order shear deformation theory, are discussed for plate geometries in the case of pure mechanical analysis. Their extensions to piezoelectric problems is almost immediate, if a linear through-the-thickness electric potential is included. Examples are given for equilibrium equations in the case of smart structures; both the Kirchhoff and Reissner–Mindlin plate/shell theories are introduced in the case of a piezoelectric layer embedded in a multilayered smart structure.

3.1 Plate/shell theories

The analysis, design, and construction of layered structures is a cumbersome task. New, different, and complicated effects have arisen to add to those that are already known for traditional one-layered isotropic structures (Carrera 2002). Of all the possible topics, the present chapter is dedicated to aspects related to two-dimensional modeling of layered plate and shell structures (Reddy 2004). Classical two-dimensional models will be extended to piezoelectric problems.

Several approaches can be used to analyze plates and shells: three-dimensional approaches; continuum-based methods; axiomatic and asymptotic two-dimensional theories. The theories most commonly employed in this work are axiomatic two-dimensional models. This chapter discusses only classical theories, while refined and advanced ones can be found in Chapters 6–9.

Plates and Shells for Smart Structures: Classical and Advanced Theories for Modeling and Analysis, First Edition.
Erasmo Carrera, Salvatore Brischetto and Pietro Nali.

3.1.1 Three-dimensional problems

A first obvious approach to multilayered plates and shells is that of three-dimensional (3D) analysis. Such a 3D analysis can be implemented by solving, in a strong or a weak form, the fundamental differential equations of 3D elasticity: equilibrium equations, compatibility equations, and physical constitutive relations (Carrera 2002). Arrays of differential operators in the equations are defined in a 3D continuum body with domain Σ and boundary Γ. The unknown quantities, such as the displacements, stresses, and strains, are defined at each point $P(x, y, z)$ of a given reference system (x, y, z). When a plate/shell problem is dealt with using the direct solution of equilibrium, compatibility, and constitutive equations, a 3D analysis has been acquired. Typical examples of 3D solutions for layered structures are given in Noor and Rarig (1974), Pagano (1969, 1970), and Pagano and Hatfiled (1972) for pure mechanical problems.

However, 3D solutions are difficult to obtain, and often cannot be given in strong form for each geometry, laminate layout, boundary or loading condition case. Moreover, if we consider a finite element implementation of 3D approaches, its computational costs often prove to be prohibitive for practical problems. For all these reasons, two-dimensional (2D) approaches have become more popular than 3D ones.

3.1.2 Two-dimensional approaches

Plates and shells are, by definition, 2D structures, because one dimension, in general the thickness h, is at least one order of magnitude lower than representative in-plane dimensions a and b that are measured on the reference plate/shell surface Ω. This fact makes it possible to reduce a 3D problem to a 2D one. Such a reduction can be seen as a transformation of the problem defined at each point $P_\Sigma(x, y, z)$ of the 3D continuum, occupied by the considered plate, into a problem defined at each point $P_\Omega(x, y)$ of a reference plate surface Ω. A typical multilayered plate is given in Figure 3.1, where (x, y, z) is an orthogonal

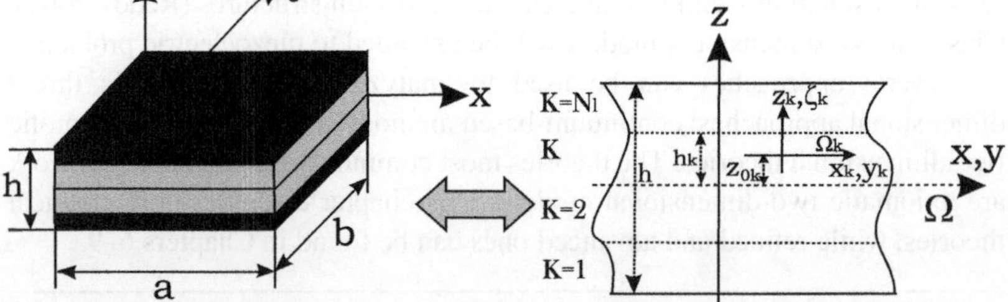

Figure 3.1 Geometrical notations for a multilayered plate.

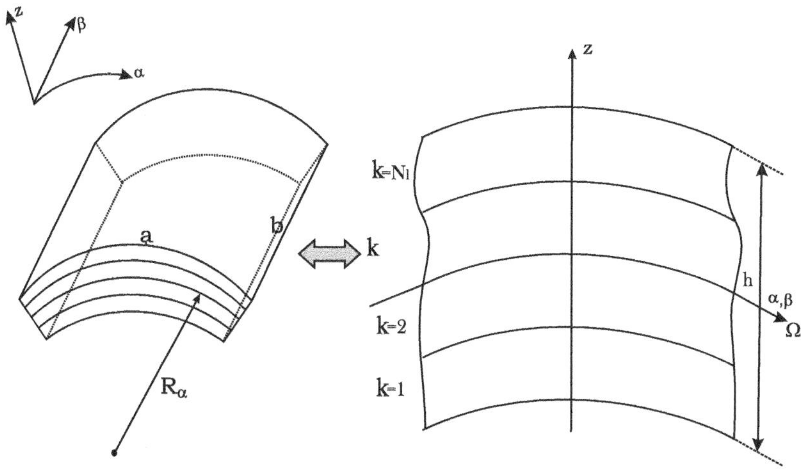

Figure 3.2 Geometrical notations for a multilayered shell.

rectilinear coordinate system, Ω is the middle reference surface of the multilayered structure, and Ω_k is the reference surface for each k layer of thickness h_k. A local orthogonal rectilinear coordinate system (x_k, y_k, z_k) can be defined for each layer.

The reduction from a 3D problem to a 2D one can also be made for shell structures; in this case, the considered structures are curved in the two in-plane curvilinear directions α and β. A typical multilayered shell is shown in Figure 3.2. The reference surface Ω is a curvilinear surface and the 2D problem is obtained by considering points $P_\Omega(\alpha, \beta)$ instead of points $P_\Sigma(\alpha, \beta, z)$. In Figure 3.2, (α, β, z) is the curvilinear orthogonal reference system. Plates can be considered as particular cases of shell geometries; however, in this book, plate and shell refined models will be considered separately in order to help readers.

The 2D modeling of plates and shells is a classical problem of the theory of structures. The elimination of the thickness coordinate z is usually performed on integration of the equilibrium equations, compatibility equations, and physical constitutive relations. The elimination of the z-coordinate can be made according to several methodologies; these methodologies lead to a significant number of approaches and techniques (Green and Naghdi 1967; Koiter 1960; Reissner 1967). A possible classification of 2D approaches, even though there is a certain degree of conflict in the literature, can be made as follows:

- continuum-based or stress resultant-based models;
- asymptotic-type approaches;
- axiomatic-type approaches.

3.1.2.1 Continuum-based or stress resultant-based models

According to Green and Naghdi (1967), plate and shell theories can be obtained from a generalized continuum using the Cosserat surface concept (Cosserat and Cosserat 1909). In this kind of approach, a 3D continuum, i.e., a shell, is considered as a surface on which stress resultants are defined. Then 2D approximations are introduced at a certain level and integration is performed in the thickness direction. The most important advantage of these models is that they allow both geometric and physical nonlinear behavior to be considered in plate/shell theories.

3.1.2.2 Asymptotic-type approaches

In asymptotic approaches, a perturbation parameter δ, which is usually the ratio between the plate/shell thickness and a characteristic length ($\delta = h/a$), is defined. The 3D governing equations are expanded in terms of δ. For instance, the equilibrium equations $\boldsymbol{E}_\Sigma$ can be written as:

$$\boldsymbol{E}_\Sigma \approx \boldsymbol{E}_\Sigma^1 \delta^{p_1} + \boldsymbol{E}_\Sigma^2 \delta^{p_2} + \cdots + \boldsymbol{E}_\Sigma^N \delta^{p_N} \tag{3.1}$$

where the exponents p_i of the perturbation parameter δ are in general real numbers. The obtained 2D theories are related to the same order in δ. Several variational statements can be used to obtain the expansion in Equation (3.1). Interesting asymptotic approaches for shell structures were given in Cicala (1959, 1965). The main advantage of an asymptotic approach is that it gives a *consistent* approximation: all the terms have the same order of magnitude as the introduced perturbation parameter δ. The 3D solutions are approached when $\delta \rightarrow 0$. The extension of asymptotic approaches to multilayered structures introduces other difficulties; for example, in order to take into account the anisotropy of composite layers, another mechanical parameter must be introduced (e.g., the orthotropic ratio of lamina E_L/E_T).

3.1.2.3 Axiomatic-type approach

The 2D models which are dealt with in detail in this book are axiomatic approaches. In this case, the displacement field and/or stress field are postulated in the thickness direction z:

$$f(\alpha, \beta, z) = f_1(\alpha, \beta)F_1(z) + f_2(\alpha, \beta)F_2(z) + \cdots + f_N(\alpha, \beta)F_N(z) \tag{3.2}$$

where the generic function f can be the vector of displacements $\boldsymbol{u} = (u, v, w)$ in the case of a *displacement formulation*, the vector of strain components $\boldsymbol{\epsilon}$ in the case of a *strain formulation*, and the vector of stress components $\boldsymbol{\sigma}$ in

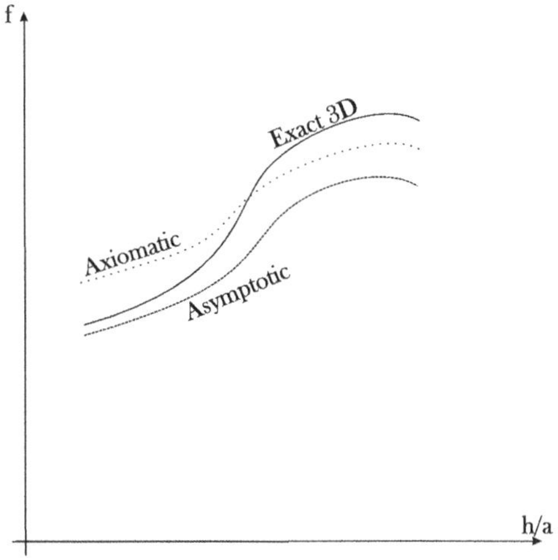

Figure 3.3 Comparison between some 2D approaches and the 3D exact solution in the case of a generic function f.

the case of a *stress formulation*. *Mixed formulations* can also be taken into account: for example, by considering both displacement components $\boldsymbol{u}$ and transverse shear/normal stress components $\boldsymbol{\sigma}_n = (\sigma_{\alpha z}, \sigma_{\beta z}, \sigma_{zz})$ as f functions. The f_i functions are the introduced unknowns that are defined on Ω, and F_i are the polynomials which have been introduced as the base functions of the expansion in z; in the case of a plate geometry, a rectilinear coordinate system (x, y, z) is considered in place of the curvilinear one (α, β, z). N is the order of expansion in the z direction. Different variational statements can be applied, depending on the formulation: a *displacement formulation* is based on the PVD (Reddy 2004), while a *mixed formulation* could use the RMVT (Reissner 1984). This book considers electromechanical problems, so other unknowns can be chosen as f functions: namely, the electric potential and the transverse normal electric displacement.

Axiomatic-type approaches offer the advantage of introducing *intuitive* approximations into plate/shell behavior (Antona 1991). Two cases of axiomatic and asymptotic approaches are compared in Figure 3.3 with respect to a 3D solution for a generic function f.

3.2 Complicating effects of layered structures

In the case of multilayered structures, new, complicating effects can arise with respect to isotropic one-layered plates and shells. These effects play a fundamental role in the development of any plate/shell theory. For these reasons, classical 2D theories are often inadequate for the analysis of such structures (Jones 1999). The main complicating effects introduced by multi-layered structures are:

- in-plane anisotropy;
- transverse anisotropy: zigzag effects and interlaminar continuity (C_z^0 requirements).

3.2.1 In-plane anisotropy

In the case of laminates made of anisotropic layers, a high in-plane anisotropy can be exhibited. This means that the structure has different mechanical–physical properties in different in-plane directions (Reddy 2004). Compared to traditional isotropic one-layered structures, multilayered composite plates/shells could show higher transverse shear/normal flexibility with respect to in-plane deformability. A consequence of this in-plane anisotropy is coupling between shear and axial strains (Jones 1999). Such a coupling leads to many complications in the solution procedure of an anisotropic structure. The 2D models must consider these effects. An example is that of the higher order shear deformation theory (HSDT); but, depending on the magnitude of the in-plane anisotropy, such theories might not be sufficient.

3.2.2 Transverse anisotropy, zigzag effects, and interlaminar continuity

A further complicating effect of multilayered structures is that of transverse anisotropy: the structures exhibit different mechanical–physical properties in the thickness direction z. Discontinuous transverse mechanical properties cause a displacement field, $\boldsymbol{u}$, in the thickness direction which can exhibit a rapid change in its slopes corresponding to each layer interface. This effect is known as the zigzag (ZZ) form of the displacement field in the thickness direction z (Carrera 2003), and it is clearly visible in the sandwich structure (two stiffer faces and a soft core) shown in Figure 3.4. In order to consider the ZZ form of displacements in deformed multilayered structures, a layer-wise approach may be necessary, as illustrated in Carrera and Brischetto (2009), or an opportune zigzag function could be added to the displacement field, as in Carrera (2004) and Demasi (2005). These topics are dealt with in more detail in the following chapters. In-plane stresses $\boldsymbol{\sigma}_p = (\sigma_{\alpha\alpha}, \sigma_{\beta\beta}, \sigma_{\alpha\beta})$ can, in general, be discontinuous at each layer interface. The transverse stresses $\boldsymbol{\sigma}_n = (\sigma_{\alpha z}, \sigma_{\beta z}, \sigma_{zz})$, instead, must be continuous at each layer interface for equilibrium reasons, as clearly

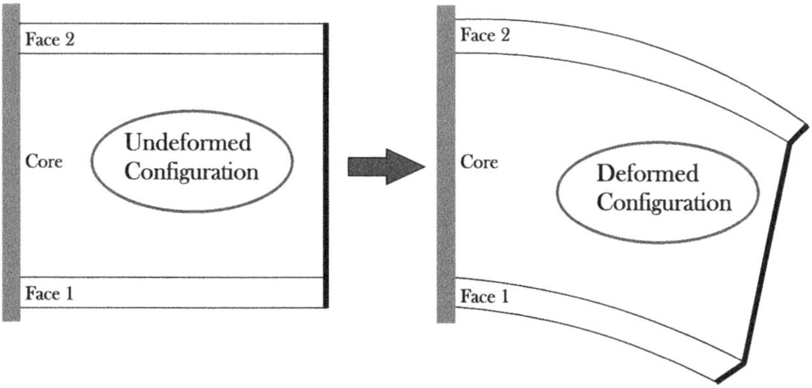

Figure 3.4 Typical zigzag effect for a sandwich structure in bending response.

illustrated in Figure 3.5 for a multilayered plate (coordinates (x, y, z) instead of the curvilinear ones (α, β, z)). In the literature, these conditions are called interlaminar continuity (IC) (Carrera 1997). The behavior that can be observed of in-plane stresses, displacements, and transverse stresses through the thickness z of a multilayered plate is clearly indicated in Figure 3.6, from a qualitative point of view. The in-plane components of stress can be discontinuous or continuous and they are only shown in the figure for comparison purposes. Displacements must be continuous in the z direction for compatibility reasons, while transverse shear/normal stresses must be continuous in the thickness z direction for equilibrium reasons, therefore $\boldsymbol{u}$ and $\boldsymbol{\sigma}_n$ are C^0-continuous functions in the z direction. Moreover, displacements and transverse stresses have discontinuous first derivatives corresponding to each interface, because the mechanical properties change in each layer (ZZ effect). In Carrera (1997) and Demasi (2008), ZZ and IC conditions are referred to as C_z^0 *requirements. The fulfillment of* C_z^0 *requirements is a crucial point in the development of appropriate 2D models for multilayered structures*. Displacement formulations must fulfill the C_z^0 requirements for the displacement components; mixed formulations must fulfill the C_z^0 requirements for both displacements and transverse stresses. In the case of multilayered anisotropic structures, classical theories, such as those based on Cauchy–Poisson–Kirchhoff–Love (Cauchy 1828; Poisson 1829; Kirchhoff 1850; Love 1906) hypotheses or Reissner–Mindlin (Reissner 1945; Mindlin 1951) hypotheses, which will be discussed in the next section, fulfill the IC conditions for displacements, but not the ZZ form of $\boldsymbol{u}$. For these reasons, they can often turn out to be inappropriate for the study of multilayered composite plates and shells. In this case, the use of refined and advanced 2D models (see Chapters 6–8) could be mandatory.

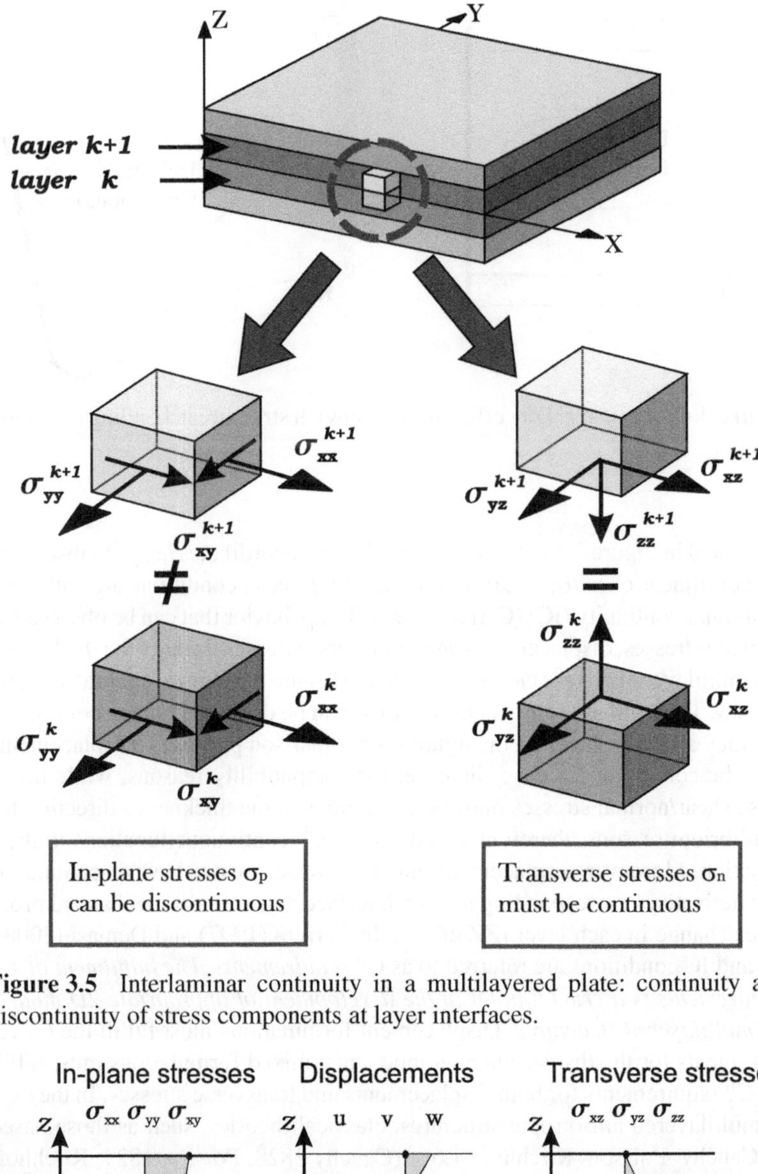

Figure 3.5 Interlaminar continuity in a multilayered plate: continuity and discontinuity of stress components at layer interfaces.

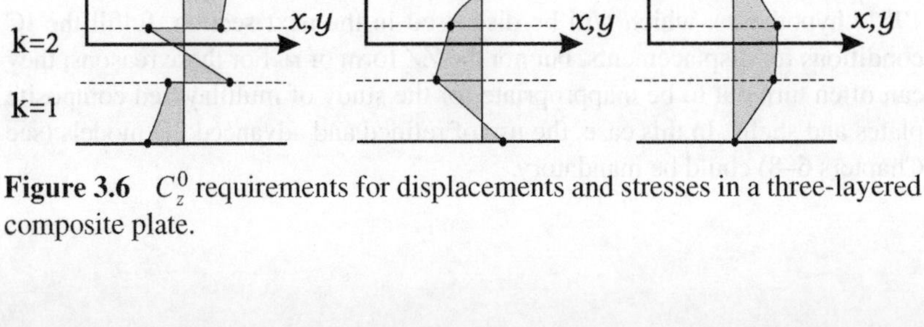

Figure 3.6 C_z^0 requirements for displacements and stresses in a three-layered composite plate.

3.3 Classical theories

Classical theories were originally developed for one-layered isotropic structures. These theories can be divided into two main groups: *Love first-approximation theories* (LFAT) and *Love second-approximation theories* (LSAT). LFAT are based on the well-known Cauchy–Poisson–Kirchhoff–Love thin shell assumptions (Cauchy 1828; Poisson 1829; Kirchhoff 1850; Love 1906): normals to the reference surface Ω remain normal in the deformed states and do not change in length. This means that transverse shear and transverse normal strains are negligible with respect to the other strains. When one or more of these LFAT postulates are removed, we obtain the so-called LSAT (Koiter 1960), which means the effects of transverse shear and/or transverse normal stresses can be taken into account. As a consequence of the introduction of multilayered structures, several LFAT and LSAT were extended to multilayered plates and shells. However, these extensions are part of the framework of equivalent single layer (ESL) theories: the layers in the multilayered structure are seen as only one equivalent plate or shell, and the 2D approximation does not consider dependency on the index layer k.

3.3.1 Classical lamination theory

A possible application of LFAT to multilayered structures is the *classical lamination theory* (CLT); see the books by Reddy (Reddy 2004) and Jones (Jones 1999). CLT is based on Kirchhoff hypotheses (Kirchhoff 1850):

- straight lines that are perpendicular to the midsurface (i.e., transverse normals) before deformation remain straight after the deformation;
- the transverse normals do not experience elongation (i.e., they are inextensible);
- the transverse normals rotate so that they remain perpendicular to the midsurface after the deformation.

These hypotheses are clearly summarized in Figure 3.7. The first two assumptions imply that transverse displacement is independent of the transverse (or thickness) coordinate and the transverse normal strain ϵ_{zz} is zero. The third assumption results in zero transverse shear strains: $\gamma_{xz} = \gamma_{yz} = 0$. The displacement field for a plate is:

$$
\begin{aligned}
u(x, y, z) &= u_0(x, y) - z\frac{\partial w_0}{\partial x} \\
v(x, y, z) &= v_0(x, y) - z\frac{\partial w_0}{\partial y} \qquad (3.3)\\
w(x, y, z) &= w_0(x, y)
\end{aligned}
$$

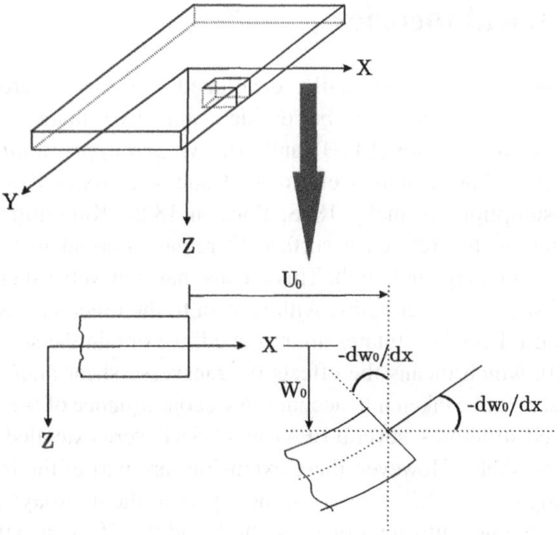

Figure 3.7 Undeformed and deformed geometry of a plate according to the Kirchhoff hypotheses.

Only three degrees of freedom are used for this 2D theory: the displacements in the three directions refer to the midsurface Ω. In CLT for pure mechanical problems, in order to avoid the Poisson locking phenomenon, the $\sigma_{zz} = 0$ condition must be enforced in the constitutive equations. For further details about this topic, readers can refer to Carrera and Brischetto (2008a,b). Typical displacements through the thickness direction z for the case of a three-layered plate are given in Figure 3.8. Figure 3.9 shows the typical behavior of in-plane displacement components u, v (linear and equivalent single layer) and transverse shear stresses (zero for all the multilayer) in the thickness direction z.

3.3.2 First-order shear deformation theory

A one of the typical LSAT for the case of multilayered structures is the *first-order shear deformation theory* (FSDT). The third part of the Kirchhoff hypotheses is removed, therefore the transverse normals do not remain perpendicular to the midsurface after deformation. In this way, transverse shear strains γ_{xz} and γ_{yz} are included in the theory. However, the inextensibility of the transverse normal remains, therefore displacement w is constant in the thickness direction z. FSDT is an extension of the so-called Reissner–Mindlin model

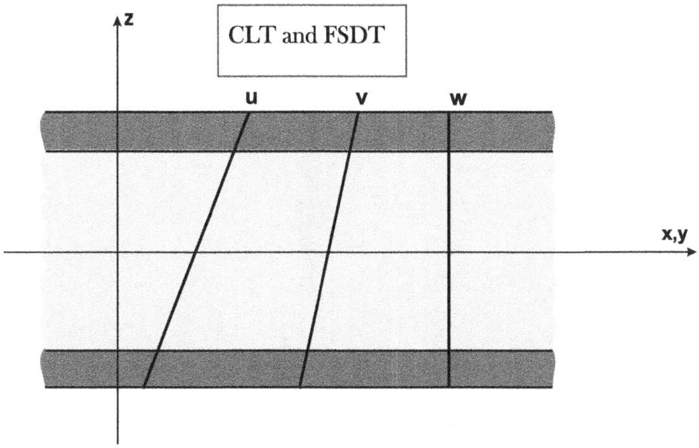

Figure 3.8 Displacement components through the thickness direction z for the case of CLT and FSDT.

(Reissner 1945; Mindlin 1951) to multilayered structures. The displacement model, in the case of FSDT for a plate, is:

$$\begin{aligned} u(x, y, z) &= u_0(x, y) + z\Phi_x(x, y) \\ v(x, y, z) &= v_0(x, y) + z\Phi_y(x, y) \\ w(x, y, z) &= w_0(x, y) \end{aligned} \tag{3.4}$$

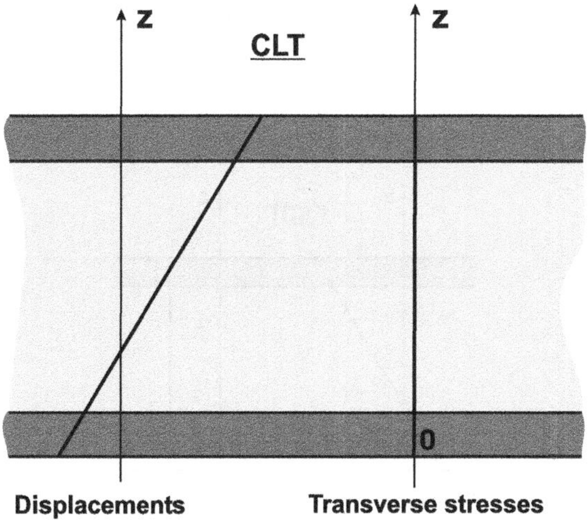

Figure 3.9 CLT: displacements u and v, and transverse shear stresses through the thickness direction z.

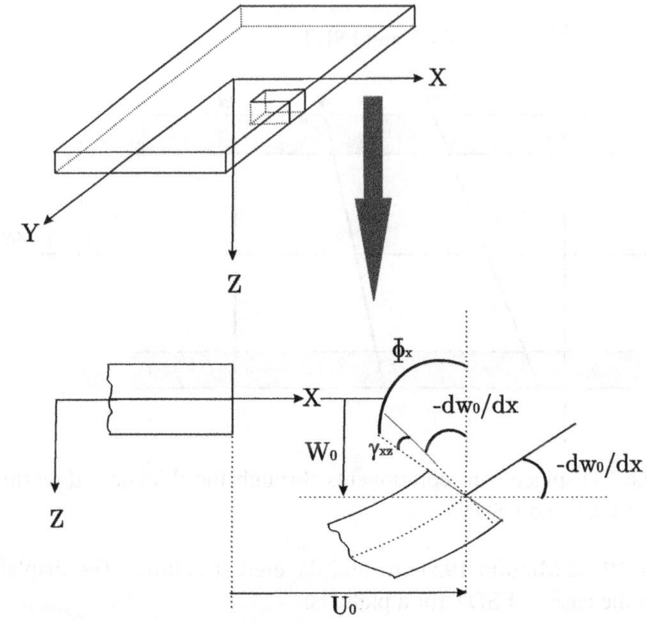

Figure 3.10 Undeformed and deformed geometries of a plate according to the Reissner–Mindlin hypotheses.

The hypotheses of FSDT are clearly shown in Figure 3.10. The displacement field of FSDT has five unknowns (there were three for CLT): the midsurface displacements (u_0, v_0, w_0) and the rotations of a transverse normal around the x- and y-axes (Φ_y, Φ_x). In the case of CLT, the rotations coincide with the derivatives $\Phi_x = -\partial w_0/\partial x$ and $\Phi_y = -\partial w_0/\partial y$. Only strain ϵ_{zz} is zero, therefore stresses σ_{xz} and σ_{yz} are different from zero. Figure 3.11 shows

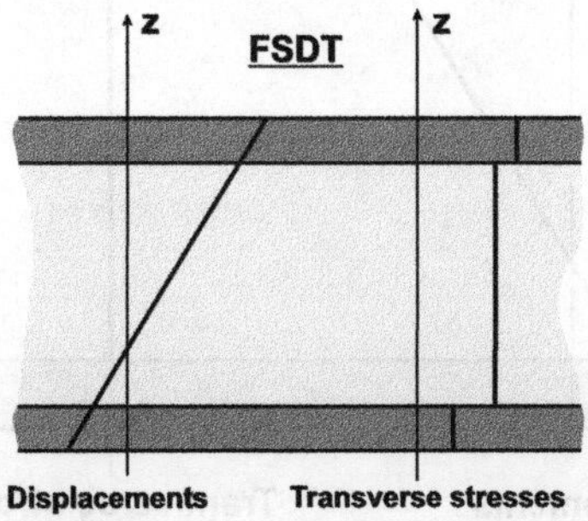

Figure 3.11 FSDT: displacements u and v, and transverse shear stresses through the thickness direction z.

the typical behavior of in-plane displacement components u, v (linear and equivalent single layer) and transverse shear stresses (constant in each layer) in the thickness direction z. Poisson locking phenomena exist because the transverse normal strain ϵ_{zz} remains zero, but it can be avoided by enforcing the $\sigma_{zz} = 0$ condition in constitutive equations, as seen in Carrera and Brischetto (2008a,b) for pure mechanical problems.

3.3.3 Vlasov–Reddy theory

A refinement of the Reissner–Mindlin theory was produced by Vlasov for the case of one-layered isotropic structures (Vlasov 1957). This theory permits the homogeneous conditions for the transverse shear stresses to be fulfilled, corresponding to the top and bottom of the plate/shell. Reddy (1984) and Reddy and Phan (1985) showed that such an inclusion leads to significant improvements compared to FSDT for layered structures (static and dynamic analysis). The resulting model is called the Vlasov–Reddy theory (VRT), and its displacement model for a plate is:

$$\begin{aligned} u(x, y, z) &= u_0(x, y) + z\Phi_x(x, y) + z^3\left(-\frac{4}{3h^2}\right)\left(\Phi_x + \frac{\partial w_0}{\partial x}\right) \\ v(x, y, z) &= v_0(x, y) + z\Phi_y(x, y) + z^3\left(-\frac{4}{3h^2}\right)\left(\Phi_y + \frac{\partial w_0}{\partial y}\right) \\ w(x, y, z) &= w_0(x, y) \end{aligned} \tag{3.5}$$

The model in Equation (3.5), like any ESL theory with transverse displacement w constant or linear in z, needs correction for the Poisson locking phenomena (Carrera and Brischetto 2008a,b) for the case of pure mechanical problems.

3.4 Classical plate theories extended to smart structures

3.4.1 CLT plate theory extended to smart structures

CLT is extended here to smart structures by assuming a linear electric potential through the thickness of the piezoelectric layer in the multilayered plate. We consider a plate of total thickness h composed of an orthotropic layer (thickness h_2) and a piezoelectric layer at the top (thickness h_1). The geometry of the plate and its coordinate systems are indicated in Figure 3.12. The global reference system is (x, y, z) and it coincides with the middle surface of the plate; in this case, we indicate the top and the bottom of the kth layer by z_k and z_{k+1}, respectively. Therefore, the thickness of the kth layer is $h_k = z_{k+1} - z_k$.

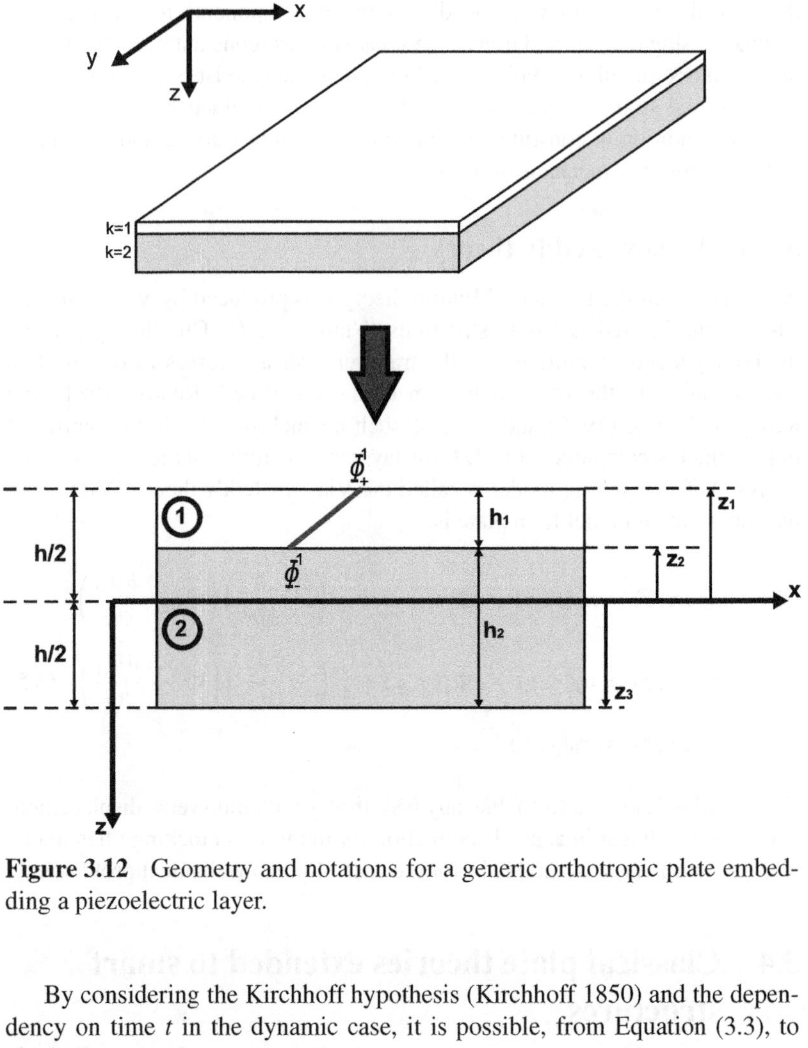

Figure 3.12 Geometry and notations for a generic orthotropic plate embedding a piezoelectric layer.

By considering the Kirchhoff hypothesis (Kirchhoff 1850) and the dependency on time t in the dynamic case, it is possible, from Equation (3.3), to obtain the extension to smart structures:

$$\begin{aligned}
u(x, y, z, t) &= u_0(x, y, t) - z\frac{\partial w_0(x, y, t)}{\partial x} \\
v(x, y, z, t) &= v_0(x, y, t) - z\frac{\partial w_0(x, y, t)}{\partial y} \\
w(x, y, z, t) &= w_0(x, y, t) \\
\Phi^k(x, y, z, t) &= \Phi_0^k(x, y, t) + (z - \bar{z}_k)\Phi_1^k(x, y, t)
\end{aligned} \tag{3.6}$$

In Equation (3.6), u_0, v_0, and w_0 are three displacement components in the middle plane of the plate; the other two unknowns are Φ_0^k and Φ_1^k and these permit the linear electric potential to be expressed in the electrical layer (in this case the electric potential is in layer 1). We can consider an electric potential at the top Φ_+^k and at the bottom Φ_-^k of the generic piezoelectric layer k. The linear electric potential through the thickness is defined by means of the mean value $\Phi_0^k = (\Phi_+^k + \Phi_-^k)/2$ and the slope $\Phi_1^k = (\Phi_-^k - \Phi_+^k)/h_k$. In order to refer the thickness coordinate to the midsurface of the kth piezoelectric layer, the value $\bar{z}_k = (z_k + z_{k+1})/2$ is defined.

The piezoelectric constitutive equations for the CLT case, referring to Equations (2.12)–(2.18), can be written for each k layer as:

$$\begin{bmatrix} \sigma_{xx}^k \\ \sigma_{yy}^k \\ \sigma_{xy}^k \end{bmatrix} = \begin{bmatrix} Q_{11}^k & Q_{12}^k & Q_{16}^k \\ Q_{12}^k & Q_{22}^k & Q_{26}^k \\ Q_{16}^k & Q_{26}^k & Q_{66}^k \end{bmatrix} \begin{bmatrix} \epsilon_{xx} \\ \epsilon_{yy} \\ \gamma_{xy} \end{bmatrix} - \begin{bmatrix} 0 & 0 & e_{31}^k \\ 0 & 0 & e_{32}^k \\ 0 & 0 & e_{36}^k \end{bmatrix} \begin{bmatrix} \mathcal{E}_x^k \\ \mathcal{E}_y^k \\ \mathcal{E}_z^k \end{bmatrix} \tag{3.7}$$

$$\begin{bmatrix} \mathcal{D}_x^k \\ \mathcal{D}_y^k \\ \mathcal{D}_z^k \end{bmatrix} = \begin{bmatrix} 0 & 0 & 0 \\ 0 & 0 & 0 \\ e_{31}^k & e_{32}^k & e_{36}^k \end{bmatrix} \begin{bmatrix} \epsilon_{xx} \\ \epsilon_{yy} \\ \gamma_{xy} \end{bmatrix} + \begin{bmatrix} \varepsilon_{11}^k & \varepsilon_{12}^k & 0 \\ \varepsilon_{12}^k & \varepsilon_{22}^k & 0 \\ 0 & 0 & \varepsilon_{33}^k \end{bmatrix} \begin{bmatrix} \mathcal{E}_x^k \\ \mathcal{E}_y^k \\ \mathcal{E}_z^k \end{bmatrix} \tag{3.8}$$

The in-plane strains ($\epsilon_{xx}, \epsilon_{yy}, \gamma_{xy}$) do not depend on the k layer because the displacements are in ESL form. No correction for the Poisson locking phenomena is made in this book for the electromechanical CLT case. Instead, the Poisson locking phenomenon clearly appears for the pure mechanical CLT case, and it can be contrasted by means of the reduced elastic coefficients given in Carrera and Brischetto (2008a,b) for plate and shell geometries, respectively.

The geometrical relations are obtained in Reddy (2004). The strain and electric field components, in the case of CLT analysis, are linked to the displacements and electric potential by means of the following relations:

$$\epsilon_{xx} = \frac{\partial u}{\partial x} = \frac{\partial u_0}{\partial x} - z\frac{\partial^2 w_0}{\partial x^2} = \epsilon_{xx}^{(0)} + z\epsilon_{xx}^{(1)} \tag{3.9}$$

$$\epsilon_{yy} = \frac{\partial v}{\partial y} = \frac{\partial v_0}{\partial y} - z\frac{\partial^2 w_0}{\partial y^2} = \epsilon_{yy}^{(0)} + z\epsilon_{yy}^{(1)} \tag{3.10}$$

$$\gamma_{xy} = \frac{\partial u}{\partial y} + \frac{\partial v}{\partial x} = \frac{\partial u_0}{\partial y} - z\frac{\partial^2 w_0}{\partial x \partial y} + \frac{\partial v_0}{\partial x} - z\frac{\partial^2 w_0}{\partial x \partial y}$$

$$= \frac{\partial u_0}{\partial y} + \frac{\partial v_0}{\partial x} - 2z\frac{\partial^2 w_0}{\partial x \partial y} = \gamma_{xy}^{(0)} + z\gamma_{xy}^{(1)} \tag{3.11}$$

$$\mathcal{E}_x^k = -\frac{\partial \Phi^k}{\partial x} = -\frac{\partial \Phi_0^k}{\partial x} + \bar{z}_k\frac{\partial \Phi_1^k}{\partial x} - z\frac{\partial \Phi_1^k}{\partial x} = \mathcal{E}_x^{k(0)} + z\mathcal{E}_x^{k(1)} \tag{3.12}$$

$$\mathcal{E}_y^k = -\frac{\partial \Phi^k}{\partial y} = -\frac{\partial \Phi_0^k}{\partial y} + \bar{z}_k \frac{\partial \Phi_1^k}{\partial y} - z\frac{\partial \Phi_1^k}{\partial y} = \mathcal{E}_y^{k(0)} + z\mathcal{E}_y^{k(1)} \tag{3.13}$$

$$\mathcal{E}_z^k = -\frac{\partial \Phi^k}{\partial z} = -\Phi_1^k = \mathcal{E}_z^{k(0)} \tag{3.14}$$

The dynamic version of the PVD, already proposed in Reddy (2004), where t is the time which goes from 0 to T, states:

$$0 = \int_0^T (\delta U + \delta V - \delta K)dt \tag{3.15}$$

where the virtual internal work δU (volume integral of δU_0), in the case of electromechanical coupling, is a summation of δU_m (virtual strain energy) and δU_e (virtual electrical internal work). The term δU_e is not considered for a partial electromechanical coupling and the electrical contribution is only considered by means of the second term in Equation (3.7). A fully electromechanical coupling is accounted for in this work. δV is the virtual work done by the applied forces, and it permits one to obtain the mechanical forces for the case of sensor applications; no mechanical forces are considered for actuator applications and the electric potential can be directly applied at the top and bottom of the considered piezoelectric actuator layer. δK is the virtual kinetic energy.

The virtual internal work δU is given by:

$$\begin{aligned}
\delta U = \delta U_m + \delta U_e &= \int_v \delta U_0 dv = \int_v (\delta U_{0m} + \delta U_{0e})\, dv \\
&= \int_{\Omega_0} \left(\int_{-h/2}^{h/2} [(\sigma_{xx}^k \delta\epsilon_{xx} + \sigma_{yy}^k \delta\epsilon_{yy} + \sigma_{xy}^k \delta\gamma_{xy}) \right. \\
&\quad \left. + (-\mathcal{D}_x^k \delta\mathcal{E}_x^k - \mathcal{D}_y^k \delta\mathcal{E}_y^k - \mathcal{D}_z^k \delta\mathcal{E}_z^k)\,]dz \right) dxdy \\
&= \int_{\Omega_0} \left(\sum_{k=1}^{N_l} \int_{-h_k/2}^{h_k/2} [\sigma_{xx}^k (\delta\epsilon_{xx}^{(0)} + z\delta\epsilon_{xx}^{(1)}) + \sigma_{yy}^k (\delta\epsilon_{yy}^{(0)} + z\delta\epsilon_{yy}^{(1)}) \right. \\
&\quad + \sigma_{xy}^k (\delta\gamma_{xy}^{(0)} + z\delta\gamma_{xy}^{(1)}) - \mathcal{D}_x^k (\delta\mathcal{E}_x^{k(0)} + z\delta\mathcal{E}_x^{k(1)}) \\
&\quad \left. - \mathcal{D}_y^k (\delta\mathcal{E}_y^{k(0)} + z\delta\mathcal{E}_y^{k(1)}) - \mathcal{D}_z^k (\delta\mathcal{E}_z^{k(0)})]dz \right) dxdy
\end{aligned} \tag{3.16}$$

where v is the volume of the plate, Ω_0 is the reference midsurface of the whole multilayered plate, k indicates the layer, and N_l is the number of embedded layers. The strains in Equation (3.16) do not depend on the k layer because the displacements are in ESL form, while the electric field instead depends on the k layer since the electric potential is in layer-wise form.

The virtual work done by applied forces δV is given by:

$$\begin{aligned}\delta V &= -\int_{\Omega_0} \left[q_b(x,y)\delta w(x,y,h/2) + q_t(x,y)\delta w(x,y,-h/2)\right] dxdy \\ &\quad - \int_{\Gamma_\sigma} \left(\int_{-h/2}^{h/2} \left[\hat{\sigma}_{nn}\delta u_n + \hat{\sigma}_{ns}\delta u_s + \hat{\sigma}_{nz}\delta w\right] dz \right) ds \\ &= -\int_{\Omega_0} \left[(q_b(x,y) + q_t(x,y))\delta w_0\right] dxdy - \int_{\Gamma_\sigma} \left(\int_{-h/2}^{h/2} \left[\hat{\sigma}_{nn}(\delta u_{0n} \right.\right. \\ &\quad \left.\left. - (z\partial\delta w_0)/\partial n) + \hat{\sigma}_{ns}(\delta u_{0s} - z(\partial\delta w_0)/\partial s) + \hat{\sigma}_{nz}\delta w_0\right] dz \right) ds \end{aligned} \tag{3.17}$$

where q_b and q_t are the distributed forces at the bottom and top of the multilayered plate, respectively. The bottom and top coordinates are $z = h/2$ and $z = -h/2$, respectively. $\hat{\sigma}_{nn}$, $\hat{\sigma}_{ns}$, and $\hat{\sigma}_{nz}$ are the stress components on portion Γ_σ of boundary Γ. The subscripts n and s indicate the normal and tangential directions, therefore δu_{0n} and δu_{0s} are the virtual displacements in the normal and tangential directions, respectively.

The virtual kinetic energy δK is given by:

$$\begin{aligned}\delta K &= \int_{\Omega_0} \left(\int_{-h/2}^{h/2} \rho_0^k \left[\dot{u}\delta\dot{u} + \dot{v}\delta\dot{v} + \dot{w}\delta\dot{w}\right] dz \right) dxdy \\ &= \int_{\Omega_0} \left(\int_{-h/2}^{h/2} \rho_0^k \left[(\dot{u}_0 - z(\partial\dot{w}_0)/\partial x)(\delta\dot{u}_0 - z\partial\delta\dot{w}_0/\partial x) \right.\right. \\ &\quad \left.\left. + (\dot{v}_0 - z(\partial\dot{w}_0)/\partial y)(\delta\dot{v}_0 - z(\partial\delta\dot{w}_0)/\partial y) + \dot{w}_0\delta\dot{w}_0\right] dz \right) dxdy \end{aligned} \tag{3.18}$$

where ρ_0^k is the mass density of the kth layer. The overdot denotes the derivative with respect to time, e.g., $\dot{u}_0 = \partial u_0/\partial t$.

We define the following integrals in the z direction of the multilayered plate:

$$\begin{bmatrix} N_{xx} \\ N_{yy} \\ N_{xy} \end{bmatrix} = \int_{-h/2}^{h/2} \begin{bmatrix} \sigma_{xx}^k \\ \sigma_{yy}^k \\ \sigma_{xy}^k \end{bmatrix} dz \qquad \begin{bmatrix} M_{xx} \\ M_{yy} \\ M_{xy} \end{bmatrix} = \int_{-h/2}^{h/2} \begin{bmatrix} \sigma_{xx}^k \\ \sigma_{yy}^k \\ \sigma_{xy}^k \end{bmatrix} z\, dz \tag{3.19}$$

$$\begin{bmatrix} \hat{N}_{nn} \\ \hat{N}_{ns} \end{bmatrix} = \int_{-h/2}^{h/2} \begin{bmatrix} \hat{\sigma}_{nn} \\ \hat{\sigma}_{ns} \end{bmatrix} dz \qquad \begin{bmatrix} \hat{M}_{nn} \\ \hat{M}_{ns} \end{bmatrix} = \int_{-h/2}^{h/2} \begin{bmatrix} \hat{\sigma}_{nn} \\ \hat{\sigma}_{ns} \end{bmatrix} z\, dz \tag{3.20}$$

$$\hat{Q}_n = \int_{-h/2}^{h/2} \hat{\sigma}_{nz}\, dz \qquad \begin{bmatrix} I_0 \\ I_1 \\ I_2 \end{bmatrix} = \int_{-h/2}^{h/2} \begin{bmatrix} 1 \\ z \\ z^2 \end{bmatrix} \rho_0^k dz \tag{3.21}$$

$$\begin{bmatrix} O_x \\ O_y \\ O_z \end{bmatrix} = \int_{-h/2}^{h/2} \begin{bmatrix} \mathcal{D}_x^k \\ \mathcal{D}_y^k \\ \mathcal{D}_z^k \end{bmatrix} dz \qquad \begin{bmatrix} P_x \\ P_y \\ P_z \end{bmatrix} = \int_{-h/2}^{h/2} \begin{bmatrix} \mathcal{D}_x^k \\ \mathcal{D}_y^k \\ \mathcal{D}_z^k \end{bmatrix} z\, dz \tag{3.22}$$

where (N_{xx}, N_{yy}, N_{xy}) are the in-plane force resultants per unit length and (M_{xx}, M_{yy}, M_{xy}) are the moment resultants per unit length in Equation (3.19). $\hat{Q}_n$ denotes the transverse force resultant, and (I_0, I_1, I_2) are the mass moments of inertia in Equation (3.21). We define (O_x, O_y, O_z) as the electric charge resultants per unit length, and (P_x, P_y, P_z) as the electric moment resultants per unit length in Equation (3.22).

Considering Equations (3.16), (3.17), and (3.18) for δU, δV, and δK, respectively, and substituting them in Equation (3.15) (considering Equations (3.19)–(3.22)), we obtain:

$$\begin{aligned} 0 = \int_0^T \Bigg(\int_{\Omega_0} \Bigg[& N_{xx}\delta\epsilon_{xx}^{(0)} + M_{xx}\delta\epsilon_{xx}^{(1)} + N_{yy}\delta\epsilon_{yy}^{(0)} + M_{yy}\delta\epsilon_{yy}^{(1)} + N_{xy}\delta\gamma_{xy}^{(0)} \\ & + M_{xy}\delta\gamma_{xy}^{(1)} - O_x\delta\mathcal{E}_x^{k(0)} - P_x\delta\mathcal{E}_x^{k(1)} - O_y\delta\mathcal{E}_y^{k(0)} - P_y\delta\mathcal{E}_y^{k(1)} - O_z\delta\mathcal{E}_z^{k(0)} \\ & - q\delta w_0 - I_0(\dot{u}_0\delta\dot{u}_0 + \dot{v}_0\delta\dot{v}_0 + \dot{w}_0\delta\dot{w}_0) + I_1\left(\frac{\partial\delta\dot{w}_0}{\partial x}\dot{u}_0 + \frac{\partial\dot{w}_0}{\partial x}\delta\dot{u}_0\right. \\ & \left. + \frac{\partial\delta\dot{w}_0}{\partial y}\dot{v}_0 + \frac{\partial\dot{w}_0}{\partial y}\delta\dot{v}_0\right) - I_2\left(\frac{\partial\dot{w}_0}{\partial x}\frac{\partial\delta\dot{w}_0}{\partial x} + \frac{\partial\dot{w}_0}{\partial y}\frac{\partial\delta\dot{w}_0}{\partial y}\right)\Bigg]dxdy \\ & - \int_{\Gamma_\sigma}\left[\hat{N}_{nn}\delta u_{0n} + \hat{N}_{ns}\delta u_{0s} - \hat{M}_{nn}\frac{\partial\delta w_0}{\partial n} - \hat{M}_{ns}\frac{\partial\delta w_0}{\partial s} + \hat{Q}_n\delta w_0\right]ds\Bigg)dt \end{aligned} \tag{3.23}$$

where the integral through the thickness direction z has been obtained by means of Equations (3.19)–(3.22), and $q = q_b + q_t$ is the total transverse load.

The virtual strain and electric field components can be written in terms of virtual displacements and virtual electric potential in the same way as for the true strain and electric field components, in terms of the true displacements and electric potential (see Equations (3.9)–(3.14)):

$$\delta\epsilon_{xx}^{(0)} = \frac{\partial\delta u_0}{\partial x}, \qquad \delta\epsilon_{xx}^{(1)} = -\frac{\partial^2\delta w_0}{\partial x^2} \tag{3.24}$$

$$\delta\epsilon_{yy}^{(0)} = \frac{\partial\delta v_0}{\partial y}, \qquad \delta\epsilon_{yy}^{(1)} = -\frac{\partial^2\delta w_0}{\partial y^2} \tag{3.25}$$

$$\delta\gamma_{xy}^{(0)} = \frac{\partial \delta u_0}{\partial y} + \frac{\partial \delta v_0}{\partial x}, \qquad \delta\gamma_{xy}^{(1)} = -2\frac{\partial^2 \delta w_0}{\partial x \partial y} \tag{3.26}$$

$$\delta\mathcal{E}_x^{k(0)} = -\frac{\partial \delta \Phi_0^k}{\partial x} + \bar{z}_k \frac{\partial \delta \Phi_1^k}{\partial x}, \qquad \delta\mathcal{E}_x^{k(1)} = -\frac{\partial \delta \Phi_1^k}{\partial x} \tag{3.27}$$

$$\delta\mathcal{E}_y^{k(0)} = -\frac{\partial \delta \Phi_0^k}{\partial y} + \bar{z}_k \frac{\partial \delta \Phi_1^k}{\partial y}, \qquad \delta\mathcal{E}_y^{k(1)} = -\frac{\partial \delta \Phi_1^k}{\partial y} \tag{3.28}$$

$$\delta\mathcal{E}_z^{k(0)} = -\delta\Phi_1^k \tag{3.29}$$

We can substitute Equations (3.24)–(3.29) into Equation (3.23) and then integrate by parts to reveal the virtual displacements (δu_0,δv_0,δw_0) and the virtual electric potential variables ($\delta\Phi_0^k$,$\delta\Phi_1^k$). In the notation, a comma followed by subscripts denotes differentiation with respect to the subscripts, e.g., $N_{xx,x} = \partial N_{xx}/\partial x$. Note that both spatial and time integrations by parts are used to obtain the final expression. n_x and n_y are the unit vectors in the x and y direction, respectively:

$$\begin{aligned}
0 = \int_0^T \Bigg\{ \int_{\Omega_0} \Bigg[& - N_{xx,x}\delta u_0 - M_{xx,xx}\delta w_0 - N_{yy,y}\delta v_0 - M_{yy,yy}\delta w_0 \\
& - N_{xy,y}\delta u_0 - N_{xy,x}\delta v_0 - 2M_{xy,xy}\delta w_0 - q\delta w_0 - O_{x,x}\delta\Phi_0^k + O_{x,x}\bar{z}_k\delta\Phi_1^k \\
& - P_{x,x}\delta\Phi_1^k - O_{y,y}\delta\Phi_0^k + O_{y,y}\bar{z}_k\delta\Phi_1^k - P_{y,y}\delta\Phi_1^k + O_z\delta\Phi_1^k + I_0(\ddot{u}_0\delta u_0 \\
& + \ddot{v}_0\delta v_0 + \ddot{w}_0\delta w_0) - I_2\left(\frac{\partial^2 \ddot{w}_0}{\partial x^2} + \frac{\partial^2 \ddot{w}_0}{\partial y^2}\right)\delta w_0 + I_1\left(\frac{\partial \ddot{u}_0}{\partial x}\delta w_0 - \frac{\partial \ddot{w}_0}{\partial x}\delta u_0 \right.\\
& \left. + \frac{\partial \ddot{v}_0}{\partial y}\delta w_0 - \frac{\partial \ddot{w}_0}{\partial y}\delta v_0\right)\Bigg] dxdy + \oint_\Gamma \Bigg[N_{xx}n_x\delta u_0 - M_{xx}n_x\frac{\partial \delta w_0}{\partial x} \\
& + M_{xx,x}n_x\delta w_0 + N_{yy}n_y\delta v_0 - M_{yy}n_y\frac{\partial \delta w_0}{\partial y} + M_{yy,y}n_y\delta w_0 + N_{xy}n_y\delta u_0 \\
& + N_{xy}n_x\delta v_0 - M_{xy}n_x\frac{\partial \delta w_0}{\partial y} + M_{xy,x}n_y\delta w_0 - M_{xy}n_y\frac{\partial \delta w_0}{\partial x} + M_{xy,y}n_x\delta w_0 \\
& + O_x n_x\delta\Phi_0^k - O_x\bar{z}_k n_x\delta\Phi_1^k + P_x n_x\delta\Phi_1^k + O_y n_y\delta\Phi_0^k - O_y\bar{z}_k n_y\delta\Phi_1^k \\
& + P_y n_y\delta\Phi_1^k\Bigg] ds - \int_{\Gamma_\sigma} \Bigg[\hat{N}_{nn}\delta u_{0n} + \hat{N}_{ns}\delta u_{0s} - \hat{M}_{nn}\frac{\partial \delta w_0}{\partial n} - \hat{M}_{ns}\frac{\partial \delta w_0}{\partial s} \\
& + \hat{Q}_n\delta w_0\Bigg] ds + \oint_\Gamma \Bigg[I_2\Big(\frac{\partial \ddot{w}_0}{\partial x}n_x + \frac{\partial \ddot{w}_0}{\partial y}n_y\Big) - I_1(\ddot{u}_0 n_x + \ddot{v}_0 n_y)\Bigg]\delta w_0 ds \Bigg\} dt
\end{aligned} \tag{3.30}$$

Grouping terms with respect to δu_0, δv_0, δw_0, $\delta\Phi_0^k$, and $\delta\Phi_1^k$, and noting that the virtual displacements are zero on Γ_u, Equation (3.30) can be written as:

$$
\begin{aligned}
0 = \int_0^T \Bigg\{ \int_{\Omega_0} \Bigg[& -\left(N_{xx,x} + N_{xy,y} - I_0\ddot{u}_0 + I_1\frac{\partial \ddot{w}_0}{\partial x}\right)\delta u_0 - \Bigg(N_{yy,y} + N_{xy,x} \\
& - I_0\ddot{v}_0 + I_1\frac{\partial \ddot{w}_0}{\partial y}\Bigg)\delta v_0 - \Bigg(M_{xx,xx} + M_{yy,yy} + 2M_{xy,xy} + q - I_0\,\ddot{w}_0 \\
& + I_2\frac{\partial^2 \ddot{w}_0}{\partial x^2} + I_2\frac{\partial^2 \ddot{w}_0}{\partial y^2} - I_1\frac{\partial \ddot{u}_0}{\partial x} - I_1\frac{\partial \ddot{v}_0}{\partial y}\Bigg)\delta w_0 - (O_{x,x} + O_{y,y})\delta\Phi_0^k \\
& - (P_{x,x} - O_{x,x}\bar{z}_k + P_{y,y} - O_{y,y}\bar{z}_k - O_z)\delta\Phi_1^k \Bigg] dxdy \\
& + \int_{\Gamma_\sigma} \Bigg[(N_{xx}n_x + N_{xy}n_y)\delta u_0 + (N_{xy}n_x + N_{yy}n_y)\delta v_0 + \Bigg(M_{xx,x}n_x \\
& + M_{yy,y}n_y + M_{xy,x}n_y + M_{xy,y}n_x + I_2\frac{\partial \ddot{w}_0}{\partial x}n_x + I_2\frac{\partial \ddot{w}_0}{\partial y}n_y - I_1\ddot{u}_0 n_x \\
& - I_1\ddot{v}_0 n_y \Bigg)\delta w_0 - (M_{xx}n_x + M_{xy}n_y)\frac{\partial \delta w_0}{\partial x} - (M_{yy}n_y + M_{xy}n_x)\frac{\partial \delta w_0}{\partial y} \\
& + (O_x n_x + O_y n_y)\delta\Phi_0^k + (P_x n_x + P_y n_y - O_x\bar{z}_k n_x - O_y\bar{z}_k n_y)\delta\Phi_1^k \Bigg] ds \\
& - \int_{\Gamma_\sigma} \Bigg[\hat{N}_{nn}\delta u_{0n} + \hat{N}_{ns}\delta u_{0s} - \hat{M}_{nn}\frac{\partial \delta w_0}{\partial n} - \hat{M}_{ns}\frac{\partial \delta w_0}{\partial s} + \hat{Q}_n\delta w_0 \Bigg] ds \Bigg\} dt
\end{aligned}
\tag{3.31}
$$

The Euler–Lagrange equations of the CLT plate theory, extended to smart structures, can be obtained by setting the coefficients δu_0, δv_0, δw_0, $\delta\Phi_0^k$, and $\delta\Phi_1^k$ in Equation (3.31) to zero separately:

$$
\begin{aligned}
\delta u_0&: \frac{\partial N_{xx}}{\partial x} + \frac{\partial N_{xy}}{\partial y} = I_0\frac{\partial^2 u_0}{\partial t^2} - I_1\frac{\partial^2}{\partial t^2}\left(\frac{\partial w_0}{\partial x}\right) \\
\delta v_0&: \frac{\partial N_{xy}}{\partial x} + \frac{\partial N_{yy}}{\partial y} = I_0\frac{\partial^2 v_0}{\partial t^2} - I_1\frac{\partial^2}{\partial t^2}\left(\frac{\partial w_0}{\partial y}\right) \\
\delta w_0&: \frac{\partial^2 M_{xx}}{\partial x^2} + 2\frac{\partial^2 M_{xy}}{\partial x\partial y} + \frac{\partial^2 M_{yy}}{\partial y^2} + q = I_0\frac{\partial^2 w_0}{\partial t^2} \\
&\quad - I_2\frac{\partial^2}{\partial t^2}\left(\frac{\partial^2 w_0}{\partial x^2} + \frac{\partial^2 w_0}{\partial y^2}\right) + I_1\frac{\partial^2}{\partial t^2}\left(\frac{\partial u_0}{\partial x} + \frac{\partial v_0}{\partial y}\right)
\end{aligned}
$$

$$\delta\Phi_0^k\colon \frac{\partial O_x}{\partial x} + \frac{\partial O_y}{\partial y} = 0$$

$$\delta\Phi_1^k\colon \frac{\partial P_x}{\partial x} - \bar{z}_k \frac{\partial O_x}{\partial x} + \frac{\partial P_y}{\partial y} - \bar{z}_k \frac{\partial O_y}{\partial y} - O_z = 0 \tag{3.32}$$

Further details about boundary conditions can be found in Reddy (2004).

In order to write Equations (3.32) in terms of displacements and electric potential, it is necessary to consider the laminate constitutive equations. The stress resultants are given by:

$$\begin{bmatrix} N_{xx} \\ N_{yy} \\ N_{xy} \end{bmatrix} = \sum_{k=1}^{N_l} \int_{z_k}^{z_{k+1}} \begin{bmatrix} \sigma_{xx}^k \\ \sigma_{yy}^k \\ \sigma_{xy}^k \end{bmatrix} dz$$

$$= \sum_{k=1}^{N_l} \int_{z_k}^{z_{k+1}} \left(\begin{bmatrix} Q_{11}^k & Q_{12}^k & Q_{16}^k \\ Q_{12}^k & Q_{22}^k & Q_{26}^k \\ Q_{16}^k & Q_{26}^k & Q_{66}^k \end{bmatrix} \begin{bmatrix} \epsilon_{xx}^{(0)} + z\epsilon_{xx}^{(1)} \\ \epsilon_{yy}^{(0)} + z\epsilon_{yy}^{(1)} \\ \gamma_{xy}^{(0)} + z\gamma_{xy}^{(1)} \end{bmatrix} \right.$$

$$\left. - \begin{bmatrix} 0 & 0 & e_{31}^k \\ 0 & 0 & e_{32}^k \\ 0 & 0 & e_{36}^k \end{bmatrix} \begin{bmatrix} \mathcal{E}_x^{k(0)} + z\mathcal{E}_x^{k(1)} \\ \mathcal{E}_y^{k(0)} + z\mathcal{E}_y^{k(1)} \\ \mathcal{E}_z^{k(0)} + z0 \end{bmatrix} \right) dz \tag{3.33}$$

The moment resultants are given by:

$$\begin{bmatrix} M_{xx} \\ M_{yy} \\ M_{xy} \end{bmatrix} = \sum_{k=1}^{N_l} \int_{z_k}^{z_{k+1}} \begin{bmatrix} \sigma_{xx}^k \\ \sigma_{yy}^k \\ \sigma_{xy}^k \end{bmatrix} zdz$$

$$= \sum_{k=1}^{N_l} \int_{z_k}^{z_{k+1}} \left(\begin{bmatrix} Q_{11}^k & Q_{12}^k & Q_{16}^k \\ Q_{12}^k & Q_{22}^k & Q_{26}^k \\ Q_{16}^k & Q_{26}^k & Q_{66}^k \end{bmatrix} \begin{bmatrix} \epsilon_{xx}^{(0)} + z\epsilon_{xx}^{(1)} \\ \epsilon_{yy}^{(0)} + z\epsilon_{yy}^{(1)} \\ \gamma_{xy}^{(0)} + z\gamma_{xy}^{(1)} \end{bmatrix} \right.$$

$$\left. - \begin{bmatrix} 0 & 0 & e_{31}^k \\ 0 & 0 & e_{32}^k \\ 0 & 0 & e_{36}^k \end{bmatrix} \begin{bmatrix} \mathcal{E}_x^{k(0)} + z\mathcal{E}_x^{k(1)} \\ \mathcal{E}_y^{k(0)} + z\mathcal{E}_y^{k(1)} \\ \mathcal{E}_z^{k(0)} + z0 \end{bmatrix} \right) zdz \tag{3.34}$$

The electric charge per unit length resultants are given by:

$$\begin{bmatrix} O_x \\ O_y \\ O_z \end{bmatrix} = \sum_{k=1}^{N_l} \int_{z_k}^{z_{k+1}} \begin{bmatrix} \mathcal{D}_x^k \\ \mathcal{D}_y^k \\ \mathcal{D}_z^k \end{bmatrix} dz$$

$$= \sum_{k=1}^{N_l} \int_{z_k}^{z_{k+1}} \left(\begin{bmatrix} 0 & 0 & 0 \\ 0 & 0 & 0 \\ e_{31}^k & e_{32}^k & e_{36}^k \end{bmatrix} \begin{bmatrix} \epsilon_{xx}^{(0)} + z\epsilon_{xx}^{(1)} \\ \epsilon_{yy}^{(0)} + z\epsilon_{yy}^{(1)} \\ \gamma_{xy}^{(0)} + z\gamma_{xy}^{(1)} \end{bmatrix} \right.$$

$$\left. + \begin{bmatrix} \varepsilon_{11}^k & \varepsilon_{12}^k & 0 \\ \varepsilon_{12}^k & \varepsilon_{22}^k & 0 \\ 0 & 0 & \varepsilon_{33}^k \end{bmatrix} \begin{bmatrix} \mathcal{E}_x^{k(0)} + z\mathcal{E}_x^{k(1)} \\ \mathcal{E}_y^{k(0)} + z\mathcal{E}_y^{k(1)} \\ \mathcal{E}_z^{k(0)} + z0 \end{bmatrix} \right) dz \tag{3.35}$$

The electric moment resultants are given by:

$$\begin{bmatrix} P_x \\ P_y \\ P_z \end{bmatrix} = \sum_{k=1}^{N_l} \int_{z_k}^{z_{k+1}} \begin{bmatrix} \mathcal{D}_x^k \\ \mathcal{D}_y^k \\ \mathcal{D}_z^k \end{bmatrix} zdz$$

$$= \sum_{k=1}^{N_l} \int_{z_k}^{z_{k+1}} \left(\begin{bmatrix} 0 & 0 & 0 \\ 0 & 0 & 0 \\ e_{31}^k & e_{32}^k & e_{36}^k \end{bmatrix} \begin{bmatrix} \epsilon_{xx}^{(0)} + z\epsilon_{xx}^{(1)} \\ \epsilon_{yy}^{(0)} + z\epsilon_{yy}^{(1)} \\ \gamma_{xy}^{(0)} + z\gamma_{xy}^{(1)} \end{bmatrix} \right.$$

$$\left. + \begin{bmatrix} \varepsilon_{11}^k & \varepsilon_{12}^k & 0 \\ \varepsilon_{12}^k & \varepsilon_{22}^k & 0 \\ 0 & 0 & \varepsilon_{33}^k \end{bmatrix} \begin{bmatrix} \mathcal{E}_x^{k(0)} + z\mathcal{E}_x^{k(1)} \\ \mathcal{E}_y^{k(0)} + z\mathcal{E}_y^{k(1)} \\ \mathcal{E}_z^{k(0)} + z0 \end{bmatrix} \right) zdz \tag{3.36}$$

The summation in Equations (3.33)–(3.36) is done for the total number of layers N_l, and the integrals in the thickness direction are done for each k layer.

A_{ij} are the extensional stiffnesses, D_{ij} are the bending stiffnesses, and B_{ij} are the bending–extensional coupling stiffnesses, which are defined as:

$$(A_{ij}, B_{ij}, D_{ij}) = \sum_{k=1}^{N_l} \int_{z_k}^{z_{k+1}} Q_{ij}^k(1, z, z^2)dz \tag{3.37}$$

E_{ij} are the electromechanical coupling extensional stiffnesses, G_{ij} are the electromechanical coupling bending stiffnesses, and F_{ij} are the electromechanical coupling bending–extensional coupling stiffnesses, which are defined as:

$$(E_{ij}, F_{ij}, G_{ij}) = \sum_{k=1}^{N_l} \int_{z_k}^{z_{k+1}} e_{ij}^k(1, z, z^2)dz \tag{3.38}$$

R_{ij} are called dielectric extensional stiffnesses, S_{ij} are the dielectric bending stiffnesses, and T_{ij} are the dielectric bending–extensional coupling stiffnesses, which are defined as:

$$(R_{ij}, S_{ij}, T_{ij}) = \sum_{k=1}^{N_l} \int_{z_k}^{z_{k+1}} \varepsilon_{ij}^k(1, z, z^2)dz \tag{3.39}$$

Using Equations (3.37)–(3.39), Equations (3.33)–(3.36) can be rewritten as:

$$\begin{bmatrix} N_{xx} \\ N_{yy} \\ N_{xy} \end{bmatrix} = \begin{bmatrix} A_{11} & A_{12} & A_{16} \\ A_{12} & A_{22} & A_{26} \\ A_{16} & A_{26} & A_{66} \end{bmatrix} \begin{bmatrix} \epsilon_{xx}^{(0)} \\ \epsilon_{yy}^{(0)} \\ \gamma_{xy}^{(0)} \end{bmatrix} + \begin{bmatrix} B_{11} & B_{12} & B_{16} \\ B_{12} & B_{22} & B_{26} \\ B_{16} & B_{26} & B_{66} \end{bmatrix} \begin{bmatrix} \epsilon_{xx}^{(1)} \\ \epsilon_{yy}^{(1)} \\ \gamma_{xy}^{(1)} \end{bmatrix}$$
$$- \begin{bmatrix} 0 & 0 & E_{31} \\ 0 & 0 & E_{32} \\ 0 & 0 & E_{36} \end{bmatrix} \begin{bmatrix} \mathcal{E}_x^{k(0)} \\ \mathcal{E}_y^{k(0)} \\ \mathcal{E}_z^{k(0)} \end{bmatrix} - \begin{bmatrix} 0 & 0 & F_{31} \\ 0 & 0 & F_{32} \\ 0 & 0 & F_{36} \end{bmatrix} \begin{bmatrix} \mathcal{E}_x^{k(1)} \\ \mathcal{E}_y^{k(1)} \\ 0 \end{bmatrix} \tag{3.40}$$

$$\begin{bmatrix} M_{xx} \\ M_{yy} \\ M_{xy} \end{bmatrix} = \begin{bmatrix} B_{11} & B_{12} & B_{16} \\ B_{12} & B_{22} & B_{26} \\ B_{16} & B_{26} & B_{66} \end{bmatrix} \begin{bmatrix} \epsilon_{xx}^{(0)} \\ \epsilon_{yy}^{(0)} \\ \gamma_{xy}^{(0)} \end{bmatrix} + \begin{bmatrix} D_{11} & D_{12} & D_{16} \\ D_{12} & D_{22} & D_{26} \\ D_{16} & D_{26} & D_{66} \end{bmatrix} \begin{bmatrix} \epsilon_{xx}^{(1)} \\ \epsilon_{yy}^{(1)} \\ \gamma_{xy}^{(1)} \end{bmatrix}$$
$$- \begin{bmatrix} 0 & 0 & F_{31} \\ 0 & 0 & F_{32} \\ 0 & 0 & F_{36} \end{bmatrix} \begin{bmatrix} \mathcal{E}_x^{k(0)} \\ \mathcal{E}_y^{k(0)} \\ \mathcal{E}_z^{k(0)} \end{bmatrix} - \begin{bmatrix} 0 & 0 & G_{31} \\ 0 & 0 & G_{32} \\ 0 & 0 & G_{36} \end{bmatrix} \begin{bmatrix} \mathcal{E}_x^{k(1)} \\ \mathcal{E}_y^{k(1)} \\ 0 \end{bmatrix} \tag{3.41}$$

$$\begin{bmatrix} O_x \\ O_y \\ O_z \end{bmatrix} = \begin{bmatrix} 0 & 0 & 0 \\ 0 & 0 & 0 \\ E_{31} & E_{32} & E_{36} \end{bmatrix} \begin{bmatrix} \epsilon_{xx}^{(0)} \\ \epsilon_{yy}^{(0)} \\ \gamma_{xy}^{(0)} \end{bmatrix} + \begin{bmatrix} 0 & 0 & 0 \\ 0 & 0 & 0 \\ F_{31} & F_{32} & F_{36} \end{bmatrix} \begin{bmatrix} \epsilon_{xx}^{(1)} \\ \epsilon_{yy}^{(1)} \\ \gamma_{xy}^{(1)} \end{bmatrix}$$
$$+ \begin{bmatrix} R_{11} & R_{12} & 0 \\ R_{11} & R_{22} & 0 \\ 0 & 0 & R_{33} \end{bmatrix} \begin{bmatrix} \mathcal{E}_x^{k(0)} \\ \mathcal{E}_y^{k(0)} \\ \mathcal{E}_z^{k(0)} \end{bmatrix} + \begin{bmatrix} S_{11} & S_{12} & 0 \\ S_{11} & S_{22} & 0 \\ 0 & 0 & S_{33} \end{bmatrix} \begin{bmatrix} \mathcal{E}_x^{k(1)} \\ \mathcal{E}_y^{k(1)} \\ 0 \end{bmatrix} \tag{3.42}$$

$$\begin{bmatrix} P_x \\ P_y \\ P_z \end{bmatrix} = \begin{bmatrix} 0 & 0 & 0 \\ 0 & 0 & 0 \\ F_{31} & F_{32} & F_{36} \end{bmatrix} \begin{bmatrix} \epsilon_{xx}^{(0)} \\ \epsilon_{yy}^{(0)} \\ \gamma_{xy}^{(0)} \end{bmatrix} + \begin{bmatrix} 0 & 0 & 0 \\ 0 & 0 & 0 \\ G_{31} & G_{32} & G_{36} \end{bmatrix} \begin{bmatrix} \epsilon_{xx}^{(1)} \\ \epsilon_{yy}^{(1)} \\ \gamma_{xy}^{(1)} \end{bmatrix}$$
$$+ \begin{bmatrix} S_{11} & S_{12} & 0 \\ S_{11} & S_{22} & 0 \\ 0 & 0 & S_{33} \end{bmatrix} \begin{bmatrix} \mathcal{E}_x^{k(0)} \\ \mathcal{E}_y^{k(0)} \\ \mathcal{E}_z^{k(0)} \end{bmatrix} + \begin{bmatrix} T_{11} & T_{12} & 0 \\ T_{11} & T_{22} & 0 \\ 0 & 0 & T_{33} \end{bmatrix} \begin{bmatrix} \mathcal{E}_x^{k(1)} \\ \mathcal{E}_y^{k(1)} \\ 0 \end{bmatrix} \tag{3.43}$$

$\{\epsilon^0\}$ and $\{\epsilon^1\}$ are the membrane and bending strains in Equations (3.40)–(3.43). $\{\mathcal{E}^{k(0)}\}$ and $\{\mathcal{E}^{k(1)}\}$ are the membrane and bending electric field components. These four vectors are defined as:

$$\{\epsilon^0\} = \begin{bmatrix} \epsilon_{xx}^{(0)} \\ \epsilon_{yy}^{(0)} \\ \gamma_{xy}^{(0)} \end{bmatrix}, \quad \{\epsilon^1\} = \begin{bmatrix} \epsilon_{xx}^{(1)} \\ \epsilon_{yy}^{(1)} \\ \gamma_{xy}^{(1)} \end{bmatrix}, \quad \{\mathcal{E}^{k(0)}\} = \begin{bmatrix} \mathcal{E}_x^{k(0)} \\ \mathcal{E}_y^{k(0)} \\ \mathcal{E}_z^{k(0)} \end{bmatrix}, \quad \{\mathcal{E}^{k(1)}\} = \begin{bmatrix} \mathcal{E}_x^{k(1)} \\ \mathcal{E}_y^{k(1)} \\ 0 \end{bmatrix} \tag{3.44}$$

The laminate constitutive equations can be written in compact form as:

$$\begin{bmatrix} \{N\} \\ \{M\} \\ \{O\} \\ \{P\} \end{bmatrix} = \begin{bmatrix} \boldsymbol{A} & \boldsymbol{B} & -\boldsymbol{E} & -\boldsymbol{F} \\ \boldsymbol{B} & \boldsymbol{D} & -\boldsymbol{F} & -\boldsymbol{G} \\ \boldsymbol{E}^T & \boldsymbol{F}^T & \boldsymbol{R} & \boldsymbol{S} \\ \boldsymbol{F}^T & \boldsymbol{G}^T & \boldsymbol{S} & \boldsymbol{T} \end{bmatrix} \begin{bmatrix} \{\epsilon^0\} \\ \{\epsilon^1\} \\ \{\mathcal{E}^{k(0)}\} \\ \{\mathcal{E}^{k(1)}\} \end{bmatrix} - \begin{bmatrix} \{N\}^p \\ \{M\}^p \end{bmatrix} \tag{3.45}$$

where T means the transpose of a matrix, and the vectors containing the mechanical loads have the superscript p.

The Euler–Lagrange equations of the CLT plate theory, extended to smart structures (see Equations (3.32)), can be written using Equations (3.40)–(3.43) and Equations (3.9)–(3.14) in terms of displacements and electric potential. These substitutions are a good exercise for interested readers.

3.4.2 FSDT plate theory extended to smart structures

FSDT, like the CLT case, can also be extended to smart structures by assuming a linear electric potential through the thickness of the piezoelectric layer of the multilayered plate.

Considering the Reissner–Mindlin hypothesis (Reissner 1945; Mindlin 1951) and the dependency on time t, it is possible to obtain the model, extended to smart structures, from Equation (3.4):

$$\begin{aligned}
u(x, y, z, t) &= u_0(x, y, t) + z\Phi_x(x, y, t)\\
v(x, y, z, t) &= v_0(x, y, t) + z\Phi_y(x, y, t)\\
w(x, y, z, t) &= w_0(x, y, t)\\
\Phi^k(x, y, z, t) &= \Phi_0^k(x, y, t) + (z - \bar{z}_k)\Phi_1^k(x, y, t)
\end{aligned} \tag{3.46}$$

There are now seven unknowns in Equation (3.46) (for the CLT case there were five); Φ_x and Φ_y are the two additional rotations that are not considered in the Kirchhoff hypotheses. The main limitation of classical theories (CLT and FSDT) extended to smart structures is the use of a linear electric potential through the thickness which gives a constant electric field; this limitation leads to the so-called electrical locking. The piezoelectric constitutive equations for the FSDT case, with reference to Equations (2.12)–(2.18), can be written for each k layer as:

$$\begin{bmatrix} \sigma_{xx}^k \\ \sigma_{yy}^k \\ \sigma_{xy}^k \\ \sigma_{yz}^k \\ \sigma_{xz}^k \end{bmatrix} = \begin{bmatrix} Q_{11}^k & Q_{12}^k & Q_{16}^k & 0 & 0 \\ Q_{12}^k & Q_{22}^k & Q_{26}^k & 0 & 0 \\ Q_{16}^k & Q_{26}^k & Q_{66}^k & 0 & 0 \\ 0 & 0 & 0 & Q_{44}^k & Q_{45}^k \\ 0 & 0 & 0 & Q_{45}^k & Q_{55}^k \end{bmatrix} \begin{bmatrix} \epsilon_{xx} \\ \epsilon_{yy} \\ \gamma_{xy} \\ \gamma_{yz} \\ \gamma_{xz} \end{bmatrix} - \begin{bmatrix} 0 & 0 & e_{31}^k \\ 0 & 0 & e_{32}^k \\ 0 & 0 & e_{36}^k \\ e_{14}^k & e_{24}^k & 0 \\ e_{15}^k & e_{25}^k & 0 \end{bmatrix} \begin{bmatrix} \mathcal{E}_x^k \\ \mathcal{E}_y^k \\ \mathcal{E}_z^k \end{bmatrix} \tag{3.47}$$

$$\begin{bmatrix} \mathcal{D}_x^k \\ \mathcal{D}_y^k \\ \mathcal{D}_z^k \end{bmatrix} = \begin{bmatrix} 0 & 0 & 0 & e_{14}^k & e_{15}^k \\ 0 & 0 & 0 & e_{24}^k & e_{25}^k \\ e_{31}^k & e_{32}^k & e_{36}^k & 0 & 0 \end{bmatrix} \begin{bmatrix} \epsilon_{xx} \\ \epsilon_{yy} \\ \gamma_{xy} \\ \gamma_{yz} \\ \gamma_{xz} \end{bmatrix} + \begin{bmatrix} \varepsilon_{11}^k & \varepsilon_{12}^k & 0 \\ \varepsilon_{12}^k & \varepsilon_{22}^k & 0 \\ 0 & 0 & \varepsilon_{33}^k \end{bmatrix} \begin{bmatrix} \mathcal{E}_x^k \\ \mathcal{E}_y^k \\ \mathcal{E}_z^k \end{bmatrix} \tag{3.48}$$

In this book, it was decided to adopt no correction of the Poisson locking for the electromechanical version of FSDT. The transverse shear strains γ_{yz} and γ_{xz} are different from zero and the geometrical relations are (see Equations (3.9)–(3.14)):

$$\epsilon_{xx} = \frac{\partial u}{\partial x} = \frac{\partial u_0}{\partial x} + z\frac{\partial \Phi_x}{\partial x} = \epsilon_{xx}^{(0)} + z\epsilon_{xx}^{(1)} \tag{3.49}$$

$$\epsilon_{yy} = \frac{\partial v}{\partial y} = \frac{\partial v_0}{\partial y} + z\frac{\partial \Phi_y}{\partial y} = \epsilon_{yy}^{(0)} + z\epsilon_{yy}^{(1)} \tag{3.50}$$

$$\begin{aligned}\gamma_{xy} &= \frac{\partial u}{\partial y} + \frac{\partial v}{\partial x} = \frac{\partial u_0}{\partial y} + z\frac{\partial \Phi_x}{\partial y} + \frac{\partial v_0}{\partial x} + z\frac{\partial \Phi_y}{\partial x} \\ &= \left(\frac{\partial u_0}{\partial y} + \frac{\partial v_0}{\partial x}\right) + z\left(\frac{\partial \Phi_x}{\partial y} + \frac{\partial \Phi_y}{\partial x}\right) = \gamma_{xy}^{(0)} + z\gamma_{xy}^{(1)}\end{aligned} \tag{3.51}$$

$$\gamma_{yz} = \frac{\partial w}{\partial y} + \frac{\partial v}{\partial z} = \frac{\partial w_0}{\partial y} + \Phi_y = \gamma_{yz}^{(0)} \tag{3.52}$$

$$\gamma_{xz} = \frac{\partial w}{\partial x} + \frac{\partial u}{\partial z} = \frac{\partial w_0}{\partial x} + \Phi_x = \gamma_{xz}^{(0)} \tag{3.53}$$

$$\mathcal{E}_x^k = -\frac{\partial \Phi^k}{\partial x} = -\frac{\partial \Phi_0^k}{\partial x} + \bar{z}_k\frac{\partial \Phi_1^k}{\partial x} - z\frac{\partial \Phi_1^k}{\partial x} = \mathcal{E}_x^{k(0)} + z\mathcal{E}_x^{k(1)} \tag{3.54}$$

$$\mathcal{E}_y^k = -\frac{\partial \Phi^k}{\partial y} = -\frac{\partial \Phi_0^k}{\partial y} + \bar{z}_k\frac{\partial \Phi_1^k}{\partial y} - z\frac{\partial \Phi_1^k}{\partial y} = \mathcal{E}_y^{k(0)} + z\mathcal{E}_y^{k(1)} \tag{3.55}$$

$$\mathcal{E}_z^k = -\frac{\partial \Phi^k}{\partial z} = -\Phi_1^k = \mathcal{E}_z^{k(0)} \tag{3.56}$$

Using the PVD given in Equation (3.15), it is possible to obtain the Euler–Lagrange equations for the FSDT plate theory extended to smart structures. The complete procedure will be presented in the next chapter, but only for the case of finite element applications.

3.5 Classical shell theories extended to smart structures

The extensions of CLT and FSDT shell theories to smart structures do not introduce any further difficulties with respect to the already proposed plate cases. For these reasons, only some information will be given in this section; the complete procedures might be a good exercise for those readers who are interested. The kinematics equations and the constitutive equations are the same as those presented in the previous section, but a curvilinear reference system (α,β,z) is employed instead of a rectilinear one (x, y, z). The main differences concern geometrical relations, which will be explained in detail in the following section.

3.5.1 CLT and FSDT shell theories extended to smart structures

The FSDT kinematic model extended to shell geometries considers a curvilinear reference system (α,β,z) instead of the rectilinear one (x, y, z) given in Equation (3.46):

$$\begin{aligned}
u(\alpha,\beta,z,t) &= u_0(\alpha,\beta,t) + z\Phi_\alpha(\alpha,\beta,t)\\
v(\alpha,\beta,z,t) &= v_0(\alpha,\beta,t) + z\Phi_\beta(\alpha,\beta,t)\\
w(\alpha,\beta,z,t) &= w_0(\alpha,\beta,t)\\
\Phi^k(\alpha,\beta,z,t) &= \Phi_0^k(\alpha,\beta,t) + (z-\bar{z}_k)\Phi_1^k(\alpha,\beta,t)
\end{aligned} \tag{3.57}$$

In the case of the FSDT shell theory, the following geometrical relations are stated (see Equations (2.31)–(2.37) and Equation (3.57)):

$$\begin{aligned}
\epsilon_{\alpha\alpha}^k &= \frac{1}{H_\alpha^k}\frac{\partial u}{\partial \alpha} + \frac{1}{H_\alpha^k R_\alpha^k} w = \frac{1}{H_\alpha^k}\frac{\partial u_0}{\partial \alpha} + \frac{z}{H_\alpha^k}\frac{\partial \Phi_\alpha}{\partial \alpha} + \frac{1}{H_\alpha^k R_\alpha^k} w_0\\
&= \left(\frac{1}{H_\alpha^k}\frac{\partial u_0}{\partial \alpha} + \frac{1}{H_\alpha^k R_\alpha^k} w_0\right) + z\left(\frac{1}{H_\alpha^k}\frac{\partial \Phi_\alpha}{\partial \alpha}\right)
\end{aligned} \tag{3.58}$$

$$\begin{aligned}
\epsilon_{\beta\beta}^k &= \frac{1}{H_\beta^k}\frac{\partial v}{\partial \beta} + \frac{1}{H_\beta^k R_\beta^k} w = \frac{1}{H_\beta^k}\frac{\partial v_0}{\partial \beta} + \frac{z}{H_\beta^k}\frac{\partial \Phi_\beta}{\partial \beta} + \frac{1}{H_\beta^k R_\beta^k} w_0\\
&= \left(\frac{1}{H_\beta^k}\frac{\partial v_0}{\partial \beta} + \frac{1}{H_\beta^k R_\beta^k} w_0\right) + z\left(\frac{1}{H_\beta^k}\frac{\partial \Phi_\beta}{\partial \beta}\right)
\end{aligned} \tag{3.59}$$

$$\begin{aligned}
\gamma_{\alpha\beta}^k &= \frac{1}{H_\beta^k}\frac{\partial u}{\partial \beta} + \frac{1}{H_\alpha^k}\frac{\partial v}{\partial \alpha} = \frac{1}{H_\beta^k}\frac{\partial u_0}{\partial \beta} + \frac{z}{H_\beta^k}\frac{\partial \Phi_\alpha}{\partial \beta} + \frac{1}{H_\alpha^k}\frac{\partial v_0}{\partial \alpha} + \frac{z}{H_\alpha^k}\frac{\partial \Phi_\beta}{\partial \alpha}\\
&= \left(\frac{1}{H_\beta^k}\frac{\partial u_0}{\partial \beta} + \frac{1}{H_\alpha^k}\frac{\partial v_0}{\partial \alpha}\right) + z\left(\frac{1}{H_\beta^k}\frac{\partial \Phi_\alpha}{\partial \beta} + \frac{1}{H_\alpha^k}\frac{\partial \Phi_\beta}{\partial \alpha}\right)
\end{aligned} \tag{3.60}$$

$$\begin{aligned}
\gamma_{\beta z}^k &= \frac{1}{H_\beta^k}\frac{\partial w}{\partial \beta} + \frac{\partial v}{\partial z} - \frac{1}{H_\beta^k R_\beta^k} v = \left(\frac{1}{H_\beta^k}\frac{\partial w_0}{\partial \beta} + \Phi_\beta - \frac{1}{H_\beta^k R_\beta^k} v_0\right)\\
&\quad - z\left(\frac{1}{H_\beta^k R_\beta^k}\Phi_\beta\right)
\end{aligned} \tag{3.61}$$

$$\begin{aligned}
\gamma_{\alpha z}^k &= \frac{1}{H_\alpha^k}\frac{\partial w}{\partial \alpha} + \frac{\partial u}{\partial z} - \frac{1}{H_\alpha^k R_\alpha^k} u = \left(\frac{1}{H_\alpha^k}\frac{\partial w_0}{\partial \alpha} + \Phi_\alpha - \frac{1}{H_\alpha^k R_\alpha^k} u_0\right)\\
&\quad - z\left(\frac{1}{H_\alpha^k R_\alpha^k}\Phi_\alpha\right)
\end{aligned} \tag{3.62}$$

$$\varepsilon_{\alpha}^{k} = -\frac{1}{H_{\alpha}^{k}}\frac{\partial \Phi^{k}}{\partial \alpha} = \left(-\frac{1}{H_{\alpha}^{k}}\frac{\partial \Phi_{0}^{k}}{\partial \alpha} + \bar{z}_{k}\frac{1}{H_{\alpha}^{k}}\frac{\partial \Phi_{1}^{k}}{\partial \alpha}\right) - z\frac{1}{H_{\alpha}^{k}}\frac{\partial \Phi_{1}^{k}}{\partial \alpha} \tag{3.63}$$

$$\varepsilon_{\beta}^{k} = -\frac{1}{H_{\beta}^{k}}\frac{\partial \Phi^{k}}{\partial \beta} = \left(-\frac{1}{H_{\beta}^{k}}\frac{\partial \Phi_{0}^{k}}{\partial \beta} + \bar{z}_{k}\frac{1}{H_{\beta}^{k}}\frac{\partial \Phi_{1}^{k}}{\partial \beta}\right) - z\frac{1}{H_{\beta}^{k}}\frac{\partial \Phi_{1}^{k}}{\partial \beta} \tag{3.64}$$

$$\varepsilon_{z}^{k} = -\frac{\partial \Phi^{k}}{\partial z} = -\Phi_{1}^{k} \tag{3.65}$$

For shell geometries, even though the displacements are in ESL form, both the strain and electric field components depend on the k layer because of the curvature.

In order to obtain the kinematic model for CLT, in the case of a shell geometry, we impose $\gamma_{\beta z}^{k} = \gamma_{\alpha z}^{k} = 0$ in Equations (3.61) and (3.62). In this way, we obtain the relations for Φ_{β} and Φ_{α}:

$$\Phi_{\beta} = \left(\frac{1}{H_{\beta}^{k}R_{\beta}^{k}}v_{0} - \frac{1}{H_{\beta}^{k}}\frac{\partial w_{0}}{\partial \beta}\right)\frac{H_{\beta}^{k}R_{\beta}^{k}}{H_{\beta}^{k}R_{\beta}^{k} - z} \tag{3.66}$$

$$\Phi_{\alpha} = \left(\frac{1}{H_{\alpha}^{k}R_{\alpha}^{k}}u_{0} - \frac{1}{H_{\alpha}^{k}}\frac{\partial w_{0}}{\partial \alpha}\right)\frac{H_{\alpha}^{k}R_{\alpha}^{k}}{H_{\alpha}^{k}R_{\alpha}^{k} - z} \tag{3.67}$$

In the CLT and FSDT cases, Φ_{β} and Φ_{α} do not depend on the k layer, if we consider the mean value at the mid-reference surface for the radii of curvature R_{α}^{k} and R_{β}^{k} of the multilayered structure.

By substituting Equations (3.66) and (3.67) into Equation (3.57), we obtain the kinematic model for CLT for the case of a shell geometry. In the same way, it is possible to write the geometrical relations for the CLT case by starting from the geometrical relations for the FSDT case (Equations (3.58)–(3.65)), where $\gamma_{\beta z} = \gamma_{\alpha z} = 0$, and Equations (3.66) and (3.67) give the rotations Φ_{β} and Φ_{α}.

References

Antona E 1991 Mathematical models and their use in engineering. *Appl. Math. Aeronaut. Sci. Eng.* **44**, 395–433.

Carrera E 1997 C_z^0 requirements – models for the two dimensional analysis of multilayered structures. *Comp. Struct.* **37**, 373–383.

Carrera E 2002 Theories and finite elements for multilayered anisotropic, composite plates and shells. *Arch. Comput. Met. Eng.* **9**, 87–140.

Carrera E 2003 Historical review of zig-zag theories for multilayered plates and shells. *Appl. Mech. Rev.* **56**, 287–309.

Carrera E 2004 On the use of Murakami's zig-zag function in the modeling of layered plates and shells. *Comput. Struct.* **82**, 541–554.

Carrera E and Brischetto S 2008a Analysis of thickness locking in classical, refined and mixed multilayered plate theories. *Comp. Struct.* **82**, 549–562.

Carrera E and Brischetto S 2008b Analysis of thickness locking in classical, refined and mixed theories for layered shells. *Comp. Struct.* **85**, 83–90.

Carrera E and Brischetto S 2009 A survey with numerical assessment of classical and refined theories for the analysis of sandwich plates. *Appl. Mech. Rev.* **62**, 1–17.

Cauchy AL 1828 Sur l'équilibre et le mouvement d'une plaque solide. *Exercise Math.* **3**, 381–412.

Cicala P 1959 Sulla teoria elastica della parete sottile. *G. Genio Civ.* **97**, 429–449.

Cicala P 1965 *Systematic Approach to Linear Shell Theory*. Levrotto & Bella.

Cosserat E and Cosserat F 1909 *Théories des Corps Déformable*. Hermann.

Demasi L 2005 Refined multilayered plate elements based on Murakami zig-zag functions. *Comp. Struct.* **70**, 308–316.

Demasi L 2008 ∞^3 hierarchy plate theories for thick and thin composite plates: the generalized unified formulation. *Comp. Struct.* **84**, 256–270.

Green AE and Naghdi PM 1967 Shells in light of generalized continua. In *Proceedings of Second Symposium on the Theory of Thin Elastic Shells.*

Jones RM 1999 *Mechanics of Composite Materials*. Taylor & Francis.

Kirchhoff G 1850 Über das Gleichgewicht und die Bewegung einer elastischen Scheibe. *J. Reine Angew. Math.* **40**, 51–88.

Koiter WT 1960 A consistent first approximation in the general theory of thin elastic shells. In *Proceedings of First Symposium on the Theory of Thin Elastic Shells.*

Love AEH 1906 *A Treatise on the Mathematical Theory of Elasticity*. Cambridge University Press.

Mindlin RD 1951 Influence of rotatory inertia and shear on flexural motions of isotropic, elastic plates. *J. Appl. Mech.* **18**, 31–38.

Noor AK and Rarig PL 1974 Three-dimensional solutions of laminated cylinders. *Comput. Methods Appl. Mech. Eng.* **3**, 319–334.

Pagano NJ 1969 Exact solutions for composite laminates in cylindrical bending. *J. Compos. Mater.* **3**, 398–411.

Pagano NJ 1970 Exact solutions for rectangular bidirectional composites and sandwich plates. *J. Compos. Mater.* **4**, 20–34.

Pagano NJ and Hatfield SJ 1972 Elastic behavior of multilayered bidirectional composites. *AIAA J.* **10**, 931–933.

Poisson SD 1829 Mémoire sur l'équilibre et le mouvement des corps élastiques. *Mém. Acad. Sci. Paris* **8**, 357–570.

Reddy JN 1984 A simple higher-order theory for laminated composite plates. *J. Appl. Mech.* **51**, 745–752.

Reddy JN 2004 *Mechanics of Laminated Composite Plates and Shells: Theory and Analysis*. CRC Press.

Reddy JN and Phan ND 1985 Stability and vibration of isotropic, orthotropic and laminated plates according to a higher order shear deformation theory. *J. Sound Vibr.* **98**, 157–170.

Reissner E 1945 The effect of transverse shear deformation on the bending of elastic plates. *J. Appl. Mech.* **12**, 69–77.

Reissner E 1967 On the foundations of generalized linear shell theory. In *Proceedings of Second Symposium on the Theory of Thin Elastic Shells.*

Reissner E 1984 On a certain mixed variational theory and a proposed application. *Int. J. Numer. Methods Eng.* **20**, 1366–1368.

Vlasov BF 1957 On the equations of bending of plates. *Dokl. Nauk Azerbeijanskoi SSR* **3**, 955–979.

4

Finite element applications

The aim of this chapter is to obtain the finite element method governing equations of the static and dynamic electromechanical analysis for multilayered plates embedding piezoelectric layers. Such equations are dealt with in detail for the FSDT case, extended to smart plates, in analogy with the previous chapter where the equations were detailed for the CLT plate case extended to smart structures.

4.1 Preliminaries

The investigations required to find the solution of generic scientific or technical problems in general make use of numerical models which are either *discrete* or *continuous*. Discrete problems involve a finite number of components with a limited number of degrees of freedom (DOFs).

Generally, only a small subset of simplified continuous problems can be solved by mathematical manipulation. The corresponding solutions are called *exact solutions*. When analytical solutions are not available, a *discretization* is commonly introduced and the problem is expressed in terms of a finite number of discrete variables by involving a finite number of DOFs. Thus, the solution of the continuous problem is approximated by solving the discrete problem. The error due to discretization can be reduced by increasing the number of discrete variables. Exact solutions and, more in general, convergence studies permit one to verify the accuracy of the approximated solution.

Plates and Shells for Smart Structures: Classical and Advanced Theories for Modeling and Analysis, First Edition.
Erasmo Carrera, Salvatore Brischetto and Pietro Nali.

Different techniques can be applied to discretize the continuous problems. Among these, finite difference approximations (Allen 1955; Southwell 1946), and weighted residual procedures (Crandall 1958; Finlayson 1972), have been proposed by mathematicians in order to find the stationary points of appropriate functionals. As an alternative, engineers introduced the analogy between continuous *subdomains* and discrete elements (Argyris 1960; Hrenikoff 1941; McHenry 1943; Newmark 1949). The term "finite element" (FE) was first used in the work by Clough (1960) according to the latter approach.

Since the 1960s, a great deal of progress has been made. The discretization procedures of continuous problems have been extensively standardized and generalized. An excellent reference on this topic is given by the work of Zienkiewicz and Taylor (1967), where a unified treatment of "standard discrete problems" is presented by defining the finite element process as *a method of approximation to continuous problems so that:*

- *the continuum is divided into a finite number of parts (elements), the behavior of which is specified by a finite number of parameters; and*
- *the solution of the complete system, as an assembly of its elements, follows precisely the same rules as those applicable to standard discrete problems.*

The above approximation technique is commonly known as the finite element method (FEM).

4.2 Finite element discretization

The PVD, as proposed in Equations (3.15) and (3.16), can also be solved by means of the FEM (Zienkiewicz and Taylor 1967), which introduces several generic elements of the surface Ω (mesh of the plate) into the plate. Each generic element can be transformed into a master element $\hat{\Omega}$ with a given number of nodes, see Figure 4.1. Each considered master element can have a certain number of nodes where the variables are considered, and these elements can have four nodes (Q4), eight nodes (Q8), or nine nodes (Q9) (Zienkiewicz and Taylor 1967). In the FEM, the unknowns are expressed in terms of their nodal values, via the shape functions N_i. The latter assume unit values in the nodes, and permit the unknowns to be expressed at points that are different from the nodes as linear combinations of the 4, 8, or 9 values in the nodes (Zienkiewicz and Taylor 1967). In Figure 4.2, the Q4, Q8, and Q9 elements are clearly indicated, and a natural coordinate system (ξ, η) is defined which always goes from -1 to $+1$. The approach employed here for the plate is an *isoparametric approach* (Newmark 1949). The shape functions N_i for the Q4, Q8, and Q9 elements are the well-known formulas given in Zienkiewicz and

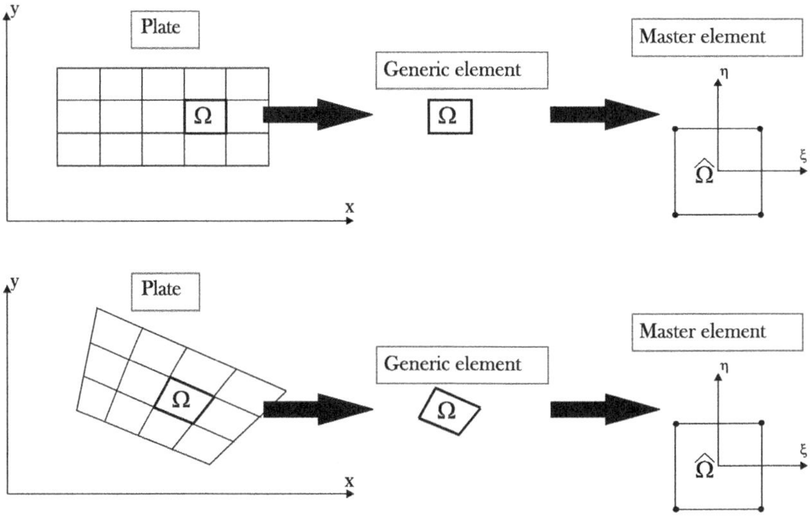

Figure 4.1 Mesh for a generic plate by means of several master elements.

Taylor (1967). Therefore, the generic variable $\boldsymbol{a}_\tau$ defined in the xy-plane, and its virtual variation $\delta\boldsymbol{a}_s$, can be expressed in terms of nodal values $\boldsymbol{q}_{\tau i}$ and $\delta\boldsymbol{q}_{sj}$ via the shape functions N_i and N_j:

$$\boldsymbol{a}_\tau(x, y) = N_i\, \boldsymbol{q}_{\tau i}, \quad \delta\boldsymbol{a}_s(x, y) = N_j\, \delta\boldsymbol{q}_s, \quad i, j = 1, 2, \ldots, N_n \quad (4.1)$$

where N_n denotes the number of nodes of the considered element.

$N_i(x, y)$ indicates the shape function for the ith node. It is convenient to recover the shape functions from the natural coordinates (ξ, η) (see Figure 4.2)

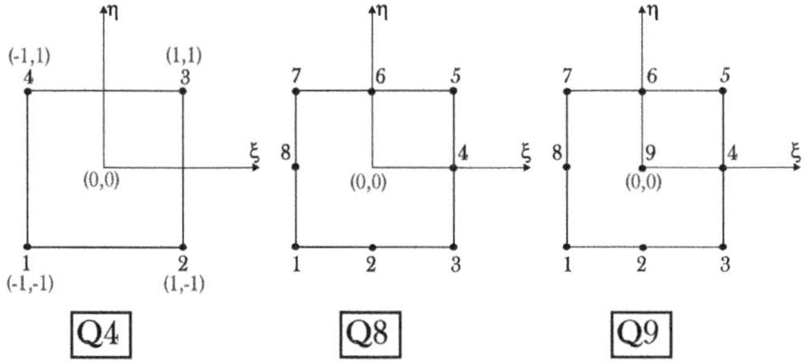

Figure 4.2 Finite element method: 4-, 8-, and 9-node master elements with natural coordinate system and node enumeration.

and not from the global ones (x, y). The shape functions of each considered element (Q4, Q8 and Q9) are obtained by means of the condition:

$$\begin{aligned} N_i &= 1 \qquad \text{if } \xi = \xi_i; \eta = \eta_i \\ N_i &= 0 \qquad \text{otherwise} \end{aligned} \tag{4.2}$$

It is clear how $N_i(\xi, \eta)$ depends on natural coordinates, because of $N_i(x(\xi, \eta), y(\xi, \eta))$. In the case of rectangular elements, as indicated in the upper part of Figure 4.1, the relation is:

$$\xi = \frac{2(x - x_1) - a}{a} \tag{4.3}$$

$$\eta = \frac{2(y - y_1) - b}{b} \tag{4.4}$$

where x_1 and y_1 are the global coordinates of the first node of the element, and a and b are the dimensions of the element in the global directions x and y, respectively.

In the case of no rectangular elements (see the bottom of Figure 4.1) the relations between global and natural coordinates are:

$$x = \sum_{i=1}^{N_{nG}} x_i \hat{N}_i(\xi, \eta) \tag{4.5}$$

$$y = \sum_{i=1}^{N_{nG}} y_i \hat{N}_i(\xi, \eta) \tag{4.6}$$

where $\hat{N}_i$ are the shape functions in the case of a generic geometry of the element (in general $\hat{N}_i \neq N_i$ when $N_{nG} \neq N_n$), and N_{nG} is the number of nodes for the change in variables (G means generic geometry). Equations (4.5) and (4.6) degenerate into Equations (4.3) and (4.4) for rectangular elements. The *isoparametric approach* is considered when $N_{nG} = N_n$ and $\hat{N}_i = N_i$.

In order to integrate along the in-plane directions, in accordance with the Gauss method, it is necessary to express everything in terms of natural coordinates (ξ, η):

$$\frac{\partial N_i(x(\xi, \eta), y(\xi, \eta))}{\partial \xi} = \frac{\partial N_i}{\partial x}\frac{\partial x}{\partial \xi} + \frac{\partial N_i}{\partial y}\frac{\partial y}{\partial \xi} \tag{4.7}$$

$$\frac{\partial N_i(x(\xi, \eta), y(\xi, \eta))}{\partial \eta} = \frac{\partial N_i}{\partial x}\frac{\partial x}{\partial \eta} + \frac{\partial N_i}{\partial y}\frac{\partial y}{\partial \eta} \tag{4.8}$$

Equations (4.7) and (4.8) can be rewritten as:

$$\begin{bmatrix} \dfrac{\partial N_i}{\partial \xi} \\ \dfrac{\partial N_i}{\partial \eta} \end{bmatrix} = \begin{bmatrix} \dfrac{\partial x}{\partial \xi} & \dfrac{\partial y}{\partial \xi} \\ \dfrac{\partial x}{\partial \eta} & \dfrac{\partial y}{\partial \eta} \end{bmatrix} \begin{bmatrix} \dfrac{\partial N_i}{\partial x} \\ \dfrac{\partial N_i}{\partial y} \end{bmatrix} \tag{4.9}$$

In Equation (4.9) the Jacobian matrix is defined as:

$$\boldsymbol{J} = \begin{bmatrix} \dfrac{\partial x}{\partial \xi} & \dfrac{\partial y}{\partial \xi} \\ \dfrac{\partial x}{\partial \eta} & \dfrac{\partial y}{\partial \eta} \end{bmatrix} \tag{4.10}$$

Considering the inverse of the Jacobian matrix in Equation (4.10), it is possible to write:

$$\begin{bmatrix} \dfrac{\partial N_i}{\partial x} \\ \dfrac{\partial N_i}{\partial y} \end{bmatrix} = [\boldsymbol{J}]^{-1} \begin{bmatrix} \dfrac{\partial N_i}{\partial \xi} \\ \dfrac{\partial N_i}{\partial \eta} \end{bmatrix} \tag{4.11}$$

where

$$[\boldsymbol{J}]^{-1} = [\boldsymbol{J}]^{*} = \begin{bmatrix} J_{11}^{*} & J_{12}^{*} \\ J_{21}^{*} & J_{22}^{*} \end{bmatrix} \tag{4.12}$$

Using Equations (4.11) and (4.12), the derivatives of the shape functions, with respect to global coordinates, can be written as linear combination of the derivatives with respect to natural coordinates (ξ,η):

$$\frac{\partial N_i}{\partial x} = N_{i,x} = J_{11}^{*} \frac{\partial N_i}{\partial \xi} + J_{12}^{*} \frac{\partial N_i}{\partial \eta} \tag{4.13}$$

$$\frac{\partial N_i}{\partial y} = N_{i,y} = J_{21}^{*} \frac{\partial N_i}{\partial \xi} + J_{22}^{*} \frac{\partial N_i}{\partial \eta} \tag{4.14}$$

In the case of integration along in-plane directions, we can consider the master element $\hat{\Omega}$ in the natural coordinates (ξ,η) or the generic element Ω in the global coordinates (x,y). The relation between the two element areas is:

$$d\Omega = dx\, dy = \det\boldsymbol{J}\, d\xi\, d\eta \tag{4.15}$$

Equations (4.13)–(4.15) will be very useful for the next section, where the governing equations will be obtained for the FSDT extended to a multilayered plate, embedding a piezoelectric layer, in the case of a FE approximation.

4.3 FSDT finite element plate theory extended to smart structures

The smart structure analyzed is the same one that was examined in Section 3.4.1, for the case of CLT extended to electromechanical problems, considering the analytical solution. Figure 4.3 shows the geometry of the plate and its coordinate systems. A linear electric potential through the thickness of the piezoelectric layer is assumed in this smart structure. We consider a plate of total thickness h composed of an orthotropic layer (thickness h_2) and a piezoelectric layer at the top (thickness h_1); the same notation as Section 3.4.1 is employed.

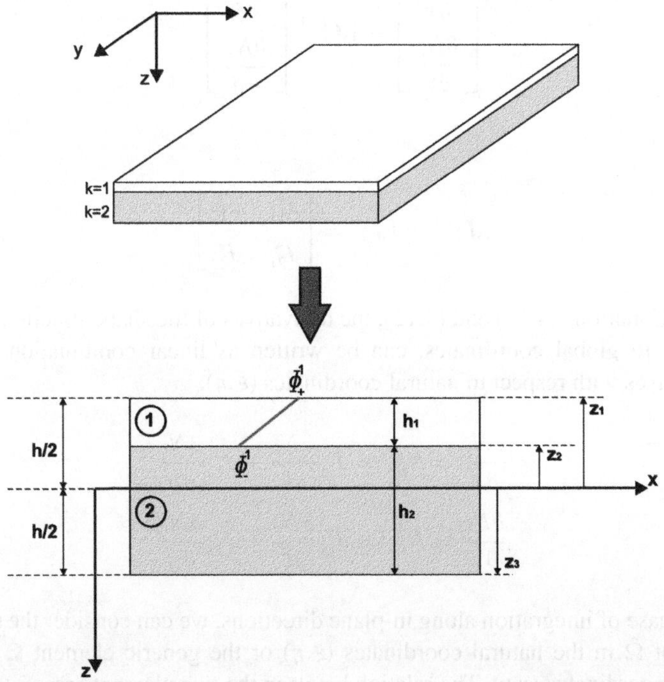

Figure 4.3 Geometry and notations for a generic orthotropic plate embedding a piezoelectric layer.

In the case of FSDT, based on the Reissner–Mindlin hypotheses (Mindlin 1951; Reissner 1945) extended to smart structures, only the last relation in Equation (3.6) remains (the electric potential is linear through the thickness in layer-wise form). Considering Equation (3.4), the model extended to smart structures becomes:

$$\begin{aligned} u(x, y, z, t) &= u_0(x, y, t) + z\Phi_x(x, y, t) \\ v(x, y, z, t) &= v_0(x, y, t) + z\Phi_y(x, y, t) \\ w(x, y, z, t) &= w_0(x, y, t) \\ \Phi^k(x, y, z, t) &= \Phi_0^k(x, y, t) + (z - \bar{z}_k)\Phi_1^k(x, y, t) \end{aligned} \tag{4.16}$$

The Poisson locking phenomenon remains, as for the analytical CLT case of the previous section, and it can be overcome, in the case of pure mechanical problems, by utilizing the plane stress conditions given in Carrera and Brischetto (2008).

u_0, v_0, and w_0 in Equation (4.16) are three displacement components in the middle plane of the plate, and Φ_x and Φ_y are the two additional rotations with respect to the CLT case. The two electrical unknowns are Φ_0^k and Φ_1^k; these allow one to express the linear electric potential through the thickness direction of the k layer where it is considered (in this case the electric potential is in layer 1). It is possible to consider an electric potential at the top Φ_+^k and at the bottom Φ_-^k of the generic piezoelectric layer k. The linear electric potential through the thickness is defined by means of the mean value $\Phi_0^k = (\Phi_+^k + \Phi_-^k)/2$ and the slope $\Phi_1^k = (\Phi_-^k - \Phi_+^k)/h_k$. The value $\bar{z}_k = (z_k + z_{k+1})/2$ is defined in order to refer the thickness coordinate to the midsurface of the kth piezoelectric layer.

If we consider the FE discretization in Equation (4.1), for the seven degrees of freedom involved in this model, it is possible to write:

$$\boldsymbol{a}_\tau(x, y) = N_i\, \boldsymbol{q}_{\tau i} \tag{4.17}$$

where i denotes the nodes of the considered element, and the two vectors are:

$$\boldsymbol{a}_\tau = \begin{bmatrix} u_0 \\ v_0 \\ w_0 \\ \Phi_x \\ \Phi_y \\ \Phi_0^k \\ \Phi_1^k \end{bmatrix}, \qquad \boldsymbol{q}_{\tau i} = \begin{bmatrix} q_{u_0 i} \\ q_{v_0 i} \\ q_{w_0 i} \\ q_{\Phi_x i} \\ q_{\Phi_y i} \\ q_{\Phi_0 i}^k \\ q_{\Phi_1 i}^k \end{bmatrix} \tag{4.18}$$

where, in Equation (4.18), the vector $\boldsymbol{q}_{\tau i}$ contains the nodal values. The same relations can also be written for the virtual variations:

$$\delta \boldsymbol{a}_s(x, y) = N_j \, \delta \boldsymbol{q}_{sj} \tag{4.19}$$

where j denotes the nodes of the considered element, and the two vectors are:

$$\delta \boldsymbol{a}_s = \begin{bmatrix} \delta u_0 \\ \delta v_0 \\ \delta w_0 \\ \delta \Phi_x \\ \delta \Phi_y \\ \delta \Phi_0^k \\ \delta \Phi_1^k \end{bmatrix}, \qquad \delta \boldsymbol{q}_{sj} = \begin{bmatrix} \delta q_{u_0 j} \\ \delta q_{v_0 j} \\ \delta q_{w_0 j} \\ \delta q_{\Phi_x j} \\ \delta q_{\Phi_y j} \\ \delta q_{\Phi_0 j}^k \\ \delta q_{\Phi_1 j}^k \end{bmatrix} \tag{4.20}$$

Using the FE discretization in Equations (4.17) and (4.18), the electromechanical model of Equation (4.16) can be written as:

$$\begin{aligned} u(x, y, z, t) &= N_i q_{u_0 i} + z N_i q_{\Phi_x i} \\ v(x, y, z, t) &= N_i q_{v_0 i} + z N_i q_{\Phi_y i} \\ w(x, y, z, t) &= N_i q_{w_0 i} \\ \Phi^k(x, y, z, t) &= N_i q_{\Phi_0 i}^k + (z - \bar{z}_k) N_i q_{\Phi_1 i}^k \end{aligned} \tag{4.21}$$

The FE discretization employed in Equation (4.21) can also be used for the virtual variations using Equations (4.19) and (4.20):

$$\begin{aligned} \delta u(x, y, z, t) &= N_j \delta q_{u_0 j} + z N_j \delta q_{\Phi_x j} \\ \delta v(x, y, z, t) &= N_j \delta q_{v_0 j} + z N_j \delta q_{\Phi_y j} \\ \delta w(x, y, z, t) &= N_j \delta q_{w_0 j} \\ \delta \Phi^k(x, y, z, t) &= N_j \delta q_{\Phi_0 j}^k + (z - \bar{z}_k) N_j \delta q_{\Phi_1 j}^k \end{aligned} \tag{4.22}$$

The piezoelectric constitutive equations for the FSDT case, with reference to Equations (2.12)–(2.18), have transverse shear strains (γ_{yz} and γ_{xz}) and stresses (σ_{yz} and σ_{xz}) in addition to the components already considered for the CLT case in Equations (3.7) and (3.8):

$$\begin{bmatrix} \sigma^k_{xx} \\ \sigma^k_{yy} \\ \sigma^k_{xy} \\ \sigma^k_{yz} \\ \sigma^k_{xz} \end{bmatrix} = \begin{bmatrix} Q^k_{11} & Q^k_{12} & Q^k_{16} & 0 & 0 \\ Q^k_{12} & Q^k_{22} & Q^k_{26} & 0 & 0 \\ Q^k_{16} & Q^k_{26} & Q^k_{66} & 0 & 0 \\ 0 & 0 & 0 & Q^k_{44} & Q^k_{45} \\ 0 & 0 & 0 & Q^k_{45} & Q^k_{55} \end{bmatrix} \begin{bmatrix} \epsilon_{xx} \\ \epsilon_{yy} \\ \gamma_{xy} \\ \gamma_{yz} \\ \gamma_{xz} \end{bmatrix} - \begin{bmatrix} 0 & 0 & e^k_{31} \\ 0 & 0 & e^k_{32} \\ 0 & 0 & e^k_{36} \\ e^k_{14} & e^k_{24} & 0 \\ e^k_{15} & e^k_{25} & 0 \end{bmatrix} \begin{bmatrix} \mathcal{E}^k_x \\ \mathcal{E}^k_y \\ \mathcal{E}^k_z \end{bmatrix} \tag{4.23}$$

$$\begin{bmatrix} \mathcal{D}^k_x \\ \mathcal{D}^k_y \\ \mathcal{D}^k_z \end{bmatrix} = \begin{bmatrix} 0 & 0 & 0 & e^k_{14} & e^k_{15} \\ 0 & 0 & 0 & e^k_{24} & e^k_{25} \\ e^k_{31} & e^k_{32} & e^k_{36} & 0 & 0 \end{bmatrix} \begin{bmatrix} \epsilon_{xx} \\ \epsilon_{yy} \\ \gamma_{xy} \\ \gamma_{yz} \\ \gamma_{xz} \end{bmatrix} + \begin{bmatrix} \varepsilon^k_{11} & \varepsilon^k_{12} & 0 \\ \varepsilon^k_{12} & \varepsilon^k_{22} & 0 \\ 0 & 0 & \varepsilon^k_{33} \end{bmatrix} \begin{bmatrix} \mathcal{E}^k_x \\ \mathcal{E}^k_y \\ \mathcal{E}^k_z \end{bmatrix} \tag{4.24}$$

The geometrical relations are obtained as indicated in Reddy (2004). Transverse shear strains are added to the strain and electric field components that were already discussed for the CLT case in Equations (3.9)–(3.14). Considering Equations (2.38)–(2.42) and introducing the derivation of the shape functions, as discussed in the previous section (Equations (4.13)–(4.14)), it is possible to write the electromechanical geometrical relation for the FSDT model, extended to multilayered piezoelectric plates, for the case of FE analysis (see Equations (3.49)–(3.57) of the analytical version of FSDT for comparison purposes):

$$\epsilon_{xx} = \frac{\partial u}{\partial x} = N_{i,x} q_{u_0 i} + z N_{i,x} q_{\Phi_x i} = \epsilon^{(0)}_{xx} + z\epsilon^{(1)}_{xx} \tag{4.25}$$

$$\epsilon_{yy} = \frac{\partial v}{\partial y} = N_{i,y} q_{v_0 i} + z N_{i,y} q_{\Phi_y i} = \epsilon^{(0)}_{yy} + z\epsilon^{(1)}_{yy} \tag{4.26}$$

$$\begin{aligned} \gamma_{xy} &= \frac{\partial u}{\partial y} + \frac{\partial v}{\partial x} = N_{i,y} q_{u_0 i} + N_{i,x} q_{v_0 i} + z(N_{i,y} q_{\Phi_x i} + N_{i,x} q_{\Phi_y i}) \\ &= \gamma^{(0)}_{xy} + z\gamma^{(1)}_{xy} \end{aligned} \tag{4.27}$$

$$\gamma_{yz} = \frac{\partial v}{\partial z} + \frac{\partial w}{\partial y} = N_i q_{\Phi_y i} + N_{i,y} q_{w_0 i} = \gamma^{(0)}_{yz} \tag{4.28}$$

$$\gamma_{xz} = \frac{\partial u}{\partial z} + \frac{\partial w}{\partial x} = N_i q_{\Phi_x i} + N_{i,x} q_{w_0 i} = \gamma^{(0)}_{xz} \tag{4.29}$$

$$\mathcal{E}_x^k = -\frac{\partial \Phi^k}{\partial x} = (-N_{i,x}q_{\Phi_0 i}^k + \bar{z}_k N_{i,x}q_{\Phi_1 i}^k) - zN_{i,x}q_{\Phi_1 i}^k = \mathcal{E}_x^{k(0)} + z\mathcal{E}_x^{k(1)} \tag{4.30}$$

$$\mathcal{E}_y^k = -\frac{\partial \Phi^k}{\partial y} = (-N_{i,y}q_{\Phi_0 i}^k + \bar{z}_k N_{i,y}q_{\Phi_1 i}^k) - zN_{i,y}q_{\Phi_1 i}^k = \mathcal{E}_y^{k(0)} + z\mathcal{E}_y^{k(1)} \tag{4.31}$$

$$\mathcal{E}_z^k = -\frac{\partial \Phi^k}{\partial z} = -N_i q_{\Phi_1 i}^k = \mathcal{E}_z^{k(0)} \tag{4.32}$$

As in Equations (4.25)–(4.32), it is possible to write the same relations for the virtual strain and electric field components (see Equations (4.19), (4.20), and (4.22)):

$$\delta\epsilon_{xx}^{(0)} = N_{j,x}\delta q_{u_0 j}, \qquad \delta\epsilon_{xx}^{(1)} = N_{j,x}\delta q_{\Phi_x j} \tag{4.33}$$

$$\delta\epsilon_{yy}^{(0)} = N_{j,y}\delta q_{v_0 j}, \qquad \delta\epsilon_{yy}^{(1)} = N_{j,y}\delta q_{\Phi_y j} \tag{4.34}$$

$$\delta\gamma_{xy}^{(0)} = N_{j,y}\delta q_{u_0 j} + N_{j,x}\delta q_{v_0 j}, \qquad \delta\gamma_{xy}^{(1)} = N_{j,y}\delta q_{\Phi_x j} + N_{j,x}\delta q_{\Phi_y j} \tag{4.35}$$

$$\delta\gamma_{yz}^{(0)} = N_j\delta q_{\Phi_y j} + N_{j,y}\delta q_{w_0 j} \tag{4.36}$$

$$\delta\gamma_{xz}^{(0)} = N_j\delta q_{\Phi_x j} + N_{j,x}\delta q_{w_0 j} \tag{4.37}$$

$$\delta\mathcal{E}_x^{k(0)} = -N_{j,x}\delta q_{\Phi_0 j}^k + \bar{z}_k N_{j,x}\delta q_{\Phi_1 j}^k, \qquad \delta\mathcal{E}_x^{k(1)} = -N_{j,x}\delta q_{\Phi_1 j}^k \tag{4.38}$$

$$\delta\mathcal{E}_y^{k(0)} = -N_{j,y}\delta q_{\Phi_0 j}^k + \bar{z}_k N_{j,y}\delta q_{\Phi_1 j}^k, \qquad \delta\mathcal{E}_y^{k(1)} = -N_{j,y}\delta q_{\Phi_1 j}^k \tag{4.39}$$

$$\delta\mathcal{E}_z^{k(0)} = -N_j\delta q_{\Phi_1 j}^k \tag{4.40}$$

The same rules of shape functions N_i are also valid for shape functions N_j as described in Equations (4.13) and (4.14).

The dynamic version of the PVD, which was first proposed in Reddy (2004), where t is the time which goes from 0 to T, states, as in the previous chapter for the CLT case of Equation (3.15), that:

$$0 = \int_0^T (\delta U + \delta V - \delta K)dt \tag{4.41}$$

where the virtual internal work δU (volume integral of δU_0) in the case of electromechanical coupling is a summation of δU_m (virtual strain energy) and δU_e (virtual electrical internal work). The term δU_e is not considered for a partial electromechanical coupling and the electrical contribution is only considered by means of the second term in Equation (4.23). In this work, a fully electro mechanical coupling is accounted for, because both δU_m and δU_e are considered. δV is the virtual work done by the applied forces, and it permits one to obtain the mechanical forces for sensor applications; no mechanical forces are considered for actuator applications and the electric potential can be applied directly to the top and bottom of the considered piezoelectric actuator layer. δK is the virtual kinetic energy.

For the sake of brevity, we only give details on the virtual internal work δU; further details concerning the virtual work done by applied forces δV and the virtual kinetic energy δK can be found in the previous chapter and in Reddy (2004). The virtual internal work δU is given by:

$$
\begin{aligned}
\delta U &= \delta U_m + \delta U_e = \int_v \delta U_0 dv = \int_v (\delta U_{0m} + \delta U_{0e}) dv \\
&= \int_{\Omega_0} \Big(\int_{-h/2}^{h/2} \Big[\Big(\sigma_{xx}^k \delta\epsilon_{xx} + \sigma_{yy}^k \delta\epsilon_{yy} + \sigma_{xy}^k \delta\gamma_{xy} + \sigma_{xz}^k \delta\gamma_{xz} + \sigma_{yz}^k \delta\gamma_{yz} \Big) \\
&\quad + \Big(- \mathcal{D}_x^k \delta\mathcal{E}_x^k - \mathcal{D}_y^k \delta\mathcal{E}_y^k - \mathcal{D}_z^k \delta\mathcal{E}_z^k \Big) \Big] dz \Big) dxdy \\
&= \int_{\Omega_0} \Big(\sum_{k=1}^{N_l} \int_{-h_k/2}^{h_k/2} \Big[\sigma_{xx}^k \Big(\delta\epsilon_{xx}^{(0)} + z\delta\epsilon_{xx}^{(1)} \Big) + \sigma_{yy}^k \Big(\delta\epsilon_{yy}^{(0)} + z\delta\epsilon_{yy}^{(1)} \Big) \\
&\quad + \sigma_{xy}^k \Big(\delta\gamma_{xy}^{(0)} + z\delta\gamma_{xy}^{(1)} \Big) + \sigma_{xz}^k \Big(\delta\gamma_{xz}^{(0)} \Big) + \sigma_{yz}^k \Big(\delta\gamma_{yz}^{(0)} \Big) \\
&\quad - \mathcal{D}_x^k \Big(\delta\mathcal{E}_x^{k(0)} + z\delta\mathcal{E}_x^{k(1)} \Big) - \mathcal{D}_y^k \Big(\delta\mathcal{E}_y^{k(0)} + z\delta\mathcal{E}_y^{k(1)} \Big) - \mathcal{D}_z^k \Big(\delta\mathcal{E}_z^{k(0)} \Big) \Big] dz \Big) dxdy
\end{aligned}
\tag{4.42}
$$

After substitution of the constitutive relations (Equations (4.23)–(4.24)), Equation (4.42) has the following form:

$$
\begin{aligned}
\delta U = \int_{\Omega_0} \Big(\sum_{k=1}^{N_l} \int_{-h_k/2}^{h_k/2} \Big[\Big(& Q_{11}^k \Big(\epsilon_{xx}^{(0)} + z\epsilon_{xx}^{(1)} \Big) + Q_{12}^k \Big(\epsilon_{yy}^{(0)} + z\epsilon_{yy}^{(1)} \Big) + Q_{16}^k \Big(\gamma_{xy}^{(0)} \\
& + z\gamma_{xy}^{(1)} \Big) - e_{31}^k \mathcal{E}_z^{k(0)} \Big) \Big(\delta\epsilon_{xx}^{(0)} + z\delta\epsilon_{xx}^{(1)} \Big) + \Big(Q_{12}^k \Big(\epsilon_{xx}^{(0)} + z\epsilon_{xx}^{(1)} \Big) \\
& + Q_{22}^k \Big(\epsilon_{yy}^{(0)} + z\epsilon_{yy}^{(1)} \Big) + Q_{26}^k \Big(\gamma_{xy}^{(0)} + z\gamma_{xy}^{(1)} \Big) - e_{32}^k \mathcal{E}_z^{k(0)} \Big) \Big(\delta\epsilon_{yy}^{(0)} + z\delta\epsilon_{yy}^{(1)} \Big) \\
& + \Big(Q_{16}^k \Big(\epsilon_{xx}^{(0)} + z\epsilon_{xx}^{(1)} \Big) + Q_{26}^k \Big(\epsilon_{yy}^{(0)} + z\epsilon_{yy}^{(1)} \Big) + Q_{66}^k \Big(\gamma_{xy}^{(0)} + z\gamma_{xy}^{(1)} \Big) \\
& - e_{36}^k \mathcal{E}_z^{k(0)} \Big) \Big(\delta\gamma_{xy}^{(0)} + z\delta\gamma_{xy}^{(1)} \Big) + \Big(Q_{44}^k \gamma_{yz}^{(0)} + Q_{45}^k \gamma_{xz}^{(0)} \\
& - e_{14}^k \Big(\mathcal{E}_x^{k(0)} + z\mathcal{E}_x^{k(1)} \Big) - e_{24}^k \Big(\mathcal{E}_y^{k(0)} + z\mathcal{E}_y^{k(1)} \Big) \Big) \Big(\delta\gamma_{yz}^{(0)} \Big) \\
& + \Big(Q_{45}^k \gamma_{yz}^{(0)} + Q_{55}^k \gamma_{xz}^{(0)} - e_{15}^k \Big(\mathcal{E}_x^{k(0)} + z\mathcal{E}_x^{k(1)} \Big)
\end{aligned}
$$

$$
\begin{aligned}
&- e_{25}^k\left(\mathcal{E}_y^{k(0)} + z\mathcal{E}_y^{k(1)}\right)\right)\left(\delta\gamma_{xz}^{(0)}\right) - \left(e_{14}^k\gamma_{yz}^{(0)} + e_{15}^k\gamma_{xz}^{(0)}\right.\\
&+ \varepsilon_{11}^k\left(\mathcal{E}_x^{k(0)} + z\mathcal{E}_x^{k(1)}\right) + \varepsilon_{12}^k\left(\mathcal{E}_y^{k(0)} + z\mathcal{E}_y^{k(1)}\right)\right)\left(\delta\mathcal{E}_x^{k(0)} + z\delta\mathcal{E}_x^{k(1)}\right)\\
&- \left(e_{24}^k\gamma_{yz}^{(0)} + e_{25}^k\gamma_{xz}^{(0)} + \varepsilon_{12}^k\left(\mathcal{E}_x^{k(0)} + z\mathcal{E}_x^{k(1)}\right)\right.\\
&+ \varepsilon_{22}^k\left(\mathcal{E}_y^{k(0)} + z\mathcal{E}_y^{k(1)}\right)\right)\left(\delta\mathcal{E}_y^{k(0)} + z\delta\mathcal{E}_y^{k(1)}\right)\\
&- \left(e_{31}^k\left(\epsilon_{xx}^{(0)} + z\epsilon_{xx}^{(1)}\right) + e_{32}^k\left(\epsilon_{yy}^{(0)} + z\epsilon_{yy}^{(1)}\right)\right.\\
&+ e_{36}^k\left(\gamma_{xy}^{(0)} + z\gamma_{xy}^{(1)}\right) + \varepsilon_{33}^k\mathcal{E}_z^{k(0)}\right)\left(\delta\mathcal{E}_z^{k(0)}\right)\Big]dz\Big)dxdy
\end{aligned}
\tag{4.43}
$$

The next step is the substitution of the geometrical relations for the electro mechanical components and their virtual variations (see Equations (4.25)–(4.32) and (4.33)–(4.40)) in the virtual internal work δU of Equation (4.43):

$$
\begin{aligned}
\delta U = \int_{\Omega_0}\Big(\sum_{k=1}^{N_l}\int_{-h_k/2}^{h_k/2}\Big[&\Big(Q_{11}^k N_{i,x}q_{u_0i} + zQ_{11}^k N_{i,x}q_{\Phi_xi} + Q_{12}^k N_{i,y}q_{v_0i}\\
&+ zQ_{12}^k N_{i,y}q_{\Phi_yi} + Q_{16}^k N_{i,y}q_{u_0i} + Q_{16}^k N_{i,x}q_{v_0i} + zQ_{16}^k N_{i,y}q_{\Phi_xi}\\
&+ zQ_{16}^k N_{i,x}q_{\Phi_yi} + e_{31}^k N_i q_{\Phi_1i}^k\Big)\Big(N_{j,x}\delta q_{u_0j} + zN_{j,x}\delta q_{\Phi_xj}\Big)\\
&+ \Big(Q_{12}^k N_{i,x}q_{u_0i} + zQ_{12}^k N_{i,x}q_{\Phi_xi} + Q_{22}^k N_{i,y}q_{v_0i} + zQ_{22}^k N_{i,y}q_{\Phi_yi}\\
&+ Q_{26}^k N_{i,y}q_{u_0i} + Q_{26}^k N_{i,x}q_{v_0i} + zQ_{26}^k N_{i,y}q_{\Phi_xi} + zQ_{26}^k N_{i,x}q_{\Phi_yi}\\
&+ e_{32}^k N_i q_{\Phi_1i}^k\Big)\Big(N_{j,y}\delta q_{v_0j} + zN_{j,y}\delta q_{\Phi_yj}\Big) + \Big(Q_{16}^k N_{i,x}q_{u_0i}\\
&+ zQ_{16}^k N_{i,x}q_{\Phi_xi} + Q_{26}^k N_{i,y}q_{v_0i} + zQ_{26}^k N_{i,y}q_{\Phi_yi} + Q_{66}^k N_{i,y}q_{u_0i}\\
&+ Q_{66}^k N_{i,x}q_{v_0i} + zQ_{66}^k N_{i,y}q_{\Phi_xi} + zQ_{66}^k N_{i,x}q_{\Phi_yi} + e_{36}^k N_i q_{\Phi_1i}^k\Big)\\
&\times\Big(N_{j,y}\delta q_{u_0j} + N_{j,x}\delta q_{v_0j} + zN_{j,y}\delta q_{\Phi_xj} + zN_{j,x}\delta q_{\Phi_yj}\Big)\\
&+ \Big(Q_{44}^k N_i q_{\Phi_yi} + Q_{44}^k N_{i,y}q_{w_0i} + Q_{45}^k N_i q_{\Phi_xi} + Q_{45}^k N_{i,x}q_{w_0i}\\
&+ e_{14}^k N_{i,x}q_{\Phi_0i}^k - e_{14}^k\bar{z}_k N_{i,x}q_{\Phi_1i}^k + e_{14}^k zN_{i,x}q_{\Phi_1i}^k + e_{24}^k N_{i,y}q_{\Phi_0i}^k\\
&- e_{24}^k\bar{z}_k N_{i,y}q_{\Phi_1i}^k + e_{24}^k zN_{i,y}q_{\Phi_1i}^k\Big)\Big(N_j\delta q_{\Phi_yj} + N_{j,y}\delta q_{w_0j}\Big)\\
&+ \Big(Q_{45}^k N_i q_{\Phi_yi} + Q_{45}^k N_{i,y}q_{w_0i} + Q_{55}^k N_i q_{\Phi_xi} + Q_{55}^k N_{i,x}q_{w_0i}\\
&+ e_{15}^k N_{i,x}q_{\Phi_0i}^k - e_{15}^k\bar{z}_k N_{i,x}q_{\Phi_1i}^k + e_{15}^k zN_{i,x}q_{\Phi_1i}^k + e_{25}^k N_{i,y}q_{\Phi_0i}^k
\end{aligned}
$$

$$
\begin{aligned}
&- e^k_{25}\bar{z}_k N_{i,y} q^k_{\Phi_1 i} + e^k_{25} z N_{i,y} q^k_{\Phi_1 i}\Big)\Big(N_j \delta q_{\Phi_x j} + N_{j,x}\delta q_{w_0 j}\Big)\\
&- \Big(e^k_{14} N_i q_{\Phi_y i} + e^k_{14} N_{i,y} q_{w_0 i} + e^k_{24} N_i q_{\Phi_x i} + e^k_{24} N_{i,x} q_{w_0 i} - \varepsilon^k_{11} N_{i,x} q^k_{\Phi_0 i}\\
&+ \varepsilon^k_{11}\bar{z}_k N_{i,x} q^k_{\Phi_1 i} - \varepsilon^k_{11} z N_{i,x} q^k_{\Phi_1 i} - \varepsilon^k_{12} N_{i,y} q^k_{\Phi_0 i} + \varepsilon^k_{12}\bar{z}_k N_{i,y} q^k_{\Phi_1 i}\\
&- \varepsilon^k_{12} z N_{i,y} q^k_{\Phi_1 i}\Big)\Big(- N_{j,x}\delta q^k_{\Phi_0 j} + \bar{z}_k N_{j,x}\delta q^k_{\Phi_1 j} - z N_{j,x}\delta q^k_{\Phi_1 j}\Big)\\
&- \Big(e^k_{24} N_i q_{\Phi_y i} + e^k_{24} N_{i,y} q_{w_0 i} + e^k_{25} N_i q_{\Phi_x i} + e^k_{25} N_{i,x} q_{w_0 i} - \varepsilon^k_{12} N_{i,x} q^k_{\Phi_0 i}\\
&+ \varepsilon^k_{12}\bar{z}_k N_{i,x} q^k_{\Phi_1 i} - \varepsilon^k_{12} z N_{i,x} q^k_{\Phi_1 i} - \varepsilon^k_{22} N_{i,y} q^k_{\Phi_0 i} + \varepsilon^k_{22}\bar{z}_k N_{i,y} q^k_{\Phi_1 i}\\
&- \varepsilon^k_{22} z N_{i,y} q^k_{\Phi_1 i}\Big)\Big(- N_{j,y}\delta q^k_{\Phi_0 j} + \bar{z}_k N_{j,y}\delta q^k_{\Phi_1 j} - z N_{j,y}\delta q^k_{\Phi_1 j}\Big)\\
&- \Big(e^k_{31} N_{i,x} q_{u_0 i} + e^k_{31} z N_{i,x} q_{\Phi_x i} + e^k_{32} N_{i,y} q_{v_0 i} + e^k_{32} z N_{i,y} q_{\Phi_y i}\\
&+ e^k_{36} N_{i,y} q_{u_0 i} + e^k_{36} N_{i,x} q_{v_0 i} + e^k_{36} z N_{i,y} q_{\Phi_x i}\\
&+ e^k_{36} z N_{i,x} q_{\Phi_y i} - \varepsilon^k_{33} N_i q^k_{\Phi_1 i}\Big)\Big(- N_j \delta q^k_{\Phi_1 j}\Big)\Big] dz\Big) dx dy
\end{aligned}
\tag{4.44}
$$

By developing the products in Equation (4.44), and collecting terms with respect to vectors $\boldsymbol{q}_{\tau i}$ and $\delta\boldsymbol{q}_{sj}$ in Equations (4.18) and (4.20), respectively, it is possible to define the components of the stiffness matrix $\boldsymbol{K}^{kij}$. The system of governing equations is:

$$
\delta\boldsymbol{q}_{sj} : \ \boldsymbol{K}^{kij}\ \boldsymbol{q}_{\tau i} = \boldsymbol{F}_j \tag{4.45}
$$

where $\boldsymbol{F}_j$ is the vector that contains the mechanical forces. In the case of an actuator configuration, the electric potential is applied directly to vector $\boldsymbol{q}_{\tau i}$. The form of the stiffness matrix $\boldsymbol{K}^{kij}$ is:

$$
\boldsymbol{K}^{kij} = \begin{bmatrix}
K^{kij}_{u_0u_0} & K^{kij}_{u_0v_0} & K^{kij}_{u_0w_0} & K^{kij}_{u_0\Phi_x} & K^{kij}_{u_0\Phi_y} & K^{kij}_{u_0\Phi_0} & K^{kij}_{u_0\Phi_1}\\
K^{kij}_{v_0u_0} & K^{kij}_{v_0v_0} & K^{kij}_{v_0w_0} & K^{kij}_{v_0\Phi_x} & K^{kij}_{v_0\Phi_y} & K^{kij}_{v_0\Phi_0} & K^{kij}_{v_0\Phi_1}\\
K^{kij}_{w_0u_0} & K^{kij}_{w_0v_0} & K^{kij}_{w_0w_0} & K^{kij}_{w_0\Phi_x} & K^{kij}_{w_0\Phi_y} & K^{kij}_{w_0\Phi_0} & K^{kij}_{w_0\Phi_1}\\
K^{kij}_{\Phi_xu_0} & K^{kij}_{\Phi_xv_0} & K^{kij}_{\Phi_xw_0} & K^{kij}_{\Phi_x\Phi_x} & K^{kij}_{\Phi_x\Phi_y} & K^{kij}_{\Phi_x\Phi_0} & K^{kij}_{\Phi_x\Phi_1}\\
K^{kij}_{\Phi_yu_0} & K^{kij}_{\Phi_yv_0} & K^{kij}_{\Phi_yw_0} & K^{kij}_{\Phi_y\Phi_x} & K^{kij}_{\Phi_y\Phi_y} & K^{kij}_{\Phi_y\Phi_0} & K^{kij}_{\Phi_y\Phi_1}\\
K^{kij}_{\Phi_0u_0} & K^{kij}_{\Phi_0v_0} & K^{kij}_{\Phi_0w_0} & K^{kij}_{\Phi_0\Phi_x} & K^{kij}_{\Phi_0\Phi_y} & K^{kij}_{\Phi_0\Phi_0} & K^{kij}_{\Phi_0\Phi_1}\\
K^{kij}_{\Phi_1u_0} & K^{kij}_{\Phi_1v_0} & K^{kij}_{\Phi_1w_0} & K^{kij}_{\Phi_1\Phi_x} & K^{kij}_{\Phi_1\Phi_y} & K^{kij}_{\Phi_1\Phi_0} & K^{kij}_{\Phi_1\Phi_1}
\end{bmatrix}
\tag{4.46}
$$

By separating the mechanical degrees of freedom $\boldsymbol{q_u} = (q_{u_0i}\ q_{v_0i}\ q_{w_0i}\ q_{\Phi_x i}\ q_{\Phi_y i})^T$ from the electric ones $\boldsymbol{q}^k_\Phi = (q^k_{\Phi_0 i}\ q^k_{\Phi_1 i})^T$, it is possible to write Equation (4.46) in a more concise form, where the pure mechanical part, the pure electrical part, and the electromechanical coupling contributions are easily noted:

$$\boldsymbol{K}^{kij} = \begin{bmatrix} \boldsymbol{K}^{kij}_{uu} & \boldsymbol{K}^{kij}_{u\Phi} \\ \boldsymbol{K}^{kij}_{\Phi u} & \boldsymbol{K}^{kij}_{\Phi\Phi} \end{bmatrix} \tag{4.47}$$

Matrix $\boldsymbol{K}^{kij}_{uu}$ has the dimension 5×5, $\boldsymbol{K}^{kij}_{u\Phi}$ is 5×2, $\boldsymbol{K}^{kij}_{\Phi u}$ is 2×5, and $\boldsymbol{K}^{kij}_{\Phi\Phi}$ has the dimension 2×2. In order to write the explicit form of each contribution, it is necessary to define the integrals along the in-plane and thickness directions, as suggested in Equations (4.48) and (4.49), respectively:

$$\triangleleft \ldots \triangleright_{\Omega_0} = \int_{\Omega_0} (\ldots) dx dy \tag{4.48}$$

$$\triangleleft \ldots \triangleright_h = \sum_{k=1}^{N_l} \int_{-h_k/2}^{h_k/2} (\ldots) dz \tag{4.49}$$

where it is necessary to recall Equation (4.15) for a generic element Ω; this equation can also be considered for the generic element Ω_0 at the middle plane. The explicit forms of each component of Equations (4.46) or (4.47) are:

$$\begin{aligned} K^{kij}_{u_0u_0} &= \triangleleft Q^k_{11} \triangleright_h \triangleleft N_{i,x} N_{j,x} \triangleright_{\Omega_0} + \triangleleft Q^k_{16} \triangleright_h \triangleleft N_{i,y} N_{j,x} \triangleright_{\Omega_0} \\ &+ \triangleleft Q^k_{16} \triangleright_h \triangleleft N_{i,x} N_{j,y} \triangleright_{\Omega_0} + \triangleleft Q^k_{66} \triangleright_h \triangleleft N_{i,y} N_{j,y} \triangleright_{\Omega_0} \end{aligned} \tag{4.50}$$

$$\begin{aligned} K^{kij}_{u_0v_0} &= \triangleleft Q^k_{12} \triangleright_h \triangleleft N_{i,y} N_{j,x} \triangleright_{\Omega_0} + \triangleleft Q^k_{26} \triangleright_h \triangleleft N_{i,y} N_{j,y} \triangleright_{\Omega_0} \\ &+ \triangleleft Q^k_{16} \triangleright_h \triangleleft N_{i,x} N_{j,x} \triangleright_{\Omega_0} + \triangleleft Q^k_{66} \triangleright_h \triangleleft N_{i,x} N_{j,y} \triangleright_{\Omega_0} \end{aligned} \tag{4.51}$$

$$K^{kij}_{u_0w_0} = 0 \tag{4.52}$$

$$\begin{aligned} K^{kij}_{u_0\Phi_x} &= \triangleleft z Q^k_{11} \triangleright_h \triangleleft N_{i,x} N_{j,x} \triangleright_{\Omega_0} + \triangleleft z Q^k_{16} \triangleright_h \triangleleft N_{i,y} N_{j,x} \triangleright_{\Omega_0} \\ &+ \triangleleft z Q^k_{16} \triangleright_h \triangleleft N_{i,x} N_{j,y} \triangleright_{\Omega_0} + \triangleleft z Q^k_{66} \triangleright_h \triangleleft N_{i,y} N_{j,y} \triangleright_{\Omega_0} \end{aligned} \tag{4.53}$$

$$\begin{aligned} K^{kij}_{u_0\Phi_y} &= \triangleleft z Q^k_{12} \triangleright_h \triangleleft N_{i,y} N_{j,x} \triangleright_{\Omega_0} + \triangleleft z Q^k_{16} \triangleright_h \triangleleft N_{i,x} N_{j,x} \triangleright_{\Omega_0} \\ &+ \triangleleft z Q^k_{26} \triangleright_h \triangleleft N_{i,y} N_{j,y} \triangleright_{\Omega_0} + \triangleleft z Q^k_{66} \triangleright_h \triangleleft N_{i,x} N_{j,y} \triangleright_{\Omega_0} \end{aligned} \tag{4.54}$$

$$\begin{aligned} K^{kij}_{v_0u_0} &= \triangleleft Q^k_{12} \triangleright_h \triangleleft N_{i,x} N_{j,y} \triangleright_{\Omega_0} + \triangleleft Q^k_{26} \triangleright_h \triangleleft N_{i,y} N_{j,y} \triangleright_{\Omega_0} \\ &+ \triangleleft Q^k_{16} \triangleright_h \triangleleft N_{i,x} N_{j,x} \triangleright_{\Omega_0} + \triangleleft Q^k_{66} \triangleright_h \triangleleft N_{i,y} N_{j,x} \triangleright_{\Omega_0} \end{aligned} \tag{4.55}$$

$$\begin{aligned} K^{kij}_{v_0v_0} &= \triangleleft Q^k_{22} \triangleright_h \triangleleft N_{i,y} N_{j,y} \triangleright_{\Omega_0} + \triangleleft Q^k_{26} \triangleright_h \triangleleft N_{i,x} N_{j,y} \triangleright_{\Omega_0} \\ &+ \triangleleft Q^k_{26} \triangleright_h \triangleleft N_{i,y} N_{j,x} \triangleright_{\Omega_0} + \triangleleft Q^k_{66} \triangleright_h \triangleleft N_{i,x} N_{j,x} \triangleright_{\Omega_0} \end{aligned} \tag{4.56}$$

$$K^{kij}_{v_0 w_0} = 0 \tag{4.57}$$

$$K^{kij}_{v_0 \Phi_x} = \lhd z Q^k_{12} \rhd_h \lhd N_{i,x} N_{j,y} \rhd_{\Omega_0} + \lhd z Q^k_{26} \rhd_h \lhd N_{i,y} N_{j,y} \rhd_{\Omega_0} + \lhd z Q^k_{16} \rhd_h \lhd N_{i,x} N_{j,y} \rhd_{\Omega_0} + \lhd z Q^k_{66} \rhd_h \lhd N_{i,y} N_{j,x} \rhd_{\Omega_0} \tag{4.58}$$

$$K^{kij}_{v_0 \Phi_y} = \lhd z Q^k_{22} \rhd_h \lhd N_{i,y} N_{j,y} \rhd_{\Omega_0} + \lhd z Q^k_{26} \rhd_h \lhd N_{i,x} N_{j,y} \rhd_{\Omega_0} + \lhd z Q^k_{26} \rhd_h \lhd N_{i,y} N_{j,x} \rhd_{\Omega_0} + \lhd z Q^k_{66} \rhd_h \lhd N_{i,x} N_{j,x} \rhd_{\Omega_0} \tag{4.59}$$

$$K^{kij}_{w_0 u_0} = 0 \tag{4.60}$$

$$K^{kij}_{w_0 v_0} = 0 \tag{4.61}$$

$$K^{kij}_{w_0 w_0} = \lhd Q^k_{44} \rhd_h \lhd N_{i,y} N_{j,y} \rhd_{\Omega_0} + \lhd Q^k_{45} \rhd_h \lhd N_{i,x} N_{j,y} \rhd_{\Omega_0} + \lhd Q^k_{45} \rhd_h \lhd N_{i,y} N_{j,x} \rhd_{\Omega_0} + \lhd Q^k_{55} \rhd_h \lhd N_{i,x} N_{j,x} \rhd_{\Omega_0} \tag{4.62}$$

$$K^{kij}_{w_0 \Phi_x} = \lhd Q^k_{45} \rhd_h \lhd N_i N_{j,y} \rhd_{\Omega_0} + \lhd Q^k_{55} \rhd_h \lhd N_i N_{j,x} \rhd_{\Omega_0} \tag{4.63}$$

$$K^{kij}_{w_0 \Phi_y} = \lhd Q^k_{44} \rhd_h \lhd N_i N_{j,y} \rhd_{\Omega_0} + \lhd Q^k_{45} \rhd_h \lhd N_i N_{j,x} \rhd_{\Omega_0} \tag{4.64}$$

$$K^{kij}_{\Phi_x u_0} = \lhd z Q^k_{11} \rhd_h \lhd N_{i,x} N_{j,x} \rhd_{\Omega_0} + \lhd z Q^k_{16} \rhd_h \lhd N_{i,y} N_{j,x} \rhd_{\Omega_0} + \lhd z Q^k_{16} \rhd_h \lhd N_{i,x} N_{j,y} \rhd_{\Omega_0} + \lhd z Q^k_{66} \rhd_h \lhd N_{i,y} N_{j,y} \rhd_{\Omega_0} \tag{4.65}$$

$$K^{kij}_{\Phi_x v_0} = \lhd z Q^k_{12} \rhd_h \lhd N_{i,y} N_{j,x} \rhd_{\Omega_0} + \lhd z Q^k_{16} \rhd_h \lhd N_{i,x} N_{j,x} \rhd_{\Omega_0} + \lhd z Q^k_{26} \rhd_h \lhd N_{i,y} N_{j,y} \rhd_{\Omega_0} + \lhd z Q^k_{66} \rhd_h \lhd N_{i,x} N_{j,y} \rhd_{\Omega_0} \tag{4.66}$$

$$K^{kij}_{\Phi_x w_0} = \lhd Q^k_{55} \rhd_h \lhd N_{i,x} N_j \rhd_{\Omega_0} + \lhd Q^k_{45} \rhd_h \lhd N_{i,y} N_j \rhd_{\Omega_0} \tag{4.67}$$

$$K^{kij}_{\Phi_x \Phi_x} = \lhd z^2 Q^k_{11} \rhd_h \lhd N_{i,x} N_{j,x} \rhd_{\Omega_0} + \lhd z^2 Q^k_{16} \rhd_h \lhd N_{i,y} N_{j,x} \rhd_{\Omega_0} + \lhd z^2 Q^k_{16} \rhd_h \lhd N_{i,x} N_{j,y} \rhd_{\Omega_0} + \lhd z^2 Q^k_{66} \rhd_h \lhd N_{i,y} N_{j,y} \rhd_{\Omega_0} + \lhd Q^k_{55} \rhd_h \lhd N_i N_j \rhd_{\Omega_0} \tag{4.68}$$

$$K^{kij}_{\Phi_x \Phi_y} = \lhd z^2 Q^k_{12} \rhd_h \lhd N_{i,y} N_{j,x} \rhd_{\Omega_0} + \lhd z^2 Q^k_{16} \rhd_h \lhd N_{i,x} N_{j,x} \rhd_{\Omega_0} + \lhd z^2 Q^k_{66} \rhd_h \lhd N_{i,x} N_{j,y} \rhd_{\Omega_0} + \lhd z^2 Q^k_{26} \rhd_h \lhd N_{i,y} N_{j,y} \rhd_{\Omega_0} + \lhd Q^k_{45} \rhd_h \lhd N_i N_j \rhd_{\Omega_0} \tag{4.69}$$

$$K^{kij}_{\Phi_y u_0} = \lhd z Q^k_{12} \rhd_h \lhd N_{i,x} N_{j,y} \rhd_{\Omega_0} + \lhd z Q^k_{26} \rhd_h \lhd N_{i,y} N_{j,y} \rhd_{\Omega_0} + \lhd z Q^k_{16} \rhd_h \lhd N_{i,x} N_{j,x} \rhd_{\Omega_0} + \lhd z Q^k_{66} \rhd_h \lhd N_{i,x} N_{j,y} \rhd_{\Omega_0} \tag{4.70}$$

$$K^{kij}_{\Phi_y v_0} = \lhd z Q^k_{22} \rhd_h \lhd N_{i,y} N_{j,y} \rhd_{\Omega_0} + \lhd z Q^k_{26} \rhd_h \lhd N_{i,x} N_{j,y} \rhd_{\Omega_0} + \lhd z Q^k_{26} \rhd_h \lhd N_{i,y} N_{j,x} \rhd_{\Omega_0} + \lhd z Q^k_{66} \rhd_h \lhd N_{i,x} N_{j,x} \rhd_{\Omega_0} \tag{4.71}$$

$$K^{kij}_{\Phi_y w_0} = \lhd Q^k_{44} \rhd_h \lhd N_{i,y} N_j \rhd_{\Omega_0} + \lhd Q^k_{45} \rhd_h \lhd N_{i,x} N_j \rhd_{\Omega_0} \tag{4.72}$$

$$
\begin{aligned}
K^{kij}_{\Phi_y\Phi_x} &= \lhd z^2 Q^k_{12}\rhd_h \lhd N_{i,x} N_{j,y}\rhd_{\Omega_0} + \lhd z^2 Q^k_{26}\rhd_h \lhd N_{i,y} N_{j,y}\rhd_{\Omega_0} \\
&\quad + \lhd z^2 Q^k_{16}\rhd_h \lhd N_{i,x} N_{j,x}\rhd_{\Omega_0} + \lhd z^2 Q^k_{66}\rhd_h \lhd N_{i,y} N_{j,x}\rhd_{\Omega_0} \\
&\quad + \lhd Q^k_{45}\rhd_h \lhd N_i N_j\rhd_{\Omega_0} &\quad (4.73)\\
K^{kij}_{\Phi_y\Phi_y} &= \lhd z^2 Q^k_{22}\rhd_h \lhd N_{i,y} N_{j,y}\rhd_{\Omega_0} + \lhd z^2 Q^k_{26}\rhd_h \lhd N_{i,x} N_{j,y}\rhd_{\Omega_0} \\
&\quad + \lhd z^2 Q^k_{26}\rhd_h \lhd N_{i,y} N_{j,x}\rhd_{\Omega_0} + \lhd z^2 Q^k_{66}\rhd_h \lhd N_{i,x} N_{j,x}\rhd_{\Omega_0} \\
&\quad + \lhd Q^k_{44}\rhd_h \lhd N_i N_j\rhd_{\Omega_0} &\quad (4.74)\\
K^{kij}_{u_0\Phi_0} &= 0 &\quad (4.75)\\
K^{kij}_{u_0\Phi_1} &= \lhd e^k_{31}\rhd_h \lhd N_i N_{i,x}\rhd_{\Omega_0} + \lhd e^k_{36}\rhd_h \lhd N_i N_{j,y}\rhd_{\Omega_0} &\quad (4.76)\\
K^{kij}_{v_0\Phi_0} &= 0 &\quad (4.77)\\
K^{kij}_{v_0\Phi_1} &= \lhd e^k_{32}\rhd_h \lhd N_i N_{j,y}\rhd_{\Omega_0} + \lhd e^k_{36}\rhd_h \lhd N_i N_{j,x}\rhd_{\Omega_0} &\quad (4.78)\\
K^{kij}_{w_0\Phi_0} &= \lhd e^k_{14}\rhd_h \lhd N_{i,x} N_{j,y}\rhd_{\Omega_0} + \lhd e^k_{24}\rhd_h \lhd N_{i,y} N_{j,y}\rhd_{\Omega_0} \\
&\quad + \lhd e^k_{15}\rhd_h \lhd N_{i,x} N_{i,x}\rhd_{\Omega_0} + \lhd e^k_{25}\rhd_h \lhd N_{i,y} N_{j,x}\rhd_{\Omega_0} &\quad (4.79)\\
K^{kij}_{w_0\Phi_1} &= - \lhd \bar{z}_k e^k_{14}\rhd_h \lhd N_{i,x} N_{j,y}\rhd_{\Omega_0} + \lhd z e^k_{14}\rhd_h \lhd N_{i,x} N_{j,y}\rhd_{\Omega_0} \\
&\quad - \lhd \bar{z}_k e^k_{24}\rhd_h \lhd N_{i,y} N_{j,y}\rhd_{\Omega_0} + \lhd z e^k_{24}\rhd_h \lhd N_{i,y} N_{j,y}\rhd_{\Omega_0} \\
&\quad - \lhd \bar{z}_k e^k_{15}\rhd_h \lhd N_{i,x} N_{j,x}\rhd_{\Omega_0} + \lhd z e^k_{15}\rhd_h \lhd N_{i,x} N_{j,x}\rhd_{\Omega_0} \\
&\quad - \lhd \bar{z}_k e^k_{25}\rhd_h \lhd N_{i,y} N_{j,x}\rhd_{\Omega_0} + \lhd z e^k_{25}\rhd_h \lhd N_{i,y} N_{j,x}\rhd_{\Omega_0} &\quad (4.80)\\
K^{kij}_{\Phi_x\Phi_0} &= \lhd e^k_{15}\rhd_h \lhd N_{i,x} N_j\rhd_{\Omega_0} + \lhd e^k_{25}\rhd_h \lhd N_{i,y} N_j\rhd_{\Omega_0} &\quad (4.81)\\
K^{kij}_{\Phi_x\Phi_1} &= \lhd z e^k_{31}\rhd_h \lhd N_i N_{j,x}\rhd_{\Omega_0} + \lhd z e^k_{36}\rhd_h \lhd N_i N_{j,y}\rhd_{\Omega_0} \\
&\quad - \lhd \bar{z}_k e^k_{15}\rhd_h \lhd N_{i,x} N_j\rhd_{\Omega_0} + \lhd z e^k_{15}\rhd_h \lhd N_{i,x} N_j\rhd_{\Omega_0} \\
&\quad - \lhd \bar{z}_k e^k_{25}\rhd_h \lhd N_{i,y} N_j\rhd_{\Omega_0} + \lhd z e^k_{25}\rhd_h \lhd N_{i,y} N_j\rhd_{\Omega_0} &\quad (4.82)\\
K^{kij}_{\Phi_y\Phi_0} &= \lhd e^k_{14}\rhd_h \lhd N_{i,x} N_j\rhd_{\Omega_0} + \lhd e^k_{24}\rhd_h \lhd N_{i,y} N_j\rhd_{\Omega_0} &\quad (4.83)\\
K^{kij}_{\Phi_y\Phi_1} &= \lhd z e^k_{32}\rhd_h \lhd N_i N_{j,y}\rhd_{\Omega_0} + \lhd z e^k_{36}\rhd_h \lhd N_i N_{j,x}\rhd_{\Omega_0} \\
&\quad - \lhd \bar{z}_k e^k_{14}\rhd_h \lhd N_{i,x} N_j\rhd_{\Omega_0} + \lhd z e^k_{14}\rhd_h \lhd N_{i,x} N_j\rhd_{\Omega_0} \\
&\quad - \lhd \bar{z}_k e^k_{24}\rhd_h \lhd N_{i,y} N_j\rhd_{\Omega_0} + \lhd z e^k_{24}\rhd_h \lhd N_{i,y} N_j\rhd_{\Omega_0} &\quad (4.84)\\
K^{kij}_{\Phi_0 u_0} &= 0 &\quad (4.85)\\
K^{kij}_{\Phi_0 v_0} &= 0 &\quad (4.86)
\end{aligned}
$$

$$K_{\Phi_0 w_0}^{kij} = \lhd e_{14}^k \rhd_h \lhd N_{i,y} N_{j,x} \rhd_{\Omega_0} + \lhd e_{24}^k \rhd_h \lhd N_{i,x} N_{j,x} \rhd_{\Omega_0} + \lhd e_{24}^k \rhd_h \lhd N_{i,y} N_{j,y} \rhd_{\Omega_0} + \lhd e_{25}^k \rhd_h \lhd N_{i,x} N_{j,y} \rhd_{\Omega_0} \quad (4.87)$$

$$K_{\Phi_0 \Phi_x}^{kij} = \lhd e_{24}^k \rhd_h \lhd N_i N_{j,x} \rhd_{\Omega_0} + \lhd e_{25}^k \rhd_h \lhd N_i N_{j,y} \rhd_{\Omega_0} \quad (4.88)$$

$$K_{\Phi_0 \Phi_y}^{kij} = \lhd e_{14}^k \rhd_h \lhd N_i N_{j,x} \rhd_{\Omega_0} + \lhd e_{24}^k \rhd_h \lhd N_i N_{j,y} \rhd_{\Omega_0} \quad (4.89)$$

$$K_{\Phi_1 u_0}^{kij} = \lhd e_{31}^k \rhd_h \lhd N_{i,x} N_j \rhd_{\Omega_0} + \lhd e_{36}^k \rhd_h \lhd N_{i,y} N_j \rhd_{\Omega_0} \quad (4.90)$$

$$K_{\Phi_1 v_0}^{kij} = \lhd e_{32}^k \rhd_h \lhd N_{i,y} N_j \rhd_{\Omega_0} + \lhd e_{36}^k \rhd_h \lhd N_{i,x} N_j \rhd_{\Omega_0} \quad (4.91)$$

$$\begin{aligned} K_{\Phi_1 w_0}^{kij} = & - \lhd \bar{z}_k e_{14}^k \rhd_h \lhd N_{i,y} N_{j,x} \rhd_{\Omega_0} - \lhd \bar{z}_k e_{24}^k \rhd_h \lhd N_{i,x} N_{j,x} \rhd_{\Omega_0} \\ & + \lhd z e_{14}^k \rhd_h \lhd N_{i,y} N_{j,x} \rhd_{\Omega_0} + \lhd z e_{24}^k \rhd_h \lhd N_{i,x} N_{j,x} \rhd_{\Omega_0} \\ & - \lhd \bar{z}_k e_{24}^k \rhd_h \lhd N_{i,y} N_{j,y} \rhd_{\Omega_0} - \lhd \bar{z}_k e_{25}^k \rhd_h \lhd N_{i,x} N_{j,y} \rhd_{\Omega_0} \\ & + \lhd z e_{24}^k \rhd_h \lhd N_{i,y} N_{j,y} \rhd_{\Omega_0} + \lhd z e_{25}^k \rhd_h \lhd N_{i,x} N_{j,y} \rhd_{\Omega_0} \end{aligned} \quad (4.92)$$

$$\begin{aligned} K_{\Phi_1 \Phi_x}^{kij} = & - \lhd \bar{z}_k e_{24}^k \rhd_h \lhd N_i N_{j,x} \rhd_{\Omega_0} + \lhd z e_{24}^k \rhd_h \lhd N_i N_{j,x} \rhd_{\Omega_0} \\ & + \lhd z e_{25}^k \rhd_h \lhd N_i N_{j,y} \rhd_{\Omega_0} + \lhd z e_{31}^k \rhd_h \lhd N_{i,x} N_j \rhd_{\Omega_0} \\ & + \lhd z e_{36}^k \rhd_h \lhd N_{i,y} N_j \rhd_{\Omega_0} \end{aligned} \quad (4.93)$$

$$\begin{aligned} K_{\Phi_1 \Phi_y}^{kij} = & - \lhd \bar{z}_k e_{14}^k \rhd_h \lhd N_i N_{j,x} \rhd_{\Omega_0} + \lhd z e_{14}^k \rhd_h \lhd N_i N_{j,x} \rhd_{\Omega_0} \\ & - \lhd \bar{z}_k e_{24}^k \rhd_h \lhd N_i N_{j,y} \rhd_{\Omega_0} - \lhd \bar{z}_k e_{25}^k \rhd_h \lhd N_i N_{j,y} \rhd_{\Omega_0} \\ & + \lhd z e_{24}^k \rhd_h \lhd N_i N_{j,y} \rhd_{\Omega_0} + \lhd z e_{32}^k \rhd_h \lhd N_{i,y} N_j \rhd_{\Omega_0} \\ & + \lhd z e_{36}^k \rhd_h \lhd N_{i,x} N_j \rhd_{\Omega_0} \end{aligned} \quad (4.94)$$

$$\begin{aligned} K_{\Phi_0 \Phi_0}^{kij} = & - \lhd \varepsilon_{11}^k \rhd_h \lhd N_{i,x} N_{j,x} \rhd_{\Omega_0} - \lhd \varepsilon_{12}^k \rhd_h \lhd N_{i,y} N_{j,x} \rhd_{\Omega_0} \\ & - \lhd \varepsilon_{12}^k \rhd_h \lhd N_{i,x} N_{j,y} \rhd_{\Omega_0} - \lhd \varepsilon_{22}^k \rhd_h \lhd N_{i,y} N_{j,y} \rhd_{\Omega_0} \end{aligned} \quad (4.95)$$

$$\begin{aligned} K_{\Phi_0 \Phi_1}^{kij} = & \lhd \bar{z}_k \varepsilon_{11}^k \rhd_h \lhd N_{i,x} N_{j,x} \rhd_{\Omega_0} - \lhd z \varepsilon_{11}^k \rhd_h \lhd N_{i,x} N_{j,x} \rhd_{\Omega_0} \\ & + \lhd \bar{z}_k \varepsilon_{12}^k \rhd_h \lhd N_{i,y} N_{j,x} \rhd_{\Omega_0} - \lhd z \varepsilon_{12}^k \rhd_h \lhd N_{i,y} N_{j,x} \rhd_{\Omega_0} \\ & + \lhd \bar{z}_k \varepsilon_{12}^k \rhd_h \lhd N_{i,x} N_{j,y} \rhd_{\Omega_0} - \lhd z \varepsilon_{12}^k \rhd_h \lhd N_{i,x} N_{j,y} \rhd_{\Omega_0} \\ & + \lhd \bar{z}_k \varepsilon_{22}^k \rhd_h \lhd N_{i,y} N_{j,y} \rhd_{\Omega_0} - \lhd z \varepsilon_{22}^k \rhd_h \lhd N_{i,y} N_{j,y} \rhd_{\Omega_0} \end{aligned} \quad (4.96)$$

$$\begin{aligned} K_{\Phi_1 \Phi_0}^{kij} = & \lhd \bar{z}_k \varepsilon_{11}^k \rhd_h \lhd N_{i,x} N_{j,x} \rhd_{\Omega_0} - \lhd z \varepsilon_{11}^k \rhd_h \lhd N_{i,x} N_{j,x} \rhd_{\Omega_0} \\ & + \lhd \bar{z}_k \varepsilon_{12}^k \rhd_h \lhd N_{i,y} N_{j,x} \rhd_{\Omega_0} - \lhd z \varepsilon_{12}^k \rhd_h \lhd N_{i,y} N_{j,x} \rhd_{\Omega_0} \\ & + \lhd \bar{z}_k \varepsilon_{12}^k \rhd_h \lhd N_{i,x} N_{j,y} \rhd_{\Omega_0} - \lhd z \varepsilon_{12}^k \rhd_h \lhd N_{i,x} N_{j,y} \rhd_{\Omega_0} \\ & + \lhd \bar{z}_k \varepsilon_{22}^k \rhd_h \lhd N_{i,y} N_{j,y} \rhd_{\Omega_0} - \lhd z \varepsilon_{22}^k \rhd_h \lhd N_{i,y} N_{j,y} \rhd_{\Omega_0} \end{aligned} \quad (4.97)$$

$$
\begin{aligned}
K^{kij}_{\Phi_1\Phi_1} = &- \triangleleft \bar{z}^2_k \varepsilon^k_{11} \triangleright_h \triangleleft N_{i,x} N_{j,x} \triangleright_{\Omega_0} + \triangleleft z \bar{z}_k \varepsilon^k_{11} \triangleright_h \triangleleft N_{i,x} N_{j,x} \triangleright_{\Omega_0} \\
&- \triangleleft \bar{z}^2_k \varepsilon^k_{12} \triangleright_h \triangleleft N_{i,y} N_{j,x} \triangleright_{\Omega_0} + \triangleleft \bar{z}_k z \varepsilon^k_{12} \triangleright_h \triangleleft N_{i,y} N_{j,x} \triangleright_{\Omega_0} \\
&+ \triangleleft z \bar{z}_k \varepsilon^k_{11} \triangleright_h \triangleleft N_{i,x} N_{j,x} \triangleright_{\Omega_0} - \triangleleft z^2 \varepsilon^k_{11} \triangleright_h \triangleleft N_{i,x} N_{j,x} \triangleright_{\Omega_0} \\
&+ \triangleleft z \bar{z}_k \varepsilon^k_{12} \triangleright_h \triangleleft N_{i,y} N_{j,x} \triangleright_{\Omega_0} - \triangleleft z^2 \varepsilon^k_{12} \triangleright_h \triangleleft N_{i,y} N_{j,x} \triangleright_{\Omega_0} \\
&+ \triangleleft z \bar{z}_k \varepsilon^k_{12} \triangleright_h \triangleleft N_{i,x} N_{j,y} \triangleright_{\Omega_0} - \triangleleft \bar{z}^2_k \varepsilon^k_{12} \triangleright_h \triangleleft N_{i,y} N_{j,y} \triangleright_{\Omega_0} \\
&+ \triangleleft z \bar{z}_k \varepsilon^k_{22} \triangleright_h \triangleleft N_{i,y} N_{j,y} \triangleright_{\Omega_0} + \triangleleft \bar{z}_k z \varepsilon^k_{12} \triangleright_h \triangleleft N_{i,x} N_{j,y} \triangleright_{\Omega_0} \\
&- \triangleleft z^2 \varepsilon^k_{12} \triangleright_h \triangleleft N_{i,x} N_{j,y} \triangleright_{\Omega_0} + \triangleleft \bar{z}_k z \varepsilon^k_{22} \triangleright_h \triangleleft N_{i,y} N_{j,y} \triangleright_{\Omega_0} \\
&- \triangleleft z^2 \varepsilon^k_{22} \triangleright_h \triangleleft N_{i,y} N_{j,y} \triangleright_{\Omega_0} - \triangleleft \varepsilon^k_{23} \triangleright_h \triangleleft N_i N_j \triangleright_{\Omega_0}
\end{aligned}
\tag{4.98}
$$

The governing equations can also be written in terms of force and moment resultants, as already done for the analytical closed-form solution in Chapter 3. It is now possible to define the following integrals in the z direction of the multilayered plate:

$$
\begin{bmatrix} N_{xx} \\ N_{yy} \\ N_{xy} \end{bmatrix} = \int_{-h/2}^{h/2} \begin{bmatrix} \sigma^k_{xx} \\ \sigma^k_{yy} \\ \sigma^k_{xy} \end{bmatrix} dz, \qquad \begin{bmatrix} M_{xx} \\ M_{yy} \\ M_{xy} \end{bmatrix} = \int_{-h/2}^{h/2} \begin{bmatrix} \sigma^k_{xx} \\ \sigma^k_{yy} \\ \sigma^k_{xy} \end{bmatrix} z\, dz \tag{4.99}
$$

$$
\begin{bmatrix} Q_x \\ Q_y \end{bmatrix} = \int_{-h/2}^{h/2} \begin{bmatrix} \sigma^k_{xz} \\ \sigma^k_{yz} \end{bmatrix} dz \tag{4.100}
$$

$$
\begin{bmatrix} O_x \\ O_y \\ O_z \end{bmatrix} = \int_{-h/2}^{h/2} \begin{bmatrix} \mathcal{D}^k_x \\ \mathcal{D}^k_y \\ \mathcal{D}^k_z \end{bmatrix} dz, \qquad \begin{bmatrix} P_x \\ P_y \\ P_z \end{bmatrix} = \int_{-h/2}^{h/2} \begin{bmatrix} \mathcal{D}^k_x \\ \mathcal{D}^k_y \\ \mathcal{D}^k_z \end{bmatrix} z\, dz \tag{4.101}
$$

As already seen for the CLT case in Chapter 3, (N_{xx}, N_{yy}, N_{xy}) are the in-plane force resultants per unit length in Equation (4.99) and (M_{xx}, M_{yy}, M_{xy}) are the moment resultants per unit length. In Equation (4.101), we define (O_x, O_y, O_z) as the electric charge resultants per unit length, and (P_x, P_y, P_z) as the electric moment resultants per unit length; the additional forces for FSDT analysis are (Q_x, Q_y), as in Equation (4.100), which denotes the transverse force resultants per unit length.

If we consider the constitutive relations in Equations (4.23) and (4.24), the explicit form of Equations (4.99)–(4.101) is obtained. The force resultants are given by:

$$\begin{bmatrix} N_{xx} \\ N_{yy} \\ N_{xy} \end{bmatrix} = \sum_{k=1}^{N_l} \int_{z_k}^{z_{k+1}} \begin{bmatrix} \sigma_{xx}^k \\ \sigma_{yy}^k \\ \sigma_{xy}^k \end{bmatrix} dz$$

$$= \sum_{k=1}^{N_l} \int_{z_k}^{z_{k+1}} \left(\begin{bmatrix} Q_{11}^k & Q_{12}^k & Q_{16}^k \\ Q_{12}^k & Q_{22}^k & Q_{26}^k \\ Q_{16}^k & Q_{26}^k & Q_{66}^k \end{bmatrix} \begin{bmatrix} \epsilon_{xx}^{(0)} + z\epsilon_{xx}^{(1)} \\ \epsilon_{yy}^{(0)} + z\epsilon_{yy}^{(1)} \\ \gamma_{xy}^{(0)} + z\gamma_{xy}^{(1)} \end{bmatrix} \right.$$

$$\left. - \begin{bmatrix} 0 & 0 & e_{31}^k \\ 0 & 0 & e_{32}^k \\ 0 & 0 & e_{36}^k \end{bmatrix} \begin{bmatrix} \mathcal{E}_x^{k(0)} + z\mathcal{E}_x^{k(1)} \\ \mathcal{E}_y^{k(0)} + z\mathcal{E}_y^{k(1)} \\ \mathcal{E}_z^{k(0)} + z0 \end{bmatrix} \right) dz \tag{4.102}$$

The moment resultants are given by:

$$\begin{bmatrix} M_{xx} \\ M_{yy} \\ M_{xy} \end{bmatrix} = \sum_{k=1}^{N_l} \int_{z_k}^{z_{k+1}} \begin{bmatrix} \sigma_{xx}^k \\ \sigma_{yy}^k \\ \sigma_{xy}^k \end{bmatrix} zdz$$

$$= \sum_{k=1}^{N_l} \int_{z_k}^{z_{k+1}} \left(\begin{bmatrix} Q_{11}^k & Q_{12}^k & Q_{16}^k \\ Q_{12}^k & Q_{22}^k & Q_{26}^k \\ Q_{16}^k & Q_{26}^k & Q_{66}^k \end{bmatrix} \begin{bmatrix} \epsilon_{xx}^{(0)} + z\epsilon_{xx}^{(1)} \\ \epsilon_{yy}^{(0)} + z\epsilon_{yy}^{(1)} \\ \gamma_{xy}^{(0)} + z\gamma_{xy}^{(1)} \end{bmatrix} \right.$$

$$\left. - \begin{bmatrix} 0 & 0 & e_{31}^k \\ 0 & 0 & e_{32}^k \\ 0 & 0 & e_{36}^k \end{bmatrix} \begin{bmatrix} \mathcal{E}_x^{k(0)} + z\mathcal{E}_x^{k(1)} \\ \mathcal{E}_y^{k(0)} + z\mathcal{E}_y^{k(1)} \\ \mathcal{E}_z^{k(0)} + z0 \end{bmatrix} \right) zdz \tag{4.103}$$

The transverse force resultants (they are in addition to the CLT case) are given by:

$$
\begin{bmatrix} Q_x \\ Q_y \end{bmatrix} = \sum_{k=1}^{N_l} \int_{z_k}^{z_{k+1}} \begin{bmatrix} \sigma_{xz}^k \\ \sigma_{yz}^k \end{bmatrix} dz
$$

$$
= \sum_{k=1}^{N_l} \int_{z_k}^{z_{k+1}} \left(\begin{bmatrix} Q_{44}^k & Q_{45}^k \\ Q_{45}^k & Q_{55}^k \end{bmatrix} \begin{bmatrix} \gamma_{yz}^{(0)} \\ \gamma_{xz}^{(0)} \end{bmatrix} - \begin{bmatrix} e_{14}^k & e_{24}^k & 0 \\ e_{15}^k & e_{25}^k & 0 \end{bmatrix} \begin{bmatrix} \mathcal{E}_x^{k(0)} + z\mathcal{E}_x^{k(1)} \\ \mathcal{E}_y^{k(0)} + z\mathcal{E}_y^{k(1)} \\ \mathcal{E}_z^{k(0)} + z0 \end{bmatrix} \right) dz \tag{4.104}
$$

The electric charge per unit length resultants are given by:

$$
\begin{bmatrix} O_x \\ O_y \\ O_z \end{bmatrix} = \sum_{k=1}^{N_l} \int_{z_k}^{z_{k+1}} \begin{bmatrix} \mathcal{D}_x^k \\ \mathcal{D}_y^k \\ \mathcal{D}_z^k \end{bmatrix} dz
$$

$$
= \sum_{k=1}^{N_l} \int_{z_k}^{z_{k+1}} \left(\begin{bmatrix} 0 & 0 & 0 & e_{14}^k & e_{15}^k \\ 0 & 0 & 0 & e_{24}^k & e_{25}^k \\ e_{31}^k & e_{32}^k & e_{36}^k & 0 & 0 \end{bmatrix} \begin{bmatrix} \epsilon_{xx}^{(0)} + z\epsilon_{xx}^{(1)} \\ \epsilon_{yy}^{(0)} + z\epsilon_{yy}^{(1)} \\ \gamma_{xy}^{(0)} + z\gamma_{xy}^{(1)} \\ \gamma_{yz}^{(0)} \\ \gamma_{xz}^{(0)} \end{bmatrix} \right.
$$

$$
\left. + \begin{bmatrix} \varepsilon_{11}^k & \varepsilon_{12}^k & 0 \\ \varepsilon_{12}^k & \varepsilon_{22}^k & 0 \\ 0 & 0 & \varepsilon_{33}^k \end{bmatrix} \begin{bmatrix} \mathcal{E}_x^{k(0)} + z\mathcal{E}_x^{k(1)} \\ \mathcal{E}_y^{k(0)} + z\mathcal{E}_y^{k(1)} \\ \mathcal{E}_z^{k(0)} + z0 \end{bmatrix} \right) dz \tag{4.105}
$$

The electric moment resultants are given by:

$$
\begin{bmatrix} P_x \\ P_y \\ P_z \end{bmatrix} = \sum_{k=1}^{N_l} \int_{z_k}^{z_{k+1}} \begin{bmatrix} \mathcal{D}_x^k \\ \mathcal{D}_y^k \\ \mathcal{D}_z^k \end{bmatrix} zdz
$$

$$
= \sum_{k=1}^{N_l} \int_{z_k}^{z_{k+1}} \left(\begin{bmatrix} 0 & 0 & 0 & e_{14}^k & e_{15}^k \\ 0 & 0 & 0 & e_{24}^k & e_{25}^k \\ e_{31}^k & e_{32}^k & e_{36}^k & 0 & 0 \end{bmatrix} \begin{bmatrix} \epsilon_{xx}^{(0)} + z\epsilon_{xx}^{(1)} \\ \epsilon_{yy}^{(0)} + z\epsilon_{yy}^{(1)} \\ \gamma_{xy}^{(0)} + z\gamma_{xy}^{(1)} \\ \gamma_{yz}^{(0)} \\ \gamma_{xz}^{(0)} \end{bmatrix} \right.
$$

$$
\left. + \begin{bmatrix} \varepsilon_{11}^k & \varepsilon_{12}^k & 0 \\ \varepsilon_{12}^k & \varepsilon_{22}^k & 0 \\ 0 & 0 & \varepsilon_{33}^k \end{bmatrix} \begin{bmatrix} \mathcal{E}_x^{k(0)} + z\mathcal{E}_x^{k(1)} \\ \mathcal{E}_y^{k(0)} + z\mathcal{E}_y^{k(1)} \\ \mathcal{E}_z^{k(0)} + z0 \end{bmatrix} \right) zdz \tag{4.106}
$$

The summation in Equations (4.102)–(4.106) is done for the total number of N_l layers.

A_{ij} are the extensional stiffnesses,, D_{ij} are the bending stiffnesses, and B_{ij} are the bending–extensional coupling stiffnesses. They are defined as:

$$
(A_{ij}, B_{ij}, D_{ij}) = \sum_{k=1}^{N_l} \int_{z_k}^{z_{k+1}} Q_{ij}^k(1, z, z^2)dz \tag{4.107}
$$

E_{ij} are the electromechanical coupling extensional stiffnesses, G_{ij} are the electro mechanical coupling bending stiffnesses, and F_{ij} are the electromechanical coupling bending–extensional stiffnesses. They are defined as:

$$
(E_{ij}, F_{ij}, G_{ij}) = \sum_{k=1}^{N_l} \int_{z_k}^{z_{k+1}} e_{ij}^k(1, z, z^2)dz \tag{4.108}
$$

R_{ij} are the dielectric extensional stiffnesses, S_{ij} are the dielectric bending stiffnesses, and T_{ij} are the dielectric bending–extensional coupling stiffnesses. They are defined as:

$$(R_{ij}, S_{ij}, T_{ij}) = \sum_{k=1}^{N_l} \int_{z_k}^{z_{k+1}} \varepsilon_{ij}^{k}(1, z, z^2)dz \tag{4.109}$$

Using Equations (4.107)–(4.109), Equations (4.102)–(4.106) can be rewritten as:

$$\begin{bmatrix} N_{xx} \\ N_{yy} \\ N_{xy} \end{bmatrix} = \begin{bmatrix} A_{11} & A_{12} & A_{16} \\ A_{12} & A_{22} & A_{26} \\ A_{16} & A_{26} & A_{66} \end{bmatrix} \begin{bmatrix} \epsilon_{xx}^{(0)} \\ \epsilon_{yy}^{(0)} \\ \gamma_{xy}^{(0)} \end{bmatrix} + \begin{bmatrix} B_{11} & B_{12} & B_{16} \\ B_{12} & B_{22} & B_{26} \\ B_{16} & B_{26} & B_{66} \end{bmatrix} \begin{bmatrix} \epsilon_{xx}^{(1)} \\ \epsilon_{yy}^{(1)} \\ \gamma_{xy}^{(1)} \end{bmatrix}$$

$$- \begin{bmatrix} 0 & 0 & E_{31} \\ 0 & 0 & E_{32} \\ 0 & 0 & E_{36} \end{bmatrix} \begin{bmatrix} \mathcal{E}_x^{k(0)} \\ \mathcal{E}_y^{k(0)} \\ \mathcal{E}_z^{k(0)} \end{bmatrix} - \begin{bmatrix} 0 & 0 & F_{31} \\ 0 & 0 & F_{32} \\ 0 & 0 & F_{36} \end{bmatrix} \begin{bmatrix} \mathcal{E}_x^{k(1)} \\ \mathcal{E}_y^{k(1)} \\ 0 \end{bmatrix} \tag{4.110}$$

$$\begin{bmatrix} Q_x \\ Q_y \end{bmatrix} = \begin{bmatrix} A_{44} & A_{45} \\ A_{45} & A_{55} \end{bmatrix} \begin{bmatrix} \gamma_{yz}^{(0)} \\ \gamma_{xz}^{(0)} \end{bmatrix} - \begin{bmatrix} E_{14} & E_{24} & 0 \\ E_{15} & E_{25} & 0 \end{bmatrix} \begin{bmatrix} \mathcal{E}_x^{k(0)} \\ \mathcal{E}_y^{k(0)} \\ \mathcal{E}_z^{k(0)} \end{bmatrix}$$

$$- \begin{bmatrix} F_{14} & F_{24} & 0 \\ F_{15} & F_{25} & 0 \end{bmatrix} \begin{bmatrix} \mathcal{E}_x^{k(1)} \\ \mathcal{E}_y^{k(1)} \\ 0 \end{bmatrix} \tag{4.111}$$

The following matrices can be defined in Equation (4.111) with the subscript s, which stands for the shear components:

$$\boldsymbol{A}_s = \begin{bmatrix} A_{44} & A_{45} \\ A_{45} & A_{55} \end{bmatrix}, \quad \boldsymbol{E}_s = \begin{bmatrix} E_{14} & E_{24} & 0 \\ E_{15} & E_{25} & 0 \end{bmatrix}, \quad \boldsymbol{F}_s = \begin{bmatrix} F_{14} & F_{24} & 0 \\ F_{15} & F_{25} & 0 \end{bmatrix} \tag{4.112}$$

The other equations for the moments and electrical contributions are:

$$\begin{bmatrix} M_{xx} \\ M_{yy} \\ M_{xy} \end{bmatrix} = \begin{bmatrix} B_{11} & B_{12} & B_{16} \\ B_{12} & B_{22} & B_{26} \\ B_{16} & B_{26} & B_{66} \end{bmatrix} \begin{bmatrix} \epsilon_{xx}^{(0)} \\ \epsilon_{yy}^{(0)} \\ \gamma_{xy}^{(0)} \end{bmatrix} + \begin{bmatrix} D_{11} & D_{12} & D_{16} \\ D_{12} & D_{22} & D_{26} \\ D_{16} & D_{26} & D_{66} \end{bmatrix} \begin{bmatrix} \epsilon_{xx}^{(1)} \\ \epsilon_{yy}^{(1)} \\ \gamma_{xy}^{(1)} \end{bmatrix}$$
$$- \begin{bmatrix} 0 & 0 & F_{31} \\ 0 & 0 & F_{32} \\ 0 & 0 & F_{36} \end{bmatrix} \begin{bmatrix} \mathcal{E}_x^{k(0)} \\ \mathcal{E}_y^{k(0)} \\ \mathcal{E}_z^{k(0)} \end{bmatrix} - \begin{bmatrix} 0 & 0 & G_{31} \\ 0 & 0 & G_{32} \\ 0 & 0 & G_{36} \end{bmatrix} \begin{bmatrix} \mathcal{E}_x^{k(1)} \\ \mathcal{E}_y^{k(1)} \\ 0 \end{bmatrix} \quad (4.113)$$

$$\begin{bmatrix} O_x \\ O_y \\ O_z \end{bmatrix} = \begin{bmatrix} 0 & 0 & 0 \\ 0 & 0 & 0 \\ E_{31} & E_{32} & E_{36} \end{bmatrix} \begin{bmatrix} \epsilon_{xx}^{(0)} \\ \epsilon_{yy}^{(0)} \\ \gamma_{xy}^{(0)} \end{bmatrix} + \begin{bmatrix} 0 & 0 & 0 \\ 0 & 0 & 0 \\ F_{31} & F_{32} & F_{36} \end{bmatrix} \begin{bmatrix} \epsilon_{xx}^{(1)} \\ \epsilon_{yy}^{(1)} \\ \gamma_{xy}^{(1)} \end{bmatrix}$$
$$+ \begin{bmatrix} E_{14} & E_{15} \\ E_{24} & E_{25} \\ 0 & 0 \end{bmatrix} \begin{bmatrix} \gamma_{yz}^{(0)} \\ \gamma_{xz}^{(0)} \end{bmatrix} + \begin{bmatrix} R_{11} & R_{12} & 0 \\ R_{11} & R_{22} & 0 \\ 0 & 0 & R_{33} \end{bmatrix} \begin{bmatrix} \mathcal{E}_x^{k(0)} \\ \mathcal{E}_y^{k(0)} \\ \mathcal{E}_z^{k(0)} \end{bmatrix}$$
$$+ \begin{bmatrix} S_{11} & S_{12} & 0 \\ S_{11} & S_{22} & 0 \\ 0 & 0 & S_{33} \end{bmatrix} \begin{bmatrix} \mathcal{E}_x^{k(1)} \\ \mathcal{E}_y^{k(1)} \\ 0 \end{bmatrix} \quad (4.114)$$

$$\begin{bmatrix} P_x \\ P_y \\ P_z \end{bmatrix} = \begin{bmatrix} 0 & 0 & 0 \\ 0 & 0 & 0 \\ F_{31} & F_{32} & F_{36} \end{bmatrix} \begin{bmatrix} \epsilon_{xx}^{(0)} \\ \epsilon_{yy}^{(0)} \\ \gamma_{xy}^{(0)} \end{bmatrix} + \begin{bmatrix} 0 & 0 & 0 \\ 0 & 0 & 0 \\ G_{31} & G_{32} & G_{36} \end{bmatrix} \begin{bmatrix} \epsilon_{xx}^{(1)} \\ \epsilon_{yy}^{(1)} \\ \gamma_{xy}^{(1)} \end{bmatrix}$$
$$+ \begin{bmatrix} F_{14} & F_{15} \\ F_{24} & F_{25} \\ 0 & 0 \end{bmatrix} \begin{bmatrix} \gamma_{yz}^{(0)} \\ \gamma_{xz}^{(0)} \end{bmatrix} + \begin{bmatrix} S_{11} & S_{12} & 0 \\ S_{11} & S_{22} & 0 \\ 0 & 0 & S_{33} \end{bmatrix} \begin{bmatrix} \mathcal{E}_x^{k(0)} \\ \mathcal{E}_y^{k(0)} \\ \mathcal{E}_z^{k(0)} \end{bmatrix}$$
$$+ \begin{bmatrix} T_{11} & T_{12} & 0 \\ T_{11} & T_{22} & 0 \\ 0 & 0 & T_{33} \end{bmatrix} \begin{bmatrix} \mathcal{E}_x^{k(1)} \\ \mathcal{E}_y^{k(1)} \\ 0 \end{bmatrix} \quad (4.115)$$

$\{\epsilon^0\}$ and $\{\epsilon^1\}$ in Equations (4.110)–(4.115) are the membrane and bending strains, and $\{\epsilon_s^0\}$ are the shear strains which are typical of FSDT applications. $\{\mathcal{E}^{k(0)}\}$ and $\{\mathcal{E}^{k(1)}\}$ are the membrane and bending electric field components. These five vectors are defined as:

$$\{\epsilon^0\} = \begin{bmatrix} \epsilon_{xx}^{(0)} \\ \epsilon_{yy}^{(0)} \\ \gamma_{xy}^{(0)} \end{bmatrix}, \quad \{\epsilon^1\} = \begin{bmatrix} \epsilon_{xx}^{(1)} \\ \epsilon_{yy}^{(1)} \\ \gamma_{xy}^{(1)} \end{bmatrix}, \quad \{\epsilon_s^0\} = \begin{bmatrix} \gamma_{yz}^{(0)} \\ \gamma_{xz}^{(0)} \end{bmatrix}$$

$$\{\mathcal{E}^{k(0)}\} = \begin{bmatrix} \mathcal{E}_x^{k(0)} \\ \mathcal{E}_y^{k(0)} \\ \mathcal{E}_z^{k(0)} \end{bmatrix}, \quad \{\mathcal{E}^{k(1)}\} = \begin{bmatrix} \mathcal{E}_x^{k(1)} \\ \mathcal{E}_y^{k(1)} \\ 0 \end{bmatrix} \tag{4.116}$$

The force, transverse force, and moment resultants for the mechanical part, and the electric charge and electric moment resultants, are grouped in the following vectors:

$$\{N\} = \begin{bmatrix} N_{xx} \\ N_{yy} \\ N_{xy} \end{bmatrix}, \quad \{Q\} = \begin{bmatrix} Q_x \\ Q_y \end{bmatrix}, \quad \{M\} = \begin{bmatrix} M_{xx} \\ M_{yy} \\ M_{xy} \end{bmatrix}$$

$$\{O\} = \begin{bmatrix} O_x \\ O_y \\ O_z \end{bmatrix}, \quad \{P\} = \begin{bmatrix} P_x \\ P_y \\ P_z \end{bmatrix}. \tag{4.117}$$

In the case of a sensor configuration, mechanical loads are applied to vectors $\{N\}^p$, $\{Q\}^p$, and $\{M\}^p$. The electric potential for the actuator configuration is imposed directly on the vector of the unknowns. The laminate constitutive equations can be written in compact form as:

$$\begin{bmatrix} \{N\} \\ \{Q\} \\ \{M\} \\ \{O\} \\ \{P\} \end{bmatrix} = \begin{bmatrix} \mathbf{A} & 0 & \mathbf{B} & -\mathbf{E} & -\mathbf{F} \\ 0 & \mathbf{A_s} & 0 & -\mathbf{E_s} & -\mathbf{F_s} \\ \mathbf{B} & 0 & \mathbf{D} & -\mathbf{F} & -\mathbf{G} \\ \mathbf{E}^T & \mathbf{E_s}^T & \mathbf{F}^T & \mathbf{R} & \mathbf{S} \\ \mathbf{F}^T & \mathbf{F_s}^T & \mathbf{G}^T & \mathbf{S} & \mathbf{T} \end{bmatrix} \begin{bmatrix} \{\epsilon^0\} \\ \{\epsilon_s^0\} \\ \{\epsilon^1\} \\ \{\mathcal{E}^{k(0)}\} \\ \{\mathcal{E}^{k(1)}\} \end{bmatrix} - \begin{bmatrix} \{N\}^p \\ \{Q\}^p \\ \{M\}^p \end{bmatrix} \tag{4.118}$$

where T means the transpose of a matrix. When the fourth and fifth columns and rows in Equation (4.118) are deleted, the pure mechanical problem is investigated. We can also consider a partial electro-mechanical problem by deleting only the fourth and fifth rows. Equation (4.118), for the FSDT case, compared to Equation (3.45), for the CLT case, also has the shear contributions due to the transverse shear strains γ_{yz} and γ_{xz}; the other components are formally the same.

References

Allen DN de G 1955 *Relaxation Methods*. McGraw-Hill.

Argyris JH 1960 *Energy Theorems and Structural Analysis*. Butterworth (reprinted from *Aircr. Eng.*, 1954–5).

Carrera E and Brischetto S 2008 Analysis of thickness locking in classical, refined and mixed multilayered plate theories. *Comp. Struct.* **82**, 549–562.

Clough RW 1960 The finite element in plane stress analysis. In *Proceedings of 2nd ASCE Conference on Electronic Computation.*

Crandall SH 1958 *Engineering Analysis*. McGraw-Hill.

Finlayson BA 1972 *The Method of Weighted Residuals and Variational Principles*. Academic Press.

Hrenikoff A 1941 Solution of problems in elasticity by the framework method. *J. Appl. Mech.* **A8**, 189–175.

McHenry D 1943 A lattice analogy for the solution of plane stress problem. *J. Inst. Civ. Eng.* **21**, 59–82.

Mindlin RD 1951 Influence of rotatory inertia and shear on flexural motions of isotropic elastic plates. *J. Appl. Mech.* **18**, 31–38.

Newmark NM 1949 *Numerical Methods in Analysis in Engineering* (ed. LE Grinter). Macmillan.

Reddy JN 2004 *Mechanics of Laminated Composite Plates and Shells; Theory and Analysis*. CRC Press.

Reissner E 1945 The effect of transverse shear deformation on the bending of elastic plates. *J. Appl. Mech.* **12**, 69–77.

Southwell RV 1946 *Relaxation Methods in Theoretical Physics*. Clarendon Press.

Zienkiewicz OC and Taylor RL 1967 *The Finite Element Method*. McGraw-Hill.

5

Numerical evaluation of classical theories and their limitations

Some preliminary results are given here in order to show the main limitations of classical theories for the static and dynamic analyses of one-layered and multilayered structures embedding piezoelectric layers. The proposed analyses compare some 3D or quasi-3D results found in the open literature to the classical theories introduced in Chapters 3 and 4 and extended there to multilayered piezoelectric plates and shells. These theories are the well-known classical lamination theory (CLT) and first order shear deformation theory (FSDT) based on the Kirchhoff and Reissner–Mindlin hypotheses, respectively, where the electric potential is described in layer-wise (LW) form with linear expansion in the thickness direction. Only analytical closed-form solutions are given and the readers can find other results (including FE analysis) in the last chapter of this book. In the present chapter, the limitations of CLT and FSDT analyses are shown for simply supported plates and shells and for harmonic loads. The extensions of CLT and FSDT to the electromechanical case are indicated as CLT($\boldsymbol{u}$, Φ) and FSDT($\boldsymbol{u}$, Φ), respectively, where the primary variables are indicated in parentheses. The limitations of the classical theories depend on the thickness ratios, the stacking lamination sequence, transverse and in-plane anisotropy, the type of loads (sensor or actuator configurations), the type of

Plates and Shells for Smart Structures: Classical and Advanced Theories for Modeling and Analysis, First Edition.
Erasmo Carrera, Salvatore Brischetto and Pietro Nali.

electromechanical variables investigated, and the order of frequencies and vibration modes in the case of dynamic analysis. The results are organized as follows: static analysis of multilayered piezoelectric plates in both sensor and actuator configurations; static analysis of one-layered and multilayered piezoelectric ring and cylindrical shells in both sensor and actuator configurations; free-vibration analysis of multilayered piezoelectric plates in closed-circuit configurations; and free-vibration analysis of multilayered piezoelectric shells in closed-circuit configurations (the electric potential is zero at the top and bottom surfaces).

5.1 Static analysis of piezoelectric plates

The considered multilayered piezoelectric plate is simply supported with harmonic loads applied to its top surface. As suggested in Figure 5.1, the sensor configuration has a bisinusoidal transverse mechanical load with amplitude $\hat{p}_z = 1$ Pa, and the actuator configuration has a bisinusoidal electric voltage with amplitude $\hat{\Phi} = 1$ V; in both cases, the imposed wave numbers are $m = n = 1$ and the electric potential at the bottom surface is set to zero. The plate is square, $a = b$, with two external piezoelectric layers in PZT-4 and two internal composite layers in graphite/epoxy (Gr/Ep) with a lamination sequence of 0°/90°. The total thickness is indicated as h, the two external piezoelectric layers have a thickness $h_1 = h_4 = 0.1h$, and the two internal composite layers have a thickness $h_2 = h_3 = 0.4h$. The material data for PZT-4 are: Young's moduli $E_1 = E_2 = 81.3$ GPa and $E_3 = 64.5$ GPa, Poisson ratios $\nu_{12} = 0.329$ and $\nu_{13} = \nu_{23} = 0.432$, shear moduli $G_{23} = G_{13} = 25.6$ GPa and $G_{12} = 30.6$ GPa, dielectric coefficients $\varepsilon_{11} = \varepsilon_{22} = 13\,060$ pC/V m and $\varepsilon_{33} = 11\,510$ pC/V m, piezoelectric coefficients $e_{15} = e_{24} = 12.72$ C/m^2, $e_{31} = e_{32} = -5.20$ C/m^2 and $e_{33} = 15.08$ C/m^2. The material data for the

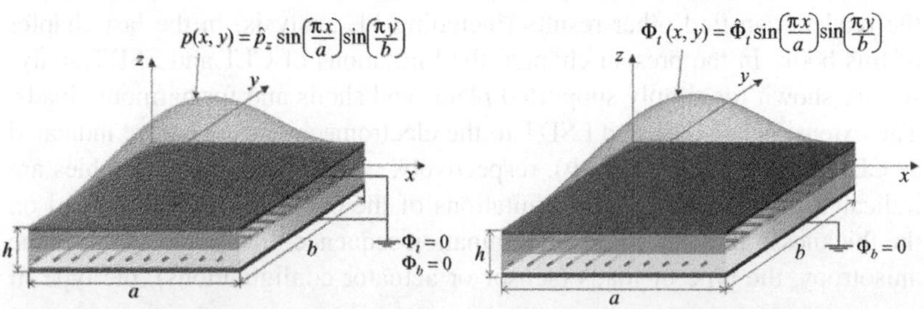

Figure 5.1 Multilayered piezoelectric plate: sensor configuration (left) and actuator configuration (right). Bi-sinusoidal distribution analyzed via analytical solution in closed form.

Table 5.1 Multilayered piezoelectric plate in sensor configuration: in-plane normal stress σ_{yy} at the top $z = h/2$.

a/h	2	4	10	100
3D	—	6.5643	—	—
Ref.	3.2208	6.5642	32.776	3142.1
CLT($\boldsymbol{u}$, Φ)	1.4726	5.8912	36.821	3682.1
FSDT($\boldsymbol{u}$, Φ)	1.4836	5.9023	36.832	3682.1

Gr/Ep are: Young's moduli $E_1 = 132.38$ GPa and $E_2 = E_3 = 10.756$ GPa, Poisson ratios $\nu_{12} = \nu_{13} = 0.24$ and $\nu_{23} = 0.49$, shear moduli $G_{23} = 3.606$ GPa and $G_{12} = G_{13} = 5.6537$ GPa, dielectric coefficients $\varepsilon_{11} = 30.9897$ pC/V m and $\varepsilon_{22} = \varepsilon_{33} = 26.563$ pC/V m. The exact 3D solution for both the actuator and sensor cases has been proposed by Heyliger (1997), while other quasi-3D solutions, based on refined models (Ref.), have been given in Carrera and Brischetto (2007) for several thickness ratio values a/h. CLT($\boldsymbol{u}$, Φ) and FSDT($\boldsymbol{u}$, Φ) are compared in Tables 5.1 and 5.2 to 3D and quasi-3D results for the sensor and actuator cases, respectively. In Table 5.1, the in-plane normal stress σ_{yy} at the top $z = h/2$ of the plate is given by the 3D solution (Heyliger, 1997) for a thick plate in a sensor configuration ($a/h = 4$). The refined theory proposed in Carrera and Brischetto (2007), by means of the CUF theory and MUL2 software, can be considered as a quasi-3D result, as will be demonstrated in Chapter 9; it gives the in-plane normal stress σ_{yy} at the top $z = h/2$ for both thick and thin plates (thickness ratios from $a/h = 2$ to $a/h = 100$). Classical theories are completely inadequate to calculate such a variable for either thick or thin plates; they calculate the electric potential to an acceptable approximation (consequently the electric field by means of geometrical relations), but the three displacement components and their derivatives are not correctly calculated to obtain the strains and, as a consequence, the input data for electromechanical constitutive equations are not adequate to

Table 5.2 Multilayered piezoelectric plate in actuator configuration: electric potential Φ at the middle $z = 0$.

a/h	2	4	10	100
3D	—	0.4477	—	—
Ref.	0.3330	0.4477	0.4910	0.4999
CLT($\boldsymbol{u}$, Φ)	0.3244	0.4469	0.4910	0.4999
FSDT($\boldsymbol{u}$, Φ)	0.3219	0.4461	0.4908	0.4999

obtain satisfactory values for in-plane stresses. The use of refined theories is mandatory to obtain the correct values of such variables, since the plate is multilayered and moderately thick with significant anisotropy. In Table 5.2, the electric potential Φ at the middle $z = 0$ of the plate has been calculated as a 3D result, according to Heyliger (1997), for a thick plate in an actuator configuration ($a/h = 4$). The proposed refined theory, based on the CUF theory and MUL2 software by Carrera and Brischetto (2007), is a quasi-3D result, as will be demonstrated in Chapter 9; it gives the electric potential Φ at the middle $z = 0$ for both thick and thin plates (thickness ratios from $a/h = 2$ to $a/h = 100$). Classical theories give satisfactory values of the electric potential (in particular for thin and moderately thin plates) because it is a primary variable in both CLT($\boldsymbol{u}$, Φ) and FSDT($\boldsymbol{u}$, Φ) analyses, which means that it is obtained directly from the governing equations in LW form by means of a linear expansion in the thickness direction for each layer k. Some differences are exhibited for thick plates, as a linear expansion in the thickness direction is not sufficient to achieve a 3D description through the multilayered structure.

5.2 Static analysis of piezoelectric shells

This section describes some results of the static analysis of one-layered and multilayered cylindrical panels and shells embedding piezoelectric layers; both geometries are considered either in sensor or actuator configuration. The proposed quasi-3D results, indicated as Ref. solutions and shown for comparison purposes, are based on refined models obtained by means of the MUL2 software in the framework of CUF theory (see the results described in Carrera and Brischetto 2007).

The first case covers a simply supported one-layered cylindrical panel in PZT-4 (see the material data in Section 5.1). The geometry, as described in Figure 5.2, has a radius of curvature $R_\alpha = 10$ m in the α direction and infinite radius of curvature R_β in the β direction. The two in-plane dimensions are $a = \pi/3R_\alpha$ in the α direction and $b = 1$ m in the β direction. The sensor configuration has a sinusoidal transverse mechanical load with amplitude $\hat{p}_z = 1$ Pa, and the actuator configuration has a sinusoidal electric voltage with amplitude $\hat{\Phi} = 1$ V; in both cases, the imposed wave numbers are $m = 1$ and $n = 0$, and the electric potential at the bottom surface is set to zero. A comparison between the Ref. solution by Carrera and Brischetto (2007) and the classical CLT($\boldsymbol{u}$, Φ) and FSDT($\boldsymbol{u}$, Φ) theories is shown in Tables 5.3 and 5.4 for the sensor and actuator configurations, respectively. The transverse displacement $\bar{w}$ is evaluated in the middle of the panel ($z = 0$) for thick and thin shells (thickness ratios from $R_\alpha/h = 2$ to $R_\alpha/h = 100$). In the sensor case shown in Table 5.3, it is clear how neither the CLT($\boldsymbol{u}$, Φ) nor FSDT($\boldsymbol{u}$, Φ) analyses achieve the 3D result, even though the shell is thin; similar conclusions can be confirmed for

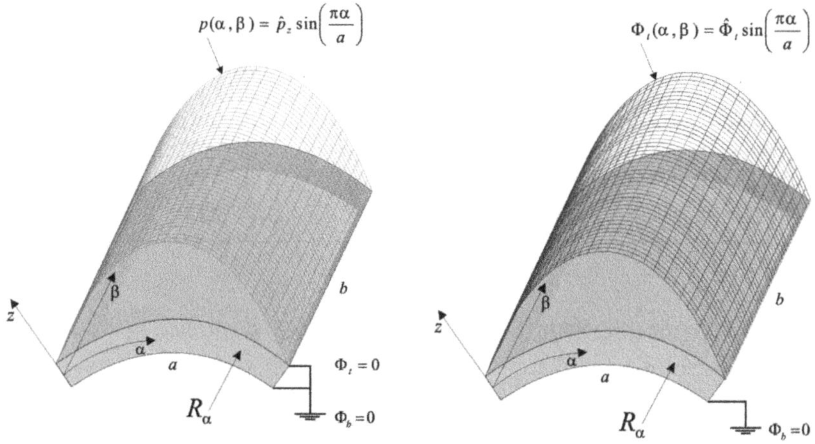

Figure 5.2 One-layered and multilayered piezoelectric cylindrical panels: sensor configuration (left) and actuator configuration (right). Cylindrical distribution analyzed via analytical solution in closed form.

the actuator case shown in Table 5.4 for the electric potential calculated in the middle of the shell where CLT($\boldsymbol{u}$, Φ) and FSDT($\boldsymbol{u}$, Φ) exhibit some difficulties because the electric potential profile is linear from the bottom, where $\Phi_b = 0$, to the top value $\Phi_t = 1$ V; for this reason, the results are acceptable for very thin shells where the electric potential is almost linear in the thickness direction. Figure 5.3 shows the differences between classical CLT($\boldsymbol{u}$, Φ) and FSDT($\boldsymbol{u}$, Φ) theories and quasi-3D descriptions (Ref.) for the evaluation of the transverse displacement through the thickness direction in the case of thick, one-layered piezoelectric panels in a sensor configuration, and a comparison between classical theories and the Ref. solution for the evaluation of the in-plane normal stress $\sigma_{\alpha\alpha}$ through the thickness direction of a moderately thick, one-layered piezoelectric shell panel in an actuator configuration. The inadequacy of

Table 5.3 One-layered piezoelectric cylindrical panel in sensor configuration: transverse normal displacement $\bar{w} = w \times 10^9$ at the middle $z = 0$.

R_α/h	2	4	10	100
Ref.	0.2740	1.4459	18.748	17508
CLT($\boldsymbol{u}$, Φ)	0.1298	0.9619	14.140	13554
FSDT($\boldsymbol{u}$, Φ)	0.2642	1.2078	14.716	13560

Table 5.4 One-layered piezoelectric cylindrical panel in actuator configuration: electric potential Φ at the middle $z = 0$.

R_α/h	2	4	10	100
Ref.	0.4213	0.4903	0.5057	0.5012
CLT($\boldsymbol{u}$, Φ)	0.5000	0.5000	0.5000	0.5000
FSDT($\boldsymbol{u}$, Φ)	0.5000	0.5000	0.5000	0.5000

classical theories shown in Tables 5.3 and 5.4 is also confirmed by the results in Figure 5.3.

The second multilayered case has the same geometry as the one-layered piezoelectric case; the panel is simply supported and both the mechanical and electrical loads (for sensor and actuator configurations, respectively) are applied at the top in sinusoidal form ($m = 1$ and $n = 0$). The stacking layer sequence is the same one that was shown in Section 5.1 for the plate geometry (two external piezoelectric PZT-4 layers and two internal composite layers with a lamination sequence of $90°/0°$ and not the $0°/90°$ of the plate case) and the geometry and loading conditions are as summarized in Figure 5.2. Table 5.5 shows the transverse displacement $\bar{w}$ at $z = 0$ for the multilayered piezoelectric sensor panel for different thickness ratios R_α/h; the combination of thick geometry and transverse anisotropy makes classical theories inadequate for such problems, even though it can be observed that FSDT($\boldsymbol{u}$, Φ) works better than CLT($\boldsymbol{u}$, Φ). These considerations can be confirmed by looking at the relative actuator case analyzed in Table 5.6. The use of refined theories for the actuator case is more desirable for thick geometries, and the electric potential for thin shells can be correctly calculated using each proposed 2D model.

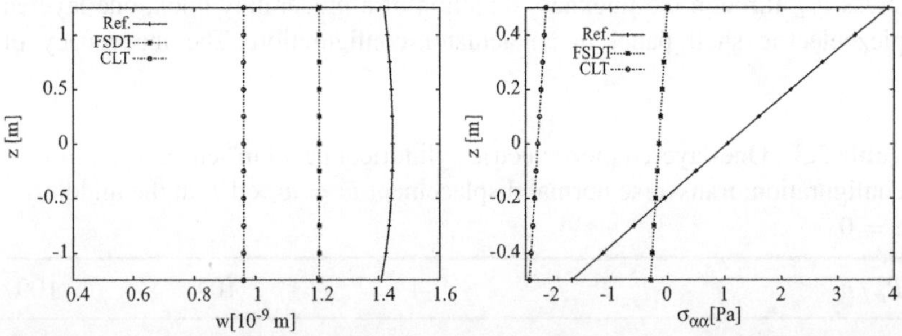

Figure 5.3 One-layered piezoelectric cylindrical panel: transverse normal displacement w vs. z for the sensor case with $R_\alpha/h = 4$ (left) and in-plane normal stress $\sigma_{\alpha\alpha}$ vs. z for the actuator case with $R_\alpha/h = 10$ (right).

Table 5.5 Multilayered piezoelectric cylindrical panel in sensor configuration: transverse normal displacement $\bar{w} = w \times 10^9$ at the middle $z = 0$.

R_α/h	2	4	10	100
Ref.	1.0304	3.3031	29.832	25 202
CLT($\boldsymbol{u}$, Φ)	0.2086	1.4620	21.015	19 957
FSDT($\boldsymbol{u}$, Φ)	0.6148	2.1808	22.682	19 973

Figure 5.4 shows the electric potential through the thickness direction z for a multilayered cylindrical sensor panel and the transverse normal electric displacement along z for a multilayered cylindrical actuator panel. The considered shell is moderately thick ($R_\alpha/h = 10$) and classical theories (CLT($\boldsymbol{u}$, Φ) and FSDT($\boldsymbol{u}$, Φ)) are compared to a refined theory (proposed as a quasi-3D solution in Carrera and Brischetto 2007). For the sensor case, classical theories satisfy the boundary conditions for the electric potential and its evaluation through the thickness, even though the correct values are only obtained by means of the Ref. model. In the actuator case, classical theories are completely inadequate for a correct evaluation of the transverse normal electric displacement, and only the quasi-3D solution gives the interlaminar continuity of the transverse normal electric displacement through the interfaces.

The third case considers the same one-layered piezoelectric configuration as the first case, embedded in a cylindrical shell geometry, as shown in Figure 5.5. The radii of curvature are $R_\alpha = 10$ m and infinite R_β with shell dimensions $a = 2\pi R_\alpha$ and $b = 40$ m. The cylindrical shell is considered to be simply supported and two different conditions are investigated: a sensor case, when a mechanical load is applied at the top, $p_z = \hat{p}_z \sin(m\pi\alpha/a)\sin(n\pi\beta/b)$, with amplitude $\hat{p}_z = 1$ Pa and wave numbers $m = 8$ and $n = 1$; and an actuator case, when an electric voltage is applied at the top, $\Phi = \hat{\Phi}\sin(m\pi\alpha/a)\sin(n\pi\beta/b)$ with amplitude $\hat{\Phi} = 1$ V and wave numbers $m = 8$ and $n = 1$. The electric potential in the sensor configuration is set to zero at the top and bottom surfaces,

Table 5.6 Multilayered piezoelectric cylindrical panel in actuator configuration: electric potential Φ at the middle $z = 0$.

R_α/h	2	4	10	100
Ref.	0.4528	0.4981	0.5056	0.5009
CLT($\boldsymbol{u}$, Φ)	0.4504	0.4979	0.5056	0.5010
FSDT($\boldsymbol{u}$, Φ)	0.4491	0.4975	0.5055	0.5010

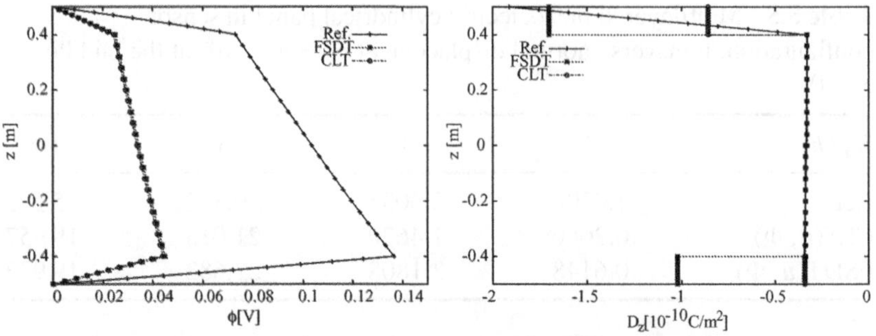

Figure 5.4 Multilayered piezoelectric cylindrical panel: electric potential Φ vs. z for the sensor case with $R_\alpha/h = 10$ (left) and transverse normal electric displacement $\mathcal{D}_z$ vs. z for the actuator case with $R_\alpha/h = 10$ (right).

while in the actuator case the electric potential is set to zero at the shell bottom. The electric potential Φ in Figure 5.6 is given through the thickness direction z of the sensor configuration (left) and actuator configuration (right). The use of the Ref. solution is mandatory for the sensor configuration since, when zero electric potential is imposed at the top and bottom of the shell, classical theories (CLT($\boldsymbol{u}$, Φ) and FSDT($\boldsymbol{u}$, Φ)), which have a linear electric potential in the thickness direction, can only give a zero electric potential all over the thickness of the shell. A linear electric potential through the thickness, which goes from 0 to 1 V, is obtained for the actuator configuration for a moderately thin one-layered shell; for this reason, it can be stated that classical theories are

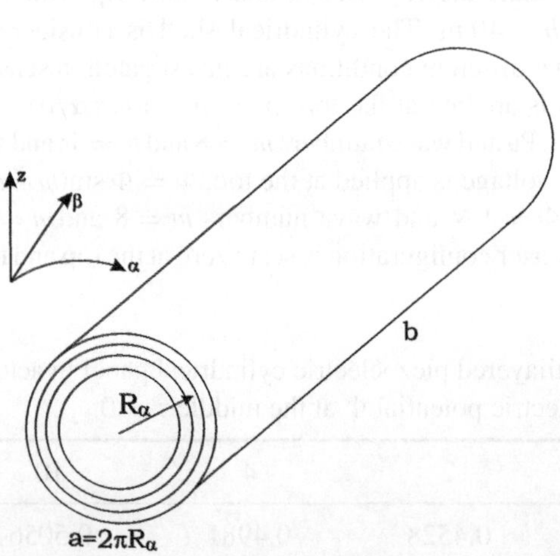

Figure 5.5 One-layered and multilayered piezoelectric cylindrical shell geometry: both sensor and actuator configurations can be considered.

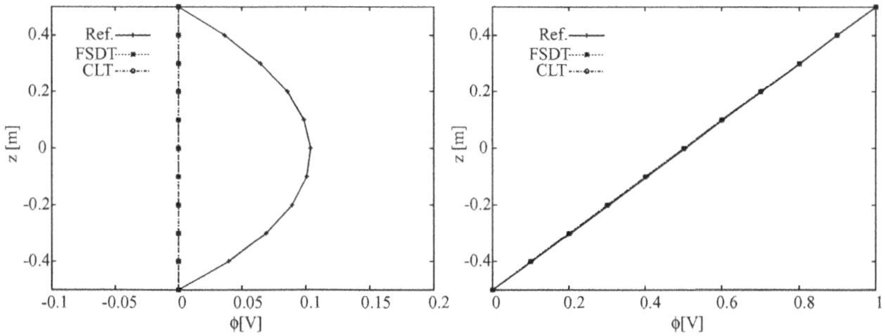

Figure 5.6 One-layered piezoelectric cylindrical shell: electric potential Φ vs. z for the sensor case with $R_\alpha/h = 10$ (left) and electric potential Φ vs. z for the actuator case with $R_\alpha/h = 10$ (right).

able to satisfy such conditions and behavior. Table 5.7 considers the transverse normal electric displacement at the top $z = h/2$ for the actuator case. Classical theories are completely inadequate for such an analysis for each considered thickness ratio R_α/h; the use of refined models, where $\mathcal{D}_z$ is a primary variable, is mandatory.

The last case considers the same cylindrical shell already described in Figure 5.5 with the same boundary conditions, geometrical parameters, and applied loads for the sensor and actuator cases already given for the third case. The only difference, with respect to the third case, is the lamination sequence; in this last case, we consider a four-layered configuration with two internal composite layers and two external piezoelectric ones, as already shown in this section for the second case. The transverse normal electric displacement is given through the thickness direction of the sensor configuration of the multilayered piezoelectric cylindrical shell on the left side of Figure 5.7. It has clearly been demonstrated how classical theories are completely inadequate and the Ref. model gives a satisfactory analysis and ensures interlaminar continuity (interlaminar continuity of a transverse electromechanical variable is only ensured by means of opportune RMVT applications as will be shown in the subsequent chapters). A

Table 5.7 One-layered piezoelectric cylindrical shell in actuator configuration: transverse normal electric displacement $\bar{\mathcal{D}}_z = \mathcal{D}_z \times 10^8$ at the top $z = h/2$.

R_α/h	2	4	10	100
Ref.	−0.5848	−0.7840	−1.6156	−16.266
CLT($\boldsymbol{u}$, Φ)	−0.2351	−0.4692	−1.1711	−10.738
FSDT($\boldsymbol{u}$, Φ)	−0.2475	−0.4822	−1.1843	−10.747

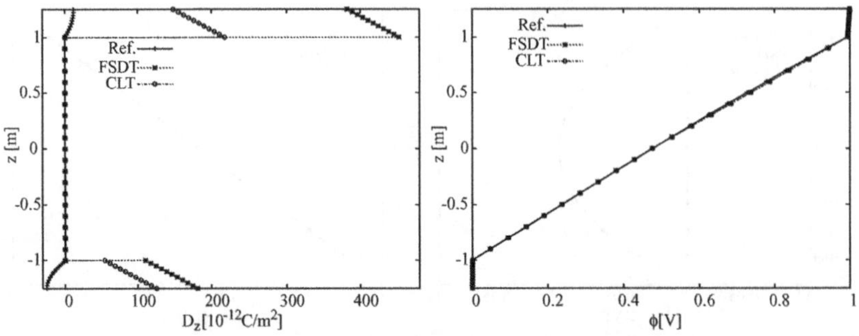

Figure 5.7 Multilayered piezoelectric cylindrical shell: transverse normal electric displacement $\mathcal{D}_z$ vs. z for the sensor case with $R_\alpha/h = 4$ (left) and electric potential Φ vs. z for the actuator case with $R_\alpha/h = 4$ (right).

correct evaluation of the electric potential through z is given on the right side of Figure 5.7 for the relative actuator configuration; each considered theory gives correct results as the electric potential is always a primary variable, and it also satisfies the boundary load conditions for an actuator configuration (electric potential equal to the imposed value at the top and zero at the bottom). Table 5.8 shows the transverse normal electric displacement $\bar{\mathcal{D}}_z$ at the top $z = h/2$ for the actuator configuration; the error obtained by means of classical theories for each thickness ratio R_α/h is clearly observed, even though such an error decreases with the thickness ratio.

5.3 Vibration analysis of piezoelectric plates

The free-vibration problem of multilayered piezoelectric plates is investigated here in order to show the main limitations of classical theories such as CLT($\boldsymbol{u}$, Φ) and FSDT($\boldsymbol{u}$, Φ). The corresponding vibration modes are obtained by imposing the wave numbers in the in-plane directions when an analytical

Table 5.8 Multilayered piezoelectric cylindrical shell in actuator configuration: transverse normal displacement $\bar{\mathcal{D}}_z = \mathcal{D}_z \times 10^{10}$ at the top $z = h/2$.

R_α/h	2	4	10	100
Ref.	−10.660	−6.6070	−3.2710	−3.6226
CLT($\boldsymbol{u}$, Φ)	−3.7634	−2.3546	−1.3225	−3.3559
FSDT($\boldsymbol{u}$, Φ)	−4.6977	−3.2782	−2.1875	−3.9044

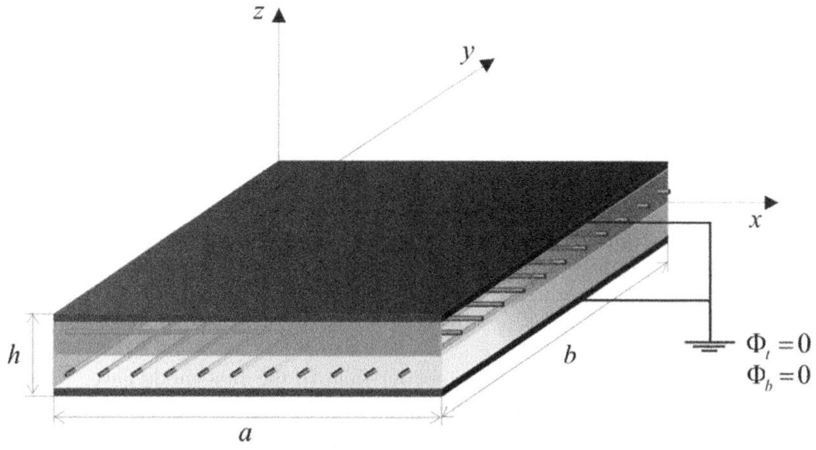

Figure 5.8 Closed-circuit configuration (electric potential Φ zero at the external surfaces) for the free-vibration analysis of a multilayered piezoelectric plate.

closed-form solution is assumed. The number of frequencies is equal to the number of degrees of freedom through the thickness, according to the considered kinematics.

A five-layered plate is considered (see Figure 5.8). The two external layers are made of piezoelectric material with a thickness $h_1 = h_5 = h/10$, while the three internal layers consist of reinforced carbon fiber layers with lamination sequence $0°/90°/0°$ and thickness $h_2 = h_3 = h_4 = \frac{4}{15}h$. The elastic and electrical properties of the multilayered plate are the same as those given in Section 5.1 for the PZT-4 and Gr/Ep materials. The 3D solution was first proposed by Heyliger and Saravanos (1995). The considered plate has a square geometry ($a = b$) in a closed-circuit configuration (electric potential applied at the top and bottom equal to zero, as indicated in Figure 5.8). In order to obtain the reference 3D solution, Heyliger and Saravanos (1995) employed a mass density $\rho = 1$ kg/m^3 for both materials; this operation does not have any physical meaning but it is nevertheless acceptable for the proposed preliminary assessment. The results are given as the first three fundamental circular frequencies $\overline{\omega} = \omega/100 = 2\pi f/100$ (for wave numbers $m = n = 1$). Two thickness ratios are investigated: a thick plate ($a/h = 4$ with $h = 0.01$ m) and a moderately thin plate ($a/h = 50$ with $h = 0.01$ m). Table 5.9 clearly shows how classical theories are completely inadequate for such an investigation. The plate is multilayered with a given transverse anisotropy. The errors given by the CLT($\boldsymbol{u}$, Φ) and FSDT($\boldsymbol{u}$, Φ) theories increase with the thickness value and also depend on the mode considered (e.g., modes 2 and 3 in Table 5.9 are in-plane modes and

Table 5.9 Multilayered piezoelectric plate: 3D results vs. classical theories. $m = n = 1$, first three modes.

	$a/h = 4$			$a/h = 50$		
	Mode 1	Mode 2	Mode 3	Mode 1	Mode 2	Mode 3
3D	57074.5	191301	250769	618.118	15681.6	21492.8
CLT($\boldsymbol{u}$, Φ)	103030	198465	286795	692.254	15877.2	22943.9
FSDT($\boldsymbol{u}$, Φ)	74105.9	198465	286795	689.870	15877.2	22943.9

in this case classical theories work better than in the first mode case, which is a thickness mode). It is clear why the use of refined models is mandatory for a complete dynamic investigation, since, when classical theories are employed, some modes are tragically lost because of the small number of degrees of freedom employed for the description through the thickness.

The case proposed in Heyliger and Saravanos (1995) is a well-known 3D benchmark; however, they assumed a mass density $\rho = 1\ \text{kg/m}^3$ for both piezoelectric and composite materials. This assumption does not have any physical meaning, therefore real mass densities $\rho = 7600\ \text{kg/m}^3$ and $\rho = 1578\ \text{kg/m}^3$ for PZT-4 and Gr/Ep materials, respectively, were considered in Carrera *et al.* (2010). The first three circular frequencies $\overline{\omega} = \omega/100$ for $m = n = 1$ are considered in Table 5.10. The most refined theory proposed in Carrera *et al.* (2010) is used as a reference solution (Ref.) for the case of real mass density as it gives a quasi-3D description of the multilayered piezoelectric plate. No further results are obtained by the introduction of different mass densities from $1\ \text{kg/m}^3$ since, although the mass density influences the frequency values, it does not add any new effects to the comparison of classical theories. For both cases (unit or real mass densities), it can be seen how classical theories obtain satisfactory results for those frequencies that are related to in-plane vibration modes.

Table 5.10 Multilayered piezoelectric plate with real mass densities: 3D results vs. classical theories. $m = n = 1$, first three modes.

	$a/h = 4$			$a/h = 50$		
	Mode 1	Mode 2	Mode 3	Mode 1	Mode 2	Mode 3
Ref.	1078.98	3460.68	4328.96	11.7167	297.176	407.510
CLT($\boldsymbol{u}$, Φ)	1393.79	3762.48	5437.02	13.0758	300.999	434.969
FSDT($\boldsymbol{u}$, Φ)	1898.84	3762.48	5437.02	13.1210	300.999	434.969

5.4 Vibration analysis of piezoelectric shells

The free-vibration problem of multilayered shells, including thickness-polarized piezoelectric layers, is now investigated. As in the plate case, the corresponding vibration modes have been obtained by imposing the wave numbers in the in-plane directions. The number of frequencies is equal to the number of degrees of freedom through the thickness of the considered kinematics model.

A two-layered ring shell and a multilayered cylindrical panel have been considered. The geometry of these shells is given in Figures 5.9 and 5.10 for the ring shell and the cylindrical panel, respectively. The free-vibration problem for both geometries has been investigated in a closed-circuit configuration, as clearly indicated in Figure 5.10: the electric potential is zero at the top and bottom of the shell ($\Phi_t = \Phi_b = 0$).

The cylindrical ring shell has two layers, an internal layer in titanium and an external one in piezoelectric PZT-4. The properties of PZT-4 have already been given in the previous sections and it has a mass density $\rho = 7600$ kg/m^3, while the properties of the titanium, which is an isotropic material, are: Young's modulus $E = 114$ GPa, Poisson ratio $\nu = 0.3$, dielectric coefficient $\varepsilon = 8.850$ pC/V m and mass density $\rho = 2768$ kg/m^3. The 3D solution was given by Heyliger *et al.* (1996): the first fundamental frequency in Hz is given by imposing $m = 0$ and $n = 4, 8, 12, 16, 20$. The thickness ratio R_β / h is equal to 72.75 (where the total thickness is $h = 0.004$ m, and the radii of curvature

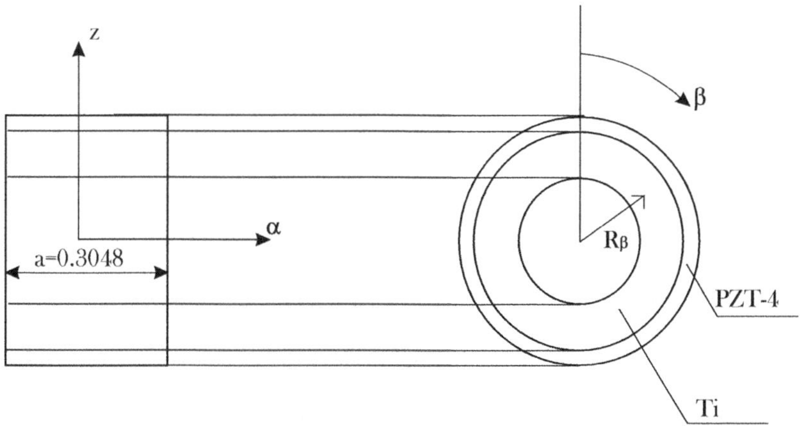

Figure 5.9 Closed-circuit configuration (electric potential Φ zero at the external surfaces) for the free-vibration analysis of a multilayered piezoelectric ring shell.

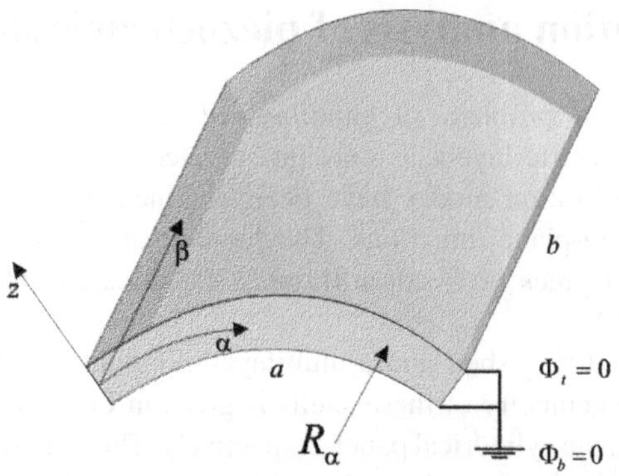

Figure 5.10 Closed-circuit configuration (electric potential Φ zero at the external surfaces) for the free-vibration analysis of a multilayered piezoelectric cylindrical panel.

at the midsurface are $R_\alpha = \infty$ and $R_\beta = 0.291$ m). The dimensions are $a = 0.3048$ m and $b = 2\pi R_\beta = 1.828\,41$ m. The PZT-4 layer has a thickness of 0.001 m and the titanium layer thickness is 0.003 m. In Table 5.11, the shell is moderately thin, so the errors, in terms of frequency, are remarkable for classical theories, and these errors increase with the wave number m. For higher values of m, the use of refined models is mandatory. When the wave numbers m and n are fixed, classical theories work quite well for the fundamental frequency, but they are inadequate for the higher frequency values.

The cylindrical panel has two external layers in piezoelectric material PTZ-4 with a thickness $h_1 = h_5 = h_{TOT}/10$ and three internal layers in graphite-epoxy Gr/Ep with a lamination sequence 0°/90°/0° and thickness $h_2 = h_3 = h_4 = \frac{4}{15} h_{TOT}$. The elastic and electric properties of these two materials have already been given in the previous sections. The dimensions are $b = 1$ m and $a = \pi/3R_\alpha = 10.471\,97$ m (radii of curvature $R_\beta = \infty$ and $R_\alpha = 10$ m). This assessment is a sort of extension of the plate case, which was proposed in the previous section, to a shell geometry. In this case, the

Table 5.11 Ring in PZT-4 and titanium, closed-circuit configuration; 3D results vs. CLT and FSDT analysis, fundamental mode for $m = 0$.

n	4	8	12	16	20
3D	31.27	170.42	407.29	745.21	1190.48
CLT($\boldsymbol{u}$, Φ)	35.54	192.83	457.44	828.11	1304.49
FSDT($\boldsymbol{u}$, Φ)	35.54	192.74	456.96	826.56	1300.69

Table 5.12 Closed-circuit vibration problem for multilayered piezoelectric cylindrical panel, fundamental frequency $\overline{\omega} = \omega\sqrt{R_\alpha{}^4(\rho)\text{PZT-4}/(E_3)\text{PZT-4}h_{TOT}^2}$ obtained by using classical theories. $m = n = 1$.

R_α/h	2	4	10	100
Ref.	30.435	51.885	127.74	472.30
CLT($\boldsymbol{u}$, Φ)	43.661	87.448	218.70	556.01
FSDT($\boldsymbol{u}$, Φ)	39.970	77.039	178.81	539.08

reference solution (Ref) is the refined quasi-3D model already discussed in Carrera *et al.* (2010). The results are given as the first fundamental circular frequency $\overline{\omega} = \omega\sqrt{R_\beta(\rho)\text{PTZ-4}/(E_3)\text{PTZ-4}h_{TOT}^2}$. Four thickness ratios are investigated: $R_\beta/h = 2, 4, 10$, and 100 (where the values of the total thickness are $h = 5, 2.5, 1$, and 0.1 m). Tables 5.12 and 5.13 give the fundamental frequencies of the shell for different thickness ratios and for wave numbers $m = n = 1$ and $m = n = 10$, respectively. Table 5.12 is for low values of the imposed wave numbers and, when the thickness ratio increases (thin shells), the results given by classical theories improve, even though the CLT($\boldsymbol{u}$, Φ) theory is always inadequate. It can be concluded that the results obtained using classical theories are inappropriate for each thickness ratio. The inadequacy of classical theories is much more notable for higher values of wave numbers, as clearly shown in Table 5.13 for $m = n = 10$. In both tables, the results obtained by means of FSDT($\boldsymbol{u}$, Φ) are better than the CLT($\boldsymbol{u}$, Φ) analysis, but they are not sufficient for a quasi-3D description, and for this reason the use of refined models must be considered.

Table 5.13 Closed-circuit vibration problem for multilayered piezoelectric cylindrical panel, fundamental frequency $\overline{\omega} = \omega\sqrt{R_\beta{}^4(\rho)\text{PZT-4}/(E_3)\text{PZT-4}h_{TOT}^2}$ obtained by using classical theories. $m = n = 10$.

R_α/h	2	4	10	100
Ref.	328.90	664.37	1763.6	12 651
CLT($\boldsymbol{u}$, Φ)	436.61	874.49	2187.2	21 875
FSDT($\boldsymbol{u}$, Φ)	418.66	834.95	2055.2	17 802

References

Carrera E and Brischetto S 2007 Reissner mixed theorem applied to static analysis of piezoelectric shells. *J. Intell. Mater. Syst. Struct.* **18**, 1083–1107.

Carrera E, Brischetto S, and Cinefra M 2010 Variable kinematics and advanced variational statements for free vibrations analysis of piezoelectric plates and shells. *Comput. Model. Eng. Sci.* **65**, 259–341.

Heyliger PR 1997 Exact solutions for simply supported laminated piezoelectric plates. *J. Appl. Mech.* **64**, 299–306.

Heyliger PR and Saravanos DA 1995 Exact free vibration analysis of laminated plates with embedded piezoelectric layers. *J. Acoust. Soc. Am.* **98**, 1547–1557.

Heyliger PR, Pei K, and Saravanos DA 1996 Layerwise mechanics and finite element model for laminated piezoelectric shells. *AIAA J.* **34**, 2353–2360.

6

Refined and advanced theories for plates

In refined and advanced models, higher orders of expansion in the thickness direction are assumed for both the electrical and mechanical components. These axiomatic 2D models can be considered in ESL or in LW form. The Carrera Unified Formulation (CUF) is a technique that permits one to handle a large variety of plate models in a unified manner. According to the CUF, the obtained theories can have an order of expansion which goes from first- to higher order values, and, depending on the thickness functions that are used, a model can be ESL or LW. The CLT and FSDT plate theories obtained in Chapters 3 and 4 can also be obtained in the CUF as particular cases of ESL theories. CLT, FSDT, ESL, and LW refined and advanced mixed theories have been implemented by means of the in-house academic code MUL2 (an acronym of MULtifield problems for MULtilayered structures).

6.1 Unified formulation: refined models

We define *refined models* as those *displacement models* where higher orders of expansion in the thickness direction z are assumed for all three displacement components. These axiomatic 2D models can be seen in ESL form when the layers included in the multilayer are considered as one equivalent structure, and in LW form when each layer embedded in the multilayer is separately

Plates and Shells for Smart Structures: Classical and Advanced Theories for Modeling and Analysis, First Edition.
Erasmo Carrera, Salvatore Brischetto and Pietro Nali.

considered in order to write the expansions in z for each layer k. In the case of electromechanical problems, *refined models* are those where the extension is made by considering the electric potential and the displacement vector as the primary variables. These models are obtained using the principle of virtual displacements (PVD) (Carrera 2002), and its extensions to multifield problems (Carrera *et al.* 2007; Ikeda 1996).

The CUF is a technique which can handle a large variety of plate/shell models in a unified manner (Carrera 1995). According to the CUF, the governing equations are written in terms of a few fundamental nuclei which do not formally depend on the order of expansion N used in the z direction and on the description of the variables (LW or ESL) (Demasi 2008a,b). The application of a 2D method for plates allows one to express the unknown variables as a set of thickness functions that only depend on the thickness coordinate z and the corresponding variable that depends on the in-plane coordinates x and y. The generic variable $\boldsymbol{f}(x, y, z)$, for instance a displacement, and its variation $\delta \boldsymbol{f}(x, y, z)$ are therefore written according to the following general expansion:

$$\boldsymbol{f}(x, y, z) = F_\tau(z)\boldsymbol{f}_\tau(x, y), \quad \delta \boldsymbol{f}(x, y, z) = F_s(z)\delta \boldsymbol{f}_s(x, y),$$
$$\text{with} \quad \tau, s = 1, \ldots, N \tag{6.1}$$

where the bold letters denote arrays, (x,y) are the in-plane coordinates, and z the thickness. The summing convention is assumed with repeated indexes τ and s. The order of expansion N goes from first- to higher order values, and, depending on the thickness functions used, a model can be either ESL, when the variable is assumed for the whole multilayer and a Taylor expansion is employed as the thickness functions $F(z)$, or LW, when the variable is considered independent in each layer and a combination of Legendre polynomials is used as the thickness functions $F(z)$. In the CUF, the maximum order of expansion N in the z direction is the fourth.

6.1.1 ESL theories

The displacement $\boldsymbol{u} = (u, v, w)$ is described according to the ESL description if the unknowns are the same for the whole plate (Librescu and Wu 1977; Librescu and Schmidt 1988). The z expansion is obtained via Taylor polynomials, that is:

$$\begin{aligned} u &= F_0\, u_0 + F_1\, u_1 + \cdots + F_N\, u_N = F_\tau\, u_\tau \\ v &= F_0\, v_0 + F_1\, v_1 + \cdots + F_N\, v_N = F_\tau\, v_\tau \\ w &= F_0\, w_0 + F_1\, w_1 + \cdots + F_N\, w_N = F_\tau\, w_\tau \end{aligned} \tag{6.2}$$

with $\tau = 0, 1, \ldots, N$; N is the order of expansion and ranges from 1 (linear) to 4:

$$F_0 = z^0 = 1, \quad F_1 = z^1 = z, \ldots, F_N = z^N \tag{6.3}$$

Equation (6.2) can be written in vectorial form:

$$\boldsymbol{u}(x, y, z) = F_\tau(z)\boldsymbol{u}_\tau(x, y), \quad \delta\boldsymbol{u}(x, y, z) = F_s(z)\delta\boldsymbol{u}_s(x, y),$$
$$\text{with} \quad \tau, s = 1, \ldots, N \tag{6.4}$$

The 2D models obtained from Equations (6.2)–(6.4) are denoted by the acronym EDN, where E indicates that an ESL approach has been employed, D indicates that the theory is a displacement formulation, and N indicates the order of expansion in the thickness direction. For example, an ED2 model has a quadratic expansion in z, an ED4 has a fourth- order of expansion in z, and so on. A typical displacement field is shown in Figure 6.1 for a three-layered structure for the case of an ED4 model. Figure 6.2 considers the displacement and the transverse stresses along the z direction for an ED2 model: displacements are quadratic in z, therefore the transverse stresses are linear (no longer constant, as in classical theories), but discontinuous at each interface. Simpler theories can be obtained from EDN models, such as those that discard the ϵ_{zz} effect; in this case, it is sufficient to impose that the transverse displacement w is constant in z. Such theories are denoted as EDNd. The ED1d model coincides with the FSDT. CLT is obtained from FSDT via an opportune penalty technique which imposes an infinite shear correction factor. It is important to recall that all the EDNd theories, which have constant transverse

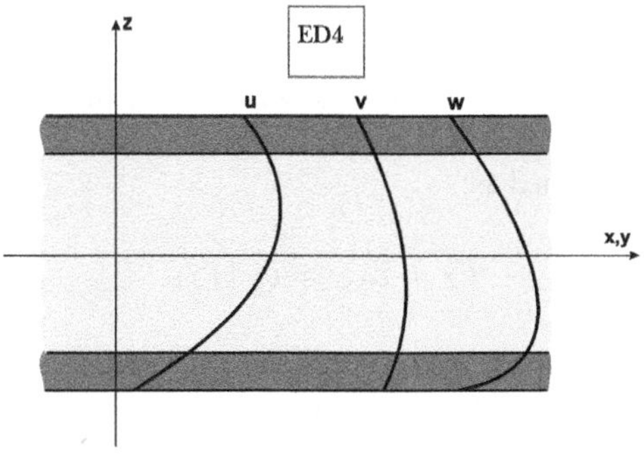

Figure 6.1 ED4: displacements u, v and w through the thickness direction z.

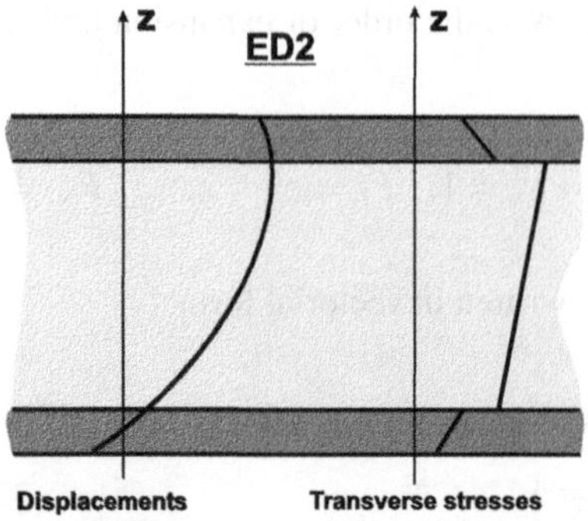

Figure 6.2 ED2: displacements and transverse shear stresses through the thickness direction z.

displacement and zero transverse normal strain ϵ_{zz}, and the ED1 model show Poisson locking phenomenon in the case of pure mechanical problems; this can be overcome via plane stress conditions in constitutive equations (Carrera and Brischetto 2008a,b).

6.1.2 Murakami zigzag function

The proposed ESL models in the previous section do not consider the zigzag (ZZ) form of displacements in the z direction, which is typical of multilayered structures with transverse anisotropy (Carrera 2003). A remedy for this limitation is the introduction of an opportune zigzag function in the ESL displacement model, in order to recover the ZZ form of displacements without the use of LW models. The latter have intrinsic ZZ behavior, but are more computationally expansive compared to ESL models (Carrera and Brischetto 2009a,b). A possible choice for the zigzag function is the so-called *Murakami zigzag function* (MZZF) (Murakami 1985, 1986). MZZF can be simply added to a displacement model and leads to remarkable improvements in the solution as it satisfies the typical ZZ form of displacements in multilayered structures.

MZZF $Z(z)$ is defined as:

$$F_Z = Z(z) = (-1)^k \zeta_k \tag{6.5}$$

with the non-dimensioned layer coordinate $\zeta_k = (2z_k)/h_k$, where z_k is the transverse thickness coordinate of the k layer and h_k is the thickness of the k layer, therefore $-1 \leq \zeta_k \leq 1$. $Z(z)$ has the following properties: it is a piecewise linear function of the layer coordinates z_k; $Z(z)$ has unit amplitude for the whole layers; and the slope $Z'(z) = dZ/dz$ assumes an opposite sign between two-adjacent layers. Its amplitude is layer thickness independent (Murakami 1986). The displacement model that includes MZZF is:

$$\begin{aligned} u &= F_0\, u_0 + F_1\, u_1 + \cdots + F_N\, u_N + F_Z\, u_Z = F_\tau\, u_\tau \\ v &= F_0\, v_0 + F_1\, v_1 + \cdots + F_N\, v_N + F_Z\, v_Z = F_\tau\, v_\tau \\ w &= F_0\, w_0 + F_1\, w_1 + \cdots + F_N\, w_N + F_Z\, w_Z = F_\tau\, w_\tau \end{aligned} \tag{6.6}$$

where $\tau = 0, 1, \ldots, (N+1)$, and N is the order of expansion ranging from 1 (linear) to 4:

$$F_0 = z^0 = 1, \quad F_1 = z^1 = z, \ldots, F_N = z^N, \quad F_{N+1} = F_Z = (-1)^k \zeta_k \tag{6.7}$$

The acronym used to indicate this kind of model is EDZN, where E stand for the ESL approach, D for the displacement formulation, and N is the order of expansion in the z direction. Z indicates that MZZF has been added (Brischetto *et al.* 2009a). The following remarks can be made: the additional degree of freedom $\boldsymbol{u}_Z$ has the meaning of displacement; the amplitude $\boldsymbol{u}_Z$ is layer independent since $\boldsymbol{u}_Z$ has an intrinsic equivalent single-layer description; and MZZF can be used for both in-plane and out-of-plane displacement components (Brischetto *et al.*,2009b,c). Figure 6.3 clearly explains the meaning of MZZF and how to add it to displacement components. The MZZF $F_Z = Z(z) = (-1)^k \zeta_k$ is considered as the $(N+1)th$ thickness function in the vectorial form of Equation (6.6):

$$\boldsymbol{u}(x, y, z) = F_\tau(z)\boldsymbol{u}_\tau(x, y), \quad \delta\boldsymbol{u}(x, y, z) = F_s(z)\delta\boldsymbol{u}_s(x, y),$$
$$\text{with} \quad \tau, s = 1, \ldots, (N+1) \tag{6.8}$$

Typical displacements and transverse shear stresses along the thickness z are shown in Figure 6.4 for an EDZ1 model: the inclusion of MZZF allows

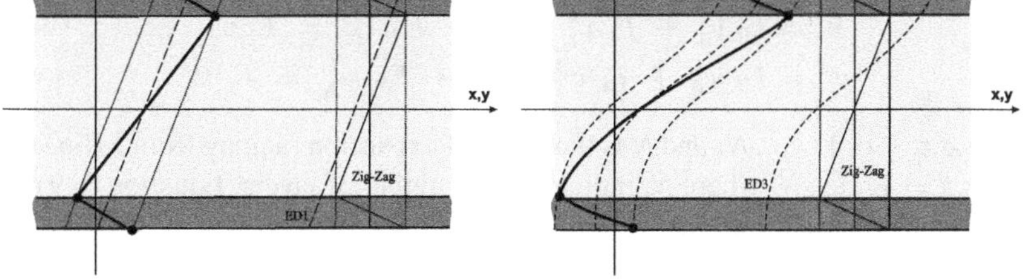

Figure 6.3 Displacement models in the EDZ1 and EDZ3 theories. Inclusion of MZZF in an ESL model.

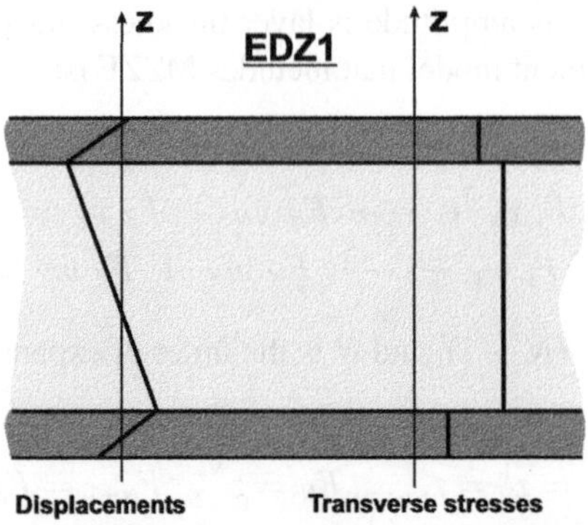

Figure 6.4 EDZ1: displacements and transverse shear stresses through the thickness direction z.

one to recover the typical ZZ form of the displacement vector for case of multilayered transverse-anisotropy structures. Like the EDN models, there is the possibility of imposing constant transverse displacements w. Such models are denoted as EDZNd models. EDZNd models require the Poisson locking phenomena to be corrected, as indicated in Carrera and Brischetto (2008a,b), for the case of pure mechanical problems.

6.1.3 LW theories

When each layer of a multilayered structure is described as an independent plate, a LW approach is necessary (Reddy 2004). The displacement $\boldsymbol{u}^k = (u, v, w)^k$ is described for each k layer. In this way, the ZZ form of displacement is easily obtained in multilayered transverse-anisotropy structures (Hsu and Wang 1970, 1971; Robbins and Reddy 1993; Srinivas 1973). The recovery of the ZZ effect via LW models is dealt with in detail in Carrera and Brischetto (2009a,b) and in Figure 6.5. The z expansion for the displacement components is made for each k layer:

$$\begin{aligned} u^k &= F_0\, u_0^k + F_1\, u_1^k + \cdots + F_N\, u_N^k = F_\tau\, u_\tau^k \\ v^k &= F_0\, v_0^k + F_1\, v_1^k + \cdots + F_N\, v_N^k = F_\tau\, v_\tau^k \\ w^k &= F_0\, w_0^k + F_1\, w_1^k + \cdots + F_N\, w_N^k = F_\tau\, w_\tau^k \end{aligned} \tag{6.9}$$

where $\tau = 0, 1, \ldots, N$, and N is the order of expansion ranging from 1 (linear) to 4; $k = 1, \ldots, N_l$ where N_l indicates the number of layers. Equation (6.9) is written in vectorial form as:

$$\boldsymbol{u}^k(x, y, z) = F_\tau(z)\boldsymbol{u}_\tau^k(x, y), \quad \delta\boldsymbol{u}^k(x, y, z) = F_s(z)\delta\boldsymbol{u}_s^k(x, y),$$
$$\text{with} \quad \tau, s = t, b, r \quad \text{and} \quad k = 1, \ldots, N_l \tag{6.10}$$

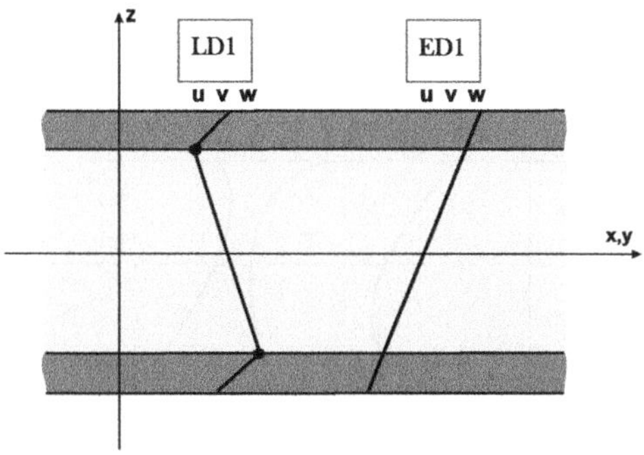

Figure 6.5 Linear expansion in the z direction for the displacement components: LW approach vs. ESL approach.

where t and b indicate the top and bottom of each k layer, respectively; N_l is the number of total layers, and r indicates the higher orders of expansion in the thickness direction: $r = 2, \ldots, N$. The thickness functions $F_\tau(\zeta_k)$ and $F_s(\zeta_k)$ have now been defined at the k-layer level, where they are a linear combination of Legendre polynomials $P_j = P_j(\zeta_k)$ of the jth-order defined in the ζ_k- domain ($\zeta_k = 2z_k/h_k$ with the z_k local coordinate and h_k thickness, both referring to the kth layer, therefore $-1 \leq \zeta_k \leq 1$). The first five Legendre polynomials are:

$$P_0 = 1, \quad P_1 = \zeta_k, \quad P_2 = \frac{(3\zeta_k{}^2 - 1)}{2}, \quad P_3 = \frac{5\zeta_k{}^3}{2} - \frac{3\zeta_k}{2},$$
$$P_4 = \frac{35\zeta_k{}^4}{8} - \frac{15\zeta_k{}^2}{4} + \frac{3}{8} \tag{6.11}$$

and their combinations for the thickness functions are:

$$F_t = F_0 = \frac{P_0 + P_1}{2}, \quad F_b = F_1 = \frac{P_0 - P_1}{2},$$
$$F_r = P_r - P_{r-2} \quad \text{with} \quad r = 2, \ldots, N \tag{6.12}$$

The chosen functions have the following interesting properties:

$$\zeta_k = 1\colon \; F_t = 1; \;\; F_b = 0; \;\; F_r = 0 \quad \text{at the top} \tag{6.13}$$

$$\zeta_k = -1\colon \; F_t = 0; \;\; F_b = 1; \;\; F_r = 0 \quad \text{at the bottom} \tag{6.14}$$

In other words, interface values of the variables are considered as variable unknowns. This fact allows one to easily impose the compatibility conditions

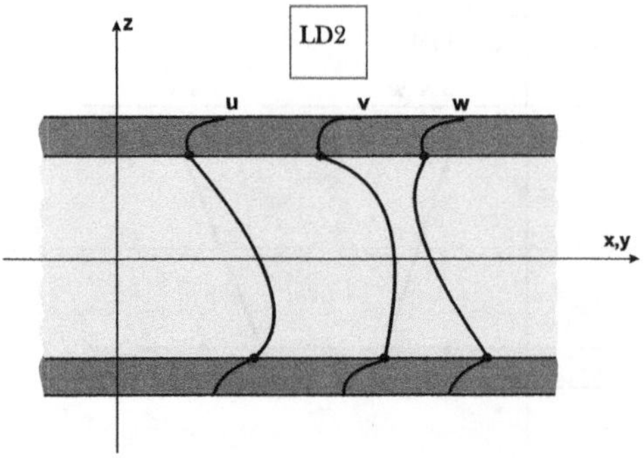

Figure 6.6 LD2: displacements u, v, and w through the thickness direction z.

for the displacements at each layer interface. The acronym to indicate such theories is LDN, where L stands for the LW approach, D indicates the displacement formulation, and N is the order of expansion in each k layer. Typical displacement behavior for a three-layered structure is indicated in Figure 6.6 for an LD2 model. Figure 6.7 indicates the displacements and transverse shear stresses for an LD3 model. The transverse shear/normal stresses are obtained via constitutive equations but this fact does not ensure interlaminar continuity (IC). IC could be enforced by a priori modeling of the transverse shear/normal stresses. In LW models, even though a linear expansion in z is considered for transverse displacement w, the Poisson locking phenomenon does not appear for a pure mechanical problem: the transverse normal strain ϵ_{zz} is piece-wise constant in the thickness direction (Carrera and Brischetto 2008a,b).

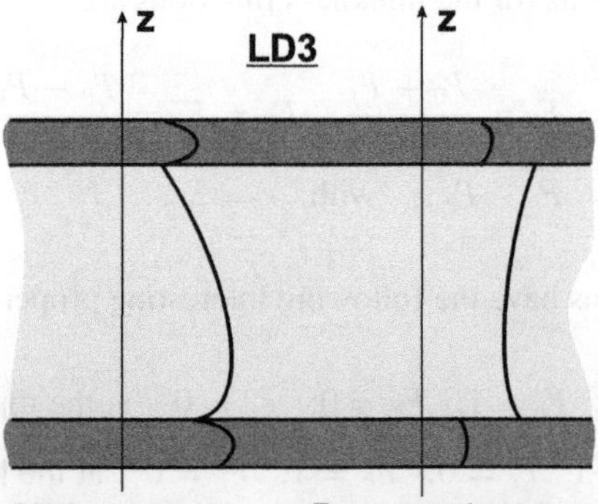

Figure 6.7 LD3: displacements and transverse shear stresses through the thickness direction z.

6.1.4 Refined models for the electromechanical case

In the case of electromechanical problems, the primary variables are the displacement vector $\boldsymbol{u} = (u, v, w)$ and the scalar electric potential Φ. Considering the higher spatial gradient of the electric potential, the variable Φ^k is always modeled as LW (Brischetto and Carrera 2009; Carrera and Brischetto 2007a,b):

$$\Phi^k(x, y, z) = F_\tau(z)\Phi^k_\tau(x, y), \quad \delta\Phi^k(x, y, z) = F_s(z)\delta\Phi^k_s(x, y)$$
$$\text{with} \quad \tau, s = t, b, r \quad \text{and} \quad k = 1, \ldots, N_l \tag{6.15}$$

where t and b indicate the top and bottom of each k layer, respectively; N_l indicates the number of total layers, and r indicates the higher orders of expansion in the thickness direction: $r = 2, \ldots, N$. The thickness functions are a combination of Legendre polynomials, as indicated in the previous section. A 2D model for electromechanical problems is defined as ESL, ESL+MZZF, or LW, depending on the choice made for the displacement vector: the electric potential is always considered LW (Ballhause *et al.* 2005; Carrera and Boscolo 2007).

6.2 Unified formulation: advanced mixed models

In the case of electromechanical problems, we define *advanced mixed models* as those 2D models that are obtained by employing the Reissner mixed variational theorem (RMVT) (Reissner 1984) and its extensions to electromechanical coupling (Carrera *et al.* 2008). These extensions allow one to a priori model some transverse quantities, which are obtained in PVD applications via post-processing. Transverse shear/normal stresses $\boldsymbol{\sigma}_n = (\sigma_{xz}, \sigma_{yz}, \sigma_{zz})$ and/or transverse normal electric displacement $\mathcal{D}_n = (\mathcal{D}_z)$ are a priori modeled and considered in LW form. The main advantage of obtaining these variables directly from the governing equations is the fulfillment of Interlaminar Continuity (IC) (Brischetto 2009; Brischetto and Carrera 2010). These advanced models are obtained by means of the CUF (Carrera 2002) which has been dealt with in detail in previous sections.

6.2.1 Transverse shear/normal stress modeling

An advanced mixed model for a pure mechanical problem considers both displacements $\boldsymbol{u} = (u, v, w)$ and transverse shear/normal stresses $\boldsymbol{\sigma}_{nM} = (\sigma_{xz}, \sigma_{yz}, \sigma_{zz})$ as the primary variables (Brischetto and Carrera 2010). The displacements can be modeled as ESL (Section 6.1.1), ESL+MZZF (Section 6.1.2), or LW (Section 6.1.3), and this choice allows one to define the considered advanced model as ESL, ESL+MZZF, or LW, respectively:

the transverse shear/normal stresses $\boldsymbol{\sigma}_{nM}$ are always LW (the subscript M means that the stresses are modeled and not obtained from the constitutive equations). The LW model for stresses is:

$$\begin{aligned}
\sigma_{xz}^k &= F_0\,\sigma_{xz0}^k + F_1\,\sigma_{xz1}^k + \cdots + F_N\,\sigma_{xzN}^k = F_\tau\,\sigma_{xz\tau}^k \\
\sigma_{yz}^k &= F_0\,\sigma_{yz0}^k + F_1\,\sigma_{yz1}^k + \cdots + F_N\,\sigma_{yzN}^k = F_\tau\,\sigma_{yz\tau}^k \\
\sigma_{zz}^k &= F_0\,\sigma_{zz0}^k + F_1\,\sigma_{zz1}^k + \cdots + F_N\,\sigma_{zzN}^k = F_\tau\,\sigma_{zz\tau}^k
\end{aligned} \tag{6.16}$$

where $\tau = 0, 1, \ldots, N$, and N is the order of expansion ranging from 1 (linear) to 4; $k = 1, \ldots, N_l$ where N_l indicates the number of layers. Equation (6.16) is written in vectorial form as:

$$\boldsymbol{\sigma}_{nM}^k(x,y,z) = F_\tau(z)\boldsymbol{\sigma}_{nM\tau}^k(x,y), \qquad \delta\boldsymbol{\sigma}_{nM}^k(x,y,z) = F_s(z)\delta\boldsymbol{\sigma}_{nMs}^k(x,y),$$
$$\text{with} \quad \tau, s = t, b, r \qquad \text{and} \quad k = 1, \ldots, N_l \tag{6.17}$$

where t and b indicate the top and bottom of each k layer, respectively; r indicates the higher orders of expansion in the thickness direction: $r = 2, \ldots, N$. The thickness functions $F_\tau(\zeta_k)$ and $F_s(\zeta_k)$ have now been defined at the k-layer level, and they are a linear combination of Legendre polynomials. The use of such thickness functions, based on the property pointed out in Equations (6.13) and (6.14), allows one to easily write IC for the transverse stresses:

$$\boldsymbol{\sigma}_{nt}^k = \boldsymbol{\sigma}_{nb}^{k+1} \quad \text{with} \quad k = 1, \ldots, (N_l - 1) \tag{6.18}$$

which means the top value of the k layer in each interface is equal to the bottom value of the $(k+1)$ layer. The same property can be used for displacements in LW form, in order to impose the compatibility conditions:

$$\boldsymbol{u}_t^k = \boldsymbol{u}_b^{k+1} \quad \text{with} \quad k = 1, \ldots, (N_l - 1) \tag{6.19}$$

Those models with displacements in the ESL form (E) and transverse stresses in the LW form are known as EMN models, where M means mixed formulation (use of RMVT), and N is the order of expansion, which is the same for both variables. EMZN models consider the displacements modeled in ESL form but they also include MZZF. LMN models consider both displacements and transverse stresses in LW form. Figure 6.8 shows the displacements and transverse stresses for an EM2 model. The displacements are considered ESL, while the transverse stresses are a priori modeled and directly obtained from the

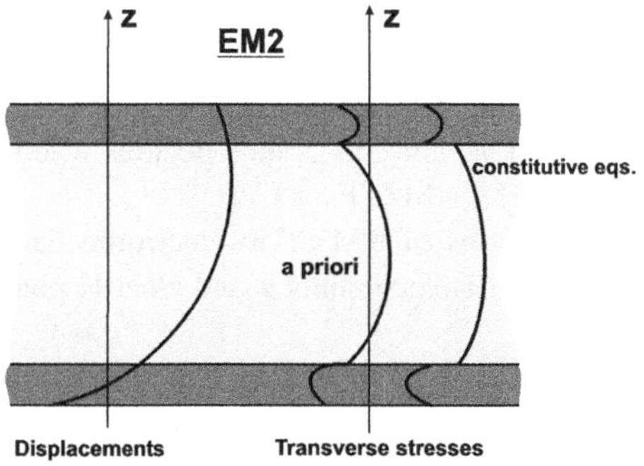

Figure 6.8 EM2: displacements and transverse shear stresses through the thickness direction z.

governing equations: they are considered in LW form, and this makes it possible to satisfy both the ZZ form and IC. If transverse stresses are obtained from constitutive equations via post-processing, IC might not be ensured. Figure 6.9 shows displacements and stresses for an LM2 model. In this case, the displacements are also LW, and the ZZ form and IC are ensured for both the displacement and transverse stress components. The transverse stresses obtained from the constitutive equations could not satisfy IC (Brischetto 2009).

6.2.2 Advanced mixed models for the electromechanical case

Several extensions of RMVT can be considered for electromechanical problems (Reissner 1984; Carrera *et al.* 2008). In such models, the $\boldsymbol{u}$ displacements and

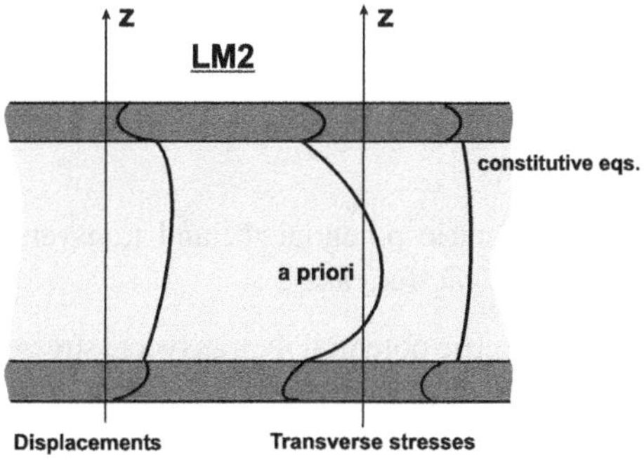

Figure 6.9 LM2: displacements and transverse shear stresses through the thickness direction z.

electric potential Φ are always considered in the governing equations, the electric potential Φ is always modeled in LW form, as discussed for the PVD case in the previous sections, the displacement components $\boldsymbol{u}$ are modeled as ESL, ESL+MZZF, or LW, and this choice decides whether the considered advanced model is ESL, ESL+MZZF, or LW.

Three different extensions of RMVT to electromechanical problems are possible and in addition to displacements $\boldsymbol{u}$ and electric potential Φ, the other modeled variables are:

1. Using only one Lagrange multiplier (Reissner 1984), the transverse stresses $\boldsymbol{\sigma}_{nM}$ are a priori modeled (LW form as described in previous sections) (Carrera and Brischetto 2007b).
2. Using only one Lagrange multiplier, the transverse normal electric displacement $\mathcal{D}_{nM} = \mathcal{D}_z$ is a priori obtained in LW form (Carrera *et al.* 2008).
3. Considering two Lagrange multipliers, both the transverse stresses and transverse normal electric displacement are a priori modeled in LW form (Carrera and Brischetto 2007a).

The LW expansion for the transverse normal electric displacement $\mathcal{D}_{nM} = \mathcal{D}_z$ is:

$$\mathcal{D}_z^k(x, y, z) = F_\tau(z)\mathcal{D}_{z\tau}^k(x, y), \qquad \delta\mathcal{D}_z^k(x, y, z) = F_s(z)\delta\mathcal{D}_{zs}^k(x, y),$$
$$\text{with} \quad \tau, s = t, b, r \quad \text{and} \quad k = 1, \ldots, N_l \tag{6.20}$$

where t and b indicate the top and bottom of each k layer, respectively, and r indicates the higher orders of expansion in the thickness direction: $r = 2, \ldots, N$.

The modeled variables for these three advanced models are:

1. displacements $\boldsymbol{u}$, transverse stresses $\boldsymbol{\sigma}_{nM}$, and electric potential Φ for case 1;
2. displacements $\boldsymbol{u}$, electric potential Φ, and transverse normal electric displacement $\mathcal{D}_{nM} = \mathcal{D}_z$ for case 2;
3. displacements $\boldsymbol{u}$, electric potential Φ, transverse stresses $\boldsymbol{\sigma}_{nM}$, and transverse normal electric displacement $\mathcal{D}_{nM} = \mathcal{D}_z$ for case 3.

The acronyms for such advanced mixed models are explained at the end of this chapter, after the discussion on the variational statements.

6.3 PVD(u, Φ) for the electromechanical plate case

The PVD has been obtained for electromechanical problems as in Equation (2.53) in Chapter 2:

$$\int_V \left(\delta\boldsymbol{\epsilon}_{pG}^T\boldsymbol{\sigma}_{pC} + \delta\boldsymbol{\epsilon}_{nG}^T\boldsymbol{\sigma}_{nC} - \delta\boldsymbol{\mathcal{E}}_G^T\boldsymbol{\mathcal{D}}_C\right)dV = \delta L_e - \delta L_{in} \tag{6.21}$$

It is not necessary to split the electric displacement for the PVD case. By considering a laminate of N_l layers, and the integral on the volume V_k of each k layer as an integral on the in-plane domain Ω_k, plus the integral in the thickness-direction domain A_k, it is possible to write:

$$\sum_{k=1}^{N_l}\int_{\Omega_k}\int_{A_k}\left\{\delta\boldsymbol{\epsilon}_{pG}^{k\ T}\boldsymbol{\sigma}_{pC}^k + \delta\boldsymbol{\epsilon}_{nG}^{k\ T}\boldsymbol{\sigma}_{nC}^k - \delta\boldsymbol{\mathcal{E}}_G^{k\ T}\boldsymbol{\mathcal{D}}_C^k\right\}d\Omega_k dz$$

$$= \sum_{k=1}^{N_l}\delta L_e^k - \sum_{k=1}^{N_l}\delta L_{in}^k \tag{6.22}$$

where δL_e^k and δL_{in}^k are the external and inertial virtual work at the k-layer level, respectively. The relative constitutive equations are those obtained in Equations (2.21)–(2.24); if the electric displacement and electric field are not split into in-plane and out-of-plane components, the relations are:

$$\boldsymbol{\sigma}_{pC}^k = \boldsymbol{Q}_{pp}^k\boldsymbol{\epsilon}_{pG}^k + \boldsymbol{Q}_{pn}^k\boldsymbol{\epsilon}_{nG}^k - \boldsymbol{e}_p^{k^T}\boldsymbol{\mathcal{E}}_G^k \tag{6.23}$$

$$\boldsymbol{\sigma}_{nC}^k = \boldsymbol{Q}_{np}^k\boldsymbol{\epsilon}_{pG}^k + \boldsymbol{Q}_{nn}^k\boldsymbol{\epsilon}_{nG}^k - \boldsymbol{e}_n^{k^T}\boldsymbol{\mathcal{E}}_G^k \tag{6.24}$$

$$\boldsymbol{\mathcal{D}}_C^k = \boldsymbol{e}_p^k\boldsymbol{\epsilon}_{pG}^k + \boldsymbol{e}_n^k\boldsymbol{\epsilon}_{nG}^k + \boldsymbol{\varepsilon}^k\boldsymbol{\mathcal{E}}_G^k \tag{6.25}$$

By substituting Equations (6.23)–(6.25), and the geometrical relations (2.38)–(2.41) in Chapter 2 for plates, in the variational statement of Equation (6.22), and considering a generic k layer (Carrera *et al.* 2007):

$$\int_{\Omega_k}\int_{A_k}\Big[\left(\boldsymbol{D}_p\delta\boldsymbol{u}^k\right)^T\left((\boldsymbol{Q}_{pp}^k\boldsymbol{D}_p + \boldsymbol{Q}_{pn}^k(\boldsymbol{D}_{np}+\boldsymbol{D}_{nz}))\boldsymbol{u}^k + \boldsymbol{e}_p^{k^T}(\boldsymbol{D}_{ep}+\boldsymbol{D}_{en})\Phi^k\right)$$

$$+\left((\boldsymbol{D}_{np}+\boldsymbol{D}_{nz})\delta\boldsymbol{u}^k\right)^T\left((\boldsymbol{Q}_{np}^k\boldsymbol{D}_p + \boldsymbol{Q}_{nn}^k(\boldsymbol{D}_{np}+\boldsymbol{D}_{nz}))\boldsymbol{u}^k + \boldsymbol{e}_n^{k^T}(\boldsymbol{D}_{ep}+\boldsymbol{D}_{en})\Phi^k\right)$$

$$+\left((\boldsymbol{D}_{ep}+\boldsymbol{D}_{en})\delta\Phi^k\right)^T\left((\boldsymbol{e}_p^k\boldsymbol{D}_p + \boldsymbol{e}_n^k(\boldsymbol{D}_{np}+\boldsymbol{D}_{nz}))\boldsymbol{u}^k - \boldsymbol{\varepsilon}^k(\boldsymbol{D}_{ep} + \boldsymbol{D}_{en})\Phi^k\right)\Big]$$

$$\times\, d\Omega_k\, dz = \delta L_e^k - \delta L_{in}^k \tag{6.26}$$

The CUF (Carrera 1995), as presented in previous sections, can be introduced into Equation (6.26) for the 2D approximation:

$$\int_{\Omega_k}\int_{A_k}\Big[\left(\boldsymbol{D}_p F_s \delta \boldsymbol{u}_s^k\right)^T \Big(\big(\boldsymbol{Q}_{pp}^k \boldsymbol{D}_p + \boldsymbol{Q}_{pn}^k (\boldsymbol{D}_{np}+\boldsymbol{D}_{nz})\big) F_\tau \boldsymbol{u}_\tau^k$$

$$+\boldsymbol{e}_p^{k^T}(\boldsymbol{D}_{ep}+\boldsymbol{D}_{en})F_\tau \Phi_\tau^k\Big) + \left((\boldsymbol{D}_{np}+\boldsymbol{D}_{nz})F_s \delta \boldsymbol{u}_s^k\right)^T$$

$$\times\Big(\big(\boldsymbol{Q}_{np}^k \boldsymbol{D}_p + \boldsymbol{Q}_{nn}^k (\boldsymbol{D}_{np}+\boldsymbol{D}_{nz})\big) F_\tau \boldsymbol{u}_\tau^k + \boldsymbol{e}_n^{k^T}(\boldsymbol{D}_{ep}+\boldsymbol{D}_{en})F_\tau \Phi_\tau^k\Big)$$

$$+\left((\boldsymbol{D}_{ep}+\boldsymbol{D}_{en})F_s \delta \Phi_s^k\right)^T \Big(\big(\boldsymbol{e}_p^k \boldsymbol{D}_p + \boldsymbol{e}_n^k (\boldsymbol{D}_{np}+\boldsymbol{D}_{nz})\big) F_\tau \boldsymbol{u}_\tau^k$$

$$-\boldsymbol{\varepsilon}^k (\boldsymbol{D}_{ep}+\boldsymbol{D}_{en})F_\tau \Phi_\tau^k\Big)\Big] d\Omega_k \, dz = \delta L_e^k - \delta L_{in}^k \tag{6.27}$$

In order to obtain a strong form of differential equations on domain Ω_k and the relative boundary conditions on edge Γ_k in Equation (6.27), integration by parts is used, which permits one to move the differential operator from the infinitesimal variation of the generic variable $\delta \boldsymbol{a}^k$ to the finite quantity $\boldsymbol{a}^k$ (Carrera 1995). For a generic variable $\boldsymbol{a}^k$, the integration by parts is:

$$\int_{\Omega_k} \left(\boldsymbol{D}_\Omega \delta \boldsymbol{a}^k\right)^T \boldsymbol{a}^k d\Omega_k = -\int_{\Omega_k} \delta \boldsymbol{a}^{k^T} \left(\boldsymbol{D}_\Omega^T \boldsymbol{a}^k\right) d\Omega_k + \int_{\Gamma_k} \delta \boldsymbol{a}^{k^T} \left(\boldsymbol{I}_\Omega^T \boldsymbol{a}^k\right) d\Gamma_k \tag{6.28}$$

where $\Omega = p, np, ep$. The matrices needed to perform the integration by parts have the following form, like the matrices for the geometrical relations:

$$\boldsymbol{I}_p = \begin{bmatrix} 1 & 0 & 0 \\ 0 & 1 & 0 \\ 1 & 1 & 0 \end{bmatrix}, \quad \boldsymbol{I}_{np} = \begin{bmatrix} 0 & 0 & 1 \\ 0 & 0 & 1 \\ 0 & 0 & 0 \end{bmatrix}, \quad \boldsymbol{I}_{ep} = \begin{bmatrix} 1 \\ 1 \end{bmatrix} \tag{6.29}$$

The governing equations have the following form (Brischetto and Carrera 2009):

$$\begin{aligned} \delta \boldsymbol{u}_s^k : \quad & \boldsymbol{K}_{uu}^{k\tau s}\, \boldsymbol{u}_\tau^k + \boldsymbol{K}_{u\Phi}^{k\tau s}\, \Phi_\tau^k = \boldsymbol{p}_{us}^k - \boldsymbol{M}_{uu}^{k\tau s}\, \ddot{\boldsymbol{u}}_\tau^k \\ \delta \Phi_s^k : \quad & \boldsymbol{K}_{\Phi u}^{k\tau s}\, \boldsymbol{u}_\tau^k + \boldsymbol{K}_{\Phi\Phi}^{k\tau s}\, \Phi_\tau^k = 0 \end{aligned} \tag{6.30}$$

where $\boldsymbol{M}_{uu}^{k\tau s}$ is the inertial contribution in the fundamental nucleus form, $\boldsymbol{u}_\tau^k$ is the vector of the degrees of freedom for the displacements, Φ_τ^k is the vector of the degrees of freedom for the electric potential, and $\ddot{\boldsymbol{u}}_\tau^k$ is the second temporal derivative of $\boldsymbol{u}_\tau^k$. The array $\boldsymbol{p}_{us}^k$ indicates the variationally consistent mechanical loading that is useful for the sensor configuration, while, for the case of an actuator configuration, the electric potential is applied directly to the vector Φ^k_τ. Along with these governing equations, the following boundary conditions on edge Γ_k of the in-plane integration domain Ω_k hold:

$$\begin{aligned}\boldsymbol{\Pi}^{k\tau s}_{uu}\ \boldsymbol{u}^k_\tau + \boldsymbol{\Pi}^{k\tau s}_{u\Phi}\ \Phi^k_\tau &= \boldsymbol{\Pi}^{k\tau s}_{uu}\ \bar{\boldsymbol{u}}^k_\tau + \boldsymbol{\Pi}^{k\tau s}_{u\Phi}\ \bar{\Phi}^k_\tau \\ \boldsymbol{\Pi}^{k\tau s}_{\Phi u}\ \boldsymbol{u}^k_\tau + \boldsymbol{\Pi}^{k\tau s}_{\Phi\Phi}\ \Phi^k_\tau &= \boldsymbol{\Pi}^{k\tau s}_{\Phi u}\ \bar{\boldsymbol{u}}^k_\tau + \boldsymbol{\Pi}^{k\tau s}_{\Phi\Phi}\ \bar{\Phi}^k_\tau\end{aligned} \tag{6.31}$$

By comparing Equation (6.27), after the integration by parts (see Equation (6.28)), to Equations (6.30) and (6.31), the following fundamental nuclei can be obtained:

$$\begin{aligned}\boldsymbol{K}^{k\tau s}_{uu} = \int_{A_k} \Big[&- \boldsymbol{D}^T_p\Big(\boldsymbol{Q}^k_{pp}\boldsymbol{D}_p + \boldsymbol{Q}^k_{pn}(\boldsymbol{D}_{np} + \boldsymbol{D}_{nz})\Big) + \Big(- \boldsymbol{D}_{np} + \boldsymbol{D}_{nz}\Big)^T \\ &\times \Big(\boldsymbol{Q}^k_{np}\boldsymbol{D}_p + \boldsymbol{Q}^k_{nn}(\boldsymbol{D}_{np} + \boldsymbol{D}_{nz})\Big)\Big] F_s F_\tau dz\end{aligned} \tag{6.32}$$

$$\begin{aligned}\boldsymbol{K}^{k\tau s}_{u\Phi} = \int_{A_k} \Big[&- \boldsymbol{D}^T_p\Big(\boldsymbol{e}^{kT}_p(\boldsymbol{D}_{ep} + \boldsymbol{D}_{en})\Big) + \Big(- \boldsymbol{D}_{np} + \boldsymbol{D}_{nz}\Big)^T \\ &\times \Big(\boldsymbol{e}^{kT}_n(\boldsymbol{D}_{ep} + \boldsymbol{D}_{en})\Big)\Big] F_s F_\tau dz\end{aligned} \tag{6.33}$$

$$\boldsymbol{K}^{k\tau s}_{\Phi u} = \int_{A_k} \Big[\Big(- \boldsymbol{D}_{ep} + \boldsymbol{D}_{en}\Big)^T \Big(\boldsymbol{e}^{kT}_p \boldsymbol{D}_p + \boldsymbol{e}^{kT}_n(\boldsymbol{D}_{np} + \boldsymbol{D}_{nz})\Big)\Big] F_s F_\tau dz \tag{6.34}$$

$$\boldsymbol{K}^{k\tau s}_{\Phi\Phi} = \int_{A_k} \Big[\Big(- \boldsymbol{D}_{ep} + \boldsymbol{D}_{en}\Big)^T \Big(- \boldsymbol{\varepsilon}^k(\boldsymbol{D}_{ep} + \boldsymbol{D}_{en})\Big)\Big] F_s F_\tau dz \tag{6.35}$$

$$\boldsymbol{M}^{k\tau s}_{uu} = \int_{A_k} (\rho^k \boldsymbol{I}) F_s F_\tau dz \tag{6.36}$$

where ρ^k is the mass density of the kth layer and $\boldsymbol{I}$ is the (3×3) identity matrix. The nuclei for the boundary conditions at edge Γ_k are (Carrera and Brischetto 2007b):

$$\begin{aligned}\boldsymbol{\Pi}^{k\tau s}_{uu} = \int_{A_k} \Big[&\boldsymbol{I}^T_p\Big(\boldsymbol{Q}^k_{pp}\boldsymbol{D}_p + \boldsymbol{Q}^k_{pn}(\boldsymbol{D}_{np} + \boldsymbol{D}_{nz})\Big) \\ &+ \boldsymbol{I}^T_{np}\Big(\boldsymbol{Q}^k_{np}\boldsymbol{D}_p + \boldsymbol{Q}^k_{nn}(\boldsymbol{D}_{np} + \boldsymbol{D}_{nz})\Big)\Big] F_s F_\tau dz\end{aligned} \tag{6.37}$$

$$\boldsymbol{\Pi}^{k\tau s}_{u\Phi} = \int_{A_k} \Big[\boldsymbol{I}^T_p \Big(\boldsymbol{e}^{kT}_p (\boldsymbol{D}_{ep} + \boldsymbol{D}_{en}) \Big) + \boldsymbol{I}^T_{np} \Big(\boldsymbol{e}^{kT}_n (\boldsymbol{D}_{ep} + \boldsymbol{D}_{en}) \Big) \Big] F_s F_\tau dz \tag{6.38}$$

$$\boldsymbol{\Pi}^{k\tau s}_{\Phi u} = \int_{A_k} \Big[\boldsymbol{I}^T_{ep} \Big(\boldsymbol{e}^{kT}_p \boldsymbol{D}_p + \boldsymbol{e}^{kT}_n (\boldsymbol{D}_{np} + \boldsymbol{D}_{nz}) \Big) \Big] F_s F_\tau dz \tag{6.39}$$

$$\boldsymbol{\Pi}^{k\tau s}_{\Phi\Phi} = \int_{A_k} \Big[\boldsymbol{I}^T_{ep} \Big(-\boldsymbol{\varepsilon}^k (\boldsymbol{D}_{ep} + \boldsymbol{D}_{en}) \Big) \Big] F_s F_\tau dz \tag{6.40}$$

In order to write the explicit form of the nuclei in Equations (6.32)–(6.36), the following integrals in the z thickness direction are defined:

$$(J^{k\tau s}, J^{k\tau_z s}, J^{k\tau s_z}, J^{k\tau_z s_z}) = \int_{A_k} \left(F_\tau F_s, \frac{\partial F_\tau}{\partial z} F_s, F_\tau \frac{\partial F_s}{\partial z}, \frac{\partial F_\tau}{\partial z} \frac{\partial F_s}{\partial z} \right) dz \tag{6.41}$$

and by developing the matrix products in Equations (6.32)–(6.36) and employing a Navier-type closed-form solution (Carrera and Brischetto 2007b), the algebraic explicit form of the nuclei can be obtained.

The fundamental nucleus $\boldsymbol{K}^{k\tau s}_{uu}$ of dimension (3×3) is:

$$\begin{aligned}
K^{k\tau s}_{uu_{11}} &= Q^k_{55} J^{k\tau_z s_z} + Q^k_{11} J^{k\tau s} \bar{\alpha}^2 + Q^k_{66} J^{k\tau s} \bar{\beta}^2 \\
K^{k\tau s}_{uu_{12}} &= J^{k\tau s} \bar{\alpha} \bar{\beta} (Q^k_{12} + Q^k_{66}) \\
K^{k\tau s}_{uu_{13}} &= Q^k_{55} J^{k\tau_z s} \bar{\alpha} - Q^k_{13} J^{k\tau s_z} \bar{\alpha} \\
K^{k\tau s}_{uu_{21}} &= J^{k\tau s} \bar{\alpha} \bar{\beta} (Q^k_{12} + Q^k_{66}) \\
K^{k\tau s}_{uu_{22}} &= Q^k_{44} J^{k\tau_z s_z} + Q^k_{22} J^{k\tau s} \bar{\beta}^2 + Q^k_{66} J^{k\tau s} \bar{\alpha}^2 \\
K^{k\tau s}_{uu_{23}} &= Q^k_{44} J^{k\tau_z s} \bar{\beta} - Q^k_{23} J^{k\tau s_z} \bar{\beta} \\
K^{k\tau s}_{uu_{31}} &= Q^k_{55} J^{k\tau s_z} \bar{\alpha} - Q^k_{13} J^{k\tau_z s} \bar{\alpha} \\
K^{k\tau s}_{uu_{32}} &= Q^k_{44} J^{k\tau s_z} \bar{\beta} - Q^k_{23} J^{k\tau_z s} \bar{\beta} \\
K^{k\tau s}_{uu_{33}} &= Q^k_{55} J^{k\tau s} \bar{\alpha}^2 + Q^k_{44} J^{k\tau s} \bar{\beta}^2 + Q^k_{33} J^{k\tau_z s_z}
\end{aligned} \tag{6.42}$$

The fundamental nucleus $\boldsymbol{K}^{k\tau s}_{u\Phi}$ is of (3×1) dimension because Φ^k is scalar:

$$\begin{aligned}
K^{k\tau s}_{u\Phi_{11}} &= \bar{\alpha}(-J^{k\tau s_z} e^k_{31} + J^{k\tau_z s} e^k_{15}) \\
K^{k\tau s}_{u\Phi_{21}} &= \bar{\beta}(J^{k\tau_z s} e^k_{24} - e^k_{32} J^{k\tau s_z}) \\
K^{k\tau s}_{u\Phi_{31}} &= \bar{\alpha}^2 e^k_{15} J^{k\tau s} + \bar{\beta}^2 e^k_{24} J^{k\tau s} + e^k_{33} J^{k\tau_z s_z}
\end{aligned} \tag{6.43}$$

The fundamental nucleus $\boldsymbol{K}^{k\tau s}_{\Phi u}$ is of (1×3) dimension:

$$
\begin{aligned}
K^{k\tau s}_{\Phi u_{11}} &= -\bar{\alpha} e^{k}_{15} J^{k\tau s_z} + \bar{\alpha} e^{k}_{31} J^{k\tau_z s} \\
K^{k\tau s}_{\Phi u_{12}} &= -\bar{\beta} e^{k}_{24} J^{k\tau s_z} + \bar{\beta} e^{k}_{32} J^{k\tau_z s} \\
K^{k\tau s}_{\Phi u_{13}} &= -\bar{\alpha}^2 e^{k}_{15} J^{k\tau s} - \bar{\beta}^2 e^{k}_{24} J^{k\tau s} - e^{k}_{33} J^{k\tau_z s_z}
\end{aligned} \tag{6.44}
$$

The fundamental nucleus $\boldsymbol{K}^{k\tau s}_{\Phi\Phi}$ is of (1×1) dimension:

$$
K^{k\tau s}_{\Phi\Phi_{11}} = J^{k\tau s} \bar{\alpha}^2 \varepsilon_{11} + J^{k\tau s} \bar{\beta}^2 \varepsilon_{22} + \varepsilon_{33} J^{k\tau_z s_z} \tag{6.45}
$$

The fundamental nucleus $\boldsymbol{M}^{k\tau s}_{uu}$ is of (3×3) dimension and only the diagonal elements are different from zero:

$$
M^{k\tau s}_{uu_{11}} = M^{k\tau s}_{uu_{22}} = M^{k\tau s}_{uu_{33}} = \rho^k J^{k\tau s} \tag{6.46}
$$

$\bar{\alpha} = m\pi/a$ and $\bar{\beta} = n\pi/b$, m and n are the wave numbers in in-plane directions, and a and b are the plate dimensions in the x and y directions, respectively. A Navier-type closed-form solution is obtained via substitution of the harmonic expressions for the displacements and electric potential and by considering the following material coefficients equal to zero: $Q_{16} = Q_{26} = Q_{36} = Q_{45} = 0$ and $e_{25} = e_{14} = e_{36} = \varepsilon_{12} = 0$. The harmonic assumptions used for the displacements and the electric potential are:

$$
u^k_\tau = \sum_{m,n} \hat{U}^k_\tau \cos\left(\frac{m\pi x}{a}\right) \sin\left(\frac{n\pi y}{b}\right), \qquad k = 1, N_l \tag{6.47}
$$

$$
v^k_\tau = \sum_{m,n} \hat{V}^k_\tau \sin\left(\frac{m\pi x}{a}\right) \cos\left(\frac{n\pi y}{b}\right), \qquad \tau = t, b, r \tag{6.48}
$$

$$
w^k_\tau = \sum_{m,n} \hat{W}^k_\tau \sin\left(\frac{m\pi x}{a}\right) \sin\left(\frac{n\pi y}{b}\right), \qquad r = 2, N \tag{6.49}
$$

$$
\Phi^k_\tau = \sum_{m,n} \hat{\Phi}^k_\tau \sin\left(\frac{m\pi x}{a}\right) \sin\left(\frac{n\pi y}{b}\right) \tag{6.50}
$$

where $\hat{U}^k_\tau$, $\hat{V}^k_\tau$, $\hat{W}^k_\tau$ are the displacement amplitudes and $\hat{\Phi}^k_\tau$ is the electric potential amplitude; k indicates the layer and N_l the total number of layers. τ is the index for the order of expansion where t and b indicate the top and bottom of the layer, respectively, while r indicates the higher orders of expansion until $N = 4$. Details on the assembly procedure of the fundamental nuclei and on the acronyms are given in Sections 6.7 and 6.8, respectively.

6.4 RMVT(u, Φ, σ_n) for the electromechanical plate case

A possible extension of the RMVT (Reissner 1984) for electromechanical coupling is that indicated in Equation (2.60) in Chapter 2, where the internal electrical work is added (Carrera and Boscolo 2007; Carrera and Brischetto 2007b):

$$\int_V \Big(\delta\epsilon_{pG}^T\sigma_{pC} + \delta\epsilon_{nG}^T\sigma_{nM} + \delta\sigma_{nM}^T(\epsilon_{nG} - \epsilon_{nC})$$

$$- \delta\mathcal{E}_{pG}^T\mathcal{D}_{pC} - \delta\mathcal{E}_{nG}^T\mathcal{D}_{nC}\Big)dV = \delta L_e - \delta L_{in} \qquad (6.51)$$

For RMVT applications, we split the electrical work into in-plane and out-of-plane contributions, and this splitting will be useful for those RMVT extensions in which the transverse normal electric displacement is a primary variable of the problem. By considering a laminate of N_l layers, and the integral on volume V_k of each k layer as an integral on the in-plane domain Ω_k, plus the integral in the thickness-direction domain A_k, it is possible to write Equation (6.51) as:

$$\sum_{k=1}^{N_l}\int_{\Omega_k}\int_{A_k}\Big\{\delta\epsilon_{pG}^{k\,T}\sigma_{pC}^k + \delta\epsilon_{nG}^{k\,T}\sigma_{nM}^k + \delta\sigma_{nM}^{k\,T}(\epsilon_{nG}^k - \epsilon_{nC}^k)$$

$$-\delta\mathcal{E}_{pG}^{k\,T}\mathcal{D}_{pC}^k - \delta\mathcal{E}_{nG}^{k\,T}\mathcal{D}_{nC}^k\Big\}$$

$$d\Omega_k dz = \sum_{k=1}^{N_l}\delta L_e^k - \sum_{k=1}^{N_l}\delta L_{in}^k \qquad (6.52)$$

where δL_e^k and δL_{in}^k are the external and inertial virtual work at the k-layer level, respectively. The relative constitutive equations are those obtained in Equations (2.61)–(2.64) and considering the transverse stresses σ_n as modeled (M) and the transverse strains ϵ_n as obtained from constitutive equations (C) (D'Ottavio and Kröplin 2006):

$$\sigma_{pC}^k = \hat{C}_{\sigma_p\epsilon_p}^k\epsilon_{pG}^k + \hat{C}_{\sigma_p\sigma_n}^k\sigma_{nM}^k + \hat{C}_{\sigma_p\mathcal{E}_p}^k\mathcal{E}_{pG}^k + \hat{C}_{\sigma_p\mathcal{E}_n}^k\mathcal{E}_{nG}^k \qquad (6.53)$$

$$\epsilon_{nC}^k = \hat{C}_{\epsilon_n\epsilon_p}^k\epsilon_{pG}^k + \hat{C}_{\epsilon_n\sigma_n}^k\sigma_{nM}^k + \hat{C}_{\epsilon_n\mathcal{E}_p}^k\mathcal{E}_{pG}^k + \hat{C}_{\epsilon_n\mathcal{E}_n}^k\mathcal{E}_{nG}^k \qquad (6.54)$$

$$\mathcal{D}_{pC}^k = \hat{C}_{\mathcal{D}_p\epsilon_p}^k\epsilon_{pG}^k + \hat{C}_{\mathcal{D}_p\sigma_n}^k\sigma_{nM}^k + \hat{C}_{\mathcal{D}_p\mathcal{E}_p}^k\mathcal{E}_{pG}^k + \hat{C}_{\mathcal{D}_p\mathcal{E}_n}^k\mathcal{E}_{nG}^k \qquad (6.55)$$

$$\mathcal{D}_{nC}^k = \hat{C}_{\mathcal{D}_n\epsilon_p}^k\epsilon_{pG}^k + \hat{C}_{\mathcal{D}_n\sigma_n}^k\sigma_{nM}^k + \hat{C}_{\mathcal{D}_n\mathcal{E}_p}^k\mathcal{E}_{pG}^k + \hat{C}_{\mathcal{D}_n\mathcal{E}_n}^k\mathcal{E}_{nG}^k \qquad (6.56)$$

The meaning of the $\hat{\boldsymbol{C}}$ coefficients has already been given in Equations (2.65). Substituting Equations (6.53)–(6.56), and the geometrical relations (2.38)-(2.41) in Chapter 2 for plates, in the variational statement of Equation (6.52), and considering a generic layer k (Carrera and Boscolo 2007; Carrera *et al.* 2007; Carrera and Brischetto 2007b), we obtain:

$$\begin{aligned}
&\int_{\Omega_k}\int_{A_k}\Big[\left(\boldsymbol{D}_p F_s \delta \boldsymbol{u}_s^k\right)^T \left(\hat{\boldsymbol{C}}^k_{\sigma_p\epsilon_p}\boldsymbol{D}_p F_\tau \boldsymbol{u}_\tau^k + \hat{\boldsymbol{C}}^k_{\sigma_p\sigma_n} F_\tau \boldsymbol{\sigma}^k_{nM\tau} - \hat{\boldsymbol{C}}^k_{\sigma_p\mathcal{E}_p}\boldsymbol{D}_{ep} F_\tau \Phi_\tau^k\right.\\
&\left.-\hat{\boldsymbol{C}}^k_{\sigma_p\mathcal{E}_n}\boldsymbol{D}_{en} F_\tau \Phi_\tau^k\right) + \left((\boldsymbol{D}_{np}+\boldsymbol{D}_{nz})F_s\delta\boldsymbol{u}_s^k\right)^T\left(F_\tau \boldsymbol{\sigma}^k_{nM\tau}\right) + \left(F_s \delta\boldsymbol{\sigma}^k_{nMs}\right)^T\\
&\times\left((\boldsymbol{D}_{np}+\boldsymbol{D}_{nz})F_\tau \boldsymbol{u}_\tau^k - \hat{\boldsymbol{C}}^k_{\epsilon_n\epsilon_p}\boldsymbol{D}_p F_\tau \boldsymbol{u}_\tau^k - \hat{\boldsymbol{C}}^k_{\epsilon_n\sigma_n}F_\tau \boldsymbol{\sigma}^k_{nM\tau} + \hat{\boldsymbol{C}}^k_{\epsilon_n\mathcal{E}_p}\boldsymbol{D}_{ep}F_\tau\Phi_\tau^k\right.\\
&\left.+\hat{\boldsymbol{C}}^k_{\epsilon_n\mathcal{E}_n}\boldsymbol{D}_{en}F_\tau\Phi_\tau^k\right) + \left(\boldsymbol{D}_{ep}F_s\delta\Phi_s^k\right)^T\left(\hat{\boldsymbol{C}}^k_{\mathcal{D}_p\epsilon_p}\boldsymbol{D}_p F_\tau \boldsymbol{u}_\tau^k + \hat{\boldsymbol{C}}^k_{\mathcal{D}_p\sigma_n}F_\tau\boldsymbol{\sigma}^k_{nM\tau}\right.\\
&\left.-\hat{\boldsymbol{C}}^k_{\mathcal{D}_p\mathcal{E}_p}\boldsymbol{D}_{ep}F_\tau\Phi_\tau^k - \hat{\boldsymbol{C}}^k_{\mathcal{D}_p\mathcal{E}_n}\boldsymbol{D}_{en}F_\tau\Phi_\tau^k\right) + \left(\boldsymbol{D}_{en}F_s\delta\Phi_s^k\right)^T\left(\hat{\boldsymbol{C}}^k_{\mathcal{D}_n\epsilon_p}\boldsymbol{D}_p F_\tau\boldsymbol{u}_\tau^k\right.\\
&\left.+\hat{\boldsymbol{C}}^k_{\mathcal{D}_n\sigma_n}F_\tau\boldsymbol{\sigma}^k_{nM\tau} - \hat{\boldsymbol{C}}^k_{\mathcal{D}_n\mathcal{E}_p}\boldsymbol{D}_{ep}F_\tau\Phi_\tau^k - \hat{\boldsymbol{C}}^k_{\mathcal{D}_n\mathcal{E}_n}\boldsymbol{D}_{en}F_\tau\Phi_\tau^k\right)\Big]d\Omega_k\,dz\\
&= \delta L_e^k - \delta L_{in}^k
\end{aligned} \tag{6.57}$$

The CUF (Carrera 1995), as presented in the previous sections for the 2D approximation, has already been introduced. In order to obtain a strong form of differential equations on domain Ω_k and the relative boundary conditions at edge Γ_k in Equation (6.57), integration by parts is used, and this permits the differential operator to be moved from the infinitesimal variation of the generic variable $\delta\boldsymbol{a}^k$ to the finite quantity $\boldsymbol{a}^k$ (Carrera 1995). For a generic variable $\boldsymbol{a}^k$, the integration by parts is given in Equation (6.28) with the matrices of Equations (6.29). The governing equations have the following form (Carrera and Boscolo 2007; Carrera *et al.* 2007; Carrera and Brischetto 2007b):

$$\begin{aligned}
\delta\boldsymbol{u}_s^k:&\quad \boldsymbol{K}^{k\tau s}_{uu}\,\boldsymbol{u}_\tau^k + \boldsymbol{K}^{k\tau s}_{u\sigma}\,\boldsymbol{\sigma}^k_{nM\tau} + \boldsymbol{K}^{k\tau s}_{u\Phi}\,\Phi_\tau^k = \boldsymbol{p}^k_{us} - \boldsymbol{M}^{k\tau s}_{uu}\,\ddot{\boldsymbol{u}}_\tau^k\\
\delta\boldsymbol{\sigma}^k_{ns}:&\quad \boldsymbol{K}^{k\tau s}_{\sigma u}\,\boldsymbol{u}_\tau^k + \boldsymbol{K}^{k\tau s}_{\sigma\sigma}\,\boldsymbol{\sigma}^k_{nM\tau} + \boldsymbol{K}^{k\tau s}_{\sigma\Phi}\,\Phi_\tau^k = 0\\
\delta\Phi_s^k:&\quad \boldsymbol{K}^{k\tau s}_{\Phi u}\,\boldsymbol{u}_\tau^k + \boldsymbol{K}^{k\tau s}_{\Phi\sigma}\,\boldsymbol{\sigma}^k_{nM\tau} + \boldsymbol{K}^{k\tau s}_{\Phi\Phi}\,\Phi_\tau^k = 0
\end{aligned} \tag{6.58}$$

where $\boldsymbol{M}^{k\tau s}_{uu}$ is the inertial contribution in the form of the fundamental nucleus, $\boldsymbol{u}_\tau^k$ is the vector of the degrees of freedom for the displacements, Φ_τ^k is the vector of the degrees of freedom for the electric potential, $\boldsymbol{\sigma}^k_{nM\tau}$ is the vector of the degrees of freedom for the transverse stresses, and $\ddot{\boldsymbol{u}}_\tau^k$ is the second temporal derivative of $\boldsymbol{u}_\tau^k$. The $\boldsymbol{p}^k_{us}$ array indicates the variationally consistent mechanical loading used for the sensor configuration. In the case of the actuator configuration, the electric potential is directly imposed in the vector Φ^k_τ. Along with these governing equations, the following boundary conditions on the edge Γ_k of the in-plane integration domain Ω_k hold:

$$
\begin{aligned}
\mathbf{\Pi}^{k\tau s}_{uu}\, \boldsymbol{u}^k_\tau + \mathbf{\Pi}^{k\tau s}_{u\sigma}\, \boldsymbol{\sigma}^k_{nM\tau} + \mathbf{\Pi}^{k\tau s}_{u\Phi}\, \Phi^k_\tau &= \mathbf{\Pi}^{k\tau s}_{uu}\, \bar{\boldsymbol{u}}^k_\tau + \mathbf{\Pi}^{k\tau s}_{u\sigma}\, \bar{\boldsymbol{\sigma}}^k_{nM\tau} + \mathbf{\Pi}^{k\tau s}_{u\Phi}\, \bar{\Phi}^k_\tau \\
\mathbf{\Pi}^{k\tau s}_{\Phi u}\, \boldsymbol{u}^k_\tau + \mathbf{\Pi}^{k\tau s}_{\Phi\sigma}\, \boldsymbol{\sigma}^k_{nM\tau} + \mathbf{\Pi}^{k\tau s}_{\Phi\Phi}\, \Phi^k_\tau &= \mathbf{\Pi}^{k\tau s}_{\Phi u}\, \bar{\boldsymbol{u}}^k_\tau + \mathbf{\Pi}^{k\tau s}_{\Phi\sigma}\, \bar{\boldsymbol{\sigma}}^k_{nM\tau} + \mathbf{\Pi}^{k\tau s}_{\Phi\Phi}\, \bar{\Phi}^k_\tau
\end{aligned}
\tag{6.59}
$$

Comparing Equation (6.57), after the integration by parts, to the Equations (6.58) and (6.59), the fundamental nuclei can be obtained:

$$\boldsymbol{K}^{k\tau s}_{uu} = \int_{A_k} \Big[-\boldsymbol{D}^T_p \hat{\boldsymbol{C}}^k_{\sigma_p\epsilon_p} \boldsymbol{D}_p \Big] F_s F_\tau dz \tag{6.60}$$

$$\boldsymbol{K}^{k\tau s}_{u\sigma} = \int_{A_k} \Big[-\boldsymbol{D}^T_p \hat{\boldsymbol{C}}^k_{\sigma_p\sigma_n} + (-\boldsymbol{D}_{np} + \boldsymbol{D}_{nz})^T \Big] F_s F_\tau dz \tag{6.61}$$

$$\boldsymbol{K}^{k\tau s}_{u\Phi} = \int_{A_k} \Big[-\boldsymbol{D}^T_p (-\hat{\boldsymbol{C}}^k_{\sigma_p\mathcal{E}_p} \boldsymbol{D}_{ep} - \hat{\boldsymbol{C}}^k_{\sigma_p\mathcal{E}_n} \boldsymbol{D}_{en}) \Big] F_s F_\tau dz \tag{6.62}$$

$$\boldsymbol{K}^{k\tau s}_{\sigma u} = \int_{A_k} \Big[(\boldsymbol{D}_{np} + \boldsymbol{D}_{nz}) - \hat{\boldsymbol{C}}^k_{\epsilon_n\epsilon_p} \boldsymbol{D}_p \Big] F_s F_\tau dz \tag{6.63}$$

$$\boldsymbol{K}^{k\tau s}_{\sigma\sigma} = \int_{A_k} \Big[-\hat{\boldsymbol{C}}^k_{\epsilon_n\sigma_n} \Big] F_s F_\tau dz \tag{6.64}$$

$$\boldsymbol{K}^{k\tau s}_{\sigma\Phi} = \int_{A_k} \Big[\hat{\boldsymbol{C}}^k_{\epsilon_n\mathcal{E}_p} \boldsymbol{D}_{ep} + \hat{\boldsymbol{C}}^k_{\epsilon_n\mathcal{E}_n} \boldsymbol{D}_{en} \Big] F_s F_\tau dz \tag{6.65}$$

$$\boldsymbol{K}^{k\tau s}_{\Phi u} = \int_{A_k} \Big[(-\boldsymbol{D}_{ep}{}^T \hat{\boldsymbol{C}}^k_{\mathcal{D}_p\epsilon_p} + \boldsymbol{D}_{en}{}^T \hat{\boldsymbol{C}}^k_{\mathcal{D}_n\epsilon_p}) \boldsymbol{D}_p \Big] F_s F_\tau dz \tag{6.66}$$

$$\boldsymbol{K}^{k\tau s}_{\Phi\sigma} = \int_{A_k} \Big[-\boldsymbol{D}_{ep}{}^T \hat{\boldsymbol{C}}^k_{\mathcal{D}_p\sigma_n} + \boldsymbol{D}_{en}{}^T \hat{\boldsymbol{C}}^k_{\mathcal{D}_n\sigma_n} \Big] F_s F_\tau dz \tag{6.67}$$

$$
\begin{aligned}
\boldsymbol{K}^{k\tau s}_{\Phi\Phi} = \int_{A_k} \Big[& -\boldsymbol{D}_{ep}{}^T (-\hat{\boldsymbol{C}}^k_{\mathcal{D}_p\mathcal{E}_p} \boldsymbol{D}_{ep} - \hat{\boldsymbol{C}}^k_{\mathcal{D}_p\mathcal{E}_n} \boldsymbol{D}_{en}) \\
& + \boldsymbol{D}_{en}{}^T (-\hat{\boldsymbol{C}}^k_{\mathcal{D}_n\mathcal{E}_p} \boldsymbol{D}_{ep} - \hat{\boldsymbol{C}}^k_{\mathcal{D}_n\mathcal{E}_n} \boldsymbol{D}_{en}) \Big] F_s F_\tau dz
\end{aligned}
\tag{6.68}
$$

The nuclei for the boundary conditions on edge Γ_k are (D'Ottavio and Kröplin 2006; Carrera and Brischetto 2007b):

$$\boldsymbol{\Pi}_{uu}^{k\tau s} = \int_{A_k} \left[\boldsymbol{I}_p^T \hat{\boldsymbol{C}}^k_{\sigma_p \epsilon_p} \boldsymbol{D}_p \right] F_s F_\tau dz \tag{6.69}$$

$$\boldsymbol{\Pi}_{u\sigma}^{k\tau s} = \int_{A_k} \left[\boldsymbol{I}_p^T \hat{\boldsymbol{C}}^k_{\sigma_p \sigma_n} + \boldsymbol{I}_{np}^T \right] F_s F_\tau dz \tag{6.70}$$

$$\boldsymbol{\Pi}_{u\Phi}^{k\tau s} = \int_{A_k} \left[\boldsymbol{I}_p^T (-\hat{\boldsymbol{C}}^k_{\sigma_p \mathcal{E}_p} \boldsymbol{D}_{ep} - \hat{\boldsymbol{C}}^k_{\sigma_p \mathcal{E}_n} \boldsymbol{D}_{en}) \right] F_s F_\tau dz \tag{6.71}$$

$$\boldsymbol{\Pi}_{\Phi u}^{k\tau s} = \int_{A_k} \left[\boldsymbol{I}_{ep}^T \hat{\boldsymbol{C}}^k_{\mathcal{D}_p \epsilon_p} \boldsymbol{D}_p \right] F_s F_\tau dz \tag{6.72}$$

$$\boldsymbol{\Pi}_{\Phi\sigma}^{k\tau s} = \int_{A_k} \left[\boldsymbol{I}_{ep}^T \hat{\boldsymbol{C}}^k_{\mathcal{D}_p \sigma_n} \right] F_s F_\tau dz \tag{6.73}$$

$$\boldsymbol{\Pi}_{\Phi\Phi}^{k\tau s} = \int_{A_k} \left[\boldsymbol{I}_{ep}^T (-\hat{\boldsymbol{C}}^k_{\mathcal{D}_p \mathcal{E}_p} \boldsymbol{D}_{ep} - \hat{\boldsymbol{C}}^k_{\mathcal{D}_p \mathcal{E}_n} \boldsymbol{D}_{en}) \right] F_s F_\tau dz \tag{6.74}$$

In order to perform the integration by parts (see Equation (6.28)), the matrices $\boldsymbol{I}_p$, $\boldsymbol{I}_{np}$, and $\boldsymbol{I}_{ep}$ which are presented in Equation (6.29) must be introduced. To write the explicit form of the nuclei in Equations (6.60)–(6.68), the integrals in the z thickness- direction are defined as in Equation (6.41). By developing the matrix products in Equations (6.60)–(6.68) and employing a Navier-type closed-form solution (Carrera and Brischetto 2007b), the explicit algebraic form of the nuclei can be obtained.

Nucleus $\boldsymbol{K}_{uu}^{k\tau s}$ of dimension (3×3) is:

$$\begin{aligned}
K_{uu_{11}}^{k\tau s} &= \bar{\alpha}^2 J^{k\tau s} \hat{C}^k_{\sigma_p \epsilon_p 11} + \bar{\beta}^2 J^{k\tau s} \hat{C}^k_{\sigma_p \epsilon_p 33} \\
K_{uu_{12}}^{k\tau s} &= J^{k\tau s} (\hat{C}^k_{\sigma_p \epsilon_p 12} + \hat{C}^k_{\sigma_p \epsilon_p 33}) \bar{\alpha} \bar{\beta} \\
K_{uu_{13}}^{k\tau s} &= 0, \quad K_{uu_{21}}^{k\tau s} = J^{k\tau s} (\hat{C}^k_{\sigma_p \epsilon_p 21} + \hat{C}^k_{\sigma_p \epsilon_p 33}) \bar{\alpha} \bar{\beta} \\
K_{uu_{22}}^{k\tau s} &= \bar{\beta}^2 J^{k\tau s} \hat{C}^k_{\sigma_p \epsilon_p 22} + \bar{\alpha}^2 J^{k\tau s} \hat{C}^k_{\sigma_p \epsilon_p 33}, \quad K_{uu_{23}}^{k\tau s} = 0 \\
K_{uu_{31}}^{k\tau s} &= 0, \quad K_{uu_{32}}^{k\tau s} = 0, \quad K_{uu_{33}}^{k\tau s} = 0
\end{aligned} \tag{6.75}$$

Nucleus $\boldsymbol{K}_{u\sigma}^{k\tau s}$ of dimension (3×3) is:

$$\begin{aligned}
K_{u\sigma_{11}}^{k\tau s} &= J^{k\tau s_z}, \quad K_{u\sigma_{12}}^{k\tau s} = 0, \quad K_{u\sigma_{13}}^{k\tau s} = -\bar{\alpha} J^{k\tau s} \hat{C}^k_{\sigma_p \sigma_n 13} \\
K_{u\sigma_{21}}^{k\tau s} &= 0, \quad K_{u\sigma_{22}}^{k\tau s} = J^{k\tau s_z}, \quad K_{u\sigma_{23}}^{k\tau s} = -\bar{\beta} J^{k\tau s} \hat{C}^k_{\sigma_p \sigma_n 23} \\
K_{u\sigma_{31}}^{k\tau s} &= \bar{\alpha} J^{k\tau s}, \quad K_{u\sigma_{32}}^{k\tau s} = \bar{\beta} J^{k\tau s}, \quad K_{u\sigma_{33}}^{k\tau s} = J^{k\tau s_z}
\end{aligned} \tag{6.76}$$

Nucleus $\boldsymbol{K}_{u\Phi}^{k\tau s}$ of dimension (3×1) is:

$$K_{u\Phi_{11}}^{k\tau s} = \bar{\alpha} J^{k\tau_z s} \hat{C}_{\sigma_p \mathcal{E}_n 11}^{k}, \quad K_{u\Phi_{21}}^{k\tau s} = \bar{\beta} J^{k\tau_z s} \hat{C}_{\sigma_p \mathcal{E}_n 21}^{k}, \quad K_{u\Phi_{31}}^{k\tau s} = 0 \tag{6.77}$$

Nucleus $\boldsymbol{K}_{\sigma u}^{k\tau s}$ of dimension (3×3) is:

$$\begin{aligned}
K_{\sigma u_{11}}^{k\tau s} &= J^{k\tau_z s}, \quad K_{\sigma u_{12}}^{k\tau s} = 0, \quad K_{\sigma u_{13}}^{k\tau s} = \bar{\alpha} J^{k\tau s} \\
K_{\sigma u_{21}}^{k\tau s} &= 0, \quad K_{\sigma u_{22}}^{k\tau s} = J^{k\tau_z s}, \quad K_{u\sigma_{23}}^{k\tau s} = \bar{\beta} J^{k\tau s}, \\
K_{\sigma u_{31}}^{k\tau s} &= \bar{\alpha} J^{k\tau s} \hat{C}_{\epsilon_n \epsilon_p 31}^{k} \\
K_{\sigma u_{32}}^{k\tau s} &= \bar{\beta} J^{k\tau s} \hat{C}_{\epsilon_n \epsilon_p 32}^{k}, \quad K_{\sigma u_{33}}^{k\tau s} = J^{k\tau_z s}
\end{aligned} \tag{6.78}$$

Nucleus $\boldsymbol{K}_{\sigma\sigma}^{k\tau s}$ of dimension (3×3) is:

$$\begin{aligned}
K_{\sigma\sigma_{11}}^{k\tau s} &= -J^{k\tau s} \hat{C}_{\epsilon_n \sigma_n 11}^{k}, \quad K_{\sigma\sigma_{12}}^{k\tau s} = 0, \quad K_{\sigma\sigma_{13}}^{k\tau s} = 0 \\
K_{\sigma\sigma_{21}}^{k\tau s} &= 0, \quad K_{\sigma\sigma_{22}}^{k\tau s} = -J^{k\tau s} \hat{C}_{\epsilon_n \sigma_n 22}^{k}, \quad K_{\sigma\sigma_{23}}^{k\tau s} = 0 \\
K_{\sigma\sigma_{31}}^{k\tau s} &= 0, \quad K_{\sigma\sigma_{32}}^{k\tau s} = 0, \quad K_{\sigma\sigma_{33}}^{k\tau s} = -J^{k\tau s} \hat{C}_{\epsilon_n \sigma_n 33}^{k}
\end{aligned} \tag{6.79}$$

Nucleus $\boldsymbol{K}_{\sigma\Phi}^{k\tau s}$ of dimension (3×1) is:

$$K_{\sigma\Phi_{11}}^{k\tau s} = \bar{\alpha} J^{k\tau s} \hat{C}_{\epsilon_n \mathcal{E}_p 11}^{k}, \quad K_{\sigma\Phi_{21}}^{k\tau s} = \bar{\beta} J^{k\tau s} \hat{C}_{\epsilon_n \mathcal{E}_p 22}^{k}, \quad K_{\sigma\Phi_{31}}^{k\tau s} = J^{k\tau_z s} \hat{C}_{\epsilon_n \mathcal{E}_n 31}^{k} \tag{6.80}$$

Nucleus $\boldsymbol{K}_{\Phi u}^{k\tau s}$ of dimension (1×3) is:

$$K_{\Phi u_{11}}^{k\tau s} = -\bar{\alpha} J^{k\tau s_z} \hat{C}_{\mathcal{D}_n \epsilon_p 11}^{k}, \quad K_{\Phi u_{12}}^{k\tau s} = -\bar{\beta} J^{k\tau s_z} \hat{C}_{\mathcal{D}_n \epsilon_p 12}^{k}, \quad K_{\Phi u_{13}}^{k\tau s} = 0 \tag{6.81}$$

Nucleus $\boldsymbol{K}_{\Phi\sigma}^{k\tau s}$ of dimension (1×3) is:

$$K_{\Phi\sigma_{11}}^{k\tau s} = \bar{\alpha} J^{k\tau s} \hat{C}_{\mathcal{D}_n \sigma_n 11}^{k}, \quad K_{\Phi\sigma_{12}}^{k\tau s} = \bar{\beta} J^{k\tau s} \hat{C}_{\mathcal{D}_n \sigma_n 22}^{k}, \quad K_{\Phi\sigma_{13}}^{k\tau s} = J^{k\tau s_z} \hat{C}_{\mathcal{D}_n \sigma_n 13}^{k} \tag{6.82}$$

Nucleus $\boldsymbol{K}_{\Phi\Phi}^{k\tau s}$ of dimension (1×1) is:

$$K_{\Phi\Phi_{11}}^{k\tau s} = -J^{k\tau_z s_z} \hat{C}_{\mathcal{D}_n \mathcal{E}_n 11}^{k} - \bar{\alpha}^2 J^{k\tau s} \hat{C}_{\mathcal{D}_p \mathcal{E}_p 11}^{k} - \bar{\beta}^2 J^{k\tau s} \hat{C}_{\mathcal{D}_p \mathcal{E}_p 22}^{k} \tag{6.83}$$

The fundamental nucleus for the inertial matrix $\boldsymbol{M}_{uu}^{k\tau s}$ and its components were given in Equations (6.36) and (6.46).

$\bar{\alpha} = m\pi/a$ and $\bar{\beta} = n\pi/b$, where m and n are the wave numbers in the in-plane directions and a and b are the plate dimensions in the x and y directions, respectively. A Navier-type closed-form solution is obtained via substitution of the harmonic expressions for the displacements, electric potential, and transverse stresses and by considering the following material coefficients to be equal to zero: $Q_{16} = Q_{26} = Q_{36} = Q_{45} = 0$ and $e_{25} = e_{14} = e_{36} = \varepsilon_{12} = 0$. The harmonic assumptions used for the displacements, the electric potential, and the transverse stresses are:

$$(u_\tau^k, \sigma_{xz\tau}^k) = \sum_{m,n} (\hat{U}_\tau^k, \hat{\sigma}_{xz\tau}^k) \cos\left(\frac{m\pi x}{a}\right) \sin\left(\frac{n\pi y}{b}\right), \quad k = 1, N_l \tag{6.84}$$

$$(v_\tau^k, \sigma_{yz\tau}^k) = \sum_{m,n} (\hat{V}_\tau^k, \hat{\sigma}_{yz\tau}^k) \sin\left(\frac{m\pi x}{a}\right) \cos\left(\frac{n\pi y}{b}\right), \quad \tau = t, b, r \tag{6.85}$$

$$(w_\tau^k, \sigma_{zz\tau}^k, \Phi_\tau^k) = \sum_{m,n} (\hat{W}_\tau^k, \hat{\sigma}_{zz\tau}^k, \hat{\Phi}_\tau^k) \sin\left(\frac{m\pi x}{a}\right) \sin\left(\frac{n\pi y}{b}\right), \quad r = 2, N \tag{6.86}$$

where $\hat{U}_\tau^k$, $\hat{V}_\tau^k$, $\hat{W}_\tau^k$ are the displacement amplitudes, $\hat{\Phi}_\tau^k$ is the electric potential amplitude, and $\hat{\sigma}_{xz\tau}^k$, $\hat{\sigma}_{yz\tau}^k$, and $\hat{\sigma}_{zz\tau}^k$ are the transverse stress amplitudes; k indicates the layer and N_l is the total number of layers. τ is the index for the order of expansion, where t and b indicate the top and bottom of the layer, respectively, while r indicates the higher orders of expansion until $N = 4$. Details on the assembly procedure of the fundamental nuclei and on the acronyms used are given in Sections 6.7 and 6.8, respectively.

The meaning of the $\hat{\boldsymbol{C}}$ coefficients for Equations (6.53)–(6.56) is given in Equations (2.65). The following equations give the explicit form of each of their components.

The $\hat{\boldsymbol{C}}_{\sigma_p\epsilon_p}^k$ array has a (3×3) dimension with components for each k layer:

$$\begin{aligned}
&\hat{C}_{\sigma_p\epsilon_p11} = Q_{11} - \frac{Q_{13}^2}{Q_{33}}, \quad \hat{C}_{\sigma_p\epsilon_p12} = Q_{12} - \frac{Q_{13}Q_{23}}{Q_{33}}, \quad \hat{C}_{\sigma_p\epsilon_p13} = 0 \\
&\hat{C}_{\sigma_p\epsilon_p21} = Q_{12} - \frac{Q_{13}Q_{23}}{Q_{33}}, \quad \hat{C}_{\sigma_p\epsilon_p22} = Q_{22} - \frac{Q_{23}^2}{Q_{33}}, \quad \hat{C}_{\sigma_p\epsilon_p23} = 0 \\
&\hat{C}_{\sigma_p\epsilon_p31} = 0 \\
&\hat{C}_{\sigma_p\epsilon_p32} = 0, \quad \hat{C}_{\sigma_p\epsilon_p33} = Q_{66}
\end{aligned} \tag{6.87}$$

The $\hat{C}^k_{\sigma_p\sigma_n}$ array has a (3 × 3) dimension with components for each k layer:

$$\begin{aligned}
&\hat{C}_{\sigma_p\sigma_n 11} = 0, \quad \hat{C}_{\sigma_p\sigma_n 12} = 0, \quad \hat{C}_{\sigma_p\sigma_n 13} = \frac{Q_{13}}{Q_{33}}, \quad \hat{C}_{\sigma_p\sigma_n 21} = 0 \\
&\hat{C}_{\sigma_p\sigma_n 22} = 0, \quad \hat{C}_{\sigma_p\sigma_n 23} = \frac{Q_{23}}{Q_{33}}, \quad \hat{C}_{\sigma_p\sigma_n 31} = 0, \quad \hat{C}_{\sigma_p\sigma_n 32} = 0 \\
&\hat{C}_{\sigma_p\sigma_n 33} = 0
\end{aligned} \tag{6.88}$$

The $\hat{C}^k_{\sigma_p\mathcal{E}_p}$ array has a (3 × 2) dimension with components for each k layer:

$$\hat{C}_{\sigma_p\mathcal{E}_p 11} = \hat{C}_{\sigma_p\mathcal{E}_p 12} = \hat{C}_{\sigma_p\mathcal{E}_p 21} = \hat{C}_{\sigma_p\mathcal{E}_p 22} = \hat{C}_{\sigma_p\mathcal{E}_p 31} = \hat{C}_{\sigma_p\mathcal{E}_p 32} = 0 \tag{6.89}$$

The $\hat{C}^k_{\sigma_p\mathcal{E}_n}$ array has a (3 × 1) dimension with components for each k layer:

$$\hat{C}_{\sigma_p\mathcal{E}_n 11} = -e_{31} + \frac{e_{33}Q_{13}}{Q_{33}}, \quad \hat{C}_{\sigma_p\mathcal{E}_n 21} = -e_{32} + \frac{e_{33}Q_{23}}{Q_{33}}, \quad \hat{C}_{\sigma_p\mathcal{E}_n 31} = 0 \tag{6.90}$$

The $\hat{C}^k_{\epsilon_n\epsilon_p}$ array has a (3 × 3) dimension with components for each k layer:

$$\begin{aligned}
&\hat{C}_{\epsilon_n\epsilon_p 11} = \hat{C}_{\epsilon_n\epsilon_p 12} = \hat{C}_{\epsilon_n\epsilon_p 13} = \hat{C}_{\epsilon_n\epsilon_p 21} = \hat{C}_{\epsilon_n\epsilon_p 22} = \hat{C}_{\epsilon_n\epsilon_p 23} = 0 \\
&\hat{C}_{\epsilon_n\epsilon_p 31} = -\frac{Q_{13}}{Q_{33}}, \quad \hat{C}_{\epsilon_n\epsilon_p 32} = -\frac{Q_{23}}{Q_{33}}, \quad \hat{C}_{\epsilon_n\epsilon_p 33} = 0
\end{aligned} \tag{6.91}$$

The $\hat{C}^k_{\epsilon_n\sigma_n}$ array has a (3 × 3) dimension with components for each k layer:

$$\begin{aligned}
&\hat{C}_{\epsilon_n\sigma_n 11} = \frac{1}{Q_{55}}, \quad \hat{C}_{\epsilon_n\sigma_n 12} = \hat{C}_{\epsilon_n\sigma_n 13} = \hat{C}_{\epsilon_n\sigma_n 21} = 0, \quad \hat{C}_{\epsilon_n\sigma_n 22} = \frac{1}{Q_{44}} \\
&\hat{C}_{\epsilon_n\sigma_n 23} = \hat{C}_{\epsilon_n\sigma_n 31} = \hat{C}_{\epsilon_n\sigma_n 32} = 0, \quad \hat{C}_{\epsilon_n\sigma_n 33} = \frac{1}{Q_{33}}
\end{aligned} \tag{6.92}$$

The $\hat{C}^k_{\epsilon_n\mathcal{E}_p}$ array has a (3 × 2) dimension with components for each k layer:

$$\begin{aligned}
&\hat{C}_{\epsilon_n\mathcal{E}_p 11} = \frac{e_{15}}{Q_{55}}, \quad \hat{C}_{\epsilon_n\mathcal{E}_p 12} = \hat{C}_{\epsilon_n\mathcal{E}_p 21} = 0, \quad \hat{C}_{\epsilon_n\mathcal{E}_p 22} = \frac{e_{24}}{Q_{44}} \\
&\hat{C}_{\epsilon_n\mathcal{E}_p 31} = \hat{C}_{\epsilon_n\mathcal{E}_p 32} = 0
\end{aligned} \tag{6.93}$$

The $\hat{\boldsymbol{C}}^k_{\epsilon_n\mathcal{E}_n}$ array has a (3 × 1) dimension with components for each k layer:

$$\hat{C}_{\epsilon_n\mathcal{E}_n 11} = \hat{C}_{\epsilon_n\mathcal{E}_n 21} = 0, \quad \hat{C}_{\epsilon_n\mathcal{E}_n 31} = \frac{e_{33}}{Q_{33}} \tag{6.94}$$

The $\hat{\boldsymbol{C}}^k_{\mathcal{D}_p\epsilon_p}$ array has a (2 × 3) dimension with components for each k layer:

$$\hat{C}_{\mathcal{D}_p\epsilon_p 11} = \hat{C}_{\mathcal{D}_p\epsilon_p 12} = \hat{C}_{\mathcal{D}_p\epsilon_p 13} = \hat{C}_{\mathcal{D}_p\epsilon_p 21} = \hat{C}_{\mathcal{D}_p\epsilon_p 22} = \hat{C}_{\mathcal{D}_p\epsilon_p 23} = 0 \tag{6.95}$$

The $\hat{\boldsymbol{C}}^k_{\mathcal{D}_p\sigma_n}$ array has a (2 × 3) dimension with components for each k layer:

$$\begin{aligned} &\hat{C}_{\mathcal{D}_p\sigma_n 11} = \frac{e_{15}}{Q_{55}}, \quad \hat{C}_{\mathcal{D}_p\sigma_n 12} = \hat{C}_{\mathcal{D}_p\sigma_n 13} = \hat{C}_{\mathcal{D}_p\sigma_n 21} = 0, \\ &\hat{C}_{\mathcal{D}_p\sigma_n 22} = \frac{e_{24}}{Q_{44}}, \hat{C}_{\mathcal{D}_p\sigma_n 23} = 0 \end{aligned} \tag{6.96}$$

The $\hat{\boldsymbol{C}}^k_{\mathcal{D}_p\mathcal{E}_p}$ array has a (2 × 2) dimension with components for each k layer:

$$\hat{C}_{\mathcal{D}_p\mathcal{E}_p 11} = \frac{e_{15}^2}{Q_{55}} + \varepsilon_{11}, \quad \hat{C}_{\mathcal{D}_p\mathcal{E}_p 12} = \hat{C}_{\mathcal{D}_p\mathcal{E}_p 21} = 0, \quad \hat{C}_{\mathcal{D}_p\mathcal{E}_p 22} = \frac{e_{24}^2}{Q_{44}} + \varepsilon_{22} \tag{6.97}$$

The $\hat{\boldsymbol{C}}^k_{\mathcal{D}_p\mathcal{E}_n}$ array has a (2 × 1) dimension with components for each k layer:

$$\hat{C}_{\mathcal{D}_p\mathcal{E}_n 11} = \hat{C}_{\mathcal{D}_p\mathcal{E}_n 21} = 0 \tag{6.98}$$

The $\hat{\boldsymbol{C}}^k_{\mathcal{D}_n\epsilon_p}$ array has a (1 × 3) dimension with components for each k layer:

$$\hat{C}_{\mathcal{D}_n\epsilon_p 11} = e_{31} - \frac{e_{33}Q_{13}}{Q_{33}}, \quad \hat{C}_{\mathcal{D}_n\epsilon_p 12} = e_{32} - \frac{e_{33}Q_{23}}{Q_{33}}, \quad \hat{C}_{\mathcal{D}_n\epsilon_p 13} = 0 \tag{6.99}$$

The $\hat{\boldsymbol{C}}^k_{\mathcal{D}_n\sigma_n}$ array has a (1 × 3) dimension with components for each k layer:

$$\hat{C}_{\mathcal{D}_n\sigma_n 11} = \hat{C}_{\mathcal{D}_n\sigma_n 12} = 0, \quad \hat{C}_{\mathcal{D}_n\sigma_n 13} = \frac{e_{33}}{Q_{33}} \tag{6.100}$$

The $\hat{\boldsymbol{C}}^k_{\mathcal{D}_n\mathcal{E}_p}$ array has a (1 × 2) dimension with components for each k layer:

$$\hat{C}_{\mathcal{D}_n\mathcal{E}_p 11} = \hat{C}_{\mathcal{D}_n\mathcal{E}_p 12} = 0 \tag{6.101}$$

The $\hat{\boldsymbol{C}}^k_{\mathcal{D}_n\mathcal{E}_n}$ array has a (1×1) dimension with components for each k layer:

$$\hat{C}_{\mathcal{D}_n\mathcal{E}_n 11} = \frac{e_{33}^2}{Q_{33}} + \varepsilon_{33} \tag{6.102}$$

6.5 RMVT($\boldsymbol{u}$, $\boldsymbol{\Phi}$, $\boldsymbol{\mathcal{D}}_n$) for the electromechanical plate case

A second possible extension of the RMVT (Reissner 1984) for electromechanical coupling is that indicated in Equation (2.74) in Chapter 2, where the internal electrical work has been considered and a Lagrange multiplier has been added for the transverse normal electric displacement (Carrera *et al.* 2010a,b):

$$\int_V \Big(\delta\boldsymbol{\epsilon}_{pG}^T\boldsymbol{\sigma}_{pC} + \delta\boldsymbol{\epsilon}_{nG}^T\boldsymbol{\sigma}_{nC} - \delta\boldsymbol{\mathcal{E}}_{pG}^T\boldsymbol{\mathcal{D}}_{pC} - \delta\boldsymbol{\mathcal{E}}_{nG}^T\boldsymbol{\mathcal{D}}_{nM}$$
$$-\delta\boldsymbol{\mathcal{D}}_{nM}^T(\boldsymbol{\mathcal{E}}_{nG} - \boldsymbol{\mathcal{E}}_{nC})\Big)dV = \delta L_e - \delta L_{in} \tag{6.103}$$

where the subscript M means a priori modeled variable. Considering a laminate of N_l layers, and the integral on the volume V_k of each k layer as an integral on the in-plane domain Ω_k, plus the integral in the thickness-direction domain A_k, it is possible to write Equation (6.103) as:

$$\sum_{k=1}^{N_l}\int_{\Omega_k}\int_{A_k}\Big\{\delta\boldsymbol{\epsilon}_{pG}^{k\ T}\boldsymbol{\sigma}_{pC}^k + \delta\boldsymbol{\epsilon}_{nG}^{k\ T}\boldsymbol{\sigma}_{nC}^k - \delta\boldsymbol{\mathcal{E}}_{pG}^{k\ T}\boldsymbol{\mathcal{D}}_{pC}^k - \delta\boldsymbol{\mathcal{E}}_{nG}^{k\ T}\boldsymbol{\mathcal{D}}_{nM}^k$$
$$-\delta\boldsymbol{\mathcal{D}}_{nM}^{k\ T}(\boldsymbol{\mathcal{E}}_{nG}^k - \boldsymbol{\mathcal{E}}_{nC}^k)\Big\}$$
$$d\Omega_k dz = \sum_{k=1}^{N_l}\delta L_e^k - \sum_{k=1}^{N_l}\delta L_{in}^k \tag{6.104}$$

where δL_e^k and δL_{in}^k are the external and inertial virtual work at the k-layer level, respectively. The relative constitutive equations are those obtained from Equations (2.75)–(2.78), in which the transverse normal electric displacement $\boldsymbol{\mathcal{D}}_n$ is modeled (M) and the transverse normal electric field $\boldsymbol{\mathcal{E}}_n$ is obtained from constitutive equations (C) (Carrera *et al.* 2010a,b):

$$\boldsymbol{\sigma}_{pC}^k = \bar{\boldsymbol{C}}_{\sigma_p\epsilon_p}^k\boldsymbol{\epsilon}_{pG}^k + \bar{\boldsymbol{C}}_{\sigma_p\epsilon_n}^k\boldsymbol{\epsilon}_{nG}^k + \bar{\boldsymbol{C}}_{\sigma_p\mathcal{E}_p}^k\boldsymbol{\mathcal{E}}_{pG}^k + \bar{\boldsymbol{C}}_{\sigma_p\mathcal{D}_n}^k\boldsymbol{\mathcal{D}}_{nM}^k \tag{6.105}$$

$$\boldsymbol{\sigma}_{nC}^k = \bar{\boldsymbol{C}}_{\sigma_n\epsilon_p}^k\boldsymbol{\epsilon}_{pG}^k + \bar{\boldsymbol{C}}_{\sigma_n\epsilon_n}^k\boldsymbol{\epsilon}_{nG}^k + \bar{\boldsymbol{C}}_{\sigma_n\mathcal{E}_p}^k\boldsymbol{\mathcal{E}}_{pG}^k + \bar{\boldsymbol{C}}_{\sigma_n\mathcal{D}_n}^k\boldsymbol{\mathcal{D}}_{nM}^k \tag{6.106}$$

$$\mathcal{D}^k_{pC} = \bar{C}^k_{\mathcal{D}_p\epsilon_p}\epsilon^k_{pG} + \bar{C}^k_{\mathcal{D}_p\epsilon_n}\epsilon^k_{nG} + \bar{C}^k_{\mathcal{D}_p\mathcal{E}_p}\mathcal{E}^k_{pG} + \bar{C}^k_{\mathcal{D}_p\mathcal{D}_n}\mathcal{D}^k_{nM} \quad (6.107)$$

$$\mathcal{E}^k_{nC} = \bar{C}^k_{\mathcal{E}_n\epsilon_p}\epsilon^k_{pG} + \bar{C}^k_{\mathcal{E}_n\epsilon_n}\epsilon^k_{nG} + \bar{C}^k_{\mathcal{E}_n\mathcal{E}_p}\mathcal{E}^k_{pG} + \bar{C}^k_{\mathcal{E}_n\mathcal{D}_n}\mathcal{D}^k_{nM} \quad (6.108)$$

The meaning of the $\bar{C}$ coefficients was given in Equations (2.79). By substituting Equations (6.105)–(6.108), and the geometrical relations (2.38)–(2.41) in Chapter 2 for plates, in the variational statement of Equation (6.104), and considering a generic layer k (Carrera *et al.* 2010a,b), we obtain:

$$\begin{aligned}
&\int_{\Omega_k}\int_{A_k}\Big[\left(\boldsymbol{D}_p F_s\delta\boldsymbol{u}^k_s\right)^T\Big(\bar{C}^k_{\sigma_p\epsilon_p}\boldsymbol{D}_p F_\tau\boldsymbol{u}^k_\tau + \bar{C}^k_{\sigma_p\epsilon_n}(\boldsymbol{D}_{np}+\boldsymbol{D}_{nz})F_\tau\boldsymbol{u}^k_\tau \\
&- \bar{C}^k_{\sigma_p\mathcal{E}_p}\boldsymbol{D}_{ep}F_\tau\Phi^k_\tau + \bar{C}^k_{\sigma_p\mathcal{D}_n}F_\tau\mathcal{D}^k_{nM\tau}\Big) + \left((\boldsymbol{D}_{np}+\boldsymbol{D}_{nz})F_s\delta\boldsymbol{u}^k_s\right)^T \\
&\times\Big(\bar{C}^k_{\sigma_n\epsilon_p}\boldsymbol{D}_p F_\tau\boldsymbol{u}^k_\tau + \bar{C}^k_{\sigma_n\epsilon_n}(\boldsymbol{D}_{np}+\boldsymbol{D}_{nz})F_\tau\boldsymbol{u}^k_\tau - \bar{C}^k_{\sigma_n\mathcal{E}_p}\boldsymbol{D}_{ep}F_\tau\Phi^k_\tau \\
&+ \bar{C}^k_{\sigma_n\mathcal{D}_n}F_\tau\mathcal{D}^k_{nM\tau}\Big) + \left(\boldsymbol{D}_{ep}F_s\delta\Phi^k_s\right)^T\Big(\bar{C}^k_{\mathcal{D}_p\epsilon_p}\boldsymbol{D}_p F_\tau\boldsymbol{u}^k_\tau + \bar{C}^k_{\mathcal{D}_p\epsilon_n}(\boldsymbol{D}_{np} \\
&+ \boldsymbol{D}_{nz})F_\tau\boldsymbol{u}^k_\tau - \bar{C}^k_{\mathcal{D}_p\mathcal{E}_p}\boldsymbol{D}_{ep}F_\tau\Phi^k_\tau + \bar{C}^k_{\mathcal{D}_p\mathcal{D}_n}F_\tau\mathcal{D}^k_{nM\tau}\Big) + \left(\boldsymbol{D}_{en}F_s\delta\Phi^k_s\right)^T \\
&\times\left(F_\tau\mathcal{D}^k_{nM\tau}\right) - \left(F_s\delta\mathcal{D}^k_{nMs}\right)^T\Big(-\boldsymbol{D}_{en}F_\tau\Phi^k_\tau - \bar{C}^k_{\mathcal{E}_n\epsilon_p}\boldsymbol{D}_p F_\tau\boldsymbol{u}^k_\tau \\
&- \bar{C}^k_{\mathcal{E}_n\epsilon_n}(\boldsymbol{D}_{np}+\boldsymbol{D}_{nz})F_\tau\delta\boldsymbol{u}^k_\tau + \bar{C}^k_{\mathcal{E}_n\mathcal{E}_p}\boldsymbol{D}_{ep}F_\tau\Phi^k_\tau - \bar{C}^k_{\mathcal{E}_n\mathcal{D}_n}F_\tau\mathcal{D}^k_{nM\tau}\Big)\Big] \\
&\times d\Omega_k\,dz = \delta L^k_e - \delta L^k_{in} \qquad (6.109)
\end{aligned}$$

The CUF (Carrera 1995), as presented in previous sections, has already been introduced for the 2D approximation. In order to obtain a strong form of differential equations on domain Ω_k and the relative boundary conditions at edge Γ_k in Equation (6.109), integration by parts is used. This permits one to move the differential operator from the infinitesimal variation of the generic variable $\delta\boldsymbol{a}^k$ to the finite quantity $\boldsymbol{a}^k$ (Carrera 1995). The integration by parts for a generic variable $\boldsymbol{a}^k$ is given in Equation (6.28); it has matrices as in Equations (6.29). The governing equations have the following form (Carrera *et al.* 2010a,b):

$$\begin{aligned}
\delta\boldsymbol{u}^k_s:&\quad \boldsymbol{K}^{k\tau s}_{uu}\,\boldsymbol{u}^k_\tau + \boldsymbol{K}^{k\tau s}_{u\mathcal{D}}\,\mathcal{D}^k_{nM\tau} + \boldsymbol{K}^{k\tau s}_{u\Phi}\,\Phi^k_\tau = \boldsymbol{p}^k_{us} - \boldsymbol{M}^{k\tau s}_{uu}\,\ddot{\boldsymbol{u}}^k_\tau \\
\delta\mathcal{D}^k_{ns}:&\quad \boldsymbol{K}^{k\tau s}_{\mathcal{D}u}\,\boldsymbol{u}^k_\tau + \boldsymbol{K}^{k\tau s}_{\mathcal{D}\mathcal{D}}\,\mathcal{D}^k_{nM\tau} + \boldsymbol{K}^{k\tau s}_{\mathcal{D}\Phi}\,\Phi^k_\tau = 0 \qquad (6.110)\\
\delta\Phi^k_s:&\quad \boldsymbol{K}^{k\tau s}_{\Phi u}\,\boldsymbol{u}^k_\tau + \boldsymbol{K}^{k\tau s}_{\Phi\mathcal{D}}\,\mathcal{D}^k_{nM\tau} + \boldsymbol{K}^{k\tau s}_{\Phi\Phi}\,\Phi^k_\tau = 0
\end{aligned}$$

where $\boldsymbol{M}_{uu}^{k\tau s}$ is the inertial contribution in the form of the fundamental nucleus, $\boldsymbol{u}_\tau^k$ is the vector of the degrees of freedom for the displacements, Φ_τ^k is the vector of the degrees of freedom for the electric potential, $\boldsymbol{\mathcal{D}}_{nM\tau}^k$ is the vector of the degrees of freedom for the transverse normal electric displacement, and $\ddot{\boldsymbol{u}}_\tau^k$ is the second temporal derivative of $\boldsymbol{u}_\tau^k$. The array $\boldsymbol{p}_{us}^k$ indicates the variationally consistent mechanical loading employed for the sensor configuration; the electric potential is directly imposed in the vector Φ_τ^k for the case of the actuator configuration. Along with these governing equations, the following boundary conditions, on the edge Γ_k of the in-plane integration domain Ω_k, hold:

$$\begin{aligned}
\boldsymbol{\Pi}_{uu}^{k\tau s}\,\boldsymbol{u}_\tau^k + \boldsymbol{\Pi}_{u\mathcal{D}}^{k\tau s}\,\boldsymbol{\mathcal{D}}_{nM\tau}^k + \boldsymbol{\Pi}_{u\Phi}^{k\tau s}\,\Phi_\tau^k &= \boldsymbol{\Pi}_{uu}^{k\tau s}\,\bar{\boldsymbol{u}}_\tau^k + \boldsymbol{\Pi}_{u\mathcal{D}}^{k\tau s}\,\bar{\boldsymbol{\mathcal{D}}}_{nM\tau}^k + \boldsymbol{\Pi}_{u\Phi}^{k\tau s}\,\bar{\Phi}_\tau^k\\
\boldsymbol{\Pi}_{\Phi u}^{k\tau s}\,\boldsymbol{u}_\tau^k + \boldsymbol{\Pi}_{\Phi\mathcal{D}}^{k\tau s}\,\boldsymbol{\mathcal{D}}_{nM\tau}^k + \boldsymbol{\Pi}_{\Phi\Phi}^{k\tau s}\,\Phi_\tau^k &= \boldsymbol{\Pi}_{\Phi u}^{k\tau s}\,\bar{\boldsymbol{u}}_\tau^k + \boldsymbol{\Pi}_{\Phi\mathcal{D}}^{k\tau s}\,\bar{\boldsymbol{\mathcal{D}}}_{nM\tau}^k + \boldsymbol{\Pi}_{\Phi\Phi}^{k\tau s}\,\bar{\Phi}_\tau^k
\end{aligned} \tag{6.111}$$

Comparing Equation (6.109), after the integration by parts, to Equations (6.110) and (6.111), the fundamental nuclei can be obtained:

$$\begin{aligned}
\boldsymbol{K}_{uu}^{k\tau s} = \int_{A_k} \Big[&-\boldsymbol{D}_p{}^T\,(\bar{\boldsymbol{C}}_{\sigma_p\epsilon_p}^k\,\boldsymbol{D}_p + \bar{\boldsymbol{C}}_{\sigma_p\epsilon_n}^k\,(\boldsymbol{D}_{np}+\boldsymbol{D}_{nz}))\\
&+(\boldsymbol{D}_{nz}-\boldsymbol{D}_{np})^T(\bar{\boldsymbol{C}}_{\sigma_n\epsilon_p}^k\,\boldsymbol{D}_p + \bar{\boldsymbol{C}}_{\sigma_n\epsilon_n}^k\,(\boldsymbol{D}_{np}+\boldsymbol{D}_{nz}))\Big]F_\tau F_s\,dz
\end{aligned} \tag{6.112}$$

$$\boldsymbol{K}_{u\mathcal{D}}^{k\tau s} = \int_{A_k} \Big[\boldsymbol{D}_p{}^T\bar{\boldsymbol{C}}_{\sigma_p\mathcal{D}_n}^k + (\boldsymbol{D}_{nz}-\boldsymbol{D}_{np})^T\bar{\boldsymbol{C}}_{\sigma_n\mathcal{D}_n}^k\Big]F_\tau F_s\,dz \tag{6.113}$$

$$\begin{aligned}
\boldsymbol{K}_{u\Phi}^{k\tau s} = \int_{A_k} &\Big[-\boldsymbol{D}_p{}^T(-\bar{\boldsymbol{C}}_{\sigma_p\mathcal{E}_n}^k\,\boldsymbol{D}_{ep}) + (\boldsymbol{D}_{nz}-\boldsymbol{D}_{np})^T(-\bar{\boldsymbol{C}}_{\sigma_n\mathcal{E}_n}^k\,\boldsymbol{D}_{ep})\Big]\\
&\times F_\tau F_s\,dz
\end{aligned} \tag{6.114}$$

$$\boldsymbol{K}_{\mathcal{D}u}^{k\tau s} = \int_{A_k} \Big[\bar{\boldsymbol{C}}_{\mathcal{E}_n\epsilon_p}^k\,\boldsymbol{D}_p + \bar{\boldsymbol{C}}_{\mathcal{E}_n\epsilon_n}^k\,(\boldsymbol{D}_{np}+\boldsymbol{D}_{nz})\Big]F_\tau F_s\,dz \tag{6.115}$$

$$\boldsymbol{K}_{\mathcal{D}\mathcal{D}}^{k\tau s} = \int_{A_k} \Big[\bar{\boldsymbol{C}}_{\mathcal{E}_n\mathcal{D}_n}^k\Big]F_\tau F_s\,dz \tag{6.116}$$

$$\boldsymbol{K}_{\mathcal{D}\Phi}^{k\tau s} = \int_{A_k} \Big[\boldsymbol{D}_{en} - \bar{\boldsymbol{C}}_{\mathcal{E}_n\mathcal{E}_p}^k\,\boldsymbol{D}_{ep}\Big]F_\tau F_s\,dz \tag{6.117}$$

$$\boldsymbol{K}_{\Phi u}^{k\tau s} = \int_{A_k} \Big[-\boldsymbol{D}_{ep}{}^T(\bar{\boldsymbol{C}}_{\mathcal{D}_p\epsilon_p}^k\,\boldsymbol{D}_p + \bar{\boldsymbol{C}}_{\mathcal{D}_p\epsilon_n}^k\,(\boldsymbol{D}_{np}+\boldsymbol{D}_{nz}))\Big]F_\tau F_s\,dz \tag{6.118}$$

$$\boldsymbol{K}_{\Phi\mathcal{D}}^{k\tau s} = \int_{A_k} \Big[-\boldsymbol{D}_{ep}{}^T\bar{\boldsymbol{C}}_{\mathcal{D}_p\mathcal{D}_n}^k + \boldsymbol{D}_{en}{}^T\Big]F_\tau F_s\,dz \tag{6.119}$$

$$\boldsymbol{K}_{\Phi\Phi}^{k\tau s} = \int_{A_k} \Big[-\boldsymbol{D}_{ep}{}^T(-\bar{\boldsymbol{C}}_{\mathcal{D}_p\mathcal{E}_p}^k\,\boldsymbol{D}_{ep})\Big]F_\tau F_s\,dz \tag{6.120}$$

The inertial arrays are no different from those of the previous sections. The nuclei for the boundary conditions are (Carrera *et al.* 2010a,b):

$$\boldsymbol{\Pi}_{uu}^{k\tau s} = \int_{A_k} \Big[\boldsymbol{I}_p^T (\bar{\boldsymbol{C}}_{\sigma_p\epsilon_p}^k \boldsymbol{D}_p + \bar{\boldsymbol{C}}_{\sigma_p\epsilon_n}^k (\boldsymbol{D}_{np} + \boldsymbol{D}_{nz})) + \boldsymbol{I}_{np}^T (\bar{\boldsymbol{C}}_{\sigma_n\epsilon_p}^k \boldsymbol{D}_p + \bar{\boldsymbol{C}}_{\sigma_n\epsilon_n}^k (\boldsymbol{D}_{np} + \boldsymbol{D}_{nz})) \Big] F_\tau F_s \, dz \tag{6.121}$$

$$\boldsymbol{\Pi}_{u\mathcal{D}}^{k\tau s} = \int_{A_k} \Big[\boldsymbol{I}_p^T \bar{\boldsymbol{C}}_{\sigma_p\mathcal{D}_n}^k + \boldsymbol{I}_{np}^T \bar{\boldsymbol{C}}_{\sigma_n\mathcal{D}_n}^k \Big] F_\tau F_s \, dz \tag{6.122}$$

$$\boldsymbol{\Pi}_{u\Phi}^{k\tau s} = \int_{A_k} \Big[\boldsymbol{I}_p^T (-\bar{\boldsymbol{C}}_{\sigma_p\mathcal{E}_n}^k \boldsymbol{D}_{ep}) + \boldsymbol{I}_{np}^T (-\bar{\boldsymbol{C}}_{\sigma_n\mathcal{E}_n}^k \boldsymbol{D}_{ep}) \Big] F_\tau F_s \, dz \tag{6.123}$$

$$\boldsymbol{\Pi}_{\Phi u}^{k\tau s} = \int_{A_k} \Big[\boldsymbol{I}_{ep}^T (\bar{\boldsymbol{C}}_{\mathcal{D}_p\epsilon_p}^k \boldsymbol{D}_p + \bar{\boldsymbol{C}}_{\mathcal{D}_p\epsilon_n}^k (\boldsymbol{D}_{np} + \boldsymbol{D}_{nz})) \Big] F_\tau F_s \, dz \tag{6.124}$$

$$\boldsymbol{\Pi}_{\Phi\mathcal{D}}^{k\tau s} = \int_{A_k} \Big[\boldsymbol{I}_{ep}^T \bar{\boldsymbol{C}}_{\mathcal{D}_p\mathcal{D}_n}^k \Big] F_\tau F_s \, dz \tag{6.125}$$

$$\boldsymbol{\Pi}_{\Phi\Phi}^{k\tau s} = \int_{A_k} \Big[\boldsymbol{I}_{ep}^T (-\bar{\boldsymbol{C}}_{\mathcal{D}_p\mathcal{E}_p}^k \boldsymbol{D}_{ep}) \Big] F_\tau F_s \, dz \tag{6.126}$$

In order to perform the integration by parts (see Equation (6.28)), the matrices $\boldsymbol{I}_p$, $\boldsymbol{I}_{np}$, and $\boldsymbol{I}_{ep}$, which are those presented in Equation (6.29), are introduced. To write the explicit form of the nuclei in Equations (6.112)–(6.120), the integrals in the z thickness direction are defined as in Equation (6.41). By developing the matrix products in Equations (6.112)–(6.120) and employing a Navier-type closed-form solution (Carrera *et al.* 2010a,b; Carrera and Brischetto 2007b) the explicit algebraic form of the nuclei can be obtained.

Nucleus $\boldsymbol{K}_{uu}^{k\tau s}$ of (3×3) dimension is:

$$\begin{aligned}
K_{uu_{11}}^{k\tau s} &= \bar{C}_{\sigma_p\epsilon_p 11}^k \bar{\alpha}^2 J^{k\tau s} + \bar{C}_{\sigma_p\epsilon_p 33}^k \bar{\beta}^2 J^{k\tau s} + \bar{C}_{\sigma_p\epsilon_n 11}^k J^{k\tau_z s_z} \\
K_{uu_{12}}^{k\tau s} &= (\bar{C}_{\sigma_p\epsilon_p 11}^k + \bar{C}_{\sigma_p\epsilon_p 33}^k)\bar{\alpha}\bar{\beta} J^{k\tau s} \\
K_{uu_{13}}^{k\tau s} &= \bar{C}_{\sigma_p\epsilon_n 13}^k \bar{\alpha} J^{k\tau s_z} + \bar{C}_{\sigma_n\epsilon_n 11}^k \bar{\alpha} J^{k\tau_z s} \\
K_{uu_{21}}^{k\tau s} &= (\bar{C}_{\sigma_p\epsilon_p 12}^k + \bar{C}_{\sigma_p\epsilon_p 33}^k)\bar{\alpha}\bar{\beta} J^{k\tau s} \\
K_{uu_{22}}^{k\tau s} &= \bar{C}_{\sigma_n\epsilon_n 22}^k J^{k\tau_z s_z} + \bar{C}_{\sigma_p\epsilon_p 22}^k \bar{\beta}^2 J^{k\tau s} + \bar{C}_{\sigma_p\epsilon_p 33}^k \bar{\alpha}^2 J^{k\tau s} \\
K_{uu_{23}}^{k\tau s} &= \bar{C}_{\sigma_n\epsilon_n 22}^k \bar{\beta} J^{k\tau_z s} - \bar{C}_{\sigma_p\epsilon_n 23}^k \bar{\beta} J^{k\tau s_z} \\
K_{uu_{31}}^{k\tau s} &= -\bar{C}_{\sigma_p\epsilon_n 13}^k \bar{\alpha} J^{k\tau_z s} + \bar{C}_{\sigma_p\epsilon_n 11}^k \bar{\alpha} J^{k\tau s_z} \\
K_{uu_{32}}^{k\tau s} &= -\bar{C}_{\sigma_p\epsilon_n 23}^k \bar{\beta} J^{k\tau_z s} + \bar{C}_{\sigma_n\epsilon_n 22}^k \bar{\beta} J^{k\tau s_z} \\
K_{uu_{33}}^{k\tau s} &= \bar{C}_{\sigma_n\epsilon_n 33}^k J^{k\tau_z s_z} + \bar{C}_{\sigma_n\epsilon_n 22}^k \bar{\beta}^2 J^{k\tau s} + \bar{C}_{\sigma_n\epsilon_n 11}^k \bar{\alpha}^2 J^{k\tau s}
\end{aligned} \tag{6.127}$$

Nucleus $\boldsymbol{K}_{u\Phi}^{k\tau s}$ of (3 × 1) dimension is:

$$
\begin{aligned}
K_{u\Phi_{11}}^{k\tau s} &= -\bar{C}_{\sigma_n\mathcal{E}_p11}^{k}\bar{\alpha}J^{k\tau_z s}, \qquad K_{u\Phi_{21}}^{k\tau s} = -\bar{C}_{\sigma_n\mathcal{E}_p21}^{k}\bar{\beta}J^{k\tau_z s}\\
K_{u\Phi_{31}}^{k\tau s} &= -\bar{C}_{\sigma_n\mathcal{E}_p11}^{k}\bar{\alpha}^2J^{k\tau s} - \bar{C}_{\sigma_n\mathcal{E}_p21}^{k}\bar{\beta}^2J^{k\tau s}
\end{aligned}
\tag{6.128}
$$

Nucleus $\boldsymbol{K}_{u\mathcal{D}}^{k\tau s}$ of (3 × 1) dimension is:

$$
\begin{aligned}
K_{u\mathcal{D}_{11}}^{k\tau s} &= -\bar{C}_{\mathcal{E}_n\epsilon_p11}^{k}\bar{\alpha}J^{k\tau s}, \qquad K_{u\mathcal{D}_{21}}^{k\tau s} = -\bar{C}_{\mathcal{E}_n\epsilon_p12}^{k}\bar{\beta}J^{k\tau s}\\
K_{u\mathcal{D}_{31}}^{k\tau s} &= -\bar{C}_{\sigma_n\mathcal{D}_n31}^{k}J^{k\tau_z s}
\end{aligned}
\tag{6.129}
$$

Nucleus $\boldsymbol{K}_{\Phi u}^{k\tau s}$ of (1 × 3) dimension is:

$$
\begin{aligned}
K_{\Phi u_{11}}^{k\tau s} &= \bar{C}_{\mathcal{D}_p\epsilon_n11}^{k}\bar{\alpha}J^{k\tau s_z}, \qquad K_{\Phi u_{12}}^{k\tau s} = \bar{C}_{\mathcal{D}_p\epsilon_n22}^{k}\bar{\beta}J^{k\tau s_z}\\
K_{\Phi u_{13}}^{k\tau s} &= \bar{C}_{\mathcal{D}_p\epsilon_n11}\bar{\alpha}^2J^{k\tau s} + \bar{C}_{\mathcal{D}_p\epsilon_n22}\bar{\beta}^2J^{k\tau s}
\end{aligned}
\tag{6.130}
$$

Nucleus $\boldsymbol{K}_{\Phi\Phi}^{k\tau s}$ of (1 × 1) dimension is:

$$
K_{\Phi\Phi_{11}}^{k\tau s} = -\bar{C}_{\mathcal{D}_p\mathcal{E}_p11}^{k}\bar{\alpha}^2J^{k\tau s} - \bar{C}_{\mathcal{D}_p\mathcal{E}_p22}^{k}\bar{\beta}^2J^{k\tau s}
\tag{6.131}
$$

Nucleus $\boldsymbol{K}_{\Phi\mathcal{D}}^{k\tau s}$ of (1 × 1) dimension is:

$$
K_{\Phi\mathcal{D}_{11}}^{k\tau s} = J^{k\tau_z s}
\tag{6.132}
$$

Nucleus $\boldsymbol{K}_{\mathcal{D}u}^{k\tau s}$ of (1 × 3) dimension is:

$$
\begin{aligned}
K_{\mathcal{D}u_{11}}^{k\tau s} &= -\bar{C}_{\mathcal{E}_n\epsilon_p11}^{k}\bar{\alpha}J^{k\tau s}, \qquad K_{\mathcal{D}u_{12}}^{k\tau s} = -\bar{C}_{\mathcal{E}_n\epsilon_p12}^{k}\bar{\beta}J^{k\tau s}\\
K_{\mathcal{D}u_{13}}^{k\tau s} &= -\bar{C}_{\mathcal{E}_n\epsilon_n13}^{k}J^{k\tau s_z}
\end{aligned}
\tag{6.133}
$$

Nucleus $\boldsymbol{K}_{\mathcal{D}\Phi}^{k\tau s}$ of (1 × 1) dimension is:

$$
K_{\mathcal{D}\Phi_{11}}^{k\tau s} = J^{k\tau s_z}
\tag{6.134}
$$

Nucleus $\boldsymbol{K}_{\mathcal{D}\mathcal{D}}^{k\tau s}$ of (1 × 1) dimension is:

$$
K_{\mathcal{D}\mathcal{D}_{11}}^{k\tau s} = \bar{C}_{\mathcal{E}_n\mathcal{D}_n11}^{k}J^{k\tau s}
\tag{6.135}
$$

The fundamental nucleus for the inertial matrix $\boldsymbol{M}_{uu}^{k\tau s}$ and its components were given in Equations (6.36) and (6.46).

$\bar{\alpha} = m\pi/a$ and $\bar{\beta} = n\pi/b$, where m and n are the wave numbers in the in-plane directions and a and b are the plate dimensions in the x and y directions, respectively. A Navier-type closed-form solution is obtained via substitution of the harmonic expressions for the displacements, electric potential, and transverse stresses and by considering the following material coefficients to be equal to zero: $Q_{16} = Q_{26} = Q_{36} = Q_{45} = 0$ and $e_{25} = e_{14} = e_{36} = \varepsilon_{12} = 0$. The harmonic assumptions used for the displacements, the electric potential, and the transverse normal electric displacement are:

$$(u_\tau^k) = \sum_{m,n} (\hat{U}_\tau^k) \cos\left(\frac{m\pi x}{a}\right) \sin\left(\frac{n\pi y}{b}\right), \quad k = 1, N_l \tag{6.136}$$

$$(v_\tau^k) = \sum_{m,n} (\hat{V}_\tau^k) \sin\left(\frac{m\pi x}{a}\right) \cos\left(\frac{n\pi y}{b}\right), \quad \tau = t, b, r \tag{6.137}$$

$$(w_\tau^k, \Phi_\tau^k, \mathcal{D}_{z\tau}^k) = \sum_{m,n} (\hat{W}_\tau^k, \hat{\Phi}_\tau^k, \hat{\mathcal{D}}_{z\tau}^k) \sin\left(\frac{m\pi x}{a}\right) \sin\left(\frac{n\pi y}{b}\right), \quad r = 2, N \tag{6.138}$$

where $\hat{U}_\tau^k$, $\hat{V}_\tau^k$, $\hat{W}_\tau^k$ are the displacement amplitudes, $\hat{\Phi}_\tau^k$ is the electric potential amplitude, and $\hat{\mathcal{D}}_{z\tau}^k$ is the transverse normal electric displacement amplitude; k indicates the layer and N_l is the total number of layers. τ is the index for the order of expansion, where t and b indicate the top and bottom of the layer, respectively, while r indicates the higher orders of expansion until $N = 4$. Details on the assembly procedure of the fundamental nuclei and on the acronyms are given in Sections 6.7 and 6.8, respectively.

The meaning of the $\bar{\boldsymbol{C}}$ coefficients for Equations (6.105)–(6.108) has been given in Equations (2.79). The following equations give the explicit form of each of their components.

The $\bar{\boldsymbol{C}}_{\sigma_p\epsilon_p}^k$ array has a (3×3) dimension with components for each k layer:

$$\begin{aligned}
&\bar{C}_{\sigma_p\epsilon_p11} = Q_{11} + \frac{e_{31}^2}{\varepsilon_{33}}, \quad \bar{C}_{\sigma_p\epsilon_p12} = Q_{12} + \frac{e_{31}e_{32}}{\varepsilon_{33}}, \quad \bar{C}_{\sigma_p\epsilon_p13} = 0\\
&\bar{C}_{\sigma_p\epsilon_p21} = Q_{12} + \frac{e_{31}e_{32}}{\varepsilon_{33}}, \quad \bar{C}_{\sigma_p\epsilon_p22} = Q_{22} + \frac{e_{32}^2}{\varepsilon_{33}}, \quad \bar{C}_{\sigma_p\epsilon_p23} = 0\\
&\bar{C}_{\sigma_p\epsilon_p31} = 0, \quad \bar{C}_{\sigma_p\epsilon_p32} = 0, \quad \bar{C}_{\sigma_p\epsilon_p33} = Q_{66}
\end{aligned} \tag{6.139}$$

The $\bar{\boldsymbol{C}}_{\sigma_p\epsilon_n}^k$ array has a (3×3) dimension with components for each k layer:

$$\begin{aligned}
&\bar{C}_{\sigma_p\epsilon_n11} = 0, \quad \bar{C}_{\sigma_p\epsilon_n12} = 0, \quad \bar{C}_{\sigma_p\epsilon_n13} = Q_{13} + \frac{e_{31}e_{33}}{\varepsilon_{33}}\\
&\bar{C}_{\sigma_p\epsilon_n21} = 0, \quad \bar{C}_{\sigma_p\epsilon_n22} = 0, \quad \bar{C}_{\sigma_p\epsilon_n23} = Q_{23} + \frac{e_{32}e_{33}}{\varepsilon_{33}}\\
&\bar{C}_{\sigma_p\epsilon_n31} = 0, \quad \bar{C}_{\sigma_p\epsilon_n32} = 0, \quad \bar{C}_{\sigma_p\epsilon_n33} = 0
\end{aligned} \tag{6.140}$$

The $\bar{\boldsymbol{C}}^k_{\sigma_p\mathcal{E}_p}$ array has a (3 × 2) dimension with components for each k layer:

$$\begin{aligned} &\bar{C}_{\sigma_p\mathcal{E}_p11} = 0, \quad \bar{C}_{\sigma_p\mathcal{E}_p12} = 0, \quad \bar{C}_{\sigma_p\mathcal{E}_p21} = 0 \\ &\bar{C}_{\sigma_p\mathcal{E}_p22} = 0, \quad \bar{C}_{\sigma_p\mathcal{E}_p31} = 0, \quad \bar{C}_{\sigma_p\mathcal{E}_p32} = 0 \end{aligned} \tag{6.141}$$

The $\bar{\boldsymbol{C}}^k_{\sigma_p\mathcal{D}_n}$ array has a (3 × 1) dimension with components for each k layer:

$$\bar{C}_{\sigma_p\mathcal{D}_n11} = -\frac{e_{31}}{\varepsilon_{33}}, \quad \bar{C}_{\sigma_p\mathcal{D}_n21} = -\frac{e_{32}}{\varepsilon_{33}}, \quad \bar{C}_{\sigma_p\mathcal{D}_n31} = 0 \tag{6.142}$$

The $\bar{\boldsymbol{C}}^k_{\sigma_n\epsilon_p}$ array has a (3 × 3) dimension with components for each k layer:

$$\begin{aligned} &\bar{C}_{\sigma_n\epsilon_p11} = 0, \quad \bar{C}_{\sigma_n\epsilon_p12} = 0, \quad \bar{C}_{\sigma_n\epsilon_p13} = 0 \\ &\bar{C}_{\sigma_n\epsilon_p21} = 0, \quad \bar{C}_{\sigma_n\epsilon_p22} = 0, \quad \bar{C}_{\sigma_n\epsilon_p23} = 0 \\ &\bar{C}_{\sigma_n\epsilon_p31} = Q_{13} + \frac{e_{31}e_{33}}{\varepsilon_{33}}, \quad \bar{C}_{\sigma_n\epsilon_p32} = Q_{23} + \frac{e_{32}e_{33}}{\varepsilon_{33}}, \quad \bar{C}_{\sigma_n\epsilon_p33} = 0 \end{aligned} \tag{6.143}$$

The $\bar{\boldsymbol{C}}^k_{\sigma_n\epsilon_n}$ array has a (3 × 3) dimension with components for each k layer:

$$\begin{aligned} &\bar{C}_{\sigma_n\epsilon_n11} = Q_{55}, \quad \bar{C}_{\sigma_n\epsilon_n12} = 0, \quad \bar{C}_{\sigma_n\epsilon_n13} = 0 \\ &\bar{C}_{\sigma_n\epsilon_n21} = 0, \quad \bar{C}_{\sigma_n\epsilon_n22} = Q_{44}, \quad \bar{C}_{\sigma_n\epsilon_n23} = 0 \\ &\bar{C}_{\sigma_n\epsilon_n31} = 0, \quad \bar{C}_{\sigma_n\epsilon_n32} = 0, \quad \bar{C}_{\sigma_n\epsilon_n33} = Q_{33} + \frac{e_{33}^2}{\varepsilon_{33}} \end{aligned} \tag{6.144}$$

The $\bar{\boldsymbol{C}}^k_{\sigma_n\mathcal{E}_p}$ array has a (3 × 2) dimension with components for each k layer:

$$\begin{aligned} &\bar{C}_{\sigma_n\mathcal{E}_p11} = -e_{15}, \quad \bar{C}_{\sigma_n\mathcal{E}_p12} = 0, \quad \bar{C}_{\sigma_n\mathcal{E}_p21} = 0 \\ &\bar{C}_{\sigma_n\mathcal{E}_p22} = -e_{24}, \quad \bar{C}_{\sigma_n\mathcal{E}_p31} = 0, \quad \bar{C}_{\sigma_n\mathcal{E}_p32} = 0 \end{aligned} \tag{6.145}$$

The $\bar{\boldsymbol{C}}^k_{\sigma_n\mathcal{D}_n}$ array has a (3 × 1) dimension with components for each k layer:

$$\bar{C}_{\sigma_n\mathcal{D}_n11} = 0, \quad \bar{C}_{\sigma_n\mathcal{D}_n21} = 0, \quad \bar{C}_{\sigma_n\mathcal{D}_n31} = -\frac{e_{33}}{\varepsilon_{33}} \tag{6.146}$$

The $\bar{\boldsymbol{C}}^k_{\mathcal{D}_p\epsilon_p}$ array has a (2 × 3) dimension with components for each k layer:

$$\begin{aligned} &\bar{C}_{\mathcal{D}_p\epsilon_p11} = 0, \quad \bar{C}_{\mathcal{D}_p\epsilon_p12} = 0, \quad \bar{C}_{\mathcal{D}_p\epsilon_p13} = 0 \\ &\bar{C}_{\mathcal{D}_p\epsilon_p21} = 0, \quad \bar{C}_{\mathcal{D}_p\epsilon_p22} = 0, \quad \bar{C}_{\mathcal{D}_p\epsilon_p23} = 0 \end{aligned} \tag{6.147}$$

The $\bar{\boldsymbol{C}}^k_{\mathcal{D}_p\epsilon_n}$ array has a (2 × 3) dimension with components for each k layer:

$$\begin{aligned} \bar{C}_{\mathcal{D}_p\epsilon_n 11} &= e_{15}, \quad \bar{C}_{\mathcal{D}_p\epsilon_n 12} = 0, \quad \bar{C}_{\mathcal{D}_p\epsilon_n 13} = 0 \\ \bar{C}_{\mathcal{D}_p\epsilon_n 21} &= 0, \quad \bar{C}_{\mathcal{D}_p\epsilon_n 22} = e_{24}, \quad \bar{C}_{\mathcal{D}_p\epsilon_n 23} = 0 \end{aligned} \tag{6.148}$$

The $\bar{\boldsymbol{C}}^k_{\mathcal{D}_p\mathcal{E}_p}$ array has a (2 × 2) dimension with components for each k layer:

$$\bar{C}_{\mathcal{D}_p\mathcal{E}_p 11} = \varepsilon_{11}, \quad \bar{C}_{\mathcal{D}_p\mathcal{E}_p 12} = 0, \quad \bar{C}_{\mathcal{D}_p\mathcal{E}_p 21} = 0, \quad \bar{C}_{\mathcal{D}_p\mathcal{E}_p 22} = \varepsilon_{22} \tag{6.149}$$

The $\bar{\boldsymbol{C}}^k_{\mathcal{D}_p\mathcal{D}_n}$ array has a (2 × 1) dimension with components for each k layer:

$$\bar{C}_{\mathcal{D}_p\mathcal{D}_n 11} = 0, \quad \bar{C}_{\mathcal{D}_p\mathcal{D}_n 21} = 0 \tag{6.150}$$

The $\bar{\boldsymbol{C}}^k_{\mathcal{E}_n\epsilon_p}$ array has a (1 × 3) dimension with components for each k layer:

$$\bar{C}_{\mathcal{E}_n\epsilon_p 11} = -\frac{e_{31}}{\varepsilon_{33}}, \quad \bar{C}_{\mathcal{E}_n\epsilon_p 12} = -\frac{e_{32}}{\varepsilon_{33}}, \quad \bar{C}_{\mathcal{E}_n\epsilon_p 13} = 0 \tag{6.151}$$

The $\bar{\boldsymbol{C}}^k_{\mathcal{E}_n\epsilon_n}$ array has a (1 × 3) dimension with components for each k layer:

$$\bar{C}_{\mathcal{E}_n\epsilon_n 11} = 0, \quad \bar{C}_{\mathcal{E}_n\epsilon_n 12} = 0, \quad \bar{C}_{\mathcal{E}_n\epsilon_n 13} = -\frac{e_{33}}{\varepsilon_{33}} \tag{6.152}$$

The $\bar{\boldsymbol{C}}^k_{\mathcal{E}_n\mathcal{E}_p}$ array has a (1 × 2) dimension with components for each k layer:

$$\bar{C}_{\mathcal{E}_n\mathcal{E}_p 11} = 0, \quad \bar{C}_{\mathcal{E}_n\mathcal{E}_p 12} = 0 \tag{6.153}$$

The $\bar{\boldsymbol{C}}^k_{\mathcal{E}_n\mathcal{D}_n}$ array has a (1 × 1) dimension with components for each k layer:

$$\bar{C}_{\mathcal{E}_n\mathcal{D}_n 11} = \frac{1}{\varepsilon_{33}} \tag{6.154}$$

6.6 RMVT($\boldsymbol{u}$, $\boldsymbol{\Phi}$, $\boldsymbol{\sigma_n}$, $\boldsymbol{\mathcal{D}_n}$) for the electromechanical plate case

The third possible extension of the RMVT (Reissner 1984) is that indicated in Equation (2.83) in Chapter 2, where the internal electrical work has been considered and two Lagrange multipliers have been added for the transverse stress components and the transverse normal electric displacement (Carrera and Brischetto 2007a; Carrera *et al.* 2008):

$$\int_V \Big(\delta\boldsymbol{\epsilon}_{pG}^T\boldsymbol{\sigma}_{pC} + \delta\boldsymbol{\epsilon}_{nG}^T\boldsymbol{\sigma}_{nM} + \delta\boldsymbol{\sigma}_{nM}^T(\boldsymbol{\epsilon}_{nG} - \boldsymbol{\epsilon}_{nC}) - \delta\boldsymbol{\mathcal{E}}_{pG}^T\boldsymbol{\mathcal{D}}_{pC} - \delta\boldsymbol{\mathcal{E}}_{nG}^T\boldsymbol{\mathcal{D}}_{nM}$$

$$-\delta\boldsymbol{\mathcal{D}}_{nM}^T(\boldsymbol{\mathcal{E}}_{nG} - \boldsymbol{\mathcal{E}}_{nC})\Big)dV = \delta L_e - \delta L_{in} \tag{6.155}$$

where the subscript M means a priori modeled variables. Considering the multilayered structure of N_l layers, the integral on the volume V in Equation (6.155) can be rewritten as:

$$\sum_{k=1}^{N_l}\int_{\Omega_k}\int_{A_k}\Big\{\delta\boldsymbol{\epsilon}_{pG}^{k\;T}\boldsymbol{\sigma}_{pC}^k + \delta\boldsymbol{\epsilon}_{nG}^{k\;T}\boldsymbol{\sigma}_{nM}^k + \delta\boldsymbol{\sigma}_{nM}^{k\;T}(\boldsymbol{\epsilon}_{nG}^k - \boldsymbol{\epsilon}_{nC}^k)$$

$$-\delta\boldsymbol{\mathcal{E}}_{pG}^{k\;T}\boldsymbol{\mathcal{D}}_{pC}^k - \delta\boldsymbol{\mathcal{E}}_{nG}^{k\;T}\boldsymbol{\mathcal{D}}_{nM}^k - \delta\boldsymbol{\mathcal{D}}_{nM}^{k\;T}(\boldsymbol{\mathcal{E}}_{nG}^k - \boldsymbol{\mathcal{E}}_{nC}^k)\Big\}d\Omega_k dz$$

$$= \sum_{k=1}^{N_l}\delta L_e^k - \sum_{k=1}^{N_l}\delta L_{in}^k \tag{6.156}$$

where δL_e^k and δL_{in}^k are the external and inertial virtual work at the k-layer level, respectively. The relative constitutive equations are those obtained from Equations (2.84)–(2.87) where the transverse normal electric displacement $\boldsymbol{\mathcal{D}}_n$ is modeled (M) and the transverse normal electric field $\boldsymbol{\mathcal{E}}_n$ is obtained from constitutive equations (C); the transverse stresses $\boldsymbol{\sigma}_n$ are modeled (M) and the transverse strains $\boldsymbol{\epsilon}_n$ are obtained from the constitutive equations (C) (Carrera and Brischetto 2007a; Carrera *et al.* 2008):

$$\boldsymbol{\sigma}_{pC}^k = \tilde{\boldsymbol{C}}_{\sigma_p\epsilon_p}^k\boldsymbol{\epsilon}_{pG}^k + \tilde{\boldsymbol{C}}_{\sigma_p\sigma_n}^k\boldsymbol{\sigma}_{nM}^k + \tilde{\boldsymbol{C}}_{\sigma_p\mathcal{E}_p}^k\boldsymbol{\mathcal{E}}_{pG}^k + \tilde{\boldsymbol{C}}_{\sigma_p\mathcal{D}_n}^k\boldsymbol{\mathcal{D}}_{nM}^k \tag{6.157}$$

$$\boldsymbol{\epsilon}_{nC}^k = \tilde{\boldsymbol{C}}_{\epsilon_n\epsilon_p}^k\boldsymbol{\epsilon}_{pG}^k + \tilde{\boldsymbol{C}}_{\epsilon_n\sigma_n}^k\boldsymbol{\sigma}_{nM}^k + \tilde{\boldsymbol{C}}_{\epsilon_n\mathcal{E}_p}^k\boldsymbol{\mathcal{E}}_{pG}^k + \tilde{\boldsymbol{C}}_{\epsilon_n\mathcal{D}_n}^k\boldsymbol{\mathcal{D}}_{nM}^k \tag{6.158}$$

$$\boldsymbol{\mathcal{D}}_{pC}^k = \tilde{\boldsymbol{C}}_{\mathcal{D}_p\epsilon_p}^k\boldsymbol{\epsilon}_{pG}^k + \tilde{\boldsymbol{C}}_{\mathcal{D}_p\sigma_n}^k\boldsymbol{\sigma}_{nM}^k + \tilde{\boldsymbol{C}}_{\mathcal{D}_p\mathcal{E}_p}^k\boldsymbol{\mathcal{E}}_{pG}^k + \tilde{\boldsymbol{C}}_{\mathcal{D}_p\mathcal{D}_n}^k\boldsymbol{\mathcal{D}}_{nM}^k \tag{6.159}$$

$$\boldsymbol{\mathcal{E}}_{nC}^k = \tilde{\boldsymbol{C}}_{\mathcal{E}_n\epsilon_p}^k\boldsymbol{\epsilon}_{pG}^k + \tilde{\boldsymbol{C}}_{\mathcal{E}_n\sigma_n}^k\boldsymbol{\sigma}_{nM}^k + \tilde{\boldsymbol{C}}_{\mathcal{E}_n\mathcal{E}_p}^k\boldsymbol{\mathcal{E}}_{pG}^k + \tilde{\boldsymbol{C}}_{\mathcal{E}_n\mathcal{D}_n}^k\boldsymbol{\mathcal{D}}_{nM}^k \tag{6.160}$$

The meaning of the $\tilde{\boldsymbol{C}}$ coefficients was given in Equations (2.88). Substituting Equations (6.157)–(6.160), and the geometrical relations (2.38)–(2.41) for plates in Chapter 2 in the variational statement of Equation (6.156), and considering a generic layer k (Carrera and Brischetto 2007a; Carrera *et al.* 2008), we obtain:

$$\begin{aligned}
&\int_{\Omega_k}\int_{A_k}\Big[\Big(\boldsymbol{D}_pF_s\delta\boldsymbol{u}_s^k\Big)^T\Big(\tilde{\boldsymbol{C}}^k_{\sigma_p\epsilon_p}\boldsymbol{D}_pF_\tau\boldsymbol{u}_\tau^k+\tilde{\boldsymbol{C}}^k_{\sigma_p\sigma_n}F_\tau\boldsymbol{\sigma}^k_{nM\tau}-\tilde{\boldsymbol{C}}^k_{\sigma_p\mathcal{E}_p}\boldsymbol{D}_{ep}F_\tau\Phi^k_\tau\\
&+\tilde{\boldsymbol{C}}^k_{\sigma_p\mathcal{D}_n}F_\tau\boldsymbol{\mathcal{D}}^k_{nM\tau}\Big)+\Big((\boldsymbol{D}_{np}+\boldsymbol{D}_{nz})F_s\delta\boldsymbol{u}_s^k\Big)^T\Big(F_\tau\boldsymbol{\sigma}^k_{nM\tau}\Big)+\Big(F_s\delta\boldsymbol{\sigma}^k_{nMs}\Big)^T\\
&\times\Big((\boldsymbol{D}_{np}+\boldsymbol{D}_{nz})F_\tau\boldsymbol{u}_\tau^k-\tilde{\boldsymbol{C}}^k_{\epsilon_n\epsilon_p}\boldsymbol{D}_pF_\tau\boldsymbol{u}_\tau^k-\tilde{\boldsymbol{C}}^k_{\epsilon_n\sigma_n}F_\tau\boldsymbol{\sigma}^k_{nM\tau}+\tilde{\boldsymbol{C}}^k_{\epsilon_n\mathcal{E}_p}\boldsymbol{D}_{ep}F_\tau\Phi^k_\tau\\
&-\tilde{\boldsymbol{C}}^k_{\epsilon_n\mathcal{D}_n}F_\tau\boldsymbol{\mathcal{D}}^k_{nM\tau}\Big)+\Big(\boldsymbol{D}_{ep}F_s\delta\Phi^k_s\Big)^T\Big(\tilde{\boldsymbol{C}}^k_{\mathcal{D}_p\epsilon_p}\boldsymbol{D}_pF_\tau\boldsymbol{u}_\tau^k+\tilde{\boldsymbol{C}}^k_{\mathcal{D}_p\sigma_n}F_\tau\boldsymbol{\sigma}^k_{nM\tau}\\
&-\tilde{\boldsymbol{C}}^k_{\mathcal{D}_p\mathcal{E}_p}\boldsymbol{D}_{ep}F_\tau\Phi^k_\tau+\tilde{\boldsymbol{C}}^k_{\mathcal{D}_p\mathcal{D}_n}F_\tau\boldsymbol{\mathcal{D}}^k_{nM\tau}\Big)+\Big(\boldsymbol{D}_{en}F_s\delta\Phi^k_s\Big)^T\Big(F_\tau\boldsymbol{\mathcal{D}}^k_{nM\tau}\Big)\\
&-\Big(F_s\delta\boldsymbol{\mathcal{D}}^k_{nMs}\Big)^T\Big(-\boldsymbol{D}_{en}F_\tau\Phi^k_\tau-\tilde{\boldsymbol{C}}^k_{\mathcal{E}_n\epsilon_p}\boldsymbol{D}_pF_\tau\boldsymbol{u}_\tau^k-\tilde{\boldsymbol{C}}^k_{\mathcal{E}_n\sigma_n}F_\tau\boldsymbol{\sigma}^k_{nM\tau}\\
&+\tilde{\boldsymbol{C}}^k_{\mathcal{E}_n\mathcal{E}_p}\boldsymbol{D}_{ep}F_\tau\Phi^k_\tau-\tilde{\boldsymbol{C}}^k_{\mathcal{E}_n\mathcal{D}_n}F_\tau\boldsymbol{\mathcal{D}}^k_{nM\tau}\Big)\Big]d\Omega_k\,dz=\delta L_e^k-\delta L_{in}^k \qquad (6.161)
\end{aligned}$$

The CUF (Carrera 1995), as presented in the previous sections, has already been introduced for the 2D approximation. In order to obtain a strong form of differential equations on the domain Ω_k and the relative boundary conditions at edge Γ_k in Equation (6.161), integration by parts is used, and this permits one to move the differential operator from the infinitesimal variation of the generic variable $\delta\boldsymbol{a}^k$ to the finite quantity $\boldsymbol{a}^k$ (Carrera 1995). The integration by parts is given in Equation (6.28) for a generic variable $\boldsymbol{a}^k$, and matrices are shown in Equations (6.29). The governing equations have the following form (Carrera and Brischetto 2007a; Carrera *et al.* 2008):

$$\begin{aligned}
\delta\boldsymbol{u}_s^k:&\quad \boldsymbol{K}^{k\tau s}_{uu}\,\boldsymbol{u}_\tau^k+\boldsymbol{K}^{k\tau s}_{u\sigma}\,\boldsymbol{\sigma}^k_{nM\tau}+\boldsymbol{K}^{k\tau s}_{u\Phi}\,\Phi^k_\tau+\boldsymbol{K}^{k\tau s}_{u\mathcal{D}}\,\boldsymbol{\mathcal{D}}^k_{nM\tau}=\boldsymbol{p}^k_{us}-\boldsymbol{M}^{k\tau s}_{uu}\,\ddot{\boldsymbol{u}}^k_\tau\\
\delta\boldsymbol{\sigma}^k_{ns}:&\quad \boldsymbol{K}^{k\tau s}_{\sigma u}\,\boldsymbol{u}_\tau^k+\boldsymbol{K}^{k\tau s}_{\sigma\sigma}\,\boldsymbol{\sigma}^k_{nM\tau}+\boldsymbol{K}^{k\tau s}_{\sigma\Phi}\,\Phi^k_\tau+\boldsymbol{K}^{k\tau s}_{\sigma\mathcal{D}}\,\boldsymbol{\mathcal{D}}^k_{nM\tau}=0\\
\delta\Phi^k_s:&\quad \boldsymbol{K}^{k\tau s}_{\Phi u}\,\boldsymbol{u}_\tau^k+\boldsymbol{K}^{k\tau s}_{\Phi\sigma}\,\boldsymbol{\sigma}^k_{nM\tau}+\boldsymbol{K}^{k\tau s}_{\Phi\Phi}\,\Phi^k_\tau+\boldsymbol{K}^{k\tau s}_{\Phi\mathcal{D}}\,\boldsymbol{\mathcal{D}}^k_{nM\tau}=0\\
\delta\boldsymbol{\mathcal{D}}^k_{ns}:&\quad \boldsymbol{K}^{k\tau s}_{\mathcal{D}u}\,\boldsymbol{u}_\tau^k+\boldsymbol{K}^{k\tau s}_{\mathcal{D}\sigma}\,\boldsymbol{\sigma}^k_{nM\tau}+\boldsymbol{K}^{k\tau s}_{\mathcal{D}\Phi}\,\Phi^k_\tau+\boldsymbol{K}^{k\tau s}_{\mathcal{D}\mathcal{D}}\,\boldsymbol{\mathcal{D}}^k_{nM\tau}=0
\end{aligned} \qquad (6.162)$$

where $\boldsymbol{M}^{k\tau s}_{uu}$ is the inertial contribution in the form of the fundamental nucleus, $\boldsymbol{u}^k_\tau$ is the vector of the degrees of freedom for the displacements, Φ^k_τ is the vector of the degrees of freedom for the electric potential, $\boldsymbol{\mathcal{D}}^k_{nM\tau}$ is the vector of the degrees of freedom for the transverse normal electric displacement, $\boldsymbol{\sigma}^k_{nM\tau}$ is the vector of the degrees of freedom for the transverse stress components, and $\ddot{\boldsymbol{u}}^k_\tau$ is the second temporal derivative of $\boldsymbol{u}^k_\tau$. The array $\boldsymbol{p}^k_{us}$ indicates the variationally consistent mechanical loading that is applied in the case of the sensor configuration; in the case of an actuator configuration, the electric potential is applied directly to the vector Φ^k_τ. Along with these governing equations, the following boundary conditions on edge Γ_k of the in-plane integration domain Ω_k hold:

$$\begin{aligned}
&\boldsymbol{\Pi}^{k\tau s}_{uu}\,\boldsymbol{u}^k_\tau + \boldsymbol{\Pi}^{k\tau s}_{u\sigma}\,\boldsymbol{\sigma}^k_{nM\tau} + \boldsymbol{\Pi}^{k\tau s}_{u\Phi}\,\Phi^k_\tau + \boldsymbol{\Pi}^{k\tau s}_{u\mathcal{D}}\,\boldsymbol{\mathcal{D}}^k_{nM\tau}\\
&\quad = \boldsymbol{\Pi}^{k\tau s}_{uu}\,\bar{\boldsymbol{u}}^k_\tau + \boldsymbol{\Pi}^{k\tau s}_{u\sigma}\,\bar{\boldsymbol{\sigma}}^k_{nM\tau} + \boldsymbol{\Pi}^{k\tau s}_{u\Phi}\,\bar{\Phi}^k_\tau + \boldsymbol{\Pi}^{k\tau s}_{u\mathcal{D}}\,\bar{\boldsymbol{\mathcal{D}}}^k_{nM\tau}\\
&\boldsymbol{\Pi}^{k\tau s}_{\Phi u}\,\boldsymbol{u}^k_\tau + \boldsymbol{\Pi}^{k\tau s}_{\Phi\sigma}\,\boldsymbol{\sigma}^k_{nM\tau} + \boldsymbol{\Pi}^{k\tau s}_{\Phi\Phi}\,\Phi^k_\tau + \boldsymbol{\Pi}^{k\tau s}_{\Phi\mathcal{D}}\,\boldsymbol{\mathcal{D}}^k_{nM\tau}\\
&\quad = \boldsymbol{\Pi}^{k\tau s}_{\Phi u}\,\bar{\boldsymbol{u}}^k_\tau + \boldsymbol{\Pi}^{k\tau s}_{\Phi\sigma}\,\bar{\boldsymbol{\sigma}}^k_{nM\tau} + \boldsymbol{\Pi}^{k\tau s}_{\Phi\Phi}\,\bar{\Phi}^k_\tau + \boldsymbol{\Pi}^{k\tau s}_{\Phi\mathcal{D}}\,\bar{\boldsymbol{\mathcal{D}}}^k_{nM\tau}
\end{aligned} \tag{6.163}$$

Comparing Equation (6.161), after the integration by parts, to Equations (6.162) and (6.163), the fundamental nuclei can be obtained:

$$\boldsymbol{K}^{k\tau s}_{uu} = \int_{A_k} \Big[-\boldsymbol{D}^T_p \tilde{\boldsymbol{C}}^k_{\sigma_p\epsilon_p}\boldsymbol{D}_p\Big] F_s F_\tau dz \tag{6.164}$$

$$\boldsymbol{K}^{k\tau s}_{u\sigma} = \int_{A_k} \Big[-\boldsymbol{D}^T_p \tilde{\boldsymbol{C}}^k_{\sigma_p\sigma_n} + (-\boldsymbol{D}_{np} + \boldsymbol{D}_{nz})^T\Big] F_s F_\tau dz \tag{6.165}$$

$$\boldsymbol{K}^{k\tau s}_{u\Phi} = \int_{A_k} \Big[-\boldsymbol{D}^T_p(-\tilde{\boldsymbol{C}}^k_{\sigma_p\mathcal{E}_p}\boldsymbol{D}_{ep})\Big] F_s F_\tau dz \tag{6.166}$$

$$\boldsymbol{K}^{k\tau s}_{u\mathcal{D}} = \int_{A_k} \Big[-\boldsymbol{D}^T_p \tilde{\boldsymbol{C}}^k_{\sigma_p\mathcal{D}_n}\Big] F_s F_\tau dz \tag{6.167}$$

$$\boldsymbol{K}^{k\tau s}_{\sigma u} = \int_{A_k} \Big[(\boldsymbol{D}_{np} + \boldsymbol{D}_{nz}) - \tilde{\boldsymbol{C}}^k_{\epsilon_n\epsilon_p}\boldsymbol{D}_p\Big] F_s F_\tau dz \tag{6.168}$$

$$\boldsymbol{K}^{k\tau s}_{\sigma\sigma} = \int_{A_k} \Big[-\tilde{\boldsymbol{C}}^k_{\epsilon_n\sigma_n}\Big] F_s F_\tau dz \tag{6.169}$$

$$\boldsymbol{K}^{k\tau s}_{\sigma\Phi} = \int_{A_k} \Big[\tilde{\boldsymbol{C}}^k_{\epsilon_n\mathcal{E}_p}\boldsymbol{D}_{ep}\Big] F_s F_\tau dz \tag{6.170}$$

$$\boldsymbol{K}^{k\tau s}_{\sigma\mathcal{D}} = \int_{A_k} \Big[-\tilde{\boldsymbol{C}}^k_{\epsilon_n\mathcal{D}_n}\Big] F_s F_\tau dz \tag{6.171}$$

$$\boldsymbol{K}^{k\tau s}_{\Phi u} = \int_{A_k} \Big[-\boldsymbol{D}^T_{ep} \tilde{\boldsymbol{C}}^k_{\mathcal{D}_p\epsilon_p}\boldsymbol{D}_p\Big] F_s F_\tau dz \tag{6.172}$$

$$\boldsymbol{K}_{\Phi\sigma}^{k\tau s} = \int_{A_k} \Big[- \boldsymbol{D}_{ep}^T \tilde{\boldsymbol{C}}_{\mathcal{D}_p\sigma_n}^k \Big] F_s F_\tau dz \tag{6.173}$$

$$\boldsymbol{K}_{\Phi\Phi}^{k\tau s} = \int_{A_k} \Big[\boldsymbol{D}_{ep}^T \tilde{\boldsymbol{C}}_{\mathcal{D}_p\mathcal{E}_p}^k \boldsymbol{D}_{ep} \Big] F_s F_\tau dz \tag{6.174}$$

$$\boldsymbol{K}_{\Phi\mathcal{D}}^{k\tau s} = \int_{A_k} \Big[- \boldsymbol{D}_{ep}^T \tilde{\boldsymbol{C}}_{\mathcal{D}_p\mathcal{D}_n}^k + \boldsymbol{D}_{en}^T \Big] F_s F_\tau dz \tag{6.175}$$

$$\boldsymbol{K}_{\mathcal{D}u}^{k\tau s} = \int_{A_k} \Big[\tilde{\boldsymbol{C}}_{\mathcal{E}_n\epsilon_p}^k \boldsymbol{D}_p^T \Big] F_s F_\tau dz \tag{6.176}$$

$$\boldsymbol{K}_{\mathcal{D}\sigma}^{k\tau s} = \int_{A_k} \Big[\tilde{\boldsymbol{C}}_{\mathcal{E}_n\sigma_n}^k \Big] F_s F_\tau dz \tag{6.177}$$

$$\boldsymbol{K}_{\mathcal{D}\Phi}^{k\tau s} = \int_{A_k} \Big[\boldsymbol{D}_{en} - \tilde{\boldsymbol{C}}_{\mathcal{E}_n\mathcal{E}_p}^k \boldsymbol{D}_{ep} \Big] F_s F_\tau dz \tag{6.178}$$

$$\boldsymbol{K}_{\mathcal{D}\mathcal{D}}^{k\tau s} = \int_{A_k} \Big[\tilde{\boldsymbol{C}}_{\mathcal{E}_n\mathcal{D}_n}^k \Big] F_s F_\tau dz \tag{6.179}$$

The inertial array is the same as that of the previous sections. The nuclei for the boundary conditions are (Carrera and Brischetto 2007a; Carrera *et al.* 2008):

$$\boldsymbol{\Pi}_{uu}^{k\tau s} = \int_{A_k} \Big[\boldsymbol{I}_p^T \tilde{\boldsymbol{C}}_{\sigma_p\epsilon_p}^k \boldsymbol{D}_p \Big] F_s F_\tau dz \tag{6.180}$$

$$\boldsymbol{\Pi}_{u\sigma}^{k\tau s} = \int_{A_k} \Big[\boldsymbol{I}_p^T \tilde{\boldsymbol{C}}_{\sigma_p\sigma_n}^k + \boldsymbol{I}_{np}^T \Big] F_s F_\tau dz \tag{6.181}$$

$$\boldsymbol{\Pi}_{u\Phi}^{k\tau s} = \int_{A_k} \Big[- \boldsymbol{I}_p^T \tilde{\boldsymbol{C}}_{\sigma_p\mathcal{E}_p}^k \boldsymbol{D}_{ep} \Big] F_s F_\tau dz \tag{6.182}$$

$$\boldsymbol{\Pi}_{u\mathcal{D}}^{k\tau s} = \int_{A_k} \Big[\boldsymbol{I}_p^T \tilde{\boldsymbol{C}}_{\sigma_p\mathcal{D}_n}^k \Big] F_s F_\tau dz \tag{6.183}$$

$$\boldsymbol{\Pi}_{\Phi u}^{k\tau s} = \int_{A_k} \Big[\boldsymbol{I}_{ep}^T \tilde{\boldsymbol{C}}_{\mathcal{D}_p\epsilon_p}^k \boldsymbol{D}_p \Big] F_s F_\tau dz \tag{6.184}$$

$$\mathbf{\Pi}^{k\tau}_{\Phi\sigma} = \int_{A_k} \left[\boldsymbol{I}^T_{ep} \tilde{\boldsymbol{C}}^k_{\mathcal{D}_p\sigma_n} \right] F_s F_\tau dz \tag{6.185}$$

$$\mathbf{\Pi}^{k\tau s}_{\Phi\Phi} = \int_{A_k} \left[-\boldsymbol{I}^T_{ep} \tilde{\boldsymbol{C}}^k_{\mathcal{D}_p\mathcal{E}_p} \boldsymbol{D}_{ep} \right] F_s F_\tau dz \tag{6.186}$$

$$\mathbf{\Pi}^{k\tau s}_{\Phi\mathcal{D}} = \int_{A_k} \left[\boldsymbol{I}^T_{ep} \tilde{\boldsymbol{C}}^k_{\mathcal{D}_p\mathcal{D}_n} \right] F_s F_\tau dz \tag{6.187}$$

In order to perform the integration by parts (see Equation (6.28)), the matrices $\boldsymbol{I}_p$, $\boldsymbol{I}_{np}$, and $\boldsymbol{I}_{ep}$, which are those presented in Equation (6.29), are introduced. To write the explicit form of the nuclei in Equations (6.164)–(6.179), the integrals in the z thickness direction are defined as in Equation (6.41). By developing the matrix products in Equations (6.164)–(6.179) and employing a Navier-type closed-form solution (Carrera and Brischetto 2007a; Carrera *et al.* 2008), the explicit algebraic form of the nuclei can be obtained.

Nucleus $\boldsymbol{K}^{k\tau s}_{uu}$ of (3 × 3) dimension is:

$$\begin{aligned}
K^{k\tau s}_{uu_{11}} &= \bar{\alpha}^2 J^{k\tau s} \tilde{C}^k_{\sigma_p\epsilon_p 11} + \bar{\beta}^2 J^{k\tau s} \tilde{C}^k_{\sigma_p\epsilon_p 33} \\
K^{k\tau s}_{uu_{12}} &= J^{k\tau s}(\tilde{C}^k_{\sigma_p\epsilon_p 12} + \tilde{C}^k_{\sigma_p\epsilon_p 33})\bar{\alpha}\bar{\beta} \\
K^{k\tau s}_{uu_{13}} &= 0, \quad K^{k\tau s}_{uu_{21}} = J^{k\tau s}(\tilde{C}^k_{\sigma_p\epsilon_p 21} + \tilde{C}^k_{\sigma_p\epsilon_p 33})\bar{\alpha}\bar{\beta} \\
K^{k\tau s}_{uu_{22}} &= \bar{\beta}^2 J^{k\tau s} \tilde{C}^k_{\sigma_p\epsilon_p 22} + \bar{\alpha}^2 J^{k\tau s} \tilde{C}^k_{\sigma_p\epsilon_p 33}, \quad K^{k\tau s}_{uu_{23}} = 0 \\
K^{k\tau s}_{uu_{31}} &= 0, \quad K^{k\tau s}_{uu_{32}} = 0, \quad K^{k\tau s}_{uu_{33}} = 0
\end{aligned} \tag{6.188}$$

Nucleus $\boldsymbol{K}^{k\tau s}_{u\sigma}$ of (3 × 3) dimension is:

$$\begin{aligned}
K^{k\tau s}_{u\sigma_{11}} &= J^{k\tau s_z}, \quad K^{k\tau s}_{u\sigma_{12}} = 0, \quad K^{k\tau s}_{u\sigma_{13}} = -\bar{\alpha} J^{k\tau s} \tilde{C}^k_{\sigma_p\sigma_n 13} \\
K^{k\tau s}_{u\sigma_{21}} &= 0, \quad K^{k\tau s}_{u\sigma_{22}} = J^{k\tau s_z}, \quad K^{k\tau s}_{u\sigma_{23}} = -\bar{\beta} J^{k\tau s} \tilde{C}^k_{\sigma_p\sigma_n 23} \\
K^{k\tau s}_{u\sigma_{31}} &= \bar{\alpha} J^{k\tau s}, \quad K^{k\tau s}_{u\sigma_{32}} = \bar{\beta} J^{k\tau s}, \quad K^{k\tau s}_{u\sigma_{33}} = J^{k\tau s_z}
\end{aligned} \tag{6.189}$$

Nucleus $\boldsymbol{K}^{k\tau s}_{u\Phi}$ of (3 × 1) dimension is:

$$K^{k\tau s}_{u\Phi_{11}} = K^{k\tau s}_{u\Phi_{21}} = K^{k\tau s}_{u\Phi_{31}} = 0 \tag{6.190}$$

Nucleus $\boldsymbol{K}^{k\tau s}_{u\mathcal{D}}$ of (3 × 1) dimension is:

$$K^{k\tau s}_{u\mathcal{D}_{11}} = -\bar{\alpha} J^{k\tau s} \tilde{C}^k_{\sigma_p\mathcal{D}_n 11}, \quad K^{k\tau s}_{u\mathcal{D}_{21}} = -\bar{\beta} J^{k\tau s} \tilde{C}^k_{\sigma_p\mathcal{D}_n 21}, \quad K^{k\tau s}_{u\mathcal{D}_{31}} = 0 \tag{6.191}$$

Nucleus $\boldsymbol{K}_{\sigma u}^{k\tau s}$ of (3 × 3) dimension is:

$$K_{\sigma u_{11}}^{k\tau s} = J^{k\tau_z s}, \quad K_{\sigma u_{12}}^{k\tau s} = 0, \quad K_{\sigma u_{13}}^{k\tau s} = \bar{\alpha} J^{k\tau s}$$
$$K_{\sigma u_{21}}^{k\tau s} = 0, \quad K_{\sigma u_{22}}^{k\tau s} = J^{k\tau_z s}, \quad K_{u\sigma_{23}}^{k\tau s} = \bar{\beta} J^{k\tau s}, \quad K_{\sigma u_{31}}^{k\tau s} = \bar{\alpha} J^{k\tau s} \tilde{C}_{\epsilon_n \epsilon_p 31}^{k}$$
$$K_{\sigma u_{32}}^{k\tau s} = \bar{\beta} J^{k\tau s} \tilde{C}_{\epsilon_n \epsilon_p 32}^{k}, \quad K_{\sigma u_{33}}^{k\tau s} = J^{k\tau_z s} \tag{6.192}$$

Nucleus $\boldsymbol{K}_{\sigma\sigma}^{k\tau s}$ of (3 × 3) dimension is:

$$\begin{aligned} K_{\sigma\sigma_{11}}^{k\tau s} &= -J^{k\tau s} \tilde{C}_{\epsilon_n \sigma_n 11}^{k}, \quad K_{\sigma\sigma_{12}}^{k\tau s} = 0, \quad K_{\sigma\sigma_{13}}^{k\tau s} = 0 \\ K_{\sigma\sigma_{21}}^{k\tau s} &= 0, \quad K_{\sigma\sigma_{22}}^{k\tau s} = -J^{k\tau s} \tilde{C}_{\epsilon_n \sigma_n 22}^{k}, \quad K_{\sigma\sigma_{23}}^{k\tau s} = 0 \\ K_{\sigma\sigma_{31}}^{k\tau s} &= 0, \quad K_{\sigma\sigma_{32}}^{k\tau s} = 0, \quad K_{\sigma\sigma_{33}}^{k\tau s} = -J^{k\tau s} \tilde{C}_{\epsilon_n \sigma_n 33}^{k} \end{aligned} \tag{6.193}$$

Nucleus $\boldsymbol{K}_{\sigma\Phi}^{k\tau s}$ of (3 × 1) dimension is:

$$K_{\sigma\Phi_{11}}^{k\tau s} = \bar{\alpha} J^{k\tau s} \tilde{C}_{\epsilon_n \mathcal{E}_p 11}^{k}, \quad K_{\sigma\Phi_{21}}^{k\tau s} = \bar{\beta} J^{k\tau s} \tilde{C}_{\epsilon_n \mathcal{E}_p 22}^{k}, \quad K_{\sigma\Phi_{31}}^{k\tau s} = 0 \tag{6.194}$$

Nucleus $\boldsymbol{K}_{\sigma\mathcal{D}}^{k\tau s}$ of (3 × 1) dimension is:

$$K_{\sigma\mathcal{D}_{11}}^{k\tau s} = K_{\sigma\mathcal{D}_{21}}^{k\tau s} = 0, \quad K_{\sigma\mathcal{D}_{31}}^{k\tau s} = -J^{k\tau s} \tilde{C}_{\epsilon_n \mathcal{D}_n 31}^{k} \tag{6.195}$$

Nucleus $\boldsymbol{K}_{\Phi u}^{k\tau s}$ of (1 × 3) dimension is:

$$K_{\Phi u_{11}}^{k\tau s} = K_{\Phi u_{12}}^{k\tau s} = K_{\Phi u_{13}}^{k\tau s} = 0 \tag{6.196}$$

Nucleus $\boldsymbol{K}_{\Phi\sigma}^{k\tau s}$ of (1 × 3) dimension is:

$$K_{\Phi\sigma_{11}}^{k\tau s} = \bar{\alpha} J^{k\tau s} \tilde{C}_{\mathcal{D}_p \sigma_n 11}^{k}, \quad K_{\Phi\sigma_{12}}^{k\tau s} = \bar{\beta} J^{k\tau s} \tilde{C}_{\mathcal{D}_p \sigma_n 22}^{k}, \quad K_{\Phi\sigma_{13}}^{k\tau s} = 0 \tag{6.197}$$

Nucleus $\boldsymbol{K}_{\Phi\Phi}^{k\tau s}$ of (1 × 1) dimension is:

$$K_{\Phi\Phi_{11}}^{k\tau s} = -\bar{\alpha}^2 J^{k\tau s} \tilde{C}_{\mathcal{D}_p \mathcal{E}_p 11}^{k} - \bar{\beta}^2 J^{k\tau s} \tilde{C}_{\mathcal{D}_p \mathcal{E}_p 22}^{k} \tag{6.198}$$

Nucleus $\boldsymbol{K}_{\Phi\mathcal{D}}^{k\tau s}$ of dimension (1 × 1) is:

$$K_{\Phi\mathcal{D}_{11}}^{k\tau s} = J^{k\tau_z s} \tag{6.199}$$

Nucleus $\boldsymbol{K}_{\mathcal{D}u}^{k\tau s}$ of (1 × 3) dimension is:

$$K_{\mathcal{D}u_{11}}^{k\tau s} = -\bar{\alpha} J^{k\tau s} \tilde{C}_{\mathcal{E}_n \epsilon_p 11}^{k}, \quad K_{\mathcal{D}u_{12}}^{k\tau s} = -\bar{\beta} J_{\alpha}^{k\tau s} \tilde{C}_{\mathcal{E}_n \epsilon_p 12}^{k}, \quad K_{\mathcal{D}u_{13}}^{k\tau s} = 0 \tag{6.200}$$

Nucleus $\boldsymbol{K}^{k\tau s}_{\mathcal{D}\sigma}$ of (1×3) dimension is:

$$K^{k\tau s}_{\mathcal{D}\sigma_{11}} = K^{k\tau s}_{\mathcal{D}\sigma_{12}} = 0, \quad K^{k\tau s}_{\mathcal{D}\sigma_{13}} = J^{k\tau s}\tilde{C}^k_{\mathcal{E}_n\sigma_n 13} \tag{6.201}$$

Nucleus $\boldsymbol{K}^{k\tau s}_{\mathcal{D}\Phi}$ of (1×1) dimension is:

$$K^{k\tau s}_{\mathcal{D}\Phi_{11}} = J^{k\tau_z s} \tag{6.202}$$

Nucleus $\boldsymbol{K}^{k\tau s}_{\mathcal{D}\mathcal{D}}$ of (1×1) dimension is:

$$K^{k\tau s}_{\mathcal{D}\mathcal{D}_{11}} = J^{k\tau s}\tilde{C}^k_{\mathcal{E}_n\mathcal{D}_n 11} \tag{6.203}$$

The fundamental nucleus for the inertial matrix $\boldsymbol{M}^{k\tau s}_{uu}$ and its components were given in Equations (6.36) and (6.46).

$\bar{\alpha} = m\pi/a$ and $\bar{\beta} = n\pi/b$, where m and n are the wave numbers in the in-plane directions and a and b are the plate dimensions in the x and y directions, respectively. A Navier-type closed-form solution is obtained via substitution of the harmonic expressions for the displacements, electric potential, transverse stresses, transverse normal electric displacement, and considering the following material coefficients to be equal to zero: $Q_{16} = Q_{26} = Q_{36} = Q_{45} = 0$ and $e_{25} = e_{14} = e_{36} = \varepsilon_{12} = 0$. The harmonic assumptions used for the displacements, the electric potential, the transverse stresses, and the transverse normal electric displacement are:

$$(u^k_\tau, \sigma^k_{xz\tau}) = \sum_{m,n} (\hat{U}^k_\tau, \hat{\sigma}^k_{xz\tau}) \cos\left(\frac{m\pi x}{a}\right) \sin\left(\frac{n\pi y}{b}\right), \quad k = 1, N_l \tag{6.204}$$

$$(v^k_\tau, \sigma^k_{yz\tau}) = \sum_{m,n} (\hat{V}^k_\tau, \hat{\sigma}^k_{yz\tau}) \sin\left(\frac{m\pi x}{a}\right) \cos\left(\frac{n\pi y}{b}\right), \quad \tau = t, b, r \tag{6.205}$$

$$(w^k_\tau, \Phi^k_\tau, \mathcal{D}^k_{z\tau}, \sigma^k_{zz\tau}) = \sum_{m,n} (\hat{W}^k_\tau, \hat{\Phi}^k_\tau, \hat{\mathcal{D}}^k_{z\tau}, \hat{\sigma}^k_{zz\tau}) \sin\left(\frac{m\pi x}{a}\right) \sin\left(\frac{n\pi y}{b}\right),$$
$$r = 2, N \tag{6.206}$$

where $\hat{U}^k_\tau$, $\hat{V}^k_\tau$, $\hat{W}^k_\tau$ are the displacement amplitudes, $\hat{\Phi}^k_\tau$ is the electric potential amplitude, $\hat{\mathcal{D}}^k_{z\tau}$ is the transverse normal electric displacement amplitude, and $\hat{\sigma}^k_{xz\tau}$, $\hat{\sigma}^k_{yz\tau}$, $\hat{\sigma}^k_{zz\tau}$ are the transverse stress amplitudes; k indicates the layer and N_l is the total number of layers. τ is the index for the order of expansion where t and b indicate the top and bottom of the layer, respectively, while r indicates the higher orders of expansion until $N = 4$. Details on the assembly procedure of the fundamental nuclei and on the acronyms are given in Sections 6.7 and 6.8, respectively.

The meaning of the $\tilde{\boldsymbol{C}}$ coefficients for Equations (6.157)–(6.160) has been given in Equations (2.88). The following equations give the explicit form of each of their components.

The $\tilde{\boldsymbol{C}}^k_{\sigma_p\epsilon_p}$ array has a (3×3) dimension with components for each k layer:

$$
\begin{aligned}
\tilde{C}_{\sigma_p\epsilon_p 11} &= \frac{e_{33}^2 Q_{11} - 2e_{31}e_{33}Q_{13} - Q_{13}^2\varepsilon_{33} + Q_{33}(e_{31}^2 + Q_{11}\varepsilon_{33})}{e_{33}^2 + Q_{33}\varepsilon_{33}} \\
\tilde{C}_{\sigma_p\epsilon_p 12} &= \\
& Q_{12} + \frac{e_{32}(-e_{33}Q_{13} + e_{31}Q_{33}) - Q_{23}(e_{31}e_{33} + Q_{13}\varepsilon_{33})}{e_{33}^2 + Q_{33}\varepsilon_{33}}, \ \tilde{C}_{\sigma_p\epsilon_p 13} = 0 \\
\tilde{C}_{\sigma_p\epsilon_p 21} &= Q_{12} + \frac{e_{32}(-e_{33}Q_{13} + e_{31}Q_{33}) - Q_{23}(e_{31}e_{33} + Q_{13}\varepsilon_{33})}{e_{33}^2 + Q_{33}\varepsilon_{33}} \\
\tilde{C}_{\sigma_p\epsilon_p 22} &= \\
& \frac{e_{33}^2 Q_{22} - 2e_{32}e_{33}Q_{23} - Q_{23}^2\varepsilon_{33} + Q_{33}(e_{32}^2 + Q_{22}\varepsilon_{33})}{e_{33}^2 + Q_{33}\varepsilon_{33}}, \ \tilde{C}_{\sigma_p\epsilon_p 23} = 0 \\
\tilde{C}_{\sigma_p\epsilon_p 31} &= \tilde{C}_{\sigma_p\epsilon_p 32} = 0, \quad \tilde{C}_{\sigma_p\epsilon_p 33} = Q_{66}
\end{aligned}
\tag{6.207}
$$

The $\tilde{\boldsymbol{C}}^k_{\sigma_p\sigma_n}$ array has a (3×3) dimension with components for each k layer:

$$
\begin{aligned}
\tilde{C}_{\sigma_p\sigma_n 11} &= \tilde{C}_{\sigma_p\sigma_n 12} = 0, \quad \tilde{C}_{\sigma_p\sigma_n 13} = \frac{e_{31}e_{33} + Q_{13}\varepsilon_{33}}{e_{33}^2 + Q_{33}\varepsilon_{33}} \\
\tilde{C}_{\sigma_p\sigma_n 21} &= \tilde{C}_{\sigma_p\sigma_n 22} = 0, \quad \tilde{C}_{\sigma_p\sigma_n 23} = \frac{e_{32}e_{33} + Q_{23}\varepsilon_{33}}{e_{33}^2 + Q_{33}\varepsilon_{33}} \\
\tilde{C}_{\sigma_p\sigma_n 31} &= \tilde{C}_{\sigma_p\sigma_n 32} = \tilde{C}_{\sigma_p\sigma_n 33} = 0
\end{aligned}
\tag{6.208}
$$

The $\tilde{\boldsymbol{C}}^k_{\sigma_p\mathcal{E}_p}$ array has a (3×2) dimension with components for each k layer:

$$
\begin{aligned}
\tilde{C}_{\sigma_p\mathcal{E}_p 11} &= \tilde{C}_{\sigma_p\mathcal{E}_p 12} = \tilde{C}_{\sigma_p\mathcal{E}_p 21} = 0 \\
\tilde{C}_{\sigma_p\mathcal{E}_p 22} &= \tilde{C}_{\sigma_p\mathcal{E}_p 31} = \tilde{C}_{\sigma_p\mathcal{E}_p 32} = 0
\end{aligned}
\tag{6.209}
$$

The $\tilde{C}^k_{\sigma_p \mathcal{D}_n}$ array has a (3×1) dimension with components for each k layer:

$$\tilde{C}_{\sigma_p \mathcal{D}_n 11} = \frac{e_{33} Q_{13} - e_{31} Q_{33}}{e_{33}^2 + Q_{33} \varepsilon_{33}}, \quad \tilde{C}_{\sigma_p \mathcal{D}_n 21} = \frac{e_{33} Q_{23} - e_{32} Q_{33}}{e_{33}^2 + Q_{33} \varepsilon_{33}}, \quad \tilde{C}_{\sigma_p \mathcal{D}_n 31} = 0 \tag{6.210}$$

The $\tilde{C}^k_{\epsilon_n \epsilon_p}$ array has a (3×3) dimension with components for each k layer:

$$\tilde{C}_{\epsilon_n \epsilon_p 11} = \tilde{C}_{\epsilon_n \epsilon_p 12} = \tilde{C}_{\epsilon_n \epsilon_p 13} = \tilde{C}_{\epsilon_n \epsilon_p 21} = \tilde{C}_{\epsilon_n \epsilon_p 22} = \tilde{C}_{\epsilon_n \epsilon_p 23} = 0$$

$$\tilde{C}_{\epsilon_n \epsilon_p 31} = \frac{-2 e_{33}^2 Q_{13} + e_{31} e_{33} Q_{33} - Q_{13} Q_{33} \varepsilon_{33}}{Q_{33}(e_{33}^2 + Q_{33} \varepsilon_{33})} \tag{6.211}$$

$$\tilde{C}_{\epsilon_n \epsilon_p 32} = \frac{-2 e_{33}^2 Q_{23} + e_{32} e_{33} Q_{33} - Q_{23} Q_{33} \varepsilon_{33}}{Q_{33}(e_{33}^2 + Q_{33} \varepsilon_{33})}, \quad \tilde{C}_{\epsilon_n \epsilon_p 33} = 0$$

The $\tilde{C}^k_{\epsilon_n \sigma_n}$ array has a (3×3) dimension with components for each k layer:

$$\tilde{C}_{\epsilon_n \sigma_n 11} = \frac{1}{Q_{55}}, \quad \tilde{C}_{\epsilon_n \sigma_n 12} = \tilde{C}_{\epsilon_n \sigma_n 13} = \tilde{C}_{\epsilon_n \sigma_n 21} = 0, \quad \tilde{C}_{\epsilon_n \sigma_n 22} = \frac{1}{Q_{44}}$$

$$\tilde{C}_{\epsilon_n \sigma_n 23} = \tilde{C}_{\epsilon_n \sigma_n 31} = \tilde{C}_{\epsilon_n \sigma_n 32} = 0, \quad \tilde{C}_{\epsilon_n \sigma_n 33} = \frac{\varepsilon_{33}}{e_{33}^2 + Q_{33} \varepsilon_{33}} \tag{6.212}$$

The $\tilde{C}^k_{\epsilon_n \mathcal{E}_p}$ array has a (3×2) dimension with components for each k layer:

$$\tilde{C}_{\epsilon_n \mathcal{E}_p 11} = \frac{e_{15}}{Q_{55}}, \quad \tilde{C}_{\epsilon_n \mathcal{E}_p 12} = \tilde{C}_{\epsilon_n \mathcal{E}_p 21} = 0$$
$$\tilde{C}_{\epsilon_n \mathcal{E}_p 22} = \frac{e_{24}}{Q_{44}}, \quad \tilde{C}_{\epsilon_n \mathcal{E}_p 31} = \tilde{C}_{\epsilon_n \mathcal{E}_p 32} = 0 \tag{6.213}$$

The $\tilde{C}^k_{\epsilon_n \mathcal{D}_n}$ array has a (3×1) dimension with components for each k layer:

$$\tilde{C}_{\epsilon_n \mathcal{D}_n 11} = \tilde{C}_{\epsilon_n \mathcal{D}_n 21} = 0, \quad \tilde{C}_{\epsilon_n \mathcal{D}_n 31} = \frac{e_{33}}{e_{33}^2 + Q_{33} \varepsilon_{33}} \tag{6.214}$$

The $\tilde{\boldsymbol{C}}^k_{\mathcal{D}_p\epsilon_p}$ array has a (2 × 3) dimension with components for each k layer:

$$\tilde{C}_{\mathcal{D}_p\epsilon_p11} = \tilde{C}_{\mathcal{D}_p\epsilon_p12} = \tilde{C}_{\mathcal{D}_p\epsilon_p13} = \tilde{C}_{\mathcal{D}_p\epsilon_p21} = \tilde{C}_{\mathcal{D}_p\epsilon_p22} = \tilde{C}_{\mathcal{D}_p\epsilon_p23} = 0 \quad (6.215)$$

The $\tilde{\boldsymbol{C}}^k_{\mathcal{D}_p\sigma_n}$ array has a (2 × 3) dimension with components for each k layer:

$$\begin{aligned} &\tilde{C}_{\mathcal{D}_p\sigma_n11} = \frac{e_{15}}{Q_{55}}, \quad \tilde{C}_{\mathcal{D}_p\sigma_n12} = \tilde{C}_{\mathcal{D}_p\sigma_n13} = 0 \\ &\tilde{C}_{\mathcal{D}_p\sigma_n21} = 0, \quad \tilde{C}_{\mathcal{D}_p\sigma_n22} = \frac{e_{24}}{Q_{44}}, \quad \tilde{C}_{\mathcal{D}_p\sigma_n23} = 0 \end{aligned} \quad (6.216)$$

The $\tilde{\boldsymbol{C}}^k_{\mathcal{D}_p\mathcal{E}_p}$ array has a (2 × 2) dimension with components for each k layer:

$$\tilde{C}_{\mathcal{D}_p\mathcal{E}_p11} = \frac{e_{15}^2}{Q_{55}} + \varepsilon_{11}, \quad \tilde{C}_{\mathcal{D}_p\mathcal{E}_p12} = \tilde{C}_{\mathcal{D}_p\mathcal{E}_p21} = 0, \quad \tilde{C}_{\mathcal{D}_p\mathcal{E}_p22} = \frac{e_{24}^2}{Q_{44}} + \varepsilon_{22} \quad (6.217)$$

The $\tilde{\boldsymbol{C}}^k_{\mathcal{D}_p\mathcal{D}_n}$ array has a (2 × 1) dimension with components for each k layer:

$$\tilde{C}_{\mathcal{D}_p\mathcal{D}_n11} = \tilde{C}_{\mathcal{D}_p\mathcal{D}_n21} = 0 \quad (6.218)$$

The $\tilde{\boldsymbol{C}}^k_{\mathcal{E}_n\epsilon_p}$ array has a (1 × 3) dimension with components for each k layer:

$$\tilde{C}_{\mathcal{E}_n\epsilon_p11} = \frac{e_{33}Q_{13} - e_{31}Q_{33}}{e_{33}^2 + Q_{33}\varepsilon_{33}}, \quad \tilde{C}_{\mathcal{E}_n\epsilon_p12} = \frac{e_{33}Q_{23} - e_{32}Q_{33}}{e_{33}^2 + Q_{33}\varepsilon_{33}}, \quad \tilde{C}_{\mathcal{E}_n\epsilon_p13} = 0 \quad (6.219)$$

The $\tilde{\boldsymbol{C}}^k_{\mathcal{E}_n\sigma_n}$ array has a (1 × 3) dimension with components for each k layer:

$$\tilde{C}_{\mathcal{E}_n\sigma_n11} = \tilde{C}_{\mathcal{E}_n\sigma_n12} = 0, \quad \tilde{C}_{\mathcal{E}_n\sigma_n13} = -\frac{e_{33}}{e_{33}^2 + Q_{33}\varepsilon_{33}} \quad (6.220)$$

The $\tilde{\boldsymbol{C}}^k_{\mathcal{E}_n\mathcal{E}_p}$ array has a (1 × 2) dimension with components for each k layer:

$$\tilde{C}_{\mathcal{E}_n\mathcal{E}_p11} = \tilde{C}_{\mathcal{E}_n\mathcal{E}_p12} = 0 \quad (6.221)$$

The $\tilde{\boldsymbol{C}}^k_{\mathcal{E}_n\mathcal{D}_n}$ array has a (1×1) dimension with components for each k layer:

$$\tilde{C}_{\mathcal{E}_n\mathcal{D}_n 11} = \frac{1}{e_{33^2}/Q_{22} + \varepsilon_{33}} \tag{6.222}$$

6.7 Assembly procedure for fundamental nuclei

The models proposed in Section 6.3 are based on the extension of the PVD to electromechanical problems, PVD($\boldsymbol{u}, \Phi$), and have two primary variables in the governing equations: the displacement vector $\boldsymbol{u}^k = (u^k, v^k, w^k)$ and the electric potential Φ^k. Three extensions of the RMVT to electromechanical problems are possible. Section 6.4 shows RMVT($\boldsymbol{u}, \Phi, \boldsymbol{\sigma}_n$) in which three primary variables are considered in the governing equations: the displacement vector $\boldsymbol{u}^k = (u^k, v^k, w^k)$, the electric potential Φ^k, and the transverse stress components vector $\boldsymbol{\sigma}^k_n = (\sigma^k_{xz}, \sigma^k_{yz}, \sigma^k_{zz})$. Section 6.5 contains RMVT($\boldsymbol{u}, \Phi, \boldsymbol{\mathcal{D}}_n$) in which the three primary variables are displacement vector $\boldsymbol{u}^k = (u^k, v^k, w^k)$, the electric potential Φ^k, and the transverse normal electric displacement $\boldsymbol{\mathcal{D}}^k_n = (\mathcal{D}^k_z)$. Finally, RMVT($\boldsymbol{u}, \Phi, \boldsymbol{\sigma}_n, \boldsymbol{\mathcal{D}}_n$) in Section 6.6 has four primary variables: the displacement vector, the electric potential, the transverse normal electric displacement, and the transverse stress components.

The choice made in this book is that the displacement $\boldsymbol{u}^k = (u^k, v^k, w^k)$ can be modeled in both ESL and LW form; the other three variables are always modeled in LW form, which means that an electromechanical model is defined as ESL or LW, depending on the choice made for the displacement unknowns. Each modeled variable, regardless of which multilayer assembly procedure is considered (ESL or LW), has the same order of expansion in the thickness direction (from linear $N = 1$ to fourth order $N = 4$). A typical Taylor expansion is used in the case of an ESL assembly procedure, while a combination of Legendre polynomials is used as thickness functions in the case of an LW assembly procedure. In the ESL approach, the multilayered plate is considered as one equivalent plate and the stiffnesses of each embedded layer are simply summed, while in the LW approach, each embedded layer is considered as an independent plate and in the global stiffness matrix each contribution is partially summed considering the compatibility and/or equilibrium conditions at each layer interface.

Fundamental nuclei $\boldsymbol{K}^{k\tau s}_{uu}$ can be assembled in ESL or LW form, as indicated in Figure 6.10. Here, an example is given for a three-layered plate. The stiffness is first obtained for each layer by expansion via the indexes τ and s, which consider the order of expansion in the thickness direction, then the three stiffnesses obtained for each layer can be assembled at the multilayer level in

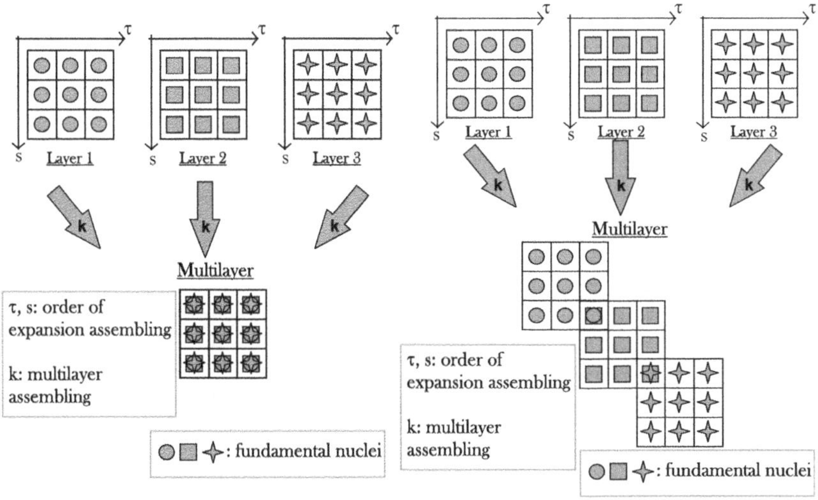

Figure 6.10 Assembly procedure for fundamental nuclei in ESL form (left) and LW form (right).

ESL form (on the left) by means of a simple summation, or assembled at the multilayer level in LW form (on the right) by considering the compatibility conditions for the displacement components at each layer interface.

Fundamental nuclei $\boldsymbol{K}_{\sigma\sigma}^{k\tau s}$, $\boldsymbol{K}_{\sigma\Phi}^{k\tau s}$, $\boldsymbol{K}_{\Phi\sigma}^{k\tau s}$, $\boldsymbol{K}_{\Phi\Phi}^{k\tau s}$, $\boldsymbol{K}_{\Phi\mathcal{D}}^{k\tau s}$, $\boldsymbol{K}_{\mathcal{D}\Phi}^{k\tau s}$, $\boldsymbol{K}_{\mathcal{D}\mathcal{D}}^{k\tau s}$, $\boldsymbol{K}_{\mathcal{D}\sigma}^{k\tau s}$, and $\boldsymbol{K}_{\sigma\mathcal{D}}^{k\tau s}$ are always assembled in LW form, as indicated on the right of Figure 6.10; in these cases, the partial summation of the stiffness matrices of each layer, for the multilayer assembly procedure, is done by imposing the continuity of the electric potential and/or the transverse stresses and/or the transverse normal electric displacement at each layer interface.

Fundamental nuclei $\boldsymbol{K}_{u\sigma}^{k\tau s}$, $\boldsymbol{K}_{u\Phi}^{k\tau s}$, and $\boldsymbol{K}_{u\mathcal{D}}^{k\tau s}$ can be assembled in LW form, as indicated on the right of Figure 6.10, by imposing the continuity of the displacements and/or transverse stresses and/or electric potential and/or transverse normal electric displacement at each layer interface. They can also be assembled in a partial ESL form, as indicated on the left of Figure 6.11, where the global stiffness matrix at the multilayer level is obtained by assembling the displacements in ESL form (see the rows) and the other variables in LW form (see the columns).

Fundamental nuclei $\boldsymbol{K}_{\sigma u}^{k\tau s}$, $\boldsymbol{K}_{\Phi u}^{k\tau s}$ and $\boldsymbol{K}_{\mathcal{D}u}^{k\tau s}$ can be assembled in LW form, as indicated on the right of Figure 6.10, by imposing the continuity of the displacements and/or transverse stresses and/or electric potential and/or transverse normal electric displacement at each layer interface. They can also be assembled in partial ESL form, as indicated on the right of Figure 6.11, where

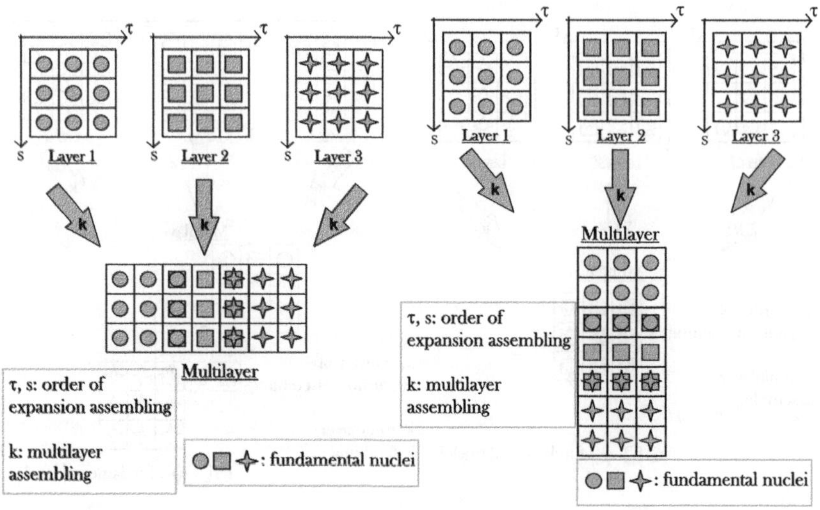

Figure 6.11 Assembly procedure for fundamental nuclei in ESL form for nuclei of type $\boldsymbol{K}_{u\sigma}^{k\tau s}$ (left) and for nuclei of type $\boldsymbol{K}_{\sigma u}^{k\tau s}$ (right).

the global stiffness matrix at the multilayer level is obtained by assembling the other variables in LW form (see the rows) and the displacements in ESL form (see the columns).

In the general fundamental nucleus $\boldsymbol{K}^{k\tau s}$, the index k for the kth layer permits the multilayer assembly procedure (both ESL and LW approaches), while the indexes τ and s permit the expansion in the thickness direction until the considered order N. In the case of the finite element (FE) approach, the general nucleus $\boldsymbol{K}^{k\tau sij}$ has two further indexes, i and j, which permit the FE assembly procedure in the plane by means of the nodes and shape functions. Details on the FE procedure can be found in Chapters 4 and 8.

6.8 Acronyms for refined and advanced models

The refined and advanced electromechanical models obtained in this chapter by means of the PVD and the three extensions of the RMVT can be defined by means of a system of acronyms that explains the multilayer approach (ESL or LW) for the displacements (the other variables are always LW), the employed variational statement (PVD or a possible extension of RMVT), and the order of expansion in the thickness direction, which is the same for all the variables (from linear to fourth order). This acronym system is shown in Figure 6.12. The letter E is used to indicate ESL displacements, while the letter L is used for the displacements in LW form; D means PVD and M stands for mixed

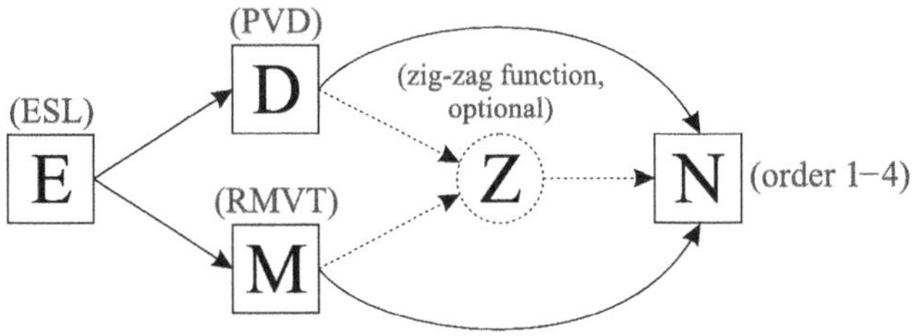

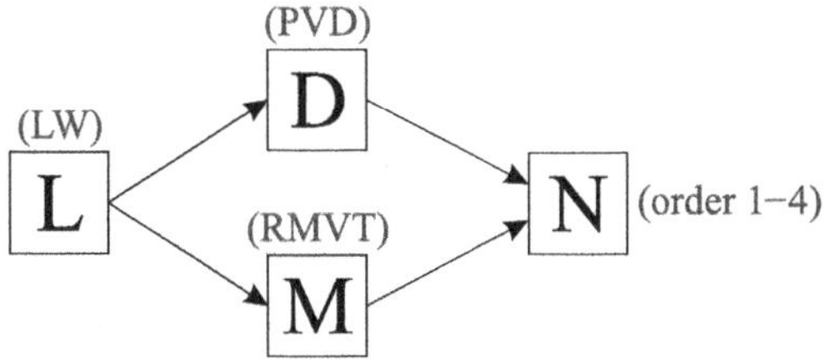

Figure 6.12 Acronyms scheme for the refined and advanced plate models.

models based on one of the three possible RMVT applications; the optional letter Z is added when the Murakami zigzag function (MZZF) is included in the ESL displacement field in order to recover the typical zigzag form of displacements in multilayered plates; a number from 1 to 4, which is the same for all the modeled variables, is used for the order of expansion in the thickness direction. The modeled a priori variables are given at the end of the acronym in parentheses. For the PVD in Section 6.3, we have $(\boldsymbol{u}, \Phi)$, while for the RMVT in Section 6.4 the variables are $(\boldsymbol{u}, \Phi, \boldsymbol{\sigma}_n)$. For the RMVT in Section 6.5 we add $(\boldsymbol{u}, \Phi, \mathcal{D}_n)$ and finally, for the RMVT in Section 6.6, the modeled variables are $(\boldsymbol{u}, \Phi, \boldsymbol{\sigma}_n, \mathcal{D}_n)$.

For example, an ESL model based on the PVD in Section 6.3 with a third order of expansion in the thickness direction for the modeled variables has the acronym ED3$(\boldsymbol{u}, \Phi)$. If we add the MZZF for the displacement to the same model, the acronym becomes EDZ3$(\boldsymbol{u}, \Phi)$. An LW model, based on the RMVT in Section 6.6, with a second order of expansion in the thickness direction for the modeled variables has the acronym LM2$(\boldsymbol{u}, \Phi, \boldsymbol{\sigma}_n, \mathcal{D}_n)$.

6.9 Pure mechanical problems as particular cases, PVD($\boldsymbol{u}$) and RMVT($\boldsymbol{u}, \boldsymbol{\sigma}_n$)

Pure mechanical models can be considered as particular cases of the electromechanical models proposed in this chapter. Pure mechanical refined models are obtained from the variational statement PVD($\boldsymbol{u}$), which is a particular case of the electromechanical PVD($\boldsymbol{u}, \Phi$) given in Section 6.3 (Carrera 2002; Carrera *et al.* 2008). Pure mechanical advanced mixed models are obtained from the variational statement RMVT($\boldsymbol{u}, \boldsymbol{\sigma}_n$), which is a particular case of the electromechanical RMVT($\boldsymbol{u}, \Phi, \boldsymbol{\sigma}_n$) shown in Section 6.4 (Carrera and Boscolo 2007; Carrera *et al.* 2008; Brischetto and Carrera 2010).

PVD($\boldsymbol{u}$) is obtained from Equation (6.22) simply by discarding the internal electrical work $\delta \boldsymbol{\mathcal{E}}_G^{k\,T} \boldsymbol{\mathcal{D}}_C^k$:

$$\sum_{k=1}^{N_l} \int_{\Omega_k} \int_{A_k} \left\{ \delta \boldsymbol{\epsilon}_{pG}^{k\,T} \boldsymbol{\sigma}_{pC}^k + \delta \boldsymbol{\epsilon}_{nG}^{k\,T} \boldsymbol{\sigma}_{nC}^k \right\} d\Omega_k dz = \sum_{k=1}^{N_l} \delta L_e^k - \sum_{k=1}^{N_l} \delta L_{in}^k \quad (6.223)$$

The relative constitutive equations are obtained from Equations (6.23)–(6.25) simply by discarding the electrical contributions and the electromechanical coupling:

$$\boldsymbol{\sigma}_{pC}^k = \boldsymbol{Q}_{pp}^k \boldsymbol{\epsilon}_{pG}^k + \boldsymbol{Q}_{pn}^k \boldsymbol{\epsilon}_{nG}^k \quad (6.224)$$

$$\boldsymbol{\sigma}_{nC}^k = \boldsymbol{Q}_{np}^k \boldsymbol{\epsilon}_{pG}^k + \boldsymbol{Q}_{nn}^k \boldsymbol{\epsilon}_{nG}^k \quad (6.225)$$

As already illustrated in Section 6.3, by substituting Equations (6.224)–(6.225) in the variational statement of Equation (6.223) and referring to the CUF for the 2D approximation (Carrera 1995), after the integration by parts the governing equations and the relative fundamental nuclei can be obtained. However, the governing equations can be obtained in a simpler way from Equations (6.30) simply by discarding the second row and column:

$$\delta \boldsymbol{u}_s^k : \quad \boldsymbol{K}_{uu}^{k\tau s}\, \boldsymbol{u}_\tau^k = \boldsymbol{p}_{us}^k - \boldsymbol{M}_{uu}^{k\tau s}\, \ddot{\boldsymbol{u}}_\tau^k \quad (6.226)$$

The fundamental nuclei $\boldsymbol{K}_{uu}^{k\tau s}$ and $\boldsymbol{M}_{uu}^{k\tau s}$ are the same as already given in Equations (6.32) and (6.36), respectively.

RMVT($\boldsymbol{u}, \boldsymbol{\sigma}_n$) is obtained from Equation (6.52) simply by discarding the internal electrical work $\delta \boldsymbol{\mathcal{E}}_G^{k\,T} \boldsymbol{\mathcal{D}}_C^k$:

$$\sum_{k=1}^{N_l} \int_{\Omega_k} \int_{A_k} \left\{ \delta \boldsymbol{\epsilon}_{pG}^{k\,T} \boldsymbol{\sigma}_{pC}^k + \delta \boldsymbol{\epsilon}_{nG}^{k\,T} \boldsymbol{\sigma}_{nM}^k + \delta \boldsymbol{\sigma}_{nM}^{k\,T} (\boldsymbol{\epsilon}_{nG}^k - \boldsymbol{\epsilon}_{nC}^k) \right\} d\Omega_k dz$$

$$= \sum_{k=1}^{N_l} \delta L_e^k - \sum_{k=1}^{N_l} \delta L_{in}^k \quad (6.227)$$

The relative constitutive equations are obtained from Equations (6.53)–(6.56) simply by discarding the electrical contributions and the electromechanical coupling:

$$\boldsymbol{\sigma}^k_{pC} = \hat{\boldsymbol{C}}^k_{\sigma_p\epsilon_p}\boldsymbol{\epsilon}^k_{pG} + \hat{\boldsymbol{C}}^k_{\sigma_p\sigma_n}\boldsymbol{\sigma}^k_{nM} \tag{6.228}$$

$$\boldsymbol{\epsilon}^k_{nC} = \hat{\boldsymbol{C}}^k_{\epsilon_n\epsilon_p}\boldsymbol{\epsilon}^k_{pG} + \hat{\boldsymbol{C}}^k_{\epsilon_n\sigma_n}\boldsymbol{\sigma}^k_{nM} \tag{6.229}$$

As already illustrated in Section 6.4, by substituting Equations (6.228)–(6.229) in the variational statement of Equation (6.227) and referring to the CUF for the 2D approximation (Carrera 1995), after the integration by parts the governing equations and the relative fundamental nuclei can be obtained. However, the governing equations can be obtained in a simpler way from Equations (6.58) simply by discarding the third row and column:

$$\begin{aligned} \delta\boldsymbol{u}^k_s: &\quad \boldsymbol{K}^{k\tau s}_{uu}\,\boldsymbol{u}^k_\tau + \boldsymbol{K}^{k\tau s}_{u\sigma}\,\boldsymbol{\sigma}^k_{nM\tau} = \boldsymbol{p}^k_{us} - \boldsymbol{M}^{k\tau s}_{uu}\,\ddot{\boldsymbol{u}}^k_\tau \\ \delta\boldsymbol{\sigma}^k_{ns}: &\quad \boldsymbol{K}^{k\tau s}_{\sigma u}\,\boldsymbol{u}^k_\tau + \boldsymbol{K}^{k\tau s}_{\sigma\sigma}\,\boldsymbol{\sigma}^k_{nM\tau} = 0 \end{aligned} \tag{6.230}$$

The fundamental nuclei $\boldsymbol{K}^{k\tau s}_{uu}$, $\boldsymbol{K}^{k\tau s}_{u\sigma}$, $\boldsymbol{K}^{k\tau s}_{\sigma u}$, $\boldsymbol{K}^{k\tau s}_{\sigma\sigma}$, and $\boldsymbol{M}^{k\tau s}_{uu}$ are the same as already given in Equations (6.60), (6.61), (6.63), (6.64), and (6.36), respectively.

6.10 Classical plate theories as particular cases of unified formulation

Pure mechanical classical plate theories as already given in Section 3.3 can also be obtained as particular cases of the CUF theory. The ED1($\boldsymbol{u}$) theory is an ESL model where the three displacement components are linear through the thickness direction z:

$$\begin{aligned} u(x,y,z) &= u_0(x,y) + z u_1(x,y) \\ v(x,y,z) &= v_0(x,y) + z v_1(x,y) \\ w(x,y,z) &= w_0(x,y) + z w_1(x,y) \end{aligned} \tag{6.231}$$

First-order shear deformation theory (FSDT) (see Equation (3.4)) can be obtained by a typical penalty technique applied to the global stiffness matrix which allows us to discard the term $z w_1(x, y)$ in Equation (6.231). Both ED1($\boldsymbol{u}$) and FSDT($\boldsymbol{u}$) have the Poisson locking phenomenon which can be overcome by means of reduced elastic coefficients in the constitutive equations imposing the condition $\sigma_{zz} = 0$ in Equations (6.224)–(6.225). Further details about this phenomenon can be found in Carrera and Brischetto (2008a).

Classical lamination theory (CLT) (see Equation (3.3)) can be considered a particular case of the above FSDT($\boldsymbol{u}$) model. In CLT($\boldsymbol{u}$) transverse shear strains γ_{xz} and γ_{yz} are zero, therefore in the FSDT($\boldsymbol{u}$) model we can penalize the coefficients Q_{55} and Q_{44} in Equations (2.22) and (2.26) in order to impose $\gamma_{xz} = \gamma_{yz} = 0$. In CLT($\boldsymbol{u}$) the Poisson locking appears too as for the FSDT($\boldsymbol{u}$) and ED1($\boldsymbol{u}$) cases and it can be corrected in the same way (Carrera and Brischetto 2008a).

In the case of electromechanical problems, CLT and FSDT extended to smart structures have been discussed in this book in Sections 3.4 and 4.3, respectively. Governing equations have been extensively written in the cases of the closed-form solution and FE method by considering a LW linear through-the-thickness electric potential. CLT($\boldsymbol{u}$, Φ) and FSDT($\boldsymbol{u}$, Φ) can also be obtained from the CUF by considering the variational statement PVD($\boldsymbol{u}$, Φ) in Equation (6.22) and the relative governing relations in Equation (6.30). They can be considered as particular cases of the ED1($\boldsymbol{u}$, Φ) model where the mechanical part is penalized as described for the pure mechanical case and the electric potential remains LW and linearly expanded through the thickness-layer direction. No Poisson locking corrections are considered for FSDT($\boldsymbol{u}$, Φ) and CLT($\boldsymbol{u}$, Φ) for the cases investigated in this book.

References

Ballhause D, D'Ottavio M, Kröplin B, and Carrera E 2005 A unified formulation to assess multilayered theories for piezoelectric plates. *Comput. Struct.* **83**, 1217–1235.

Brischetto S 2009 Classical and mixed advanced models for sandwich plates embedding functionally graded cores. *J. Mech. Mater. Struct.* **4**, 13–33.

Brischetto S and Carrera E 2009 Refined 2D models for the analysis of functionally graded piezoelectric plates. *J. Intell. Mater. Syst. Struct.* **20**, 1783–1797.

Brischetto S and Carrera E 2010 Advanced mixed theories for bending analysis of functionally graded plates. *Comput. Struct.* **88**, 1474–1483.

Brischetto S, Carrera E, and Demasi L 2009a Improved bending analysis of sandwich plates using a zig-zag function. *Comp. Struct.* **89**, 408–415.

Brischetto S, Carrera E, and Demasi L 2009b Free vibration of sandwich plates and shells by using zig-zag function. *Shock Vib.* **16**, 495–503.

Brischetto S, Carrera E, and Demasi L 2009c Improved response of unsymmetrically laminated sandwich plates by using zig-zag functions. *J. Sandwich Struct. Mater.* **11**, 257–267.

Carrera E 1995 A class of two-dimensional theories for anisotropic multilayered plates analysis. *Accad. Sci. Torino, Mem. Sci. Fis.* **19–20**, 1–39.

Carrera E 2002 Theories and finite elements for multilayered anisotropic, composite plates and shells. *Arch. Comput. Meter. Eng.* **9**, 87–140.

Carrera E 2003 Historical review of zig-zag theories for multilayered plates and shells. *App. Mech. Rev.* **56**, 287–309.

Carrera E and Boscolo M 2007 Classical and mixed finite elements for static and dynamic analysis of piezoelectric plates. *Int. J. Num. Meth. Eng.* **70**, 1135–1181.

Carrera E, Boscolo M, and Robaldo A 2007 Hierarchic multilayered plate elements for coupled multifield problems of piezoelectric adaptice structures: formulation and numerical assessment. *Arch. Comput. Methods Eng.* **14**, 383–430.

Carrera E and Brischetto S 2007a Piezoelectric shell theories with "a priori" continuous transverse electromechanical variables. *J. Mech. Mater. Struct.* **2**, 377–398.

Carrera E and Brischetto S 2007b Reissner mixed theorem applied to static analysis of piezoelectric shells. *J. Intell. Mater. Syst. Struct.* **18**, 1083–1107.

Carrera E and Brischetto S 2008a Analysis of thickness locking in classical, refined and mixed multilayered plate theories. *Comp. Struct.* **82**, 549–562.

Carrera E and Brischetto S 2008b Analysis of thickness locking in classical, refined and mixed theories for layered shells. *Comp. Struct.* **85**, 83–90.

Carrera E and Brischetto S 2009a A survey with numerical assessment of classical and refined theories for the analysis of sandwich plates. *Appl. Mech. Rev.* **62**, 1–17.

Carrera E and Brischetto S 2009b A comparison of various kinematic models for sandwich shell panels with soft core. *J. Compos. Mater.* **43**, 2201–2221.

Carrera E, Brischetto S, and Nali P 2008 Variational statements and computational models for multifield problems and multilayered structures. *Mech. Adv. Mater. Struct.* **15**, 182–198.

Carrera E, Brischetto S, and Cinefra M 2010a Variable kinematics and advanced variational statements for free vibrations analysis of piezoelectric plates and shells. *Comput. Model. Eng. Sci.* **65**, 259–341.

Carrera E, Nali P, Brischetto S, and Cinefra M 2010b Hierarchic plate and shell theories with direct evaluation of transverse electric displacement. In *Proceedings of 17th AIAA/ASME/AHS Adaptive Strutctures Conference*.

Demasi L 2008a ∞^3 hierarchy plate theories for thick and thin composite plates: the generalized unified formulation. *Comp. Struct.* **84**, 256–270.

Demasi L 2008b 2D, quasi 3D and 3D exact solutions for bending of thick and thin sandwich plates. *J. Sand. Struct. Mater.* **10**, 271–310.

D'Ottavio M and Kröplin B 2006 An extension of Reissner mixed variational theorem to piezoelectric laminates. *Mech. Adv. Mater. Struct.* **13**, 139–150.

Hsu T and Wang JT 1970 A theory of laminated cylindrical shells consisting of layers of orthotropic laminae. *AIAA J.* **8**, 2141–2146.

Hsu T and Wang JT 1971 Rotationally symmetric vibrations of orthotropic layered cylindrical shells. *J. Sound Vib.* **16**, 473–487.

Ikeda T 1996 *Fundamentals of Piezoelectricity*. Oxford University Press.

Librescu L and Schmidt R 1988 Refined theories of elastic anisotropic shells accounting for small strains and moderate rotations. *Int. J. Non-linear. Mech.* **23**, 217–229.

Librescu L and Wu EM 1977 A higher-order theory of plate deformation. Part 2: laminated plates. *J. Appl. Mech.* **44**, 669–676.

Murakami H 1985 Laminated composite plate theory with improved in-plane responses. In *ASME Proceedings of Pressure Vessels & Piping Conference.*

Murakami H 1986 Laminated composite plate theory with improved in-plane responses. *J. Appl. Mech.* **53**, 661–666.

Reddy JN 2004 *Mechanics of Laminated Composite Plates and Shells; Theory and Analysis*. CRC Press.

Reissner E 1984 On a certain mixed variational theory and a proposed application. *Int. J. Numer. Methods. Eng.* **20**, 1366–1368.

Robbins DH Jr. and Reddy JN 1993 Modeling of thick composites using a layer-wise theory. *Int. J. Numer. Methods. Eng.* **36**, 655–677.

Srinivas S 1973 A refined analysis of composite laminates. *J. Sound Vib.* **30**, 495–507.

7

Refined and advanced theories for shells

Higher orders of expansion in the thickness direction are assumed in refined and advanced models for shells for both the electrical and mechanical variables. These axiomatic 2D models can be considered in ESL or in LW form. The CUF is a technique which allows one to handle a large variety of shell models in a unified manner. According to the CUF, the obtained theories can have an order of expansion which goes from first- to higher order values, and, depending on the thickness functions used, a model can be ESL or LW. The CLT and FSDT shell theories discussed in Chapter 3 can also be obtained in the CUF as particular cases of the ESL theories. CLT, FSDT, ESL, and LW refined and advanced mixed theories have been implemented by means of the in-house academic code MUL2. The proposed shell models consider the curvature effect that is not included in the plate cases proposed in Chapter 6.

7.1 Unified formulation: refined models

Refined models for multilayered shells are those *displacements models* in which higher orders of expansion in the thickness direction z are assumed for all three displacement components. These axiomatic 2D models can be seen in ESL form, when the layers included in the multilayered shell are considered as one equivalent structure, and in LW form, when each layer embedded in the

Plates and Shells for Smart Structures: Classical and Advanced Theories for Modeling and Analysis, First Edition.
Erasmo Carrera, Salvatore Brischetto and Pietro Nali.

multilayered shell is separately considered in order to write the expansions in z for each layer k. Refined models for electromechanical problems have, as the primary variables, the electric potential in addition to the displacement vector. These models are obtained by using the PVD (Carrera 2002) and its extensions to multifield problems (Carrera *et al.* 2007; Ikeda 1996).

The CUF is a technique which handles a large variety of shell models in a unified manner (Carrera 1995) (for plate models readers should refer to Chapter 6). According to the CUF, the governing equations are written in terms of a few fundamental nuclei which do not formally depend on the order of expansion N used in the z direction and on the description of variables (LW or ESL) (Carrera 1999a,b). The application of a 2D method for shells allows one to express the unknown variables as a set of thickness functions that depend only on the thickness coordinate z and the corresponding variable that depends on the curved in-plane coordinates α and β. Therefore, the generic variable $\boldsymbol{f}(\alpha,\beta,z)$, for instance a displacement, and its variation $\delta\boldsymbol{f}(\alpha,\beta,z)$, are written according to the following general expansion:

$$\boldsymbol{f}(\alpha,\beta,z)=F_\tau(z)\boldsymbol{f}_\tau(\alpha,\beta),\qquad \delta\boldsymbol{f}(\alpha,\beta,z)=F_s(z)\delta\boldsymbol{f}_s(\alpha,\beta),$$
$$\text{with}\quad \tau,s=1,\ldots,N \tag{7.1}$$

where the bold letters denote arrays, (α,β) are the in-plane coordinates, and z the thickness one. A summing convention, with repeated indexes τ and s, is assumed. The order of expansion N goes from first- to higher order values, and, depending on the thickness functions used, a model can be either ESL, when the variable is assumed for the whole multilayer and a Taylor expansion is employed as thickness functions $F(z)$, or LW, when the variable is considered independent in each layer and a combination of Legendre polynomials used as the thickness functions $F(z)$. In the CUF, the maximum order of expansion N in the z direction is the fourth.

7.1.1 ESL theories

The displacement $\boldsymbol{u}=(u,v,w)$ is described according to the ESL description, if the unknowns are the same for the whole shell (Librescu and Wu 1977; Librescu and Schmidt 1988). The z expansion is obtained via Taylor polynomials, that is:

$$\begin{aligned}
u&=F_0\,u_0+F_1\,u_1+\cdots+F_N\,u_N=F_\tau\,u_\tau\\
v&=F_0\,v_0+F_1\,v_1+\cdots+F_N\,v_N=F_\tau\,v_\tau\\
w&=F_0\,w_0+F_1\,w_1+\cdots+F_N\,w_N=F_\tau\,w_\tau
\end{aligned} \tag{7.2}$$

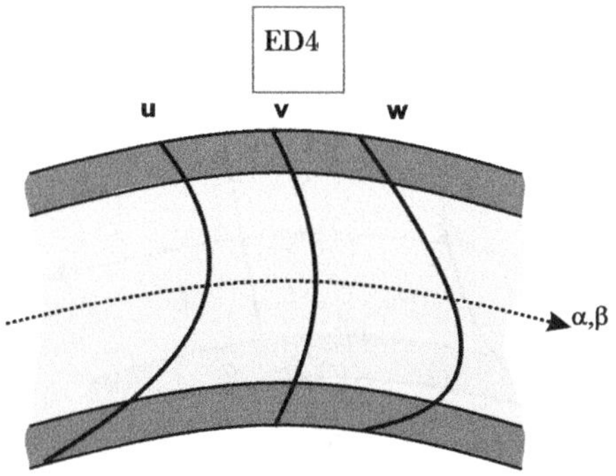

Figure 7.1 ED4: displacements u, v, and w through the thickness direction z.

with $\tau = 0, 1, \ldots, N$; N is the order of expansion and ranges from 1 (linear) to 4:

$$F_0 = z^0 = 1, \quad F_1 = z^1 = z, \ldots, \quad F_N = z^N \tag{7.3}$$

Equation (7.2) can be written in vectorial form:

$$\boldsymbol{u}(\alpha, \beta, z) = F_\tau(z)\boldsymbol{u}_\tau(\alpha, \beta), \qquad \delta\boldsymbol{u}(\alpha, \beta, z) = F_s(z)\delta\boldsymbol{u}_s(\alpha, \beta),$$
$$\text{with} \quad \tau, s = 1, \ldots, N \tag{7.4}$$

The 2D models obtained from Equations (7.2)–(7.4) are denoted by the acronym EDN, where E indicates that an ESL approach has been employed, D indicates that the theory is a displacement formulation, and N indicates the order of expansion in the thickness direction. For example, an ED2 model has a quadratic expansion in z, an ED4 has a fourth order of expansion in z, and so on. A typical displacement field is indicated in Figure 7.1 for a three-layered shell for the case of an ED4 model. Figure 7.2 considers the displacement and the transverse stresses along the z direction of the shell for an ED2 model: the displacements are quadratic in z, therefore the transverse stresses are linear (no longer constant, as in classical theories) but discontinuous at each interface. Simpler theories can be obtained from EDN models, such as those which discard the ϵ_{zz} effect; in this case, it is sufficient to impose that the transverse displacement w is constant in z. Such theories are denoted as EDNd. The ED1d model coincides with FSDT. CLT is obtained from FSDT via an opportune penalty technique, which imposes an infinite shear correction factor. It is important to recall that all the EDNd theories which have constant transverse displacement and zero

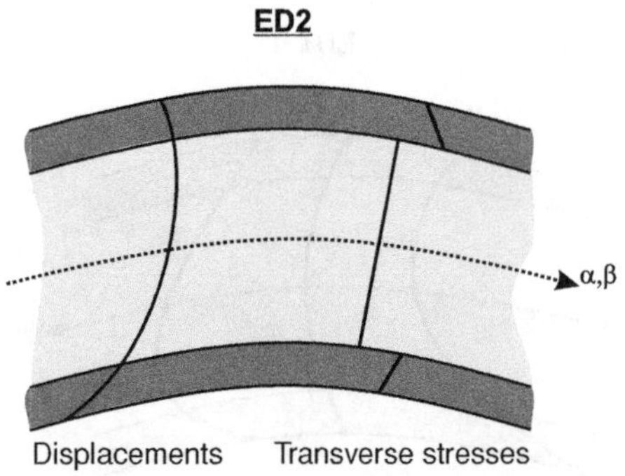

Figure 7.2 ED2: displacements and transverse shear stresses through the thickness direction z.

transverse normal strain ϵ_{zz}, the ED1 model and FSDT and CLT theories, show the Poisson locking phenomenon for pure mechanical problems; this can be overcome via plane stress conditions in constitutive equations (Carrera and Brischetto 2008a,b).

7.1.2 Murakami zigzag function

The ESL models proposed in the previous section do not consider the typical zigzag (ZZ) form of displacements in the z direction, which is typical of multilayered structures with transverse anisotropy (Carrera 2003). A remedy for this limitation is the introduction of an opportune zigzag function in the ESL displacement model, in order to recover the ZZ form of the displacements without the use of LW models. The latter have intrinsic ZZ behavior, but are more computationally expansive than ESL models (Carrera and Brischetto 2009a,b; Reddy 2004). A possible choice for the zigzag function is the so-called *Murakami zigzag function* (MZZF) (Murakami 1985, 1986). MZZF can be simply added to a displacement model and leads to remarkable improvements in the solution by satisfying the typical ZZ form of displacements in multilayered shells.

MZZF $Z(z)$ is defined as:

$$F_Z = Z(z) = (-1)^k \zeta_k \tag{7.5}$$

with the non-dimensioned layer coordinate $\zeta_k = (2z_k)/h_k$, where z_k is the transverse thickness coordinate in the k layer and h_k is the thickness of the k layer, therefore $-1 \leq \zeta_k \leq 1$. $Z(z)$ has the following properties: it is a piecewise linear function of the layer coordinates z_k; $Z(z)$ has unit amplitude for the whole layers; the slope $Z'(z) = dZ/dz$ assumes an opposite sign between two-adjacent layers. The amplitude of MZZF is layer thickness independent (Murakami 1986). The displacement model that includes MZZF is:

$$\begin{aligned} u &= F_0\, u_0 + F_1\, u_1 + \cdots + F_N\, u_N + F_Z\, u_Z &= F_\tau\, u_\tau \\ v &= F_0\, v_0 + F_1\, v_1 + \cdots + F_N\, v_N + F_Z\, v_Z &= F_\tau\, v_\tau \\ w &= F_0\, w_0 + F_1\, w_1 + \cdots + F_N\, w_N + F_Z\, w_Z &= F_\tau\, w_\tau \end{aligned} \tag{7.6}$$

where $\tau = 0, 1, \ldots, (N+1)$, and N is the order of expansion, which ranges from 1 (linear) to 4:

$$F_0 = z^0 = 1, \quad F_1 = z^1 = z, \ldots, \quad F_N = z^N, \quad F_{N+1} = F_Z = (-1)^k \zeta_k \tag{7.7}$$

The acronym to indicate such models is EDZN, where E stands for the ESL approach, D for displacement formulation, and N the order of expansion in the z direction. Z indicates that MZZF has been added (Brischetto *et al.* 2009a). The following remarks can be made: the additional degree of freedom $\boldsymbol{u}_Z$ has the meaning of displacement; the amplitude $\boldsymbol{u}_Z$ is layer independent since $\boldsymbol{u}_Z$ has an intrinsic ESL description; MZZF can be used for both in-plane and out-of-plane displacement components (Brischetto *et al.* 2009b,c). Figure 7.3 clearly shows the meaning of MZZF and how to add it to displacement components.

The MZZF $F_Z = Z(z) = (-1)^k \zeta_k$ is the $(N+1)$th thickness function in order to write the vectorial form of Equation (7.6):

$$\boldsymbol{u}(\alpha, \beta, z) = F_\tau(z)\boldsymbol{u}_\tau(\alpha, \beta), \qquad \delta\boldsymbol{u}(\alpha, \beta, z) = F_s(z)\delta\boldsymbol{u}_s(\alpha, \beta),$$
$$\text{with} \quad \tau, s = 1, \ldots, (N+1) \tag{7.8}$$

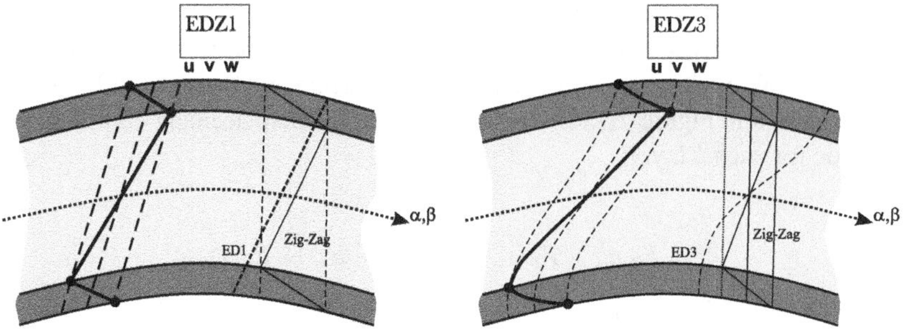

Figure 7.3 Displacements models in the EDZ1 and EDZ3 theories. Inclusion of MZZF in an ESL model.

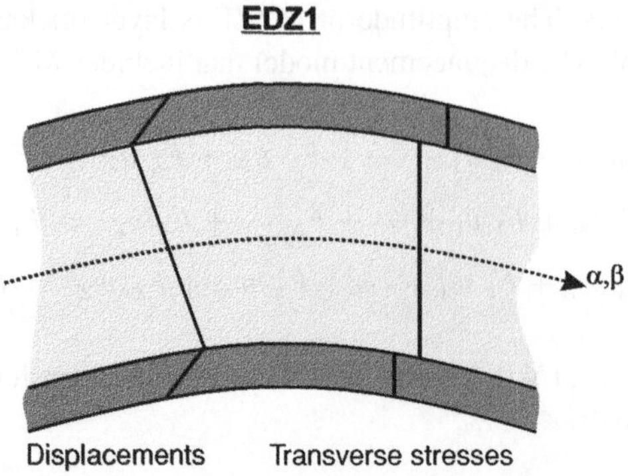

Figure 7.4 EDZ1: displacements and transverse shear stresses through the thickness direction z.

Some typical displacements and transverse shear stresses along the thickness z are shown in Figure 7.4 for an EDZ1 model: the inclusion of MZZF allows one to recover the typical ZZ form of the displacement vector in the case of multilayered transverse-anisotropic shells. Like the EDN models, it is possible to impose constant transverse displacements w. Such models are denoted as EDZNd models. It is necessary in EDZNd models to correct the Poisson locking phenomena, as indicated in Carrera and Brischetto (2008a,b) for pure mechanical problems.

7.1.3 LW theories

When each layer of a multilayered structure is described as an independent shell, a LW approach is considered (Reddy 2004). The displacement $\boldsymbol{u}^k = (u, v, w)^k$ is described for each k layer, and in this way, the ZZ form of the displacements in multilayered transverse-anisotropy shells is easily obtained (Hsu and Wang 1970, 1971; Srinivas 1973; Robbins and Reddy 1993). The recovery of the ZZ effect via LW models is dealt with in detail in Carrera and Brischetto (2009a,b) and is shown in Figure 7.5. The z expansion for the displacement components is made for each k layer:

$$
\begin{aligned}
u^k &= F_0\, u_0^k + F_1\, u_1^k + \cdots + F_N\, u_N^k = F_\tau\, u_\tau^k \\
v^k &= F_0\, v_0^k + F_1\, v_1^k + \cdots + F_N\, v_N^k = F_\tau\, v_\tau^k \\
w^k &= F_0\, w_0^k + F_1\, w_1^k + \cdots + F_N\, w_N^k = F_\tau\, w_\tau^k
\end{aligned}
\qquad (7.9)
$$

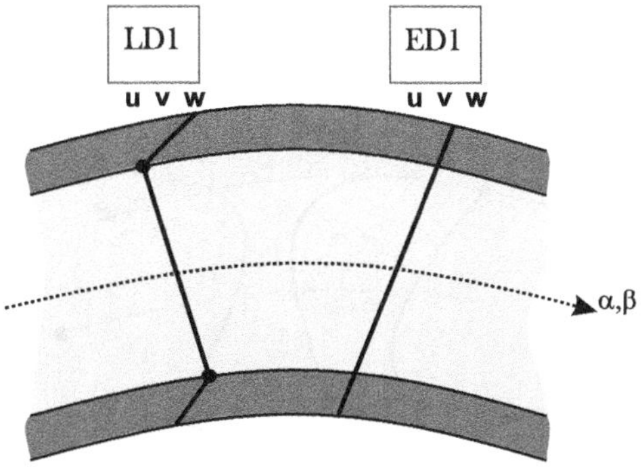

Figure 7.5 Linear expansion in the z direction for displacement components: LW approach vs. ESL approach.

with $\tau = 0, 1, \ldots, N$, and N the order of expansion, which ranges from 1 (linear) to 4; $k = 1, \ldots, N_l$ where N_l indicates the number of layers. Equation (7.9), written in vectorial form, is:

$$\boldsymbol{u}^k(\alpha, \beta, z) = F_\tau(z)\boldsymbol{u}_\tau^k(\alpha, \beta), \qquad \delta\boldsymbol{u}^k(\alpha, \beta, z) = F_s(z)\delta\boldsymbol{u}_s^k(\alpha, \beta),$$
$$\text{with} \quad \tau, s = t, b, r \qquad \text{and} \quad k = 1, \ldots, N_l \tag{7.10}$$

where t and b indicate the top and bottom of each k layer, respectively, N_l is the number of total layers, and r indicates the higher orders of expansion in the thickness direction: $r = 2, \ldots, N$. The thickness functions $F_\tau(\zeta_k)$ and $F_s(\zeta_k)$ have now been defined at the k-layer level, and are a linear combination of the Legendre polynomials $P_j = P_j(\zeta_k)$ of the jth order defined in the ζ_k domain ($\zeta_k = 2z_k/h_k$ where z_k is the local coordinate and h_k is the thickness, both referring to the kth layer, therefore $-1 \leq \zeta_k \leq 1$). The first five Legendre polynomials are:

$$P_0 = 1, \quad P_1 = \zeta_k, \quad P_2 = \frac{(3{\zeta_k}^2 - 1)}{2}, \quad P_3 = \frac{5{\zeta_k}^3}{2} - \frac{3\zeta_k}{2},$$
$$P_4 = \frac{35{\zeta_k}^4}{8} - \frac{15{\zeta_k}^2}{4} + \frac{3}{8} \tag{7.11}$$

and their combinations for the thickness functions are:

$$F_t = F_0 = \frac{P_0 + P_1}{2}, \quad F_b = F_1 = \frac{P_0 - P_1}{2}, \quad F_r = P_r - P_{r-2},$$
$$\text{with} \quad r = 2, \ldots, N \tag{7.12}$$

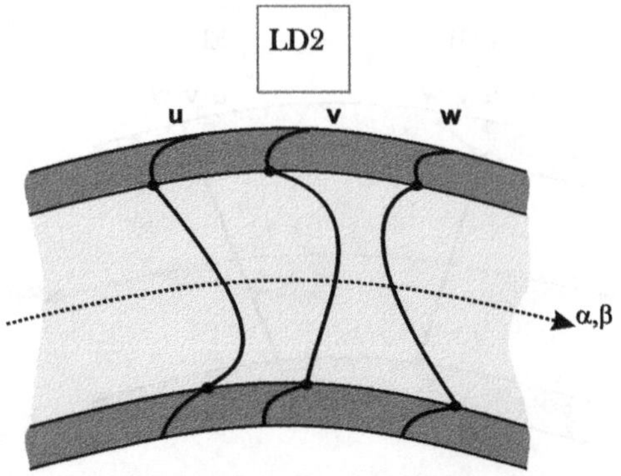

Figure 7.6 LD2: u, v, and w displacements through the thickness direction z.

The chosen functions have the following interesting properties:

$$\zeta_k = 1 : F_t = 1; F_b = 0; F_r = 0 \text{ at the top} \tag{7.13}$$

$$\zeta_k = -1 : F_t = 0; F_b = 1; F_r = 0 \text{ at the bottom} \tag{7.14}$$

In other words, interface values of the variables are considered as the variable unknowns. This fact permits one to easily impose the compatibility conditions for the displacements at each layer interface. The acronym to indicate such theories is LDN, where L stands for the LW approach, D indicates the displacement formulation, and N is the order of expansion in each k layer. Typical displacement behavior for a three-layered shell is given in Figure 7.6 for LD2 model. Figure 7.7 indicates the displacements and transverse shear stresses for a LD3 model. The transverse shear/normal stresses are obtained via the

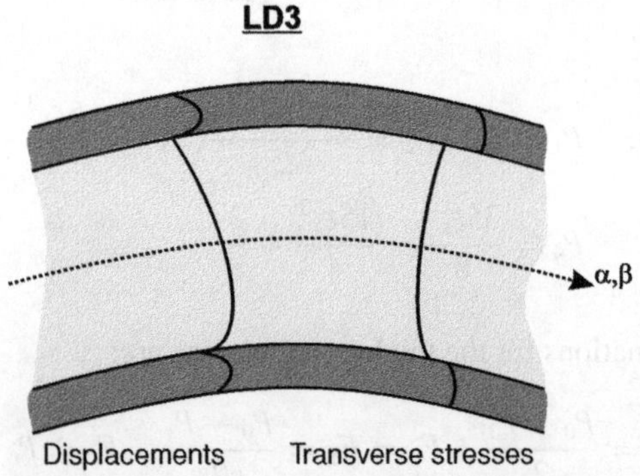

Figure 7.7 LD3: displacements and transverse shear stresses through the thickness direction z.

constitutive equations but this does not ensure interlaminar continuity (IC). IC could be enforced by "a priori" modeling of the transverse shear/normal stresses. In LW models, even though a linear expansion in z is considered for transverse displacement w, the Poisson locking phenomenon does not appear for a pure mechanical problem: the transverse normal strain ϵ_{zz} is piece-wise constant in the thickness direction (Carrera and Brischetto 2008a,b).

7.1.4 Refined models for the electromechanical case

The primary variables in electromechanical problems are the displacement vector $\boldsymbol{u} = (u, v, w)$ and the scalar electric potential Φ. Considering the higher spatial gradient of the electric potential, the variable Φ^k is always modelled as LW (Carrera and Brischetto 2007a,b; Brischetto and Carrera 2009):

$$\Phi^k(\alpha, \beta, z) = F_\tau(z)\Phi^k_\tau(\alpha, \beta), \qquad \delta\Phi^k(\alpha, \beta, z) = F_s(z)\delta\Phi^k_s(\alpha, \beta),$$
$$\text{with} \quad \tau, s = t, b, r \qquad \text{and} \quad k = 1, \ldots, N_l \tag{7.15}$$

where t and b indicate the top and bottom of each k layer, respectively; N_l indicates the number of total layers, and r indicates the higher orders of expansion in the thickness direction: $r = 2, \ldots, N$. The thickness functions are a combination of Legendre polynomials, as indicated in the previous section. A 2D model for electromechanical problems is defined as ESL, ESL+MZZF, or LW depending on the choice made concerning the displacement vector: the electric potential is always considered LW (Ballhause *et al.* 2005; Carrera and Boscolo 2007).

7.2 Unified formulation: advanced mixed models

In the case of electromechanical analysis of multilayered shells, advanced mixed models are defined as those 2D models that are obtained by employing the Reissner mixed variational theorem (RMVT) (Reissner 1984) and its extensions to electromechanical coupling (Carrera *et al.* 2008). These extensions allow one to "a priori" model some transverse quantities that are only obtained via post-processing in PVD applications. Transverse shear/normal stresses $\boldsymbol{\sigma}_n = (\sigma_{\alpha z}, \sigma_{\beta z}, \sigma_{zz})$ and/or transverse normal electric displacement $\mathcal{D}_n = (\mathcal{D}_z)$ are a priori modeled and considered in LW form. The main advantage of obtaining these variables directly from the governing equations is the fulfillment of Interlaminar Continuity (IC) (Brischetto 2009; Brischetto and Carrera 2010). These advanced models are obtained by means of the CUF (Carrera 2002) as explained in detail in previous sections.

7.2.1 Transverse shear/normal stress modeling

In the case of a pure mechanical problem, an advanced mixed model considers both displacements $\boldsymbol{u} = (u, v, w)$ and transverse shear/normal stresses $\boldsymbol{\sigma}_{nM} = (\sigma_{\alpha z}, \sigma_{\beta z}, \sigma_{zz})$ as the primary variables (Brischetto and Carrera 2010). The displacements can be modeled as ESL (Section 7.1.1), ESL+MZZF (Section 7.1.2), and LW (Section 7.1.3), and this choice allows one to define the considered advanced model as ESL, ESL+MZZF, or LW, respectively; the transverse shear/normal stresses $\boldsymbol{\sigma}_{nM}$ are always LW (the subscript M means that the stresses are modelled and not obtained from the constitutive equations). The LW model for stresses is:

$$\begin{aligned}
\sigma^k_{\alpha z} &= F_0\,\sigma^k_{\alpha z0} + F_1\,\sigma^k_{\alpha z1} + \cdots + F_N\,\sigma^k_{\alpha zN} = F_\tau\,\sigma^k_{\alpha z\tau}\\
\sigma^k_{\beta z} &= F_0\,\sigma^k_{\beta z0} + F_1\,\sigma^k_{\beta z1} + \cdots + F_N\,\sigma^k_{\beta zN} = F_\tau\,\sigma^k_{\beta z\tau}\\
\sigma^k_{zz} &= F_0\,\sigma^k_{zz0} + F_1\,\sigma^k_{zz1} + \cdots + F_N\,\sigma^k_{zzN} = F_\tau\,\sigma^k_{zz\tau}
\end{aligned} \tag{7.16}$$

with $\tau = 0, 1, \ldots, N$, and N the order of expansion ranging from 1 (linear) to 4; $k = 1, \ldots, N_l$, where N_l indicates the number of layers. Equation (7.16) is written in vectorial form as:

$$\boldsymbol{\sigma}^k_{nM}(\alpha, \beta, z) = F_\tau(z)\boldsymbol{\sigma}^k_{nM\tau}(\alpha, \beta), \qquad \delta\boldsymbol{\sigma}^k_{nM}(\alpha, \beta, z) = F_s(z)\delta\boldsymbol{\sigma}^k_{nMs}(\alpha, \beta),$$
$$\text{with} \quad \tau, s = t, b, r \quad \text{and} \quad k = 1, \ldots, N_l \tag{7.17}$$

where t and b indicate the top and bottom of each layer k, respectively; r indicates the higher orders of expansion in the thickness direction: $r = 2, \ldots, N$. The thickness functions $F_\tau(\zeta_k)$ and $F_s(\zeta_k)$ have now been defined at the k-layer level, and are a linear combination of Legendre polynomials. The use of such thickness functions, based on the property pointed out in Equations (7.13) and (7.14), permits one to easily write the IC for the transverse stresses:

$$\boldsymbol{\sigma}^k_{nt} = \boldsymbol{\sigma}^{k+1}_{nb} \quad \text{with} \quad k = 1, \ldots, (N_l - 1) \tag{7.18}$$

which means: the top value of the k layer in each interface is equal to the bottom value of the layer $(k+1)$. The same property can be used for displacements in LW form, in order to impose the compatibility conditions:

$$\boldsymbol{u}^k_t = \boldsymbol{u}^{k+1}_b \quad \text{with} \quad k = 1, \ldots, (N_l - 1) \tag{7.19}$$

We define EMN models as those models which have displacements in the ESL form (E) and transverse stresses in the LW form, where M means mixed formulation (use of RMVT), and N is the order of expansion, which is the same

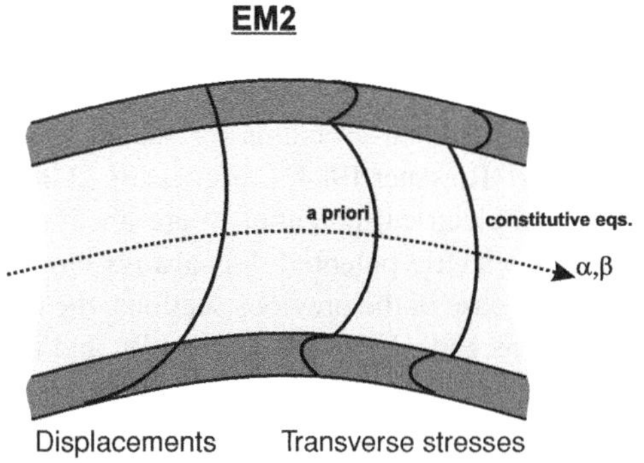

Figure 7.8 EM2: displacements and transverse shear stresses through the thickness direction z.

for both variables. EMZN models consider the displacements modelled in ESL form with the inclusion of MZZF. LMN models consider both displacements and transverse stresses in LW form. Figure 7.8 shows the displacements and transverse stresses for an EM2 model. The displacements are considered ESL, and the transverse stresses are a priori modelled and obtained directly from the governing equations: they are considered in LW form, and this allows one to satisfy both the ZZ form and IC. If transverse stresses are obtained from the constitutive equations via post-processing, IC might be not ensured. Figure 7.9 shows the displacements and stresses for LM2 model. In this case, the displacements are also LW, and the ZZ form and IC are ensured for both the displacement and transverse stress components. The transverse stresses obtained from the constitutive equations might not satisfy IC (Brischetto 2009).

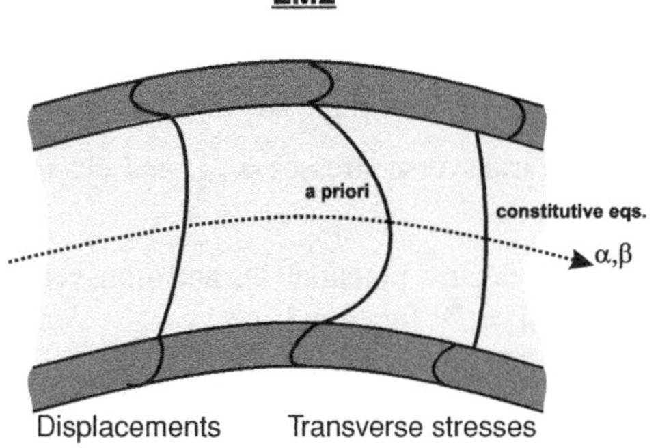

Figure 7.9 LM2: displacements and transverse shear stresses through the thickness direction z.

7.2.2 Advanced mixed models for the electromechanical case

For the case of electromechanical problems for shells, several extensions of RMVT can be considered (Reissner 1984; Carrera *et al.* 2008). In such models, the displacements $\boldsymbol{u}$ and electrical potential Φ are always considered in the governing equations, the electric potential Φ is always modelled in LW form, as discussed for the PVD case in the previous sections, the displacement components $\boldsymbol{u}$ are modelled as ESL, ESL+MZZF, or LW, and this choice defines the considered advanced model as ESL, ESL+MZZF, or LW.

Three different extensions of RMVT to electromechanical problems can be considered. In addition to displacements $\boldsymbol{u}$ and electric potential Φ, the other modelled variables are:

1. Using only one Lagrange multiplier (Reissner 1984), the transverse stresses $\boldsymbol{\sigma}_{nM}$ are a priori modelled (LW form as described in the previous sections) (Carrera and Brischetto 2007b).
2. Using only one Lagrange multiplier, the transverse normal electric displacement $\mathcal{D}_{nM} = \mathcal{D}_z$ is a priori obtained in LW form (Carrera *et al.* 2010a).
3. Considering two Lagrange multipliers, both the transverse stresses and the transverse normal electric displacement are a priori modelled in LW form (Carrera and Brischetto 2007a).

The LW expansion for the transverse normal electric displacement $\mathcal{D}_{nM} = \mathcal{D}_z$ is:

$$\mathcal{D}_z^k(\alpha, \beta, z) = F_\tau(z)\mathcal{D}_{z\tau}^k(\alpha, \beta), \qquad \delta\mathcal{D}_z^k(\alpha, \beta, z) = F_s(z)\delta\mathcal{D}_{zs}^k(\alpha, \beta),$$
$$\text{with} \quad \tau, s = t, b, r \quad \text{and} \quad k = 1, \ldots, N_l \tag{7.20}$$

where t and b indicate the top and bottom of each k layer, respectively; r indicates the higher orders of expansion in the thickness direction: $r = 2, \ldots, N$.

The modelled variables of these three advanced models are:

1. displacements $\boldsymbol{u}$, transverse stresses $\boldsymbol{\sigma}_{nM}$, and electric potential Φ for case 1;
2. displacements $\boldsymbol{u}$, electric potential Φ, and transverse normal electric displacement $\mathcal{D}_{nM} = \mathcal{D}_z$ for case 2;
3. displacements $\boldsymbol{u}$, electric potential Φ, transverse stresses $\boldsymbol{\sigma}_{nM}$, and transverse normal electric displacement $\mathcal{D}_{nM} = \mathcal{D}_z$ for case 3.

The acronyms for such advanced mixed models are explained in detail at the end of this chapter, after a discussion of the variational statements.

7.3 PVD(u, Φ) for the electromechanical shell case

The PVD has been obtained for the case of electromechanical problems as in Equations (2.53) and (6.21) in Chapters 2 and 6, respectively:

$$\int_V \left(\delta\boldsymbol{\epsilon}_{pG}^T\boldsymbol{\sigma}_{pC} + \delta\boldsymbol{\epsilon}_{nG}^T\boldsymbol{\sigma}_{nC} - \delta\boldsymbol{\mathcal{E}}_G^T\boldsymbol{\mathcal{D}}_C\right)dV = \delta L_e - \delta L_{in} \tag{7.21}$$

It is not necessary to split the electric displacement for the PVD case. Considering a laminated shell of N_l layers and the integral on the volume V_k of each layer as an integral on the in-plane curved domain Ω_k, plus the integral in the thickness-direction domain A_k, it is possible to write:

$$\sum_{k=1}^{N_l}\int_{\Omega_k}\int_{A_k}\left\{\delta\boldsymbol{\epsilon}_{pG}^{k\ T}\boldsymbol{\sigma}_{pC}^k + \delta\boldsymbol{\epsilon}_{nG}^{k\ T}\boldsymbol{\sigma}_{nC}^k - \delta\boldsymbol{\mathcal{E}}_G^{k\ T}\boldsymbol{\mathcal{D}}_C^k\right\}d\Omega_k dz = \sum_{k=1}^{N_l}\delta L_e^k - \sum_{k=1}^{N_l}\delta L_{in}^k \tag{7.22}$$

where δL_e^k and δL_{in}^k are the external and inertial virtual work at the k-layer level, respectively. The relative constitutive equations are those obtained in Equations (2.21)–(2.24), with the components in a curvilinear reference system (α, β, z) for the stress, strain, electric displacement, and electric field vectors given in Equations (2.19) and (2.20); if splitting of the electric displacement and electric field in the in-plane and out-of-plane components is not considered, the relations are the same as those already discussed in Equations (6.23)–(6.25):

$$\boldsymbol{\sigma}_{pC}^k = \boldsymbol{Q}_{pp}^k\boldsymbol{\epsilon}_{pG}^k + \boldsymbol{Q}_{pn}^k\boldsymbol{\epsilon}_{nG}^k - \boldsymbol{e}_p^{k^T}\boldsymbol{\mathcal{E}}_G^k \tag{7.23}$$

$$\boldsymbol{\sigma}_{nC}^k = \boldsymbol{Q}_{np}^k\boldsymbol{\epsilon}_{pG}^k + \boldsymbol{Q}_{nn}^k\boldsymbol{\epsilon}_{nG}^k - \boldsymbol{e}_n^{k^T}\boldsymbol{\mathcal{E}}_G^k \tag{7.24}$$

$$\boldsymbol{\mathcal{D}}_C^k = \boldsymbol{e}_p^k\boldsymbol{\epsilon}_{pG}^k + \boldsymbol{e}_n^k\boldsymbol{\epsilon}_{nG}^k + \boldsymbol{\varepsilon}^k\boldsymbol{\mathcal{E}}_G^k \tag{7.25}$$

By substituting Equations (7.23)–(7.25), and the geometrical relations (2.31)–(2.37) in Chapter 2 for the shells in the variational statement in Equation (7.22), and considering a generic k layer (Carrera *et al.* 2007):

$$\int_{\Omega_k}\int_{A_k}\Big[\left((\boldsymbol{D}_p^k + \boldsymbol{A}_p^k)\delta\boldsymbol{u}^k\right)^T\Big((\boldsymbol{Q}_{pp}^k(\boldsymbol{D}_p^k + \boldsymbol{A}_p^k) + \boldsymbol{Q}_{pn}^k(\boldsymbol{D}_{np}^k + \boldsymbol{D}_{nz}^k - \boldsymbol{A}_n^k))\boldsymbol{u}^k$$
$$+\boldsymbol{e}_p^{k^T}(\boldsymbol{D}_{ep}^k + \boldsymbol{D}_{en}^k)\Phi^k\Big) + \left((\boldsymbol{D}_{np}^k + \boldsymbol{D}_{nz}^k - \boldsymbol{A}_n^k)\delta\boldsymbol{u}^k\right)^T\Big((\boldsymbol{Q}_{np}^k(\boldsymbol{D}_p^k + \boldsymbol{A}_p^k)$$
$$+\boldsymbol{Q}_{nn}^k(\boldsymbol{D}_{np}^k + \boldsymbol{D}_{nz}^k - \boldsymbol{A}_n^k))\boldsymbol{u}^k + \boldsymbol{e}_n^{k^T}(\boldsymbol{D}_{ep}^k + \boldsymbol{D}_{en}^k)\Phi^k\Big) + \left((\boldsymbol{D}_{ep}^k + \boldsymbol{D}_{en}^k)\delta\Phi^k\right)^T$$

$$\times\Big(\big(e_p^k(D_p^k + A_p^k) + e_n^k(D_{np}^k + D_{nz}^k - A_n^k)\big)u^k - \varepsilon^k(D_{ep}^k + D_{en}^k)\Phi^k\Big)\Big]$$
$$\times\, d\Omega_k\, dz = \delta L_e^k - \delta L_{in}^k \tag{7.26}$$

The CUF (Carrera 1995) can be introduced into Equation (7.26) for the 2D approximation, as presented in Sections 7.1 and 7.2:

$$\int_{\Omega_k}\int_{A_k}\Big[\big((D_p^k + A_p^k)F_s\delta u_s^k\big)^T\Big(\big(Q_{pp}^k(D_p^k + A_p^k) + Q_{pn}^k(D_{np}^k + D_{nz}^k - A_n^k)\big)F_\tau u_\tau^k$$
$$+ e_p^{k^T}(D_{ep}^k + D_{en}^k)F_\tau\Phi_\tau^k\Big) + \big((D_{np}^k + D_{nz}^k - A_n^k)F_s\delta u_s^k\big)^T\Big(\big(Q_{np}^k(D_p^k + A_p^k)$$
$$+ Q_{nn}^k(D_{np}^k + D_{nz}^k - A_n^k)\big)F_\tau u_\tau^k + e_n^{k^T}(D_{ep}^k + D_{en}^k)F_\tau\Phi_\tau^k\Big) + \big((D_{ep}^k + D_{en}^k)F_s\delta\Phi_s^k\big)^T$$
$$\times\Big(\big(e_p^k(D_p^k + A_p^k) + e_n^k(D_{np}^k + D_{nz}^k - A_n^k)\big)F_\tau u_\tau^k - \varepsilon^k(D_{ep}^k + D_{en}^k)F_\tau\Phi_\tau^k\Big)\Big]$$
$$\times\, d\Omega_k\, dz = \delta L_e^k - \delta L_{in}^k \tag{7.27}$$

In Equation (7.27), in order to obtain a strong form of the differential equations on domain Ω_k and the relative boundary conditions on edge Γ_k, integration by parts is used, which allows one to move the differential operator from the infinitesimal variation of the generic variable δa^k to the finite quantity a^k (Carrera 1995). For a generic variable a^k, the integration by parts is:

$$\int_{\Omega_k}(D_\Omega^k\delta a^k)^T\, a^k d\Omega_k = -\int_{\Omega_k}\delta a^{k^T}\,(D_\Omega^{kT}a^k)\,d\Omega_k + \int_{\Gamma_k}\delta a^{k^T}\,(I_\Omega^{kT}a^k)\,d\Gamma_k \tag{7.28}$$

where $\Omega = p, np, ep$, and the matrices of the differential operators depend on the k layer in the case of a shell geometry; the matrices to perform the integration by parts have the following form, in analogy with the matrices of the geometrical relations of the shells in Equations (2.33) and (2.37):

$$I_p^k = \begin{bmatrix} \frac{1}{H_\alpha^k} & 0 & 0 \\ 0 & \frac{1}{H_\beta^k} & 0 \\ \frac{1}{H_\beta^k} & \frac{1}{H_\alpha^k} & 0 \end{bmatrix}, \quad I_{np}^k = \begin{bmatrix} 0 & 0 & \frac{1}{H_\alpha^k} \\ 0 & 0 & \frac{1}{H_\beta^k} \\ 0 & 0 & 0 \end{bmatrix}, \quad I_{ep}^k = \begin{bmatrix} \frac{1}{H_\alpha^k} \\ \frac{1}{H_\beta^k} \end{bmatrix} \tag{7.29}$$

where the metric coefficients $H_\alpha^k = (1 + z^k/R_\alpha^k)$ and $H_\beta^k = (1 + z^k/R_\beta^k)$ (with radii of curvature R_α^k and R_β^k in the α and β in-plane directions of the shell, respectively) were introduced in Section 2.3 and will be dealt with at the end of this chapter.

The governing equations and the boundary conditions have the same form as the plate case discussed in Section 6.3, though the meaning of the involved fundamental nuclei changes. As discussed in Brischetto and Carrera (2009), the governing equations are:

$$
\begin{aligned}
\delta \boldsymbol{u}_s^k : \quad & \boldsymbol{K}_{uu}^{k\tau s}\, \boldsymbol{u}_\tau^k + \boldsymbol{K}_{u\Phi}^{k\tau s} \Phi_\tau^k = \boldsymbol{p}_{us}^k - \boldsymbol{M}_{uu}^{k\tau s}\, \ddot{\boldsymbol{u}}_\tau^k \\
\delta \Phi_s^k : \quad & \boldsymbol{K}_{\Phi u}^{k\tau s}\, \boldsymbol{u}_\tau^k + \boldsymbol{K}_{\Phi\Phi}^{k\tau s}\, \Phi_\tau^k = 0
\end{aligned}
\tag{7.30}
$$

where $\boldsymbol{M}_{uu}^{k\tau s}$ is the inertial contribution in the form of the fundamental nucleus, $\boldsymbol{u}_\tau^k$ is the vector of the degrees of freedom for the displacements, Φ_τ^k is the vector of the degrees of freedom for the electric potential, $\ddot{\boldsymbol{u}}_\tau^k$ is the second temporal derivative of $\boldsymbol{u}_\tau^k$. The array $\boldsymbol{p}_{us}^k$ indicates the variationally consistent mechanical loading for the case of a sensor configuration; the electric potential is imposed directly in vector Φ_τ^k in the actuator application. Along with these governing equations, the following boundary conditions on edge Γ_k of the in-plane integration domain Ω_k hold:

$$
\begin{aligned}
\boldsymbol{\Pi}_{uu}^{k\tau s}\, \boldsymbol{u}_\tau^k + \boldsymbol{\Pi}_{u\Phi}^{k\tau s}\, \Phi_\tau^k &= \boldsymbol{\Pi}_{uu}^{k\tau s}\, \bar{\boldsymbol{u}}_\tau^k + \boldsymbol{\Pi}_{u\Phi}^{k\tau s}\, \bar{\Phi}_\tau^k \\
\boldsymbol{\Pi}_{\Phi u}^{k\tau s}\, \boldsymbol{u}_\tau^k + \boldsymbol{\Pi}_{\Phi\Phi}^{k\tau s}\, \Phi_\tau^k &= \boldsymbol{\Pi}_{\Phi u}^{k\tau s}\, \bar{\boldsymbol{u}}_\tau^k + \boldsymbol{\Pi}_{\Phi\Phi}^{k\tau s}\, \bar{\Phi}_\tau^k
\end{aligned}
\tag{7.31}
$$

By comparing Equation (7.27), after the integration by parts (see Equation (7.28)), to Equations (7.30) and (7.31), the fundamental nuclei can be obtained:

$$
\begin{aligned}
\boldsymbol{K}_{uu}^{k\tau s} = \int_{A_k} & \Big[\Big(-\boldsymbol{D}_p^k + \boldsymbol{A}_p^k\Big)^T \Big(\boldsymbol{Q}_{pp}^k(\boldsymbol{D}_p^k + \boldsymbol{A}_p^k) + \boldsymbol{Q}_{pn}^k(\boldsymbol{D}_{np}^k + \boldsymbol{D}_{nz}^k - \boldsymbol{A}_n^k)\Big) \\
& + \Big(-\boldsymbol{D}_{np}^k + \boldsymbol{D}_{nz}^k - \boldsymbol{A}_n^k\Big)^T \Big(\boldsymbol{Q}_{np}^k(\boldsymbol{D}_p^k + \boldsymbol{A}_p^k) + \boldsymbol{Q}_{nn}^k(\boldsymbol{D}_{np}^k + \boldsymbol{D}_{nz}^k - \boldsymbol{A}_n^k)\Big)\Big] \\
& \times F_s F_\tau H_\alpha^k H_\beta^k dz
\end{aligned}
\tag{7.32}
$$

$$
\begin{aligned}
\boldsymbol{K}_{u\Phi}^{k\tau s} = \int_{A_k} & \Big[\Big(-\boldsymbol{D}_p^k + \boldsymbol{A}_p^k\Big)^T \Big(\boldsymbol{e}_p^{kT}(\boldsymbol{D}_{ep}^k + \boldsymbol{D}_{en}^k)\Big) + \Big(-\boldsymbol{D}_{np}^k + \boldsymbol{D}_{nz}^k - \boldsymbol{A}_n^k\Big)^T \\
& \times \Big(\boldsymbol{e}_n^{kT}(\boldsymbol{D}_{ep}^k + \boldsymbol{D}_{en}^k)\Big)\Big] F_s F_\tau H_\alpha^k H_\beta^k dz
\end{aligned}
\tag{7.33}
$$

$$
\begin{aligned}
\boldsymbol{K}_{\Phi u}^{k\tau s} = \int_{A_k} & \Big[\Big(-\boldsymbol{D}_{ep}^k + \boldsymbol{D}_{en}^k\Big)^T \Big(\boldsymbol{e}_p^{kT}(\boldsymbol{D}_p^k + \boldsymbol{A}_p^k) + \boldsymbol{e}_n^{kT}(\boldsymbol{D}_{np}^k + \boldsymbol{D}_{nz}^k - \boldsymbol{A}_n^k)\Big)\Big] \\
& \times F_s F_\tau H_\alpha^k H_\beta^k dz
\end{aligned}
\tag{7.34}
$$

$$\boldsymbol{K}^{k\tau s}_{\Phi\Phi} = \int_{A_k} \Big[\Big(- \boldsymbol{D}^k_{ep} + \boldsymbol{D}^k_{en}\Big)^T \Big(- \boldsymbol{\varepsilon}^k(\boldsymbol{D}^k_{ep} + \boldsymbol{D}^k_{en})\Big) \Big] F_s F_\tau H^k_\alpha H^k_\beta dz \tag{7.35}$$

$$\boldsymbol{M}^{k\tau s}_{uu} = \int_{A_k} (\rho^k \boldsymbol{I}) F_s F_\tau H^k_\alpha H^k_\beta dz \tag{7.36}$$

ρ^k is the mass density of the kth layer and $\boldsymbol{I}$ is the (3×3) identity matrix. The nuclei for the boundary conditions on edge Γ_k are (Carrera and Brischetto 2007b):

$$\boldsymbol{\Pi}^{k\tau s}_{uu} = \int_{A_k} \Big[\boldsymbol{I}^{kT}_p \Big(\boldsymbol{Q}^k_{pp}(\boldsymbol{D}^k_p + \boldsymbol{A}^k_p) + \boldsymbol{Q}^k_{pn}(\boldsymbol{D}^k_{np} + \boldsymbol{D}^k_{nz} - \boldsymbol{A}^k_n)\Big)$$
$$+ \boldsymbol{I}^{kT}_{np} \Big(\boldsymbol{Q}^k_{np}(\boldsymbol{D}^k_p + \boldsymbol{A}^k_p) + \boldsymbol{Q}^k_{nn}(\boldsymbol{D}^k_{np} + \boldsymbol{D}^k_{nz} - \boldsymbol{A}^k_n)\Big)\Big] F_s F_\tau H^k_\alpha H^k_\beta dz \tag{7.37}$$

$$\boldsymbol{\Pi}^{k\tau s}_{u\Phi} = \int_{A_k} \Big[\boldsymbol{I}^{kT}_p \Big(\boldsymbol{e}^{kT}_p (\boldsymbol{D}^k_{ep} + \boldsymbol{D}^k_{en})\Big) + \boldsymbol{I}^{kT}_{np} \Big(\boldsymbol{e}^{kT}_n (\boldsymbol{D}^k_{ep} + \boldsymbol{D}^k_{en})\Big)\Big] F_s F_\tau H^k_\alpha H^k_\beta dz \tag{7.38}$$

$$\boldsymbol{\Pi}^{k\tau s}_{\Phi u} = \int_{A_k} \Big[\boldsymbol{I}^{kT}_{ep} \Big(\boldsymbol{e}^{kT}_p (\boldsymbol{D}^k_p + \boldsymbol{A}^k_p) + \boldsymbol{e}^{kT}_n (\boldsymbol{D}^k_{np} + \boldsymbol{D}^k_{nz} - \boldsymbol{A}^k_n)\Big)\Big]$$
$$\times F_s F_\tau H^k_\alpha H^k_\beta dz \tag{7.39}$$

$$\boldsymbol{\Pi}^{k\tau s}_{\Phi\Phi} = \int_{A_k} \Big[\boldsymbol{I}^{kT}_{ep} \Big(- \boldsymbol{\varepsilon}^k(\boldsymbol{D}^k_{ep} + \boldsymbol{D}^k_{en})\Big) \Big] F_s F_\tau H^k_\alpha H^k_\beta dz \tag{7.40}$$

The geometrical matrices and those necessary to perform integration by parts depend on the k layer for the case of a shell geometry, since they contain the parametric coefficients. They do not depend on the layer k for the plate geometry.

In order to write the explicit form of the nuclei in Equations (7.32)–(7.36), the following integrals in the z thickness-direction are defined:

$$\Big(J^{k\tau s}, J^{k\tau s}_\alpha, J^{k\tau s}_\beta, J^{k\tau s}_{\alpha/\beta}, J^{k\tau s}_{\beta/\alpha}, J^{k\tau s}_{\alpha\beta}\Big) = \int_{A_k} F_\tau F_s \Big(1, H^k_\alpha, H^k_\beta, \frac{H^k_\alpha}{H^k_\beta}, \frac{H^k_\beta}{H^k_\alpha}, H^k_\alpha H^k_\beta\Big) dz$$

$$\Big(J^{k\tau_z s}, J^{k\tau_z s}_\alpha, J^{k\tau_z s}_\beta, J^{k\tau_z s}_{\alpha/\beta}, J^{k\tau_z s}_{\beta/\alpha}, J^{k\tau_z s}_{\alpha\beta}\Big) = \int_{A_k} \frac{\partial F_\tau}{\partial z} F_s \Big(1, H^k_\alpha, H^k_\beta, \frac{H^k_\alpha}{H^k_\beta}, \frac{H^k_\beta}{H^k_\alpha}, H^k_\alpha H^k_\beta\Big) dz$$

$$\left(J^{k\tau s_z}, J_\alpha^{k\tau s_z}, J_\beta^{k\tau s_z}, J_{\alpha/\beta}^{k\tau s_z}, J_{\beta/\alpha}^{k\tau s_z}, J_{\alpha\beta}^{k\tau s_z}\right) = \int_{A_k} F_\tau \frac{\partial F_s}{\partial z}\left(1, H_\alpha^k, H_\beta^k, \frac{H_\alpha^k}{H_\beta^k}, \frac{H_\beta^k}{H_\alpha^k}, H_\alpha^k H_\beta^k\right) dz$$

$$\left(J^{k\tau_z s_z}, J_\alpha^{k\tau_z s_z}, J_\beta^{k\tau_z s_z}, J_{\alpha/\beta}^{k\tau_z s_z}, J_{\beta/\alpha}^{k\tau_z s_z}, J_{\alpha\beta}^{k\tau_z s_z}\right)$$

$$= \int_{A_k} \frac{\partial F_\tau}{\partial z}\frac{\partial F_s}{\partial z}\left(1, H_\alpha^k, H_\beta^k, \frac{H_\alpha^k}{H_\beta^k}, \frac{H_\beta^k}{H_\alpha^k}, H_\alpha^k H_\beta^k\right) dz \tag{7.41}$$

By developing the matrix products in Equations (7.32)–(7.36) and employing a Navier-type closed-form solution (Carrera and Brischetto 2007b), the explicit algebraic form of the nuclei can be obtained.

The fundamental nucleus $\boldsymbol{K}_{uu}^{k\tau s}$ of dimension (3 × 3) is:

$$\begin{aligned}
K_{uu_{11}}^{k\tau s} &= Q_{55}^k J_{\alpha\beta}^{k\tau_z s_z} + \frac{1}{R_\alpha^k} Q_{55}^k\left(-J_\beta^{k\tau_z s} - J_\beta^{k\tau s_z} + \frac{1}{R_\alpha^k} J_{\beta/\alpha}^{k\tau s}\right) + Q_{11}^k J_{\beta/\alpha}^{k\tau s}\bar{\alpha}^2 + Q_{66}^k J_{\alpha/\beta}^{k\tau s}\bar{\beta}^2\\
K_{uu_{12}}^{k\tau s} &= J^{k\tau s}\bar{\alpha}\bar{\beta}(Q_{12}^k + Q_{66}^k)\\
K_{uu_{13}}^{k\tau s} &= Q_{55}^k\left(J_\beta^{k\tau_z s}\bar{\alpha} - \frac{1}{R_\alpha^k} J_{\beta/\alpha}^{k\tau s}\bar{\alpha}\right) - Q_{13}^k J_\beta^{k\tau s_z}\bar{\alpha} - \frac{1}{R_\alpha^k} Q_{11}^k J_{\beta/\alpha}^{k\tau s}\bar{\alpha} - Q_{12}^k J^{k\tau s}\bar{\alpha}\frac{1}{R_\beta^k}\\
K_{uu_{21}}^{k\tau s} &= J^{k\tau s}\bar{\alpha}\bar{\beta}(Q_{12}^k + Q_{66}^k) \qquad (7.42)\\
K_{uu_{22}}^{k\tau s} &= Q_{44}^k J_{\alpha\beta}^{k\tau_z s_z} + \frac{1}{R_\beta^k} Q_{44}^k\left(-J_\alpha^{k\tau_z s} - J_\alpha^{k\tau s_z} + \frac{1}{R_\beta^k} J_{\alpha/\beta}^{k\tau s}\right) + Q_{22}^k J_{\alpha/\beta}^{k\tau s}\bar{\beta}^2 + Q_{66}^k J_{\beta/\alpha}^{k\tau s}\bar{\alpha}^2\\
K_{uu_{23}}^{k\tau s} &= Q_{44}^k\left(J_\alpha^{k\tau_z s}\bar{\beta} - \frac{1}{R_\beta^k} J_{\alpha/\beta}^{k\tau s}\bar{\beta}\right) - Q_{23}^k J_\alpha^{k\tau s_z}\bar{\beta} - \frac{1}{R_\beta^k} Q_{22}^k J_{\alpha/\beta}^{k\tau s}\bar{\beta} - \frac{1}{R_\alpha^k} Q_{12}^k J^{k\tau s}\bar{\beta}\\
K_{uu_{31}}^{k\tau s} &= Q_{55}^k J_\beta^{k\tau s_z}\bar{\alpha} - Q_{55}^k \frac{1}{R_\alpha^k} J_{\beta/\alpha}^{k\tau s}\bar{\alpha} - Q_{13}^k J_\beta^{k\tau_z s}\bar{\alpha} - \frac{1}{R_\alpha^k} Q_{11}^k J_{\beta/\alpha}^{k\tau s}\bar{\alpha} - \frac{1}{R_\beta^k} Q_{12}^k J^{k\tau s}\bar{\alpha}\\
K_{uu_{32}}^{k\tau s} &= Q_{44}^k\left(J_\alpha^{k\tau s_z}\bar{\beta} - \frac{1}{R_\beta^k} J_{\alpha/\beta}^{k\tau s}\bar{\beta}\right) - Q_{23}^k J_\alpha^{k\tau_z s}\bar{\beta} - \frac{1}{R_\beta^k} Q_{22}^k J_{\alpha/\beta}^{k\tau s}\bar{\beta} - \frac{1}{R_\alpha^k} Q_{12}^k J^{k\tau s}\bar{\beta},\\
K_{uu_{33}}^{k\tau s} &= Q_{55}^k J_{\beta/\alpha}^{k\tau s}\bar{\alpha}^2 + Q_{44}^k J_{\alpha/\beta}^{k\tau s}\bar{\beta}^2 + Q_{33}^k J_{\alpha\beta}^{k\tau_z s_z} + \frac{1}{R_\alpha^k}\left(\frac{1}{R_\alpha^k} Q_{11}^k J_{\beta/\alpha}^{k\tau s} + Q_{13}^k J_\beta^{k\tau s_z}\right.\\
&\quad \left. + Q_{13}^k J_\beta^{k\tau_z s}\right) + \frac{2}{R_\alpha^k R_\beta^k} J^{k\tau s} Q_{12}^k + \frac{1}{R_\beta^k}\left(\frac{1}{R_\beta^k} Q_{22}^k J_{\alpha/\beta}^{k\tau s} + Q_{23}^k J_\alpha^{k\tau_z s} + Q_{23}^k J_\alpha^{k\tau s_z}\right)
\end{aligned}$$

The fundamental nucleus $\boldsymbol{K}_{u\Phi}^{k\tau s}$ is of (3 × 1) dimension because Φ^k is scalar:

$$\begin{aligned}
K_{u\Phi_{11}}^{k\tau s} &= \bar{\alpha}\left(-J_\beta^{k\tau s_z} e_{31}^k + J_\beta^{k\tau_z s} e_{15}^k - \frac{1}{R_\alpha^k} J_{\beta/\alpha}^{k\tau s} e_{15}^k\right)\\
K_{u\Phi_{21}}^{k\tau s} &= \bar{\beta}\left(J_\alpha^{k\tau_z s} e_{24}^k - \frac{1}{R_\beta^k} J_{\alpha/\beta}^{k\tau s} e_{24}^k - e_{32}^k J_\alpha^{k\tau s_z}\right) \qquad (7.43)
\end{aligned}$$

$$K_{u\Phi_{31}}^{k\tau s} = \bar{\alpha}^2 e_{15}^k J_{\beta/\alpha}^{k\tau s} + \bar{\beta}^2 e_{24}^k J_{\alpha/\beta}^{k\tau s} + e_{33}^k J_{\alpha\beta}^{k\tau_z s_z} + \frac{1}{R_\alpha^k} e_{31}^k J_\beta^{k\tau s_z} + \frac{1}{R_\beta^k} e_{32}^k J_\alpha^{k\tau s_z}$$

The fundamental nucleus $\boldsymbol{K}_{\Phi u}^{k\tau s}$ is of (1×3) dimension:

$$\begin{aligned}
K_{\Phi u_{11}}^{k\tau s} &= -\bar{\alpha} e_{15}^k \left(J_\beta^{k\tau s_z} - J_{\beta/\alpha}^{k\tau s} \frac{1}{R_\alpha^k} \right) + \bar{\alpha} e_{31}^k J_\beta^{k\tau_z s} \\
K_{\Phi u_{12}}^{k\tau s} &= -\bar{\beta} e_{24}^k \left(J_\alpha^{k\tau s_z} - J_{\alpha/\beta}^{k\tau s} \frac{1}{R_\beta^k} \right) + \bar{\beta} e_{32}^k J_\alpha^{k\tau_z s} \\
K_{\Phi u_{13}}^{k\tau s} &= -\bar{\alpha}^2 e_{15}^k J_{\beta/\alpha}^{k\tau s} - \bar{\beta}^2 e_{24}^k J_{\alpha/\beta}^{k\tau s} - e_{33}^k J_{\alpha\beta}^{k\tau_z s_z} - e_{31}^k J_\beta^{k\tau_z s} \frac{1}{R_\alpha^k} - e_{32}^k J_\alpha^{k\tau_z s} \frac{1}{R_\beta^k}
\end{aligned} \tag{7.44}$$

The fundamental nucleus $\boldsymbol{K}_{\Phi\Phi}^{k\tau s}$ is of (1×1) dimension:

$$K_{\Phi\Phi_{11}}^{k\tau s} = J_{\beta/\alpha}^{k\tau s} \bar{\alpha}^2 \varepsilon_{11} + J_{\alpha/\beta}^{k\tau s} \bar{\beta}^2 \varepsilon_{22} + \varepsilon_{33} J_{\alpha\beta}^{k\tau_z s_z} \tag{7.45}$$

The fundamental nucleus $\boldsymbol{M}_{uu}^{k\tau s}$ is of (3×3) dimension with only the diagonal elements being different from zero:

$$M_{uu_{11}}^{k\tau s} = M_{uu_{22}}^{k\tau s} = M_{uu_{33}}^{k\tau s} = \rho^k J_{\alpha\beta}^{k\tau s} \tag{7.46}$$

$\bar{\alpha} = m\pi/a$ and $\bar{\beta} = n\pi/b$, where m and n are the wave numbers in the in-plane directions, and a and b are the shell dimensions in the α and β directions, respectively. A Navier-type closed-form solution is obtained via substitution of the harmonic expressions for the displacements and electric potential and by considering the following material coefficients to be equal to zero: $Q_{16} = Q_{26} = Q_{36} = Q_{45} = 0$ and $e_{25} = e_{14} = e_{36} = \varepsilon_{12} = 0$. The harmonic assumptions used for the displacements and the electric potential are:

$$u_\tau^k = \sum_{m,n} \hat{U}_\tau^k \cos\left(\frac{m\pi\alpha}{a}\right) \sin\left(\frac{n\pi\beta}{b}\right), \qquad k = 1, N_l \tag{7.47}$$

$$v_\tau^k = \sum_{m,n} \hat{V}_\tau^k \sin\left(\frac{m\pi\alpha}{a}\right) \cos\left(\frac{n\pi\beta}{b}\right), \qquad \tau = t, b, r \tag{7.48}$$

$$w_\tau^k = \sum_{m,n} \hat{W}_\tau^k \sin\left(\frac{m\pi\alpha}{a}\right) \sin\left(\frac{n\pi\beta}{b}\right), \qquad r = 2, N \tag{7.49}$$

$$\Phi_\tau^k = \sum_{m,n} \hat{\Phi}_\tau^k \sin\left(\frac{m\pi\alpha}{a}\right) \sin\left(\frac{n\pi\beta}{b}\right) \tag{7.50}$$

where $\hat{U}^k_\tau$, $\hat{V}^k_\tau$, $\hat{W}^k_\tau$ are the displacement amplitudes and $\hat{\Phi}^k_\tau$ is the electric potential amplitude; k indicates the layer and N_l is the total number of layers. τ is the index for the order of expansion, where t and b indicate the top and bottom of the layer, respectively, while r indicates the higher orders of expansion until $N = 4$. Details on the assembly procedure of the fundamental nuclei and on the acronyms are given in Sections 7.7 and 7.8, respectively, in analogy with the plate case dealt with in Sections 6.7 and 6.8.

7.4 RMVT(u, Φ, σ_n) for the electromechanical shell case

A first possible extension of the RMVT (Reissner 1984) to electromechanical problems is that indicated in Equations (2.60) and (6.51) in Chapters 2 and 6, respectively. In this case, the internal electric work is simply added, as shown in D'Ottavio and Kröplin (2006), Carrera and Boscolo (2007), and Carrera and Brischetto (2007b):

$$\int_V \Big(\delta \boldsymbol{\epsilon}_{pG}^T \boldsymbol{\sigma}_{pC} + \delta \boldsymbol{\epsilon}_{nG}^T \boldsymbol{\sigma}_{nM} + \delta \boldsymbol{\sigma}_{nM}^T (\boldsymbol{\epsilon}_{nG} - \boldsymbol{\epsilon}_{nC}) - \delta \boldsymbol{\mathcal{E}}_{pG}^T \boldsymbol{\mathcal{D}}_{pC} - \delta \boldsymbol{\mathcal{E}}_{nG}^T \boldsymbol{\mathcal{D}}_{nC}\Big) \times dV = \delta L_e - \delta L_{in} \tag{7.51}$$

The electrical work is split into in-plane and out-of-plane contributions. This splitting will be useful for those RMVT extensions in which the transverse normal electric displacement is a primary variable of the problem. By considering a laminated shell of N_l layers, and the integral on the volume V_k of each k layer as an integral on the in-plane curved domain Ω_k, plus the integral in the thickness-direction domain A_k, it is possible to write Equation (7.51) as:

$$\sum_{k=1}^{N_l} \int_{\Omega_k} \int_{A_k} \Big\{ \delta \boldsymbol{\epsilon}_{pG}^{k\ T} \boldsymbol{\sigma}_{pC}^k + \delta \boldsymbol{\epsilon}_{nG}^{k\ T} \boldsymbol{\sigma}_{nM}^k + \delta \boldsymbol{\sigma}_{nM}^{k\ T} (\boldsymbol{\epsilon}_{nG}^k - \boldsymbol{\epsilon}_{nC}^k) - \delta \boldsymbol{\mathcal{E}}_{pG}^{k\ T} \boldsymbol{\mathcal{D}}_{pC}^k - \delta \boldsymbol{\mathcal{E}}_{nG}^{k\ T} \boldsymbol{\mathcal{D}}_{nC}^k \Big\} d\Omega_k dz = \sum_{k=1}^{N_l} \delta L_e^k - \sum_{k=1}^{N_l} \delta L_{in}^k \tag{7.52}$$

where δL_e^k and δL_{in}^k are the external and inertial virtual work at the k-layer level, respectively. The relative constitutive equations are those obtained in Equations (2.61)–(2.64), with the components as in the curvilinear reference system (α, β, z) for the stress, strain, electric displacement, and electric field vectors given in Equations (2.19) and (2.20); the same constitutive equations have already been proposed for the plate case in Equations (6.53)–(6.56), where the transverse stresses $\boldsymbol{\sigma}_n$ are a priori modelled "M" and the meaning of coefficients $\hat{C}$ was given in Equations (2.65) and (6.87)–(6.102):

$$\boldsymbol{\sigma}^k_{pC} = \hat{\boldsymbol{C}}^k_{\sigma_p\epsilon_p}\boldsymbol{\epsilon}^k_{pG} + \hat{\boldsymbol{C}}^k_{\sigma_p\sigma_n}\boldsymbol{\sigma}^k_{nM} + \hat{\boldsymbol{C}}^k_{\sigma_p\mathcal{E}_p}\boldsymbol{\mathcal{E}}^k_{pG} + \hat{\boldsymbol{C}}^k_{\sigma_p\mathcal{E}_n}\boldsymbol{\mathcal{E}}^k_{nG} \tag{7.53}$$

$$\boldsymbol{\epsilon}^k_{nC} = \hat{\boldsymbol{C}}^k_{\epsilon_n\epsilon_p}\boldsymbol{\epsilon}^k_{pG} + \hat{\boldsymbol{C}}^k_{\epsilon_n\sigma_n}\boldsymbol{\sigma}^k_{nM} + \hat{\boldsymbol{C}}^k_{\epsilon_n\mathcal{E}_p}\boldsymbol{\mathcal{E}}^k_{pG} + \hat{\boldsymbol{C}}^k_{\epsilon_n\mathcal{E}_n}\boldsymbol{\mathcal{E}}^k_{nG} \tag{7.54}$$

$$\boldsymbol{\mathcal{D}}^k_{pC} = \hat{\boldsymbol{C}}^k_{\mathcal{D}_p\epsilon_p}\boldsymbol{\epsilon}^k_{pG} + \hat{\boldsymbol{C}}^k_{\mathcal{D}_p\sigma_n}\boldsymbol{\sigma}^k_{nM} + \hat{\boldsymbol{C}}^k_{\mathcal{D}_p\mathcal{E}_p}\boldsymbol{\mathcal{E}}^k_{pG} + \hat{\boldsymbol{C}}^k_{\mathcal{D}_p\mathcal{E}_n}\boldsymbol{\mathcal{E}}^k_{nG} \tag{7.55}$$

$$\boldsymbol{\mathcal{D}}^k_{nC} = \hat{\boldsymbol{C}}^k_{\mathcal{D}_n\epsilon_p}\boldsymbol{\epsilon}^k_{pG} + \hat{\boldsymbol{C}}^k_{\mathcal{D}_n\sigma_n}\boldsymbol{\sigma}^k_{nM} + \hat{\boldsymbol{C}}^k_{\mathcal{D}_n\mathcal{E}_p}\boldsymbol{\mathcal{E}}^k_{pG} + \hat{\boldsymbol{C}}^k_{\mathcal{D}_n\mathcal{E}_n}\boldsymbol{\mathcal{E}}^k_{nG} \tag{7.56}$$

By substituting Equations (7.53)–(7.56), and the geometrical relations (2.31)–(2.37) in Chapter 2 for shells in the variational statement of Equation (7.52), and considering a generic k layer (Carrera and Brischetto 2007a,b):

$$\begin{aligned}
&\int_{\Omega_k}\int_{A_k}\Bigg[\left((\boldsymbol{D}^k_p + \boldsymbol{A}^k_p)F_s\delta\boldsymbol{u}^k_s\right)^T\Big(\hat{\boldsymbol{C}}^k_{\sigma_p\epsilon_p}(\boldsymbol{D}^k_p + \boldsymbol{A}^k_p)F_\tau\boldsymbol{u}^k_\tau + \hat{\boldsymbol{C}}^k_{\sigma_p\sigma_n}F_\tau\boldsymbol{\sigma}^k_{nM\tau}\\
&- \hat{\boldsymbol{C}}^k_{\sigma_p\mathcal{E}_p}\boldsymbol{D}^k_{ep}F_\tau\Phi^k_\tau - \hat{\boldsymbol{C}}^k_{\sigma_p\mathcal{E}_n}\boldsymbol{D}^k_{en}F_\tau\Phi^k_\tau\Big) + \left((\boldsymbol{D}^k_{np} + \boldsymbol{D}^k_{nz} - \boldsymbol{A}^k_n)F_s\delta\boldsymbol{u}^k_s\right)^T\left(F_\tau\boldsymbol{\sigma}^k_{nM\tau}\right)\\
&+\left(F_s\delta\boldsymbol{\sigma}^k_{nMs}\right)^T\Big((\boldsymbol{D}^k_{np} + \boldsymbol{D}^k_{nz} - \boldsymbol{A}^k_n)F_\tau\boldsymbol{u}^k_\tau - \hat{\boldsymbol{C}}^k_{\epsilon_n\epsilon_p}(\boldsymbol{D}^k_p + \boldsymbol{A}^k_p)F_\tau\boldsymbol{u}^k_\tau\\
&- \hat{\boldsymbol{C}}^k_{\epsilon_n\sigma_n}F_\tau\boldsymbol{\sigma}^k_{nM\tau} + \hat{\boldsymbol{C}}^k_{\epsilon_n\mathcal{E}_p}\boldsymbol{D}^k_{ep}F_\tau\Phi^k_\tau + \hat{\boldsymbol{C}}^k_{\epsilon_n\mathcal{E}_n}\boldsymbol{D}^k_{en}F_\tau\Phi^k_\tau\Big) + \left(\boldsymbol{D}^k_{ep}F_s\delta\Phi^k_s\right)^T\\
&\times\Big(\hat{\boldsymbol{C}}^k_{\mathcal{D}_p\epsilon_p}(\boldsymbol{D}^k_p + \boldsymbol{A}^k_p)F_\tau\boldsymbol{u}^k_\tau + \hat{\boldsymbol{C}}^k_{\mathcal{D}_p\sigma_n}F_\tau\boldsymbol{\sigma}^k_{nM\tau} - \hat{\boldsymbol{C}}^k_{\mathcal{D}_p\mathcal{E}_p}\boldsymbol{D}^k_{ep}F_\tau\Phi^k_\tau - \hat{\boldsymbol{C}}^k_{\mathcal{D}_p\mathcal{E}_n}\boldsymbol{D}^k_{en}F_\tau\Phi^k_\tau\Big)\\
&+\left(\boldsymbol{D}^k_{en}F_s\delta\Phi^k_s\right)^T\Big(\hat{\boldsymbol{C}}^k_{\mathcal{D}_n\epsilon_p}(\boldsymbol{D}^k_p + \boldsymbol{A}^k_p)F_\tau\boldsymbol{u}^k_\tau + \hat{\boldsymbol{C}}^k_{\mathcal{D}_n\sigma_n}F_\tau\boldsymbol{\sigma}^k_{nM\tau} - \hat{\boldsymbol{C}}^k_{\mathcal{D}_n\mathcal{E}_p}\boldsymbol{D}^k_{ep}F_\tau\Phi^k_\tau\\
&- \hat{\boldsymbol{C}}^k_{\mathcal{D}_n\mathcal{E}_n}\boldsymbol{D}^k_{en}F_\tau\Phi^k_\tau\Big)\Bigg]d\Omega_k\,dz = \delta L^k_e - \delta L^k_{in}
\end{aligned} \tag{7.57}$$

The CUF (Carrera 1995) for the 2D approximation of shells, as presented in the previous sections, has already been introduced. In Equation (7.57), in order to obtain a strong form of the differential equations on domain Ω_k and the relative boundary conditions on edge Γ_k, integration by parts is used (see Equation (7.28) and the matrices in Equation (7.29)), which allows one to move the differential operator from the infinitesimal variation of the generic variable $\delta\boldsymbol{a}^k$ to the finite quantity $\boldsymbol{a}^k$ (Carrera 1995). The governing equations and the boundary conditions have the same form as the plate case discussed in Section 6.4, though the meaning of the fundamental nuclei changes. As discussed in Carrera and Boscolo (2007) and Carrera and Brischetto (2007b), the governing equations state:

$$
\begin{aligned}
\delta \boldsymbol{u}_s^k : \quad & \boldsymbol{K}_{uu}^{k\tau s}\, \boldsymbol{u}_\tau^k + \boldsymbol{K}_{u\sigma}^{k\tau s}\, \boldsymbol{\sigma}_{nM\tau}^k + \boldsymbol{K}_{u\Phi}^{k\tau s}\, \Phi_\tau^k = \boldsymbol{p}_{us}^k - \boldsymbol{M}_{uu}^{k\tau s}\, \ddot{\boldsymbol{u}}_\tau^k \\
\delta \boldsymbol{\sigma}_{ns}^k : \quad & \boldsymbol{K}_{\sigma u}^{k\tau s}\, \boldsymbol{u}_\tau^k + \boldsymbol{K}_{\sigma\sigma}^{k\tau s}\, \boldsymbol{\sigma}_{nM\tau}^k + \boldsymbol{K}_{\sigma\Phi}^{k\tau s}\, \Phi_\tau^k = 0 \\
\delta \Phi_s^k : \quad & \boldsymbol{K}_{\Phi u}^{k\tau s}\, \boldsymbol{u}_\tau^k + \boldsymbol{K}_{\Phi\sigma}^{k\tau s}\, \boldsymbol{\sigma}_{nM\tau}^k + \boldsymbol{K}_{\Phi\Phi}^{k\tau s}\, \Phi_\tau^k = 0
\end{aligned}
\tag{7.58}
$$

where $\boldsymbol{M}_{uu}^{k\tau s}$ is the inertial contribution in the form of the fundamental nucleus, $\boldsymbol{u}_\tau^k$ is the vector of the degrees of freedom for the displacements, Φ_τ^k is the vector of the degrees of freedom for the electric potential, $\boldsymbol{\sigma}_{nM\tau}^k$ is the vector of the degrees of freedom for the transverse stresses, and $\ddot{\boldsymbol{u}}_\tau^k$ is the second temporal derivative of $\boldsymbol{u}_\tau^k$. The array $\boldsymbol{p}_{us}^k$ indicates the variationally consistent mechanical loading for the case of the sensor configuration; in the case of the actuator configuration, the electric potential is applied directly to vector Φ_τ^k. Along with these governing equations, the following boundary conditions on edge Γ_k of the in-plane integration domain Ω_k hold:

$$
\begin{aligned}
\boldsymbol{\Pi}_{uu}^{k\tau s}\, \boldsymbol{u}_\tau^k + \boldsymbol{\Pi}_{u\sigma}^{k\tau s}\, \boldsymbol{\sigma}_{nM\tau}^k + \boldsymbol{\Pi}_{u\Phi}^{k\tau s}\, \Phi_\tau^k &= \boldsymbol{\Pi}_{uu}^{k\tau s}\, \bar{\boldsymbol{u}}_\tau^k + \boldsymbol{\Pi}_{u\sigma}^{k\tau s}\, \bar{\boldsymbol{\sigma}}_{nM\tau}^k + \boldsymbol{\Pi}_{u\Phi}^{k\tau s}\, \bar{\Phi}_\tau^k \\
\boldsymbol{\Pi}_{\Phi u}^{k\tau s}\, \boldsymbol{u}_\tau^k + \boldsymbol{\Pi}_{\Phi\sigma}^{k\tau s}\, \boldsymbol{\sigma}_{nM\tau}^k + \boldsymbol{\Pi}_{\Phi\Phi}^{k\tau s}\, \Phi_\tau^k &= \boldsymbol{\Pi}_{\Phi u}^{k\tau s}\, \bar{\boldsymbol{u}}_\tau^k + \boldsymbol{\Pi}_{\Phi\sigma}^{k\tau s}\, \bar{\boldsymbol{\sigma}}_{nM\tau}^k + \boldsymbol{\Pi}_{\Phi\Phi}^{k\tau s}\, \bar{\Phi}_\tau^k
\end{aligned}
\tag{7.59}
$$

By comparing Equation (7.57), after the integration by parts (see Equation (7.28)), to Equations (7.58) and (7.59), the fundamental nuclei can be obtained:

$$
\boldsymbol{K}_{uu}^{k\tau s} = \int_{A_k} \left[\left(-\boldsymbol{D}_p^k + \boldsymbol{A}_p^k \right)^T \left(\hat{\boldsymbol{C}}_{\sigma_p \epsilon_p}^k (\boldsymbol{D}_p^k + \boldsymbol{A}_p^k) \right) \right] F_s F_\tau H_\alpha^k H_\beta^k dz \tag{7.60}
$$

$$
\boldsymbol{K}_{u\sigma}^{k\tau s} = \int_{A_k} \left[\left(-\boldsymbol{D}_p^k + \boldsymbol{A}_p^k \right)^T \left(\hat{\boldsymbol{C}}_{\sigma_p \sigma_n}^k \right) + \left(-\boldsymbol{D}_{np}^k + \boldsymbol{D}_{nz}^k - \boldsymbol{A}_n^k \right)^T \right] F_s F_\tau H_\alpha^k H_\beta^k dz \tag{7.61}
$$

$$
\boldsymbol{K}_{u\Phi}^{k\tau s} = \int_{A_k} \left[\left(-\boldsymbol{D}_p^k + \boldsymbol{A}_p^k \right)^T \left(-\hat{\boldsymbol{C}}_{\sigma_p \mathcal{E}_p}^k \boldsymbol{D}_{ep}^k - \hat{\boldsymbol{C}}_{\sigma_p \mathcal{E}_n}^k \boldsymbol{D}_{en}^k \right) \right] F_s F_\tau H_\alpha^k H_\beta^k dz \tag{7.62}
$$

$$
\boldsymbol{K}_{\sigma u}^{k\tau s} = \int_{A_k} \left[\left(\boldsymbol{D}_{np}^k + \boldsymbol{D}_{nz}^k - \boldsymbol{A}_n^k \right) - \left(\hat{\boldsymbol{C}}_{\epsilon_n \epsilon_p}^k \right) \left(\boldsymbol{D}_p^k + \boldsymbol{A}_p^k \right) \right] F_s F_\tau H_\alpha^k H_\beta^k dz \tag{7.63}
$$

$$
\boldsymbol{K}_{\sigma\sigma}^{k\tau s} = \int_{A_k} \left[-\hat{\boldsymbol{C}}_{\epsilon_n \sigma_n}^k \right] F_s F_\tau H_\alpha^k H_\beta^k dz \tag{7.64}
$$

$$\boldsymbol{K}^{k\tau s}_{\sigma\Phi} = \int_{A_k} \left[\hat{\boldsymbol{C}}^k_{\epsilon_n\mathcal{E}_p} \boldsymbol{D}^k_{ep} + \hat{\boldsymbol{C}}^k_{\epsilon_n\mathcal{E}_n} \boldsymbol{D}^k_{en}\right] F_s F_\tau H^k_\alpha H^k_\beta dz \tag{7.65}$$

$$\boldsymbol{K}^{k\tau s}_{\Phi u} = \int_{A_k} \left[\left(-\boldsymbol{D}^{k\,T}_{ep} \hat{\boldsymbol{C}}^k_{\mathcal{D}_p\epsilon_p} + \boldsymbol{D}^{k\,T}_{en} \hat{\boldsymbol{C}}^k_{\mathcal{D}_n\epsilon_p}\right)\left(\boldsymbol{D}^k_p + \boldsymbol{A}^k_p\right)\right] F_s F_\tau H^k_\alpha H^k_\beta dz \tag{7.66}$$

$$\boldsymbol{K}^{k\tau s}_{\Phi\sigma} = \int_{A_k} \left[-\boldsymbol{D}^{k\,T}_{ep} \hat{\boldsymbol{C}}^k_{\mathcal{D}_p\sigma_n} + \boldsymbol{D}^{k\,T}_{en} \hat{\boldsymbol{C}}^k_{\mathcal{D}_n\sigma_n}\right] F_s F_\tau H^k_\alpha H^k_\beta dz \tag{7.67}$$

$$\boldsymbol{K}^{k\tau s}_{\Phi\Phi} = \int_{A_k} \left[-\boldsymbol{D}^{k\,T}_{ep}\left(-\hat{\boldsymbol{C}}^k_{\mathcal{D}_p\mathcal{E}_p} \boldsymbol{D}^k_{ep} - \hat{\boldsymbol{C}}^k_{\mathcal{D}_p\mathcal{E}_n} \boldsymbol{D}^k_{en}\right) + \boldsymbol{D}^{k\,T}_{en}\left(-\hat{\boldsymbol{C}}^k_{\mathcal{D}_n\mathcal{E}_p} \boldsymbol{D}^k_{ep}\right.\right.$$
$$\left.\left. - \hat{\boldsymbol{C}}^k_{\mathcal{D}_n\mathcal{E}_n} \boldsymbol{D}^k_{en}\right)\right] F_s F_\tau H^k_\alpha H^k_\beta dz \tag{7.68}$$

The fundamental nucleus for the inertial matrix $\boldsymbol{M}^{k\tau s}_{uu}$ is the same one that was given for the PVD case in Equation (7.36). The nuclei for the boundary conditions on edge Γ_k are (Carrera and Brischetto 2007b):

$$\boldsymbol{\Pi}^{k\tau s}_{uu} = \int_{A_k} \left[\boldsymbol{I}^{k\,T}_p \hat{\boldsymbol{C}}^k_{\sigma_p\epsilon_p}\left(\boldsymbol{D}^k_p + \boldsymbol{A}^k_p\right)\right] F_s F_\tau H^k_\alpha H^k_\beta dz \tag{7.69}$$

$$\boldsymbol{\Pi}^{k\tau s}_{u\sigma} = \int_{A_k} \left[\boldsymbol{I}^{k\,T}_p \hat{\boldsymbol{C}}^k_{\sigma_p\sigma_n} + \boldsymbol{I}^{k\,T}_{np}\right] F_s F_\tau H^k_\alpha H^k_\beta dz \tag{7.70}$$

$$\boldsymbol{\Pi}^{k\tau s}_{u\Phi} = \int_{A_k} \left[\boldsymbol{I}^{k\,T}_p\left(-\hat{\boldsymbol{C}}^k_{\sigma_p\mathcal{E}_p} \boldsymbol{D}^k_{ep} - \hat{\boldsymbol{C}}^k_{\sigma_p\mathcal{E}_n} \boldsymbol{D}^k_{en}\right)\right] F_s F_\tau H^k_\alpha H^k_\beta dz \tag{7.71}$$

$$\boldsymbol{\Pi}^{k\tau s}_{\Phi u} = \int_{A_k} \left[\boldsymbol{I}^{k\,T}_{ep}\left(\hat{\boldsymbol{C}}^k_{\mathcal{D}_p\epsilon_p}\left(\boldsymbol{D}^k_p + \boldsymbol{A}^k_p\right)\right)\right] F_s F_\tau H^k_\alpha H^k_\beta dz \tag{7.72}$$

$$\boldsymbol{\Pi}^{k\tau s}_{\Phi\sigma} = \int_{A_k} \left[\boldsymbol{I}^{k\,T}_{ep} \hat{\boldsymbol{C}}^k_{\mathcal{D}_p\sigma_n}\right] F_s F_\tau H^k_\alpha H^k_\beta dz \tag{7.73}$$

$$\boldsymbol{\Pi}^{k\tau s}_{\Phi\Phi} = \int_{A_k} \left[\boldsymbol{I}^{k\,T}_{ep}\left(-\hat{\boldsymbol{C}}^k_{\mathcal{D}_p\mathcal{E}_p} \boldsymbol{D}^k_{ep} - \hat{\boldsymbol{C}}^k_{\mathcal{D}_p\mathcal{E}_n} \boldsymbol{D}^k_{en}\right)\right] F_s F_\tau H^k_\alpha H^k_\beta dz \tag{7.74}$$

The geometrical matrices and those for the integration by parts depend on the k layer, in the case of shells, because of the parametric coefficients and radii of curvature.

In order to write the nuclei in Equations (7.60)–(7.68) in explicit form, the integrals in the z thickness direction are defined as in Equations (7.41); by developing the matrix products in Equations (7.60)–(7.68) and employing a

Navier-type closed-form solution (Carrera and Brischetto 2007b), the explicit algebraic form of the nuclei can be obtained.

The fundamental nucleus $\boldsymbol{K}_{uu}^{k\tau s}$ of (3 × 3) dimension is:

$$
\begin{aligned}
K_{uu_{11}}^{k\tau s} &= \bar{\alpha}^2 J_{\beta/\alpha}^{k\tau s}\hat{C}_{\sigma_p\epsilon_p 11}^{k} + \bar{\beta}^2 J_{\alpha/\beta}^{k\tau s}\hat{C}_{\sigma_p\epsilon_p 33}^{k}\\
K_{uu_{12}}^{k\tau s} &= J^{k\tau s}(\hat{C}_{\sigma_p\epsilon_p 12}^{k} + \hat{C}_{\sigma_p\epsilon_p 33}^{k})\bar{\alpha}\bar{\beta}\\
K_{uu_{13}}^{k\tau s} &= -\frac{1}{R_\alpha^k} J_{\beta/\alpha}^{k\tau s}\bar{\alpha}\hat{C}_{\sigma_p\epsilon_p 11}^{k} - \frac{1}{R_\beta^k} J^{k\tau s}\bar{\alpha}\hat{C}_{\sigma_p\epsilon_p 12}^{k}\\
K_{uu_{21}}^{k\tau s} &= J^{k\tau s}(\hat{C}_{\sigma_p\epsilon_p 21}^{k} + \hat{C}_{\sigma_p\epsilon_p 33}^{k})\bar{\alpha}\bar{\beta}\\
K_{uu_{22}}^{k\tau s} &= \bar{\beta}^2 J_{\alpha/\beta}^{k\tau s}\hat{C}_{\sigma_p\epsilon_p 22}^{k} + \bar{\alpha}^2 J_{\beta/\alpha}^{k\tau s}\hat{C}_{\sigma_p\epsilon_p 33}^{k}\\
K_{uu_{23}}^{k\tau s} &= -\frac{1}{R_\alpha^k} J^{k\tau s}\bar{\beta}\hat{C}_{\sigma_p\epsilon_p 21}^{k} - \frac{1}{R_\beta^k} J_{\alpha/\beta}^{k\tau s}\bar{\alpha}\hat{C}_{\sigma_p\epsilon_p 22}^{k}\\
K_{uu_{31}}^{k\tau s} &= -\frac{1}{R_\alpha^k} J_{\beta/\alpha}^{k\tau s}\bar{\alpha}\hat{C}_{\sigma_p\epsilon_p 11}^{k} - \frac{1}{R_\beta^k} J^{k\tau s}\bar{\alpha}\hat{C}_{\sigma_p\epsilon_p 21}^{k}\\
K_{uu_{32}}^{k\tau s} &= -\frac{1}{R_\alpha^k} J^{k\tau s}\bar{\beta}\hat{C}_{\sigma_p\epsilon_p 12}^{k} - \frac{1}{R_\beta^k} J_{\alpha/\beta}^{k\tau s}\bar{\beta}\hat{C}_{\sigma_p\epsilon_p 22}^{k}\\
K_{uu_{33}}^{k\tau s} &= \frac{1}{R_\alpha^{k^2}} J_{\beta/\alpha}^{k\tau s}\hat{C}_{\sigma_p\epsilon_p 11}^{k} + \frac{1}{R_\alpha^k R_\beta^k} J^{k\tau s}(\hat{C}_{\sigma_p\epsilon_p 12}^{k} + \hat{C}_{\sigma_p\epsilon_p 21}^{k}) + \frac{1}{R_\beta^{k^2}} J_{\alpha/\beta}^{k\tau s}\hat{C}_{\sigma_p\epsilon_p 22}^{k}
\end{aligned}
\tag{7.75}
$$

Nucleus $\boldsymbol{K}_{u\sigma}^{k\tau s}$ of (3 × 3) dimension is:

$$
\begin{aligned}
&K_{u\sigma_{11}}^{k\tau s} = -\frac{1}{R_\alpha^k} J_\beta^{k\tau s} + J_{\alpha\beta}^{k\tau s_z}, \quad K_{u\sigma_{12}}^{k\tau s} = 0, \quad K_{u\sigma_{13}}^{k\tau s} = -\bar{\alpha} J_\beta^{k\tau s}\hat{C}_{\sigma_p\sigma_n 13}^{k}\\
&K_{u\sigma_{21}}^{k\tau s} = 0 \quad K_{u\sigma_{22}}^{k\tau s} = -\frac{1}{R_\beta^k} J_\alpha^{k\tau s} + J_{\alpha\beta}^{k\tau s_z}, \quad K_{u\sigma_{23}}^{k\tau s} = -\bar{\beta} J_\alpha^{k\tau s}\hat{C}_{\sigma_p\sigma_n 23}^{k}\\
&K_{u\sigma_{31}}^{k\tau s} = \bar{\alpha} J_\beta^{k\tau s}, \quad K_{u\sigma_{32}}^{k\tau s} = \bar{\beta} J_\alpha^{k\tau s}\\
&K_{u\sigma_{33}}^{k\tau s} = J_{\alpha\beta}^{k\tau s_z} + \frac{1}{R_\alpha^k} J_\beta^{k\tau s}\hat{C}_{\sigma_p\sigma_n 13}^{k} + \frac{1}{R_\beta^k} J_\alpha^{k\tau s}\hat{C}_{\sigma_p\sigma_n 23}^{k}
\end{aligned}
\tag{7.76}
$$

Nucleus $\boldsymbol{K}_{u\Phi}^{k\tau s}$ of (3 × 1) dimension is:

$$
\begin{aligned}
&K_{u\Phi_{11}}^{k\tau s} = \bar{\alpha} J_\beta^{k\tau_z s}\hat{C}_{\sigma_p\mathcal{E}_n 11}^{k}, \quad K_{u\Phi_{21}}^{k\tau s} = \bar{\beta} J_\alpha^{k\tau_z s}\hat{C}_{\sigma_p\mathcal{E}_n 21}^{k}\\
&K_{u\Phi_{31}}^{k\tau s} = -\frac{1}{R_\alpha^k} J_\beta^{k\tau_z s}\hat{C}_{\sigma_p\mathcal{E}_n 11}^{k} - \frac{1}{R_\beta^k} J_\alpha^{k\tau_z s}\hat{C}_{\sigma_p\mathcal{E}_n 21}^{k}
\end{aligned}
\tag{7.77}
$$

Nucleus $\boldsymbol{K}^{k\tau s}_{\sigma u}$ of (3×3) dimension is:

$$
\begin{aligned}
&K^{k\tau s}_{\sigma u_{11}} = -\frac{1}{R^k_\alpha} J^{k\tau s}_\beta + J^{k\tau_z s}_{\alpha\beta}, \quad K^{k\tau s}_{\sigma u_{12}} = 0, \quad K^{k\tau s}_{\sigma u_{13}} = \bar{\alpha} J^{k\tau s}_\beta \\
&K^{k\tau s}_{\sigma u_{21}} = 0 \quad K^{k\tau s}_{\sigma u_{22}} = -\frac{1}{R^k_\beta} J^{k\tau s}_\alpha + J^{k\tau_z s}_{\alpha\beta}, \quad K^{k\tau s}_{u\sigma_{23}} = \bar{\beta} J^{k\tau s}_\alpha \\
&K^{k\tau s}_{\sigma u_{31}} = \bar{\alpha} J^{k\tau s}_\beta \hat{C}^k_{\epsilon_n\epsilon_p 31} \\
&K^{k\tau s}_{\sigma u_{32}} = \bar{\beta} J^{k\tau s}_\alpha \hat{C}^k_{\epsilon_n\epsilon_p 32}, \quad K^{k\tau s}_{\sigma u_{33}} = J^{k\tau_z s}_{\alpha\beta} - \frac{1}{R^k_\alpha} J^{k\tau s}_\beta \hat{C}^k_{\epsilon_n\epsilon_p 31} - \frac{1}{R^k_\beta} J^{k\tau s}_\alpha \hat{C}^k_{\epsilon_n\epsilon_p 32}
\end{aligned}
\tag{7.78}
$$

Nucleus $\boldsymbol{K}^{k\tau s}_{\sigma\sigma}$ of (3×3) dimension is:

$$
\begin{aligned}
&K^{k\tau s}_{\sigma\sigma_{11}} = -J^{k\tau s}_{\alpha\beta} \hat{C}^k_{\epsilon_n\sigma_n 11} \quad K^{k\tau s}_{\sigma\sigma_{12}} = 0, \quad K^{k\tau s}_{\sigma\sigma_{13}} = 0 \\
&K^{k\tau s}_{\sigma\sigma_{21}} = 0 \quad K^{k\tau s}_{\sigma\sigma_{22}} = -J^{k\tau s}_{\alpha\beta} \hat{C}^k_{\epsilon_n\sigma_n 22} \quad K^{k\tau s}_{\sigma\sigma_{23}} = 0 \\
&K^{k\tau s}_{\sigma\sigma_{31}} = 0 \quad K^{k\tau s}_{\sigma\sigma_{32}} = 0 \quad K^{k\tau s}_{\sigma\sigma_{33}} = -J^{k\tau s}_{\alpha\beta} \hat{C}^k_{\epsilon_n\sigma_n 33}
\end{aligned}
\tag{7.79}
$$

Nucleus $\boldsymbol{K}^{k\tau s}_{\sigma\Phi}$ of (3×1) dimension is:

$$
K^{k\tau s}_{\sigma\Phi_{11}} = \bar{\alpha} J^{k\tau s}_\beta \hat{C}^k_{\epsilon_n\mathcal{E}_p 11} \quad K^{k\tau s}_{\sigma\Phi_{21}} = \bar{\beta} J^{k\tau s}_\alpha \hat{C}^k_{\epsilon_n\mathcal{E}_p 22} \quad K^{k\tau s}_{\sigma\Phi_{31}} = J^{k\tau_z s}_{\alpha\beta} \hat{C}^k_{\epsilon_n\mathcal{E}_n 31}
\tag{7.80}
$$

Nucleus $\boldsymbol{K}^{k\tau s}_{\Phi u}$ of (1×3) dimension is:

$$
\begin{aligned}
&K^{k\tau s}_{\Phi u_{11}} = -\bar{\alpha} J^{k\tau s_z}_\beta \hat{C}^k_{\mathcal{D}_n\epsilon_p 11} \quad K^{k\tau s}_{\Phi u_{12}} = -\bar{\beta} J^{k\tau s_z}_\alpha \hat{C}^k_{\mathcal{D}_n\epsilon_p 12} \\
&K^{k\tau s}_{\Phi u_{13}} = \frac{1}{R^k_\alpha} J^{k\tau s_z}_\beta \hat{C}^k_{\mathcal{D}_n\epsilon_p 11} + \frac{1}{R^k_\beta} J^{k\tau s_z}_\alpha \hat{C}^k_{\mathcal{D}_n\epsilon_p 12}
\end{aligned}
\tag{7.81}
$$

Nucleus $\boldsymbol{K}^{k\tau s}_{\Phi\sigma}$ of (1×3) dimension is:

$$
K^{k\tau s}_{\Phi\sigma_{11}} = \bar{\alpha} J^{k\tau s}_\beta \hat{C}^k_{\mathcal{D}_n\sigma_n 11} \quad K^{k\tau s}_{\Phi\sigma_{12}} = \bar{\beta} J^{k\tau s}_\alpha \hat{C}^k_{\mathcal{D}_n\sigma_n 22} \quad K^{k\tau s}_{\Phi\sigma_{13}} = J^{k\tau s_z}_{\alpha\beta} \hat{C}^k_{\mathcal{D}_n\sigma_n 13}
\tag{7.82}
$$

Nucleus $\boldsymbol{K}^{k\tau s}_{\Phi\Phi}$ of (1×1) dimension is:

$$
K^{k\tau s}_{\Phi\Phi_{11}} = -J^{k\tau_z s_z}_{\alpha\beta} \hat{C}^k_{\mathcal{D}_n\mathcal{E}_n 11} - \bar{\alpha}^2 J^{k\tau s}_{\beta/\alpha} \hat{C}^k_{\mathcal{D}_p\mathcal{E}_p 11} - \bar{\beta}^2 J^{k\tau s}_{\alpha/\beta} \hat{C}^k_{\mathcal{D}_p\mathcal{E}_p 22}
\tag{7.83}
$$

The fundamental nucleus $\boldsymbol{M}^{k\tau s}_{uu}$ is of (3×3) dimension with the diagonal elements as already given in Equation (7.46).

$\bar{\alpha} = m\pi/a$ and $\bar{\beta} = n\pi/b$, where m and n are the wave numbers in the in-plane directions and a and b are the shell dimensions in the α and β directions, respectively. The explicit form of coefficients $\hat{\boldsymbol{C}}$ and their components are given in Equations (2.65) and (6.87)–(6.102), respectively.

A Navier-type closed-form solution is obtained via substitution of the harmonic expressions for the displacements, electric potential, and transverse stresses and by considering the following material coefficients to be equal to zero: $Q_{16} = Q_{26} = Q_{36} = Q_{45} = 0$ and $e_{25} = e_{14} = e_{36} = \varepsilon_{12} = 0$. The harmonic assumptions used for the displacements, the electric potential, and the transverse stresses are:

$$(u_\tau^k, \sigma_{\alpha z\tau}^k) = \sum_{m,n} (\hat{U}_\tau^k, \hat{\sigma}_{\alpha z\tau}^k) \cos\left(\frac{m\pi\alpha}{a}\right) \sin\left(\frac{n\pi\beta}{b}\right), \quad k = 1, N_l \tag{7.84}$$

$$(v_\tau^k, \sigma_{\beta z\tau}^k) = \sum_{m,n} (\hat{V}_\tau^k, \hat{\sigma}_{\beta z\tau}^k) \sin\left(\frac{m\pi\alpha}{a}\right) \cos\left(\frac{n\pi\beta}{b}\right), \quad \tau = t, b, r \tag{7.85}$$

$$(w_\tau^k, \sigma_{zz\tau}^k, \Phi_\tau^k) = \sum_{m,n} (\hat{W}_\tau^k, \hat{\sigma}_{zz\tau}^k, \hat{\Phi}_\tau^k) \sin\left(\frac{m\pi\alpha}{a}\right) \sin\left(\frac{n\pi\beta}{b}\right), \quad r = 2, N \tag{7.86}$$

where $\hat{U}_\tau^k$, $\hat{V}_\tau^k$, $\hat{W}_\tau^k$ are the displacement amplitudes, $\hat{\Phi}_\tau^k$ is the electric potential amplitude, and $\hat{\sigma}_{\alpha z\tau}^k$, $\hat{\sigma}_{\beta z\tau}^k$, $\hat{\sigma}_{zz\tau}^k$ are the transverse stress amplitudes; k indicates the layer and N_l is the total number of layers. τ is the index for the order of expansion, where t and b indicate the top and bottom of the layer, respectively, while r indicates the higher orders of expansion until $N = 4$. Details on the assembling procedure of the fundamental nuclei and on the acronyms are given in Sections 7.7 and 7.8, respectively, as already mentioned for the plate case in Sections 6.7 and 6.8.

7.5 RMVT(u, Φ, $\mathcal{D}_n$) for the electromechanical shell case

The second possible extension of the RMVT (Reissner 1984) to electromechanical problems is that indicated in Equations (2.74) and (6.103) of Chapters 2 and 6, respectively. In this case, the internal electrical work is simply added and the transverse normal electric displacement is modeled via the Lagrange multiplier (see Carrera *et al.* 2010a,b; 2008):

$$\int_V \left(\delta\epsilon^T_{pG}\sigma_{pC} + \delta\epsilon^T_{nG}\sigma_{nC} - \delta\mathcal{E}^T_{pG}\mathcal{D}_{pC} - \delta\mathcal{E}^T_{nG}\mathcal{D}_{nM} - \delta\mathcal{D}^T_{nM}(\mathcal{E}_{nG} - \mathcal{E}_{nC})\right)dV$$
$$= \delta L_e - \delta L_{in} \tag{7.87}$$

where the subscript M means a priori modeled variable. Considering a laminate of N_l layers, and the integral on the volume V_k of each layer k as an integral on the in-plane domain Ω_k, plus the integral in the thickness-direction domain A_k, it is possible to write Equation (7.87) as:

$$\sum_{k=1}^{N_l} \int_{\Omega_k}\int_{A_k} \left\{\delta\epsilon^k_{pG}{}^T \sigma^k_{pC} + \delta\epsilon^k_{nG}{}^T \sigma^k_{nC} - \delta\mathcal{E}^k_{pG}{}^T \mathcal{D}^k_{pC} - \delta\mathcal{E}^k_{nG}{}^T \mathcal{D}^k_{nM}\right.$$
$$\left. - \delta\mathcal{D}^k_{nM}{}^T (\mathcal{E}^k_{nG} - \mathcal{E}^k_{nC})\right\} \tag{7.88}$$
$$d\Omega_k dz = \sum_{k=1}^{N_l} \delta L^k_e - \sum_{k=1}^{N_l} \delta L^k_{in}$$

where δL^k_e and δL^k_{in} are the external and inertial virtual work at the k-layer level, respectively. The relative constitutive equations are those obtained in Equations (2.75)–(2.78) with components in the curvilinear reference system (α,β,z) for the stress, strain, electric displacement, and electric field vectors given in Equations (2.19) and (2.20); the same constitutive equations were proposed for the plate case in Equations (6.105)–(6.108) where the transverse normal electric displacement $\mathcal{D}_n$ is a priori modeled "M" and the meaning of coefficients $\bar{C}$ was given in Equations (2.79) and (6.139)–(6.154) (Carrera *et al.* 2010a,b):

$$\sigma^k_{pC} = \bar{C}^k_{\sigma_p\epsilon_p}\epsilon^k_{pG} + \bar{C}^k_{\sigma_p\epsilon_n}\epsilon^k_{nG} + \bar{C}^k_{\sigma_p\mathcal{E}_p}\mathcal{E}^k_{pG} + \bar{C}^k_{\sigma_p\mathcal{D}_n}\mathcal{D}^k_{nM} \tag{7.89}$$

$$\sigma^k_{nC} = \bar{C}^k_{\sigma_n\epsilon_p}\epsilon^k_{pG} + \bar{C}^k_{\sigma_n\epsilon_n}\epsilon^k_{nG} + \bar{C}^k_{\sigma_n\mathcal{E}_p}\mathcal{E}^k_{pG} + \bar{C}^k_{\sigma_n\mathcal{D}_n}\mathcal{D}^k_{nM} \tag{7.90}$$

$$\mathcal{D}^k_{pC} = \bar{C}^k_{\mathcal{D}_p\epsilon_p}\epsilon^k_{pG} + \bar{C}^k_{\mathcal{D}_p\epsilon_n}\epsilon^k_{nG} + \bar{C}^k_{\mathcal{D}_p\mathcal{E}_p}\mathcal{E}^k_{pG} + \bar{C}^k_{\mathcal{D}_p\mathcal{D}_n}\mathcal{D}^k_{nM} \tag{7.91}$$

$$\mathcal{E}^k_{nC} = \bar{C}^k_{\mathcal{E}_n\epsilon_p}\epsilon^k_{pG} + \bar{C}^k_{\mathcal{E}_n\epsilon_n}\epsilon^k_{nG} + \bar{C}^k_{\mathcal{E}_n\mathcal{E}_p}\mathcal{E}^k_{pG} + \bar{C}^k_{\mathcal{E}_n\mathcal{D}_n}\mathcal{D}^k_{nM} \tag{7.92}$$

By substituting Equations (7.89)–(7.92), and the geometrical relations (2.31)–(2.37) of Chapter 2 for shells, in the variational statement of Equation (7.88), and considering a generic k layer (Carrera *et al.* 2010a,b):

$$
\begin{aligned}
&\int_{\Omega_k}\int_{A_k}\Bigg[\left((\boldsymbol{D}_p^k + \boldsymbol{A}_p^k)F_s\delta\boldsymbol{u}_s^k\right)^T \left(\bar{\boldsymbol{C}}^k_{\sigma_p\epsilon_p}(\boldsymbol{D}_p^k + \boldsymbol{A}_p^k)F_\tau\boldsymbol{u}_\tau^k + \bar{\boldsymbol{C}}^k_{\sigma_p\epsilon_n}(\boldsymbol{D}_{np}^k + \boldsymbol{D}_{nz}^k\right.\\
&\left.-\boldsymbol{A}_n^k)F_\tau\boldsymbol{u}_\tau^k - \bar{\boldsymbol{C}}^k_{\sigma_p\mathcal{E}_p}\boldsymbol{D}_{ep}^kF_\tau\Phi_\tau^k + \bar{\boldsymbol{C}}^k_{\sigma_p\mathcal{D}_n}F_\tau\boldsymbol{\mathcal{D}}_{nM\tau}^k\right) + \left((\boldsymbol{D}_{np}^k + \boldsymbol{D}_{nz}^k - \boldsymbol{A}_n^k)F_s\delta\boldsymbol{u}_s^k\right)^T\\
&\times\left(\bar{\boldsymbol{C}}^k_{\sigma_n\epsilon_p}(\boldsymbol{D}_p^k + \boldsymbol{A}_p^k)F_\tau\boldsymbol{u}_\tau^k + \bar{\boldsymbol{C}}^k_{\sigma_n\epsilon_n}(\boldsymbol{D}_{np}^k + \boldsymbol{D}_{nz}^k - \boldsymbol{A}_n^k)F_\tau\boldsymbol{u}_\tau^k - \bar{\boldsymbol{C}}^k_{\sigma_n\mathcal{E}_p}\boldsymbol{D}_{ep}^kF_\tau\Phi_\tau^k\right.\\
&\left.+\bar{\boldsymbol{C}}^k_{\sigma_n\mathcal{D}_n}F_\tau\boldsymbol{\mathcal{D}}_{nM\tau}^k\right) + \left(\boldsymbol{D}_{ep}^kF_s\delta\Phi_s^k\right)^T\left(\bar{\boldsymbol{C}}^k_{\mathcal{D}_p\epsilon_p}(\boldsymbol{D}_p^k + \boldsymbol{A}_p^k)F_\tau\boldsymbol{u}_\tau^k + \bar{\boldsymbol{C}}^k_{\mathcal{D}_p\epsilon_n}(\boldsymbol{D}_{np}^k\right.\\
&\left.+\boldsymbol{D}_{nz}^k - \boldsymbol{A}_n^k)F_\tau\boldsymbol{u}_\tau^k - \bar{\boldsymbol{C}}^k_{\mathcal{D}_p\mathcal{E}_p}\boldsymbol{D}_{ep}^kF_\tau\Phi_\tau^k + \bar{\boldsymbol{C}}^k_{\mathcal{D}_p\mathcal{D}_n}F_\tau\boldsymbol{\mathcal{D}}_{nM\tau}^k\right) + \left(\boldsymbol{D}_{en}^kF_s\delta\Phi_s^k\right)^T\\
&\times\left(F_\tau\boldsymbol{\mathcal{D}}_{nM\tau}^k\right) - \left(F_s\delta\boldsymbol{\mathcal{D}}_{nMs}^k\right)^T\left(-\boldsymbol{D}_{en}^kF_\tau\Phi_\tau^k - \bar{\boldsymbol{C}}^k_{\mathcal{E}_n\epsilon_p}(\boldsymbol{D}_p^k + \boldsymbol{A}_p^k)F_\tau\boldsymbol{u}_\tau^k - \bar{\boldsymbol{C}}^k_{\mathcal{E}_n\epsilon_n}(\boldsymbol{D}_{np}^k\right.\\
&\left.+\boldsymbol{D}_{nz}^k - \boldsymbol{A}_n^k)F_\tau\delta\boldsymbol{u}_\tau^k + \bar{\boldsymbol{C}}^k_{\mathcal{E}_n\mathcal{E}_p}\boldsymbol{D}_{ep}^kF_\tau\Phi_\tau^k - \bar{\boldsymbol{C}}^k_{\mathcal{E}_n\mathcal{D}_n}F_\tau\boldsymbol{\mathcal{D}}_{nM\tau}^k\right)\Bigg]d\Omega_k\,dz = \delta L_e^k - \delta L_{in}^k
\end{aligned}
\tag{7.93}
$$

The CUF (Carrera 1995), as presented in the previous sections for the 2D approximation of shells, has already been introduced. In order to obtain a strong form of the differential equations on domain Ω_k and the relative boundary conditions on edge Γ_k in Equation (7.93), integration by parts is used (see Equation (7.28) and the matrices in Equation (7.29)), and this allows one to move the differential operator from the infinitesimal variation of the generic variable $\delta\boldsymbol{a}^k$ to the finite quantity $\boldsymbol{a}^k$ (Carrera 1995). The governing equations and the boundary conditions have the same form as the plate case discussed in Section 6.5, though, the meaning of the fundamental nuclei changes. As discussed in Carrera *et al.* (2010a,b; 2008), the governing equations state:

$$
\begin{aligned}
\delta\boldsymbol{u}_s^k:&\quad \boldsymbol{K}_{uu}^{k\tau s}\,\boldsymbol{u}_\tau^k + \boldsymbol{K}_{u\mathcal{D}}^{k\tau s}\,\boldsymbol{\mathcal{D}}_{nM\tau}^k + \boldsymbol{K}_{u\Phi}^{k\tau s}\,\Phi_\tau^k = \boldsymbol{p}_{us}^k - \boldsymbol{M}_{uu}^{k\tau s}\,\ddot{\boldsymbol{u}}_\tau^k\\
\delta\boldsymbol{\mathcal{D}}_{ns}^k:&\quad \boldsymbol{K}_{\mathcal{D}u}^{k\tau s}\,\boldsymbol{u}_\tau^k + \boldsymbol{K}_{\mathcal{D}\mathcal{D}}^{k\tau s}\,\boldsymbol{\mathcal{D}}_{nM\tau}^k + \boldsymbol{K}_{\mathcal{D}\Phi}^{k\tau s}\,\Phi_\tau^k = 0\\
\delta\Phi_s^k:&\quad \boldsymbol{K}_{\Phi u}^{k\tau s}\,\boldsymbol{u}_\tau^k + \boldsymbol{K}_{\Phi\mathcal{D}}^{k\tau s}\,\boldsymbol{\mathcal{D}}_{nM\tau}^k + \boldsymbol{K}_{\Phi\Phi}^{k\tau s}\,\Phi_\tau^k = 0
\end{aligned}
\tag{7.94}
$$

where $\boldsymbol{M}_{uu}^{k\tau s}$ is the inertial contribution in the form of the fundamental nucleus, $\boldsymbol{u}_\tau^k$ is the vector of the degrees of freedom for the displacements, Φ_τ^k is the vector of the degrees of freedom for the electric potential, $\boldsymbol{\mathcal{D}}_{nM\tau}^k$ is the vector of the degrees of freedom for the transverse normal electric displacement, and $\ddot{\boldsymbol{u}}_\tau^k$ is the second temporal derivative of $\boldsymbol{u}_\tau^k$. The array $\boldsymbol{p}_{us}^k$ indicates the variationally consistent mechanical loading for the case of a sensor application; for the actuator case, the electric voltage is applied directly to the vector Φ_τ^k. Along with these governing equations, the following boundary conditions on edge Γ_k of the in-plane integration domain Ω_k hold:

$$\begin{aligned}
&\boldsymbol{\Pi}^{k\tau s}_{uu}\,\boldsymbol{u}^k_\tau + \boldsymbol{\Pi}^{k\tau s}_{u\mathcal{D}}\,\boldsymbol{\mathcal{D}}^k_{nM\tau} + \boldsymbol{\Pi}^{k\tau s}_{u\Phi}\,\Phi^k_\tau = \boldsymbol{\Pi}^{k\tau s}_{uu}\,\bar{\boldsymbol{u}}^k_\tau + \boldsymbol{\Pi}^{k\tau s}_{u\mathcal{D}}\,\bar{\boldsymbol{\mathcal{D}}}^k_{nM\tau} + \boldsymbol{\Pi}^{k\tau s}_{u\Phi}\,\bar{\Phi}^k_\tau \\
&\boldsymbol{\Pi}^{k\tau s}_{\Phi u}\,\boldsymbol{u}^k_\tau + \boldsymbol{\Pi}^{k\tau s}_{\Phi\mathcal{D}}\,\boldsymbol{\mathcal{D}}^k_{nM\tau} + \boldsymbol{\Pi}^{k\tau s}_{\Phi\Phi}\,\Phi^k_\tau = \boldsymbol{\Pi}^{k\tau s}_{\Phi u}\,\bar{\boldsymbol{u}}^k_\tau + \boldsymbol{\Pi}^{k\tau s}_{\Phi\mathcal{D}}\,\bar{\boldsymbol{\mathcal{D}}}^k_{nM\tau} + \boldsymbol{\Pi}^{k\tau s}_{\Phi\Phi}\,\bar{\Phi}^k_\tau
\end{aligned} \tag{7.95}$$

By comparing Equation (7.93), after the integration by parts (see Equation (7.28)), to Equations (7.94) and (7.95), the fundamental nuclei can be obtained:

$$\begin{aligned}
\boldsymbol{K}^{k\tau s}_{uu} = \int_{A_k} \Big[&(-\boldsymbol{D}^k_p + \boldsymbol{A}^k_p)^T\,(\bar{\boldsymbol{C}}^k_{\sigma_p\epsilon_p}\,(\boldsymbol{D}^k_p + \boldsymbol{A}^k_p) + \bar{\boldsymbol{C}}^k_{\sigma_p\epsilon_n}\,(\boldsymbol{D}^k_{np} + \boldsymbol{D}^k_{nz} - \boldsymbol{A}^k_n)) \\
&+ (\boldsymbol{D}^k_{nz} - \boldsymbol{D}^k_{np} - \boldsymbol{A}^k_n)^T(\bar{\boldsymbol{C}}^k_{\sigma_n\epsilon_p}\,(\boldsymbol{D}^k_p + \boldsymbol{A}^k_p) + \bar{\boldsymbol{C}}^k_{\sigma_n\epsilon_n}\,(\boldsymbol{D}^k_{np} + \boldsymbol{D}^k_{nz} - \boldsymbol{A}^k_n))\Big] \\
&\times F_\tau F_s H^k_\alpha H^k_\beta\, dz
\end{aligned} \tag{7.96}$$

$$\begin{aligned}
\boldsymbol{K}^{k\tau s}_{u\mathcal{D}} = \int_{A_k} &\Big[(-\boldsymbol{D}^k_p + \boldsymbol{A}^k_p)^T\bar{\boldsymbol{C}}^k_{\sigma_p\mathcal{D}_n} + (\boldsymbol{D}^k_{nz} - \boldsymbol{D}^k_{np} - \boldsymbol{A}^k_n)^T\bar{\boldsymbol{C}}^k_{\sigma_n\mathcal{D}_n}\Big] \\
&\times F_\tau F_s H^k_\alpha H^k_\beta\, dz
\end{aligned} \tag{7.97}$$

$$\begin{aligned}
\boldsymbol{K}^{k\tau s}_{u\Phi} = \int_{A_k} &\Big[(-\boldsymbol{D}^k_p + \boldsymbol{A}^k_p)^T(-\bar{\boldsymbol{C}}^k_{\sigma_p\mathcal{E}_n}\,\boldsymbol{D}^k_{ep}) + (\boldsymbol{D}^k_{nz} - \boldsymbol{D}^k_{np} - \boldsymbol{A}^k_n)^T(-\bar{\boldsymbol{C}}^k_{\sigma_n\mathcal{E}_n}\,\boldsymbol{D}^k_{ep})\Big] \\
&\times F_\tau F_s H^k_\alpha H^k_\beta\, dz
\end{aligned} \tag{7.98}$$

$$\begin{aligned}
\boldsymbol{K}^{k\tau s}_{\mathcal{D}u} = \int_{A_k} &\Big[\bar{\boldsymbol{C}}^k_{\mathcal{E}_n\epsilon_p}\,(\boldsymbol{D}^k_p + \boldsymbol{A}^k_p) + \bar{\boldsymbol{C}}^k_{\mathcal{E}_n\epsilon_n}\,(\boldsymbol{D}^k_{np} + \boldsymbol{D}^k_{nz} - \boldsymbol{A}^k_n)\Big] \\
&\times F_\tau F_s H^k_\alpha H^k_\beta\, dz
\end{aligned} \tag{7.99}$$

$$\boldsymbol{K}^{k\tau s}_{\mathcal{D}\mathcal{D}} = \int_{A_k} \Big[\bar{\boldsymbol{C}}^k_{\mathcal{E}_n\mathcal{D}_n}\Big] F_\tau F_s H^k_\alpha H^k_\beta\, dz \tag{7.100}$$

$$\boldsymbol{K}^{k\tau s}_{\mathcal{D}\Phi} = \int_{A_k} \Big[\boldsymbol{D}^k_{en} - \bar{\boldsymbol{C}}^k_{\mathcal{E}_n\mathcal{E}_p}\,\boldsymbol{D}^k_{ep}\Big] F_\tau F_s H^k_\alpha H^k_\beta\, dz \tag{7.101}$$

$$\begin{aligned}
\boldsymbol{K}^{k\tau s}_{\Phi u} = \int_{A_k} &\Big[-(\boldsymbol{D}^k_{ep})^T(\bar{\boldsymbol{C}}^k_{\mathcal{D}_p\epsilon_p}\,(\boldsymbol{D}^k_p + \boldsymbol{A}^k_p) + \bar{\boldsymbol{C}}^k_{\mathcal{D}_p\epsilon_n}\,(\boldsymbol{D}^k_{np} + \boldsymbol{D}^k_{nz} - \boldsymbol{A}^k_n))\Big] \\
&\times F_\tau F_s H^k_\alpha H^k_\beta\, dz
\end{aligned} \tag{7.102}$$

$$\boldsymbol{K}^{k\tau s}_{\Phi\mathcal{D}} = \int_{A_k} \Big[(-\boldsymbol{D}^k_{ep})^T\bar{\boldsymbol{C}}^k_{\mathcal{D}_p\mathcal{D}_n} + (\boldsymbol{D}^k_{en})^T\Big] F_\tau F_s H^k_\alpha H^k_\beta\, dz \tag{7.103}$$

$$\boldsymbol{K}^{k\tau s}_{\Phi\Phi} = \int_{A_k} \Big[(-\boldsymbol{D}^k_{ep})^T(-\bar{\boldsymbol{C}}^k_{\mathcal{D}_p\mathcal{E}_p}\,\boldsymbol{D}^k_{ep})\Big] F_\tau F_s H^k_\alpha H^k_\beta\, dz \tag{7.104}$$

The fundamental nucleus for the inertial matrix $\boldsymbol{M}_{uu}^{k\tau s}$ is the one that was given for the PVD case in Equation (7.36). The nuclei for the boundary conditions on edge Γ_k are (Carrera *et al.* 2010a,b; 2008):

$$\boldsymbol{\Pi}_{uu}^{k\tau s} = \int_{A_k} \Big[\boldsymbol{I}_p^{kT} \, (\bar{\boldsymbol{C}}_{\sigma_p\epsilon_p}^k \, (\boldsymbol{D}_p^k + \boldsymbol{A}_p^k) + \bar{\boldsymbol{C}}_{\sigma_p\epsilon_n}^k \, (\boldsymbol{D}_{np}^k + \boldsymbol{D}_{nz}^k - \boldsymbol{A}_n^k)) + \boldsymbol{I}_{np}^{kT} \times (\bar{\boldsymbol{C}}_{\sigma_n\epsilon_p}^k \, (\boldsymbol{D}_p^k + \boldsymbol{A}_p^k) + \bar{\boldsymbol{C}}_{\sigma_n\epsilon_n}^k \, (\boldsymbol{D}_{np}^k + \boldsymbol{D}_{nz}^k - \boldsymbol{A}_n^k)) \Big] F_\tau F_s H_\alpha^k H_\beta^k \, dz \tag{7.105}$$

$$\boldsymbol{\Pi}_{u\mathcal{D}}^{k\tau s} = \int_{A_k} \Big[\boldsymbol{I}_p^{kT} \, \bar{\boldsymbol{C}}_{\sigma_p\mathcal{D}_n}^k + \boldsymbol{I}_{np}^{kT} \, \bar{\boldsymbol{C}}_{\sigma_n\mathcal{D}_n}^k \Big] F_\tau F_s H_\alpha^k H_\beta^k \, dz \tag{7.106}$$

$$\boldsymbol{\Pi}_{u\Phi}^{k\tau s} = \int_{A_k} \Big[\boldsymbol{I}_p^{kT} \, (-\bar{\boldsymbol{C}}_{\sigma_p\mathcal{E}_n}^k \, \boldsymbol{D}_{ep}^k) + \boldsymbol{I}_{np}^{kT} \, (-\bar{\boldsymbol{C}}_{\sigma_n\mathcal{E}_n}^k \, \boldsymbol{D}_{ep}^k) \Big] F_\tau F_s H_\alpha^k H_\beta^k \, dz \tag{7.107}$$

$$\boldsymbol{\Pi}_{\Phi u}^{k\tau s} = \int_{A_k} \Big[\boldsymbol{I}_{ep}^{kT} \, (\bar{\boldsymbol{C}}_{\mathcal{D}_p\epsilon_p}^k \, (\boldsymbol{D}_p^k + \boldsymbol{A}_p^k) + \bar{\boldsymbol{C}}_{\mathcal{D}_p\epsilon_n}^k \, (\boldsymbol{D}_{np}^k + \boldsymbol{D}_{nz}^k - \boldsymbol{A}_n^k)) \Big] \times F_\tau F_s H_\alpha^k H_\beta^k \, dz \tag{7.108}$$

$$\boldsymbol{\Pi}_{\Phi\mathcal{D}}^{k\tau s} = \int_{A_k} \Big[\boldsymbol{I}_{ep}^{kT} \, \bar{\boldsymbol{C}}_{\mathcal{D}_p\mathcal{D}_n}^k \Big] F_\tau F_s H_\alpha^k H_\beta^k \, dz \tag{7.109}$$

$$\boldsymbol{\Pi}_{\Phi\Phi}^{k\tau s} = \int_{A_k} \Big[\boldsymbol{I}_{ep}^{kT} \, (-\bar{\boldsymbol{C}}_{\mathcal{D}_p\mathcal{E}_p}^k \, \boldsymbol{D}_{ep}^k) \Big] F_\tau F_s H_\alpha^k H_\beta^k \, dz \tag{7.110}$$

In order to write the explicit form of the nuclei in Equations (7.96)–(7.104), the integrals in the z thickness direction are defined, as in Equations (7.41). The explicit algebraic form of the nuclei can be obtained by developing the matrix products in Equations (7.96)–(7.104) and employing a Navier-type closed form solution (Carrera *et al.* 2010a,b; 2008). The fundamental nucleus $\boldsymbol{K}_{uu}^{k\tau s}$ of (3×3) dimension is:

$$K_{uu_{11}}^{k\tau s} = \bar{C}_{\sigma_p\epsilon_p 11}^k \bar{\alpha}^2 J_{\beta/\alpha}^{k\tau s} + \bar{C}_{\sigma_p\epsilon_p 33}^k \bar{\beta}^2 J_{\alpha/\beta}^{k\tau s} + \bar{C}_{\sigma_n\epsilon_n 11}^k J_{\alpha\beta}^{k\tau_z s_z} + \bar{C}_{\sigma_n\epsilon_n 11}^k \times \left(\frac{1}{R_\alpha^{k\,2}} J_{\beta/\alpha}^{k\tau s} - \frac{1}{R_\alpha^k} J_\beta^{k\tau_z s} - \frac{1}{R_\alpha^k} J_\beta^{k\tau s_z} \right)$$

$$K_{uu_{12}}^{k\tau s} = (\bar{C}_{\sigma_p\epsilon_p 11}^k + \bar{C}_{\sigma_p\epsilon_p 33}^k) \bar{\alpha}\bar{\beta} J^{k\tau s}$$

$$K_{uu_{13}}^{k\tau s} = -\bar{C}_{\sigma_p\epsilon_n 13}^k \bar{\alpha} J_\beta^{k\tau s_z} + \bar{C}_{\sigma_n\epsilon_n 11}^k \bar{\alpha} J_\beta^{k\tau_z s} - \bar{C}_{\sigma_n\epsilon_n 11}^k \bar{\alpha} \frac{1}{R_\alpha^k} J_{\beta/\alpha}^{k\tau s} - \bar{C}_{\sigma_p\epsilon_p 11}^k \bar{\alpha} \frac{1}{R_\alpha^k} J_{\beta/\alpha}^{k\tau s} - \bar{C}_{\sigma_p\epsilon_p 12}^k \bar{\alpha} \frac{1}{R_\beta^k} J^{k\tau s}$$

$$K^{k\tau s}_{uu_{21}} = (\bar{C}^k_{\sigma_p\epsilon_p 12} + \bar{C}^k_{\sigma_p\epsilon_p 33})\bar{\alpha}\bar{\beta}J^{k\tau s}$$

$$K^{k\tau s}_{uu_{22}} = \bar{C}^k_{\sigma_n\epsilon_n 22}\left(\frac{1}{R^{k\,2}_\beta}J^{k\tau s}_{\alpha/\beta} - \frac{1}{R^k_\beta}J^{k\tau_z s}_\alpha - \frac{1}{R^k_\beta}J^{k\tau s_z}_\alpha + J^{k\tau_z s_z}_{\alpha\beta}\right)$$

$$+\bar{C}^k_{\sigma_p\epsilon_p 22}\bar{\beta}^2 J^{k\tau s}_{\alpha/\beta} + \bar{C}^k_{\sigma_p\epsilon_p 33}\bar{\alpha}^2 J^{k\tau s}_{\beta/\alpha} \qquad (7.111)$$

$$K^{k\tau s}_{uu_{23}} = \bar{C}^k_{\sigma_n\epsilon_n 22}\left(\bar{\beta}J^{k\tau_z s}_\alpha - \bar{\beta}\frac{1}{R^k_\beta}J^{k\tau s}_{\alpha/\beta}\right) - \bar{C}^k_{\sigma_n\epsilon_p 23}\bar{\beta}J^{k\tau s_z}_\alpha - \bar{C}^k_{\sigma_p\epsilon_p 12}\bar{\beta}\frac{1}{R^k_\alpha}J^{k\tau s}$$

$$\times \bar{C}^k_{\sigma_p\epsilon_p 22}\bar{\beta}\frac{1}{R^k_\beta}J^{k\tau s}_{\alpha/\beta}$$

$$K^{k\tau s}_{uu_{31}} = -\bar{C}^k_{\sigma_p\epsilon_n 13}\bar{\alpha}J^{k\tau_z s}_\beta - \bar{C}^k_{\sigma_p\epsilon_p 11}\bar{\alpha}\frac{1}{R^k_\alpha}J^{k\tau s}_{\beta/\alpha} - \bar{C}^k_{\sigma_p\epsilon_p 12}\bar{\alpha}\frac{1}{R^k_\beta}J^{k\tau s}$$

$$+\bar{C}^k_{\sigma_n\epsilon_n 11}(\bar{\alpha}J^{k\tau s_z}_\beta - \bar{\alpha}\frac{1}{R^k_\alpha}J^{k\tau s}_{\beta/\alpha})$$

$$K^{k\tau s}_{uu_{32}} = -\bar{C}^k_{\sigma_p\epsilon_n 23}\bar{\beta}J^{k\tau_z s}_\alpha + \bar{C}^k_{\sigma_n\epsilon_n 22}\left(\bar{\beta}J^{k\tau s_z}_\alpha - \bar{\beta}\frac{1}{R^k_\beta}J^{k\tau s}_{\alpha/\beta}\right) - \bar{C}^k_{\sigma_p\epsilon_p 22}$$

$$\times \bar{\beta}\frac{1}{R^k_\beta}J^{k\tau s}_{\alpha/\beta} - \bar{C}^k_{\sigma_p\epsilon_p 12}\bar{\beta}\frac{1}{R^k_\alpha}J^{k\tau s}$$

$$K^{k\tau s}_{uu_{33}} = \bar{C}^k_{\sigma_n\epsilon_n 33}J^{k\tau_z s_z}_{\alpha\beta} + \bar{C}^k_{\sigma_n\epsilon_n 22}\bar{\beta}^2 J^{k\tau s}_{\alpha/\beta} + \bar{C}^k_{\sigma_n\epsilon_n 11}\bar{\alpha}^2 J^{k\tau s}_{\beta/\alpha} + \bar{C}^k_{\sigma_p\epsilon_p 12}\frac{2}{R_{\alpha^k}R^k_\beta}J^{k\tau s}$$

$$+\bar{C}^k_{\sigma_p\epsilon_p 11}\frac{1}{R^{k\,2}_\alpha}J^{k\tau s}_{\beta/\alpha} + \bar{C}^k_{\sigma_p\epsilon_p 22}\frac{1}{R^{k\,2}_\beta}J^{k\tau s}_{\alpha/\beta} + \bar{C}^k_{\sigma_n\epsilon_p 31}\frac{1}{R^k_\alpha}(J^{k\tau s_z}_\beta + J^{k\tau_z s}_\beta)$$

$$+\bar{C}^k_{\sigma_n\epsilon_p 32}\frac{1}{R^k_\beta}(J^{k\tau s_z}_\alpha + J^{k\tau_z s}_\alpha)$$

Nucleus $\boldsymbol{K}^{k\tau s}_{u\Phi}$ of (3×1) dimension is:

$$K^{k\tau s}_{u\Phi_{11}} = \bar{C}^k_{\sigma_n\mathcal{E}_p 11}\left(\bar{\alpha}J^{k\tau_z s}_\beta - \bar{\alpha}\frac{1}{R^k_\alpha}J^{k\tau s}_{\beta/\alpha}\right)$$

$$K^{k\tau s}_{u\Phi_{21}} = \bar{C}^k_{\sigma_n\mathcal{E}_p 21}\left[-\bar{\beta}J^{k\tau_z s}_\alpha + \bar{\beta}\frac{1}{R^k_\beta}J^{k\tau s}_{\alpha/\beta}\right] \qquad (7.112)$$

$$K^{k\tau s}_{u\Phi_{31}} = -\bar{C}^k_{\sigma_n\mathcal{E}_p 11}\bar{\alpha}^2 J^{k\tau s}_{\beta/\alpha} - \bar{C}^k_{\sigma_n\mathcal{E}_p 21}\bar{\beta}^2 J^{k\tau s}_{\alpha/\beta}$$

Nucleus $\boldsymbol{K}_{u\mathcal{D}}^{k\tau s}$ of (3 × 1) dimension is:

$$
\begin{aligned}
K_{u\mathcal{D}_{11}}^{k\tau s} &= -\bar{C}_{\mathcal{E}_n\epsilon_p 11}^{k}\bar{\alpha}J_{\beta}^{k\tau s} \qquad K_{u\mathcal{D}_{21}}^{k\tau s} = -\bar{C}_{\mathcal{E}_n\epsilon_p 12}^{k}\bar{\beta}J_{\alpha}^{k\tau s} \\
K_{u\mathcal{D}_{31}}^{k\tau s} &= -\bar{C}_{\sigma_n\mathcal{D}_n 31}^{k}J_{\alpha\beta}^{k\tau_z s} - \bar{C}_{\mathcal{E}_n\epsilon_p 11}^{k}\frac{1}{R_{\alpha}^{k}}J_{\beta}^{k\tau s} - \bar{C}_{\mathcal{E}_n\epsilon_p 12}^{k}\frac{1}{R_{\beta}^{k}}J_{\alpha}^{k\tau s}
\end{aligned}
\tag{7.113}
$$

Nucleus $\boldsymbol{K}_{\Phi u}^{k\tau s}$ of (1 × 3) dimension is:

$$
\begin{aligned}
K_{\Phi u_{11}}^{k\tau s} &= \bar{C}_{\mathcal{D}_p\epsilon_n 11}^{k}\left(\bar{\alpha}J_{\beta}^{k\tau s_z} - \bar{\alpha}\frac{1}{R_{\alpha}^{k}}J_{\beta/\alpha}^{k\tau s}\right) \\
K_{\Phi u_{12}}^{k\tau s} &= \bar{C}_{\mathcal{D}_p\epsilon_n 22}^{k}\left[\bar{\beta}J_{\alpha}^{k\tau s_z} - \bar{\beta}\frac{1}{R_{\beta}^{k}}J_{\alpha/\beta}^{k\tau s}\right] \\
K_{\Phi u_{13}}^{k\tau s} &= \bar{C}_{\mathcal{D}_p\epsilon_n 11}\bar{\alpha}^2 J_{\beta/\alpha}^{k\tau s} + \bar{C}_{\mathcal{D}_p\epsilon_n 22}\bar{\beta}^2 J_{\alpha/\beta}^{k\tau s}
\end{aligned}
\tag{7.114}
$$

Nucleus $\boldsymbol{K}_{\Phi\Phi}^{k\tau s}$ of (1 × 1) dimension is:

$$
K_{\Phi\Phi_{11}}^{k\tau s} = -\bar{C}_{\mathcal{D}_p\mathcal{E}_p 11}^{k}\bar{\alpha}^2 J_{\beta/\alpha}^{k\tau s} - \bar{C}_{\mathcal{D}_p\mathcal{E}_p 22}^{k}\bar{\beta}^2 J_{\alpha/\beta}^{k\tau s}
\tag{7.115}
$$

Nucleus $\boldsymbol{K}_{\Phi\mathcal{D}}^{k\tau s}$ of (1 × 1) dimension is:

$$
K_{\Phi\mathcal{D}_{11}}^{k\tau s} = J_{\alpha\beta}^{k\tau_z s}
\tag{7.116}
$$

Nucleus $\boldsymbol{K}_{\mathcal{D}u}^{k\tau s}$ of (1 × 3) dimension is:

$$
\begin{aligned}
K_{\mathcal{D}u_{11}}^{k\tau s} &= -\bar{C}_{\mathcal{E}_n\epsilon_p 11}^{k}\bar{\alpha}J_{\beta}^{k\tau s}, \qquad K_{\mathcal{D}u_{12}}^{k\tau s} = -\bar{C}_{\mathcal{E}_n\epsilon_p 12}^{k}\bar{\beta}J_{\alpha}^{k\tau s} \\
K_{\mathcal{D}u_{13}}^{k\tau s} &= -\bar{C}_{\mathcal{E}_n\epsilon_n 13}^{k}J_{\alpha\beta}^{k\tau s_z} + \bar{C}_{\mathcal{E}_n\epsilon_p 11}^{k}\frac{1}{R_{\alpha}^{k}}J_{\beta}^{k\tau s} + \bar{C}_{\mathcal{E}_n\epsilon_p 12}^{k}\frac{1}{R_{\beta}^{k}}J_{\alpha}^{k\tau s}
\end{aligned}
\tag{7.117}
$$

Nucleus $\boldsymbol{K}_{\mathcal{D}\Phi}^{k\tau s}$ of (1 × 1) dimension is:

$$
K_{\mathcal{D}\Phi_{11}}^{k\tau s} = J_{\alpha\beta}^{k\tau s_z}
\tag{7.118}
$$

Nucleus $\boldsymbol{K}_{\mathcal{D}\mathcal{D}}^{k\tau s}$ of (1 × 1) dimension is:

$$
K_{\mathcal{D}\mathcal{D}_{11}}^{k\tau s} = \bar{C}_{\mathcal{E}_n\mathcal{D}_n 11}^{k}J_{\alpha\beta}^{k\tau s}
\tag{7.119}
$$

The fundamental nucleus $\boldsymbol{M}_{uu}^{k\tau s}$ is of (3 × 3) dimension with the diagonal elements as already given in Equation (7.46).

$\bar{\alpha} = m\pi/a$ and $\bar{\beta} = n\pi/b$, where m and n are the wave numbers in the in-plane directions and a and b are the shell dimensions in the α and β directions, respectively. The explicit form of coefficients $\bar{\boldsymbol{C}}$ and their components are given in Equations (2.79) and (6.139)–(6.154), respectively.

A Navier-type closed-form solution is obtained via substitution of the harmonic expressions for the displacements, electric potential, and transverse normal electric displacement and by considering the following material coefficients to be equal to zero: $Q_{16} = Q_{26} = Q_{36} = Q_{45} = 0$ and $e_{25} = e_{14} = e_{36} = \varepsilon_{12} = 0$. The harmonic assumptions used for the displacements, the electric potential, and the transverse normal electric displacement are:

$$u_\tau^k = \sum_{m,n} \hat{U}_\tau^k \cos\left(\frac{m\pi\alpha}{a}\right) \sin\left(\frac{n\pi\beta}{b}\right), \quad k = 1, N_l \tag{7.120}$$

$$v_\tau^k = \sum_{m,n} \hat{V}_\tau^k \sin\left(\frac{m\pi\alpha}{a}\right) \cos\left(\frac{n\pi\beta}{b}\right), \quad \tau = t, b, r \tag{7.121}$$

$$(w_\tau^k, \Phi_\tau^k, \mathcal{D}_{z\tau}^k) = \sum_{m,n} (\hat{W}_\tau^k, \hat{\Phi}_\tau^k, \hat{\mathcal{D}}_{z\tau}^k) \sin\left(\frac{m\pi\alpha}{a}\right) \sin\left(\frac{n\pi\beta}{b}\right), \quad r = 2, N \tag{7.122}$$

where $\hat{U}_\tau^k$, $\hat{V}_\tau^k$, $\hat{W}_\tau^k$ are the displacement amplitudes; $\hat{\Phi}_\tau^k$ indicates potential amplitude and $\hat{\mathcal{D}}_{z\tau}^k$ is the transverse normal electric displacement amplitude; k indicates the layer and N_l is the total number of layers. τ is the index for the order of expansion, where t and b indicate the top and bottom of the layer, respectively, while r indicates the higher orders of expansion until $N = 4$. Details on the assembling procedure of the fundamental nuclei and on the acronyms are given in Sections 7.7 and 7.8, respectively, as already discussed for the plate case in Sections 6.7 and 6.8.

7.6 RMVT(u, Φ, σ_n, $\mathcal{D}_n$) for the electromechanical shell case

The third possible extension of the RMVT (Reissner 1984) to electromechanical problems is that indicated in Equations (2.83) and (6.155) of Chapters 2 and 6, respectively. In this last case, the internal electrical work has been considered and two Lagrange multipliers have been added for the transverse stress components and the transverse normal electric displacement (Carrera and Brischetto 2007a; Carrera *et al.* 2008):

$$\int_V \Big(\delta\boldsymbol{\epsilon}_{pG}^T\boldsymbol{\sigma}_{pC} + \delta\boldsymbol{\epsilon}_{nG}^T\boldsymbol{\sigma}_{nM} + \delta\boldsymbol{\sigma}_{nM}^T(\boldsymbol{\epsilon}_{nG} - \boldsymbol{\epsilon}_{nC}) - \delta\boldsymbol{\mathcal{E}}_{pG}^T\boldsymbol{\mathcal{D}}_{pC} - \delta\boldsymbol{\mathcal{E}}_{nG}^T\boldsymbol{\mathcal{D}}_{nM}$$
$$- \delta\boldsymbol{\mathcal{D}}_{nM}^T(\boldsymbol{\mathcal{E}}_{nG} - \boldsymbol{\mathcal{E}}_{nC})\Big)dV = \delta L_e - \delta L_{in} \tag{7.123}$$

where the subscript M indicates the two a priori modeled variables. By considering a laminated shell of N_l layers, and the integral on the volume V_k of each layer k as an integral on the in-plane domain Ω_k, plus the integral in the thickness-direction domain A_k, it is possible to write Equation (7.123) as:

$$\sum_{k=1}^{N_l}\int_{\Omega_k}\int_{A_k}\Big\{\delta\boldsymbol{\epsilon}_{pG}^{k\ T}\boldsymbol{\sigma}_{pC}^k + \delta\boldsymbol{\epsilon}_{nG}^{k\ T}\boldsymbol{\sigma}_{nM}^k + \delta\boldsymbol{\sigma}_{nM}^{k\ T}(\boldsymbol{\epsilon}_{nG}^k - \boldsymbol{\epsilon}_{nC}^k) - \delta\boldsymbol{\mathcal{E}}_{pG}^{k\ T}\boldsymbol{\mathcal{D}}_{pC}^k$$
$$- \delta\boldsymbol{\mathcal{E}}_{nG}^{k\ T}\boldsymbol{\mathcal{D}}_{nM}^k - \delta\boldsymbol{\mathcal{D}}_{nM}^{k\ T}(\boldsymbol{\mathcal{E}}_{nG}^k - \boldsymbol{\mathcal{E}}_{nC}^k)\Big\}d\Omega_k dz = \sum_{k=1}^{N_l}\delta L_e^k - \sum_{k=1}^{N_l}\delta L_{in}^k \tag{7.124}$$

where δL_e^k and δL_{in}^k are the external and inertial virtual work at the k-layer level, respectively. The relative constitutive equations are those obtained in Equations (2.84)–(2.87) with the components in the curvilinear reference system (α,β,z) for the stress, strain, electric displacement, and electric field vectors given in Equations (2.19) and (2.20); the same constitutive equations were proposed for the plate case in Equations (6.157)–(6.160), where the transverse normal electric displacement $\boldsymbol{\mathcal{D}}_n$ and the transverse stresses $\boldsymbol{\sigma}_n$ were a priori modeled "M" and the meaning of the coefficients $\tilde{\boldsymbol{C}}$ was given in Equations (2.88) and (6.207)–(6.222) (Carrera and Brischetto 2007a; Carrera *et al.* 2008):

$$\boldsymbol{\sigma}_{pC}^k = \tilde{\boldsymbol{C}}_{\sigma_p\epsilon_p}^k\boldsymbol{\epsilon}_{pG}^k + \tilde{\boldsymbol{C}}_{\sigma_p\sigma_n}^k\boldsymbol{\sigma}_{nM}^k + \tilde{\boldsymbol{C}}_{\sigma_p\mathcal{E}_p}^k\boldsymbol{\mathcal{E}}_{pG}^k + \tilde{\boldsymbol{C}}_{\sigma_p\mathcal{D}_n}^k\boldsymbol{\mathcal{D}}_{nM}^k \tag{7.125}$$

$$\boldsymbol{\epsilon}_{nC}^k = \tilde{\boldsymbol{C}}_{\epsilon_n\epsilon_p}^k\boldsymbol{\epsilon}_{pG}^k + \tilde{\boldsymbol{C}}_{\epsilon_n\sigma_n}^k\boldsymbol{\sigma}_{nM}^k + \tilde{\boldsymbol{C}}_{\epsilon_n\mathcal{E}_p}^k\boldsymbol{\mathcal{E}}_{pG}^k + \tilde{\boldsymbol{C}}_{\epsilon_n\mathcal{D}_n}^k\boldsymbol{\mathcal{D}}_{nM}^k \tag{7.126}$$

$$\boldsymbol{\mathcal{D}}_{pC}^k = \tilde{\boldsymbol{C}}_{\mathcal{D}_p\epsilon_p}^k\boldsymbol{\epsilon}_{pG}^k + \tilde{\boldsymbol{C}}_{\mathcal{D}_p\sigma_n}^k\boldsymbol{\sigma}_{nM}^k + \tilde{\boldsymbol{C}}_{\mathcal{D}_p\mathcal{E}_p}^k\boldsymbol{\mathcal{E}}_{pG}^k + \tilde{\boldsymbol{C}}_{\mathcal{D}_p\mathcal{D}_n}^k\boldsymbol{\mathcal{D}}_{nM}^k \tag{7.127}$$

$$\boldsymbol{\mathcal{E}}_{nC}^k = \tilde{\boldsymbol{C}}_{\mathcal{E}_n\epsilon_p}^k\boldsymbol{\epsilon}_{pG}^k + \tilde{\boldsymbol{C}}_{\mathcal{E}_n\sigma_n}^k\boldsymbol{\sigma}_{nM}^k + \tilde{\boldsymbol{C}}_{\mathcal{E}_n\mathcal{E}_p}^k\boldsymbol{\mathcal{E}}_{pG}^k + \tilde{\boldsymbol{C}}_{\mathcal{E}_n\mathcal{D}_n}^k\boldsymbol{\mathcal{D}}_{nM}^k \tag{7.128}$$

By substituting Equations (7.125)–(7.128), and the geometrical relations (2.31)–(2.37) in Chapter 2 for shells in the variational statement of

Equation (7.124), and considering a generic layer k (Carrera and Brischetto 2007a; Carrera *et al.* 2008):

$$\begin{aligned}
&\int_{\Omega_k}\int_{A_k}\Bigg[\left((\boldsymbol{D}_p^k+\boldsymbol{A}_p^k)F_s\delta\boldsymbol{u}_s^k\right)^T\left(\tilde{\boldsymbol{C}}^k_{\sigma_p\epsilon_p}(\boldsymbol{D}_p^k+\boldsymbol{A}_p^k)F_\tau\boldsymbol{u}_\tau^k+\tilde{\boldsymbol{C}}^k_{\sigma_p\sigma_n}F_\tau\boldsymbol{\sigma}^k_{nM\tau}\right.\\
&\left.-\tilde{\boldsymbol{C}}^k_{\sigma_p\mathcal{E}_p}\boldsymbol{D}^k_{ep}F_\tau\Phi^k_\tau+\tilde{\boldsymbol{C}}^k_{\sigma_p\mathcal{D}_n}F_\tau\boldsymbol{\mathcal{D}}^k_{nM\tau}\right)+\left((\boldsymbol{D}^k_{np}+\boldsymbol{D}^k_{nz}-\boldsymbol{A}^k_n)F_s\delta\boldsymbol{u}^k_s\right)^T\left(F_\tau\boldsymbol{\sigma}^k_{nM\tau}\right)\\
&+\left(F_s\delta\boldsymbol{\sigma}^k_{nMs}\right)^T\left((\boldsymbol{D}^k_{np}+\boldsymbol{D}^k_{nz}-\boldsymbol{A}^k_n)F_\tau\boldsymbol{u}^k_\tau-\tilde{\boldsymbol{C}}^k_{\epsilon_n\epsilon_p}(\boldsymbol{D}^k_p+\boldsymbol{A}^k_p)F_\tau\boldsymbol{u}^k_\tau\right.\\
&\left.-\tilde{\boldsymbol{C}}^k_{\epsilon_n\sigma_n}F_\tau\boldsymbol{\sigma}^k_{nM\tau}+\tilde{\boldsymbol{C}}^k_{\epsilon_n\mathcal{E}_p}\boldsymbol{D}^k_{ep}F_\tau\Phi^k_\tau-\tilde{\boldsymbol{C}}^k_{\epsilon_n\mathcal{D}_n}F_\tau\boldsymbol{\mathcal{D}}^k_{nM\tau}\right)+\left(\boldsymbol{D}^k_{ep}F_s\delta\Phi^k_s\right)^T\\
&\left(\tilde{\boldsymbol{C}}^k_{\mathcal{D}_p\epsilon_p}(\boldsymbol{D}^k_p+\boldsymbol{A}^k_p)F_\tau\boldsymbol{u}^k_\tau+\tilde{\boldsymbol{C}}^k_{\mathcal{D}_p\sigma_n}F_\tau\boldsymbol{\sigma}^k_{nM\tau}-\tilde{\boldsymbol{C}}^k_{\mathcal{D}_p\mathcal{E}_p}\boldsymbol{D}^k_{ep}F_\tau\Phi^k_\tau+\tilde{\boldsymbol{C}}^k_{\mathcal{D}_p\mathcal{D}_n}F_\tau\boldsymbol{\mathcal{D}}^k_{nM\tau}\right)\\
&+\left(\boldsymbol{D}^k_{en}F_s\delta\Phi^k_s\right)^T\left(F_\tau\boldsymbol{\mathcal{D}}^k_{nM\tau}\right)-\left(F_s\delta\boldsymbol{\mathcal{D}}^k_{nMs}\right)^T\left(-\boldsymbol{D}^k_{en}F_\tau\Phi^k_\tau-\tilde{\boldsymbol{C}}^k_{\mathcal{E}_n\epsilon_p}(\boldsymbol{D}^k_p+\boldsymbol{A}^k_p)F_\tau\boldsymbol{u}^k_\tau\right.\\
&\left.-\tilde{\boldsymbol{C}}^k_{\mathcal{E}_n\sigma_n}F_\tau\boldsymbol{\sigma}^k_{nM\tau}+\tilde{\boldsymbol{C}}^k_{\mathcal{E}_n\mathcal{E}_p}\boldsymbol{D}^k_{ep}F_\tau\Phi^k_\tau-\tilde{\boldsymbol{C}}^k_{\mathcal{E}_n\mathcal{D}_n}F_\tau\boldsymbol{\mathcal{D}}^k_{nM\tau}\right)\Bigg]d\Omega_k\,dz=\delta L^k_e-\delta L^k_{in}
\end{aligned} \tag{7.129}$$

The CUF (Carrera 1995), as presented in previous sections for the 2D approximation of shells, has already been introduced. In order to obtain a strong form of differential equations on the domain Ω_k and the relative boundary conditions on edge Γ_k in Equation (7.129), integration by parts is used (see Equation (7.28) and the matrices in Equation (7.29)), which allows one to move the differential operator from the infinitesimal variation of the generic variable $\delta\boldsymbol{a}^k$ to the finite quantity $\boldsymbol{a}^k$ (Carrera 1995). The governing equations and the boundary conditions have the same form as the plate case discussed in Section 6.6, though the meaning of the fundamental nuclei changes. As discussed in Carrera and Brischetto (2007a) and Carrera *et al.* (2008), the governing equations state:

$$\begin{aligned}
\delta\boldsymbol{u}^k_s &: \boldsymbol{K}^{k\tau s}_{uu}\,\boldsymbol{u}^k_\tau+\boldsymbol{K}^{k\tau s}_{u\sigma}\,\boldsymbol{\sigma}^k_{nM\tau}+\boldsymbol{K}^{k\tau s}_{u\Phi}\,\Phi^k_\tau+\boldsymbol{K}^{k\tau s}_{u\mathcal{D}}\,\boldsymbol{\mathcal{D}}^k_{nM\tau}=\boldsymbol{p}^k_{us}-\boldsymbol{M}^{k\tau s}_{uu}\,\ddot{\boldsymbol{u}}^k_\tau\\
\delta\boldsymbol{\sigma}^k_{ns} &: \boldsymbol{K}^{k\tau s}_{\sigma u}\,\boldsymbol{u}^k_\tau+\boldsymbol{K}^{k\tau s}_{\sigma\sigma}\,\boldsymbol{\sigma}^k_{nM\tau}+\boldsymbol{K}^{k\tau s}_{\sigma\Phi}\,\Phi^k_\tau+\boldsymbol{K}^{k\tau s}_{\sigma\mathcal{D}}\,\boldsymbol{\mathcal{D}}^k_{nM\tau}=0\\
\delta\Phi^k_s &: \boldsymbol{K}^{k\tau s}_{\Phi u}\,\boldsymbol{u}^k_\tau+\boldsymbol{K}^{k\tau s}_{\Phi\sigma}\,\boldsymbol{\sigma}^k_{nM\tau}+\boldsymbol{K}^{k\tau s}_{\Phi\Phi}\,\Phi^k_\tau+\boldsymbol{K}^{k\tau s}_{\Phi\mathcal{D}}\,\boldsymbol{\mathcal{D}}^k_{nM\tau}=0\\
\delta\boldsymbol{\mathcal{D}}^k_{ns} &: \boldsymbol{K}^{k\tau s}_{\mathcal{D}u}\,\boldsymbol{u}^k_\tau+\boldsymbol{K}^{k\tau s}_{\mathcal{D}\sigma}\,\boldsymbol{\sigma}^k_{nM\tau}+\boldsymbol{K}^{k\tau s}_{\mathcal{D}\Phi}\,\Phi^k_\tau+\boldsymbol{K}^{k\tau s}_{\mathcal{D}\mathcal{D}}\,\boldsymbol{\mathcal{D}}^k_{nM\tau}=0
\end{aligned} \tag{7.130}$$

where $\boldsymbol{M}^{k\tau s}_{uu}$ is the inertial contribution in the form of the fundamental nucleus, $\boldsymbol{u}^k_\tau$ is the vector of the degrees of freedom for the displacements, Φ^k_τ is the vector of the degrees of freedom for the electric potential, $\boldsymbol{\mathcal{D}}^k_{nM\tau}$ is the vector of the degrees of freedom for the transverse normal electric displacement, $\boldsymbol{\sigma}^k_{nM\tau}$ is the vector of the degrees of freedom for the transverse stress components, and $\ddot{\boldsymbol{u}}^k_\tau$ is the second temporal derivative of $\boldsymbol{u}_\tau^k$. The array $\boldsymbol{p}_{us}^k$ indicates the variationally consistent mechanical loading used for the case of sensor applications; the electric potential is directly applied in vector Φ_τ^k for the actuator case. Along with these governing equations, the following boundary conditions on edge Γ_k of the in-plane integration domain Ω_k hold:

$$\begin{aligned}
&\boldsymbol{\Pi}_{uu}^{k\tau s}\ \boldsymbol{u}_\tau^k + \boldsymbol{\Pi}_{u\sigma}^{k\tau s}\ \boldsymbol{\sigma}_{nM\tau}^k + \boldsymbol{\Pi}_{u\Phi}^{k\tau s}\ \Phi_\tau^k + \boldsymbol{\Pi}_{u\mathcal{D}}^{k\tau s}\ \boldsymbol{\mathcal{D}}_{nM\tau}^k \\
&\quad = \boldsymbol{\Pi}_{uu}^{k\tau s}\ \bar{\boldsymbol{u}}_\tau^k + \boldsymbol{\Pi}_{u\sigma}^{k\tau s}\ \bar{\boldsymbol{\sigma}}_{nM\tau}^k + \boldsymbol{\Pi}_{u\Phi}^{k\tau s}\ \bar{\Phi}_\tau^k + \boldsymbol{\Pi}_{u\mathcal{D}}^{k\tau s}\ \bar{\boldsymbol{\mathcal{D}}}_{nM\tau}^k \\
&\boldsymbol{\Pi}_{\Phi u}^{k\tau s}\ \boldsymbol{u}_\tau^k + \boldsymbol{\Pi}_{\Phi\sigma}^{k\tau s}\ \boldsymbol{\sigma}_{nM\tau}^k + \boldsymbol{\Pi}_{\Phi\Phi}^{k\tau s}\ \Phi_\tau^k + \boldsymbol{\Pi}_{\Phi\mathcal{D}}^{k\tau s}\ \boldsymbol{\mathcal{D}}_{nM\tau}^k \\
&\quad = \boldsymbol{\Pi}_{\Phi u}^{k\tau s}\ \bar{\boldsymbol{u}}_\tau^k + \boldsymbol{\Pi}_{\Phi\sigma}^{k\tau s}\ \bar{\boldsymbol{\sigma}}_{nM\tau}^k + \boldsymbol{\Pi}_{\Phi\Phi}^{k\tau s}\ \bar{\Phi}_\tau^k + \boldsymbol{\Pi}_{\Phi\mathcal{D}}^{k\tau s}\ \bar{\boldsymbol{\mathcal{D}}}_{nM\tau}^k
\end{aligned} \tag{7.131}$$

By comparing Equation (7.129), after the integration by parts (see Equation (7.28)), to Equations (7.130) and (7.131), the fundamental nuclei can be obtained:

$$\boldsymbol{K}_{uu}^{k\tau s} = \int_{A_k} \left[(-\boldsymbol{D}_p^k + \boldsymbol{A}_p^k)^T(\tilde{\boldsymbol{C}}_{\sigma_p\epsilon_p}^k(\boldsymbol{D}_p^k + \boldsymbol{A}_p^k))\right] F_s F_\tau H_\alpha^k H_\beta^k dz \tag{7.132}$$

$$\begin{aligned}
\boldsymbol{K}_{u\sigma}^{k\tau s} = \int_{A_k} &\left[(-\boldsymbol{D}_p^k + \boldsymbol{A}_p^k)^T(\tilde{\boldsymbol{C}}_{\sigma_p\sigma_n}^k) + (-\boldsymbol{D}_{np}^k + \boldsymbol{D}_{nz}^k - \boldsymbol{A}_n^k)^T\right] \\
&\times F_s F_\tau H_\alpha^k H_\beta^k dz
\end{aligned} \tag{7.133}$$

$$\boldsymbol{K}_{u\Phi}^{k\tau s} = \int_{A_k} \left[(-\boldsymbol{D}_p^k + \boldsymbol{A}_p^k)^T(-\tilde{\boldsymbol{C}}_{\sigma_p\mathcal{E}_p}^k \boldsymbol{D}_{ep}^k)\right] F_s F_\tau H_\alpha^k H_\beta^k dz \tag{7.134}$$

$$\boldsymbol{K}_{u\mathcal{D}}^{k\tau s} = \int_{A_k} \left[(-\boldsymbol{D}_p^k + \boldsymbol{A}_p^k)^T(\tilde{\boldsymbol{C}}_{\sigma_p\mathcal{D}_n}^k)\right] F_s F_\tau H_\alpha^k H_\beta^k dz \tag{7.135}$$

$$\boldsymbol{K}_{\sigma u}^{k\tau s} = \int_{A_k} \left[(\boldsymbol{D}_{np}^k + \boldsymbol{D}_{nz}^k - \boldsymbol{A}_n^k) - (\tilde{\boldsymbol{C}}_{\epsilon_n\epsilon_p}^k)(\boldsymbol{D}_p^k + \boldsymbol{A}_p^k)\right] F_s F_\tau H_\alpha^k H_\beta^k dz \tag{7.136}$$

$$\boldsymbol{K}_{\sigma\sigma}^{k\tau s} = \int_{A_k} \left[-\tilde{\boldsymbol{C}}_{\epsilon_n\sigma_n}^k\right] F_s F_\tau H_\alpha^k H_\beta^k dz \tag{7.137}$$

$$\boldsymbol{K}_{\sigma\Phi}^{k\tau s} = \int_{A_k} \left[\tilde{\boldsymbol{C}}_{\epsilon_n\mathcal{E}_p}^k \boldsymbol{D}_{ep}^k\right] F_s F_\tau H_\alpha^k H_\beta^k dz \tag{7.138}$$

$$\boldsymbol{K}_{\sigma\mathcal{D}}^{k\tau s} = \int_{A_k} \left[-\tilde{\boldsymbol{C}}_{\epsilon_n\mathcal{D}_n}^k\right] F_s F_\tau H_\alpha^k H_\beta^k dz \tag{7.139}$$

$$\boldsymbol{K}_{\Phi u}^{k\tau s} = \int_{A_k} \left[-\boldsymbol{D}_{ep}^{k\ T} \tilde{\boldsymbol{C}}_{\mathcal{D}_p \epsilon_p}^k (\boldsymbol{D}_p^k + \boldsymbol{A}_p^k) \right] F_s F_\tau H_\alpha^k H_\beta^k dz \tag{7.140}$$

$$\boldsymbol{K}_{\Phi \sigma}^{k\tau s} = \int_{A_k} \left[-\boldsymbol{D}_{ep}^{k\ T} \tilde{\boldsymbol{C}}_{\mathcal{D}_p \sigma_n}^k \right] F_s F_\tau H_\alpha^k H_\beta^k dz \tag{7.141}$$

$$\boldsymbol{K}_{\Phi \Phi}^{k\tau s} = \int_{A_k} \left[\boldsymbol{D}_{ep}^{k\ T} \tilde{\boldsymbol{C}}_{\mathcal{D}_p \mathcal{E}_p}^k \boldsymbol{D}_{ep}^k \right] F_s F_\tau H_\alpha^k H_\beta^k dz \tag{7.142}$$

$$\boldsymbol{K}_{\Phi \mathcal{D}}^{k\tau s} = \int_{A_k} \left[-\boldsymbol{D}_{ep}^{k\ T} \tilde{\boldsymbol{C}}_{\mathcal{D}_p \mathcal{D}_n}^k + \boldsymbol{D}_{en}^{k\ T} \right] F_s F_\tau H_\alpha^k H_\beta^k dz \tag{7.143}$$

$$\boldsymbol{K}_{\mathcal{D} u}^{k\tau s} = \int_{A_k} \left[\tilde{\boldsymbol{C}}_{\mathcal{E}_n \epsilon_p}^k (\boldsymbol{D}_p^k + \boldsymbol{A}_p^k)^T \right] F_s F_\tau H_\alpha^k H_\beta^k dz \tag{7.144}$$

$$\boldsymbol{K}_{\mathcal{D} \sigma}^{k\tau s} = \int_{A_k} \left[\tilde{\boldsymbol{C}}_{\mathcal{E}_n \sigma_n}^k \right] F_s F_\tau H_\alpha^k H_\beta^k dz \tag{7.145}$$

$$\boldsymbol{K}_{\mathcal{D} \Phi}^{k\tau s} = \int_{A_k} \left[\boldsymbol{D}_{en}^k - \tilde{\boldsymbol{C}}_{\mathcal{E}_n \mathcal{E}_p}^k \boldsymbol{D}_{ep}^k \right] F_s F_\tau H_\alpha^k H_\beta^k dz \tag{7.146}$$

$$\boldsymbol{K}_{\mathcal{D} \mathcal{D}}^{k\tau s} = \int_{A_k} \left[\tilde{\boldsymbol{C}}_{\mathcal{E}_n \mathcal{D}_n}^k \right] F_s F_\tau H_\alpha^k H_\beta^k dz \tag{7.147}$$

The fundamental nucleus for the inertial matrix $\boldsymbol{M}_{uu}^{k\tau s}$ is the one that was given for the PVD case in Equation (7.36). The nuclei for the boundary conditions on edge Γ_k are (Carrera and Brischetto 2007a; Carrera *et al.* 2008):

$$\boldsymbol{\Pi}_{uu}^{k\tau s} = \int_{A_k} \left[\boldsymbol{I}_p^{k\ T} \tilde{\boldsymbol{C}}_{\sigma_p \epsilon_p}^k (\boldsymbol{D}_p^k + \boldsymbol{A}_p^k) \right] F_s F_\tau H_\alpha^k H_\beta^k dz \tag{7.148}$$

$$\boldsymbol{\Pi}_{u\sigma}^{k\tau s} = \int_{A_k} \left[\boldsymbol{I}_p^{k\ T} \tilde{\boldsymbol{C}}_{\sigma_p \sigma_n}^k + \boldsymbol{I}_{np}^{k\ T} \right] F_s F_\tau H_\alpha^k H_\beta^k dz \tag{7.149}$$

$$\boldsymbol{\Pi}_{u\Phi}^{k\tau s} = \int_{A_k} \left[-\boldsymbol{I}_p^{k\ T} \tilde{\boldsymbol{C}}_{\sigma_p \mathcal{E}_p}^k \boldsymbol{D}_{ep}^k \right] F_s F_\tau H_\alpha^k H_\beta^k dz \tag{7.150}$$

$$\boldsymbol{\Pi}_{u\mathcal{D}}^{k\tau s} = \int_{A_k} \left[\boldsymbol{I}_p^{k\ T} \tilde{\boldsymbol{C}}_{\sigma_p \mathcal{D}_n}^k \right] F_s F_\tau H_\alpha^k H_\beta^k dz \tag{7.151}$$

$$\boldsymbol{\Pi}_{\Phi u}^{k\tau s} = \int_{A_k} \Big[\boldsymbol{I}_{ep}^{k\ T} \tilde{\boldsymbol{C}}_{\mathcal{D}_p\epsilon_p}^{k} (\boldsymbol{D}_p^k + \boldsymbol{A}_p^k) \Big] F_s F_\tau H_\alpha^k H_\beta^k dz \tag{7.152}$$

$$\boldsymbol{\Pi}_{\Phi \sigma}^{k\tau s} = \int_{A_k} \Big[\boldsymbol{I}_{ep}^{k\ T} \tilde{\boldsymbol{C}}_{\mathcal{D}_p\sigma_n}^{k} \Big] F_s F_\tau H_\alpha^k H_\beta^k dz \tag{7.153}$$

$$\boldsymbol{\Pi}_{\Phi \Phi}^{k\tau s} = \int_{A_k} \Big[-\boldsymbol{I}_{ep}^{k\ T} \tilde{\boldsymbol{C}}_{\mathcal{D}_p\mathcal{E}_p}^{k} \boldsymbol{D}_{ep}^k \Big] F_s F_\tau H_\alpha^k H_\beta^k dz \tag{7.154}$$

$$\boldsymbol{\Pi}_{\Phi \mathcal{D}}^{k\tau s} = \int_{A_k} \Big[\boldsymbol{I}_{ep}^{k\ T} \tilde{\boldsymbol{C}}_{\mathcal{D}_p\mathcal{D}_n}^{k} \Big] F_s F_\tau H_\alpha^k H_\beta^k dz \tag{7.155}$$

In order to write the nuclei in Equations (7.132)–(7.147) in explicit form, the integrals in the z thickness direction are defined as in Equations (7.41); by developing the matrix products in Equations (7.132)–(7.147) and employing a Navier-type closed-form solution (Carrera and Brischetto 2007a; Carrera *et al.* 2008), the explicit algebraic form of the nuclei can be obtained. The fundamental nucleus $\boldsymbol{K}_{uu}^{k\tau s}$ of (3×3) dimension is:

$$\begin{aligned}
K_{uu_{11}}^{k\tau s} &= \bar{\alpha}^2 J_{\beta/\alpha}^{k\tau s} \tilde{C}_{\sigma_p\epsilon_p 11}^{k} + \bar{\beta}^2 J_{\alpha/\beta}^{k\tau s} \tilde{C}_{\sigma_p\epsilon_p 33}^{k} \\
K_{uu_{12}}^{k\tau s} &= J^{k\tau s} (\tilde{C}_{\sigma_p\epsilon_p 12}^{k} + \tilde{C}_{\sigma_p\epsilon_p 33}^{k}) \bar{\alpha}\bar{\beta} \\
K_{uu_{13}}^{k\tau s} &= -\frac{1}{R_\alpha^k} J_{\beta/\alpha}^{k\tau s} \bar{\alpha} \tilde{C}_{\sigma_p\epsilon_p 11}^{k} - \frac{1}{R_\beta^k} J^{k\tau s} \bar{\alpha} \tilde{C}_{\sigma_p\epsilon_p 12}^{k} \\
K_{uu_{21}}^{k\tau s} &= J^{k\tau s} (\tilde{C}_{\sigma_p\epsilon_p 21}^{k} + \tilde{C}_{\sigma_p\epsilon_p 33}^{k}) \bar{\alpha}\bar{\beta} \\
K_{uu_{22}}^{k\tau s} &= \bar{\beta}^2 J_{\alpha/\beta}^{k\tau s} \tilde{C}_{\sigma_p\epsilon_p 22}^{k} + \bar{\alpha}^2 J_{\beta/\alpha}^{k\tau s} \tilde{C}_{\sigma_p\epsilon_p 33}^{k} \\
K_{uu_{23}}^{k\tau s} &= -\frac{1}{R_\alpha^k} J^{k\tau s} \bar{\beta} \tilde{C}_{\sigma_p\epsilon_p 21}^{k} - \frac{1}{R_\beta^k} J_{\alpha/\beta}^{k\tau s} \bar{\alpha} \tilde{C}_{\sigma_p\epsilon_p 22}^{k} \\
K_{uu_{31}}^{k\tau s} &= -\frac{1}{R_\alpha^k} J_{\beta/\alpha}^{k\tau s} \bar{\alpha} \tilde{C}_{\sigma_p\epsilon_p 11}^{k} - \frac{1}{R_\beta^k} J^{k\tau s} \bar{\alpha} \tilde{C}_{\sigma_p\epsilon_p 21}^{k} \\
K_{uu_{32}}^{k\tau s} &= -\frac{1}{R_\alpha^k} J^{k\tau s} \bar{\beta} \tilde{C}_{\sigma_p\epsilon_p 12}^{k} - \frac{1}{R_\beta^k} J_{\alpha/\beta}^{k\tau s} \bar{\beta} \tilde{C}_{\sigma_p\epsilon_p 22}^{k} \\
K_{uu_{33}}^{k\tau s} &= \frac{1}{R_\alpha^{k\,2}} J_{\beta/\alpha}^{k\tau s} \tilde{C}_{\sigma_p\epsilon_p 11}^{k} + \frac{1}{R_\alpha^k R_\beta^k} J^{k\tau s} (\tilde{C}_{\sigma_p\epsilon_p 12}^{k} + \tilde{C}_{\sigma_p\epsilon_p 21}^{k}) + \frac{1}{R_\beta^{k\,2}} J_{\alpha/\beta}^{k\tau s} \tilde{C}_{\sigma_p\epsilon_p 22}^{k}
\end{aligned} \tag{7.156}$$

Nucleus $\boldsymbol{K}_{u\sigma}^{k\tau s}$ of (3 × 3) dimension is:

$$
\begin{aligned}
K_{u\sigma_{11}}^{k\tau s} &= -\frac{1}{R_\alpha^k} J_\beta^{k\tau s} + J_{\alpha\beta}^{k\tau s_z}, \quad K_{u\sigma_{12}}^{k\tau s} = 0, \quad K_{u\sigma_{13}}^{k\tau s} = -\bar{\alpha} J_\beta^{k\tau s} \tilde{C}_{\sigma_p\sigma_n 13}^k \\
K_{u\sigma_{21}}^{k\tau s} &= 0, \quad K_{u\sigma_{22}}^{k\tau s} = -\frac{1}{R_\beta^k} J_\alpha^{k\tau s} + J_{\alpha\beta}^{k\tau s_z}, \quad K_{u\sigma_{23}}^{k\tau s} = -\bar{\beta} J_\alpha^{k\tau s} \tilde{C}_{\sigma_p\sigma_n 23}^k \\
K_{u\sigma_{31}}^{k\tau s} &= \bar{\alpha} J_\beta^{k\tau s}, \quad K_{u\sigma_{32}}^{k\tau s} = \bar{\beta} J_\alpha^{k\tau s}, \qquad (7.157) \\
K_{u\sigma_{33}}^{k\tau s} &= J_{\alpha\beta}^{k\tau s_z} + \frac{1}{R_\alpha^k} J_\beta^{k\tau s} \tilde{C}_{\sigma_p\sigma_n 13}^k + \frac{1}{R_\beta^k} J_\alpha^{k\tau s} \tilde{C}_{\sigma_p\sigma_n 23}^k
\end{aligned}
$$

Nucleus $\boldsymbol{K}_{u\Phi}^{k\tau s}$ of (3 × 1) dimension is:

$$
K_{u\Phi_{11}}^{k\tau s} = K_{u\Phi_{21}}^{k\tau s} = K_{u\Phi_{31}}^{k\tau s} = 0 \qquad (7.158)
$$

Nucleus $\boldsymbol{K}_{u\mathcal{D}}^{k\tau s}$ of (3 × 1) dimension is:

$$
\begin{aligned}
K_{u\mathcal{D}_{11}}^{k\tau s} &= -\bar{\alpha} J_\beta^{k\tau s} \tilde{C}_{\sigma_p\mathcal{D}_n 11}^k, \quad K_{u\mathcal{D}_{21}}^{k\tau s} = -\bar{\beta} J_\alpha^{k\tau s} \tilde{C}_{\sigma_p\mathcal{D}_n 21}^k \qquad (7.159) \\
K_{u\mathcal{D}_{31}}^{k\tau s} &= \frac{1}{R_\alpha^k} J_\beta^{k\tau s} \tilde{C}_{\sigma_p\mathcal{D}_n 11}^k + \frac{1}{R_\beta^k} J_\alpha^{k\tau s} \tilde{C}_{\sigma_p\mathcal{D}_n 21}^k
\end{aligned}
$$

Nucleus $\boldsymbol{K}_{\sigma u}^{k\tau s}$ of (3 × 3) dimension is:

$$
\begin{aligned}
K_{\sigma u_{11}}^{k\tau s} &= -\frac{1}{R_\alpha^k} J_\beta^{k\tau s} + J_{\alpha\beta}^{k\tau_z s}, \quad K_{\sigma u_{12}}^{k\tau s} = 0, \quad K_{\sigma u_{13}}^{k\tau s} = \bar{\alpha} J_\beta^{k\tau s}, \quad K_{\sigma u_{21}}^{k\tau s} = 0 \\
& \qquad (7.160) \\
K_{\sigma u_{22}}^{k\tau s} &= -\frac{1}{R_\beta^k} J_\alpha^{k\tau s} + J_{\alpha\beta}^{k\tau_z s}, \quad K_{u\sigma_{23}}^{k\tau s} = \bar{\beta} J_\alpha^{k\tau s}, \quad K_{\sigma u_{31}}^{k\tau s} = \bar{\alpha} J_\beta^{k\tau s} \tilde{C}_{\epsilon_n\epsilon_p 31}^k \\
K_{\sigma u_{32}}^{k\tau s} &= \bar{\beta} J_\alpha^{k\tau s} \tilde{C}_{\epsilon_n\epsilon_p 32}^k, \quad K_{\sigma u_{33}}^{k\tau s} = J_{\alpha\beta}^{k\tau_z s} - \frac{1}{R_\alpha^k} J_\beta^{k\tau s} \tilde{C}_{\epsilon_n\epsilon_p 31}^k - \frac{1}{R_\beta^k} J_\alpha^{k\tau s} \tilde{C}_{\epsilon_n\epsilon_p 32}^k
\end{aligned}
$$

Nucleus $\boldsymbol{K}_{\sigma\sigma}^{k\tau s}$ of (3 × 3) dimension is:

$$\begin{aligned} K_{\sigma\sigma_{11}}^{k\tau s} &= -J_{\alpha\beta}^{k\tau s}\tilde{C}_{\epsilon_n\sigma_n 11}^{k}, \quad K_{\sigma\sigma_{12}}^{k\tau s} = 0, \quad K_{\sigma\sigma_{13}}^{k\tau s} = 0 \\ K_{\sigma\sigma_{21}}^{k\tau s} &= 0, \quad K_{\sigma\sigma_{22}}^{k\tau s} = -J_{\alpha\beta}^{k\tau s}\tilde{C}_{\epsilon_n\sigma_n 22}^{k}, \quad K_{\sigma\sigma_{23}}^{k\tau s} = 0 \\ K_{\sigma\sigma_{31}}^{k\tau s} &= 0, \quad K_{\sigma\sigma_{32}}^{k\tau s} = 0, \quad K_{\sigma\sigma_{33}}^{k\tau s} = -J_{\alpha\beta}^{k\tau s}\tilde{C}_{\epsilon_n\sigma_n 33}^{k} \end{aligned} \tag{7.161}$$

Nucleus $\boldsymbol{K}_{\sigma\Phi}^{k\tau s}$ of (3 × 1) dimension is:

$$K_{\sigma\Phi_{11}}^{k\tau s} = \bar{\alpha} J_{\beta}^{k\tau s}\tilde{C}_{\epsilon_n\mathcal{E}_p 11}^{k}, \quad K_{\sigma\Phi_{21}}^{k\tau s} = \bar{\beta} J_{\alpha}^{k\tau s}\tilde{C}_{\epsilon_n\mathcal{E}_p 22}^{k}, \quad K_{\sigma\Phi_{31}}^{k\tau s} = 0 \tag{7.162}$$

Nucleus $\boldsymbol{K}_{\sigma\mathcal{D}}^{k\tau s}$ of (3 × 1) dimension is:

$$K_{\sigma\mathcal{D}_{11}}^{k\tau s} = K_{\sigma\mathcal{D}_{21}}^{k\tau s} = 0, \quad K_{\sigma\mathcal{D}_{31}}^{k\tau s} = -J_{\alpha\beta}^{k\tau s}\tilde{C}_{\epsilon_n\mathcal{D}_n 31}^{k} \tag{7.163}$$

Nucleus $\boldsymbol{K}_{\Phi u}^{k\tau s}$ of (1 × 3) dimension is:

$$K_{\Phi u_{11}}^{k\tau s} = K_{\Phi u_{12}}^{k\tau s} = K_{\Phi u_{13}}^{k\tau s} = 0 \tag{7.164}$$

Nucleus $\boldsymbol{K}_{\Phi\sigma}^{k\tau s}$ of (1 × 3) dimension is:

$$K_{\Phi\sigma_{11}}^{k\tau s} = \bar{\alpha} J_{\beta}^{k\tau s}\tilde{C}_{\mathcal{D}_p\sigma_n 11}^{k}, \quad K_{\Phi\sigma_{12}}^{k\tau s} = \bar{\beta} J_{\alpha}^{k\tau s}\tilde{C}_{\mathcal{D}_p\sigma_n 22}^{k}, \quad K_{\Phi\sigma_{13}}^{k\tau s} = 0 \tag{7.165}$$

Nucleus $\boldsymbol{K}_{\Phi\Phi}^{k\tau s}$ of (1 × 1) dimension is:

$$K_{\Phi\Phi_{11}}^{k\tau s} = -\bar{\alpha}^2 J_{\beta/\alpha}^{k\tau s}\tilde{C}_{\mathcal{D}_p\mathcal{E}_p 11}^{k} - \bar{\beta}^2 J_{\alpha/\beta}^{k\tau s}\tilde{C}_{\mathcal{D}_p\mathcal{E}_p 22}^{k} \tag{7.166}$$

Nucleus $\boldsymbol{K}_{\Phi\mathcal{D}}^{k\tau s}$ of (1 × 1) dimension is:

$$K_{\Phi\mathcal{D}_{11}}^{k\tau s} = J_{\alpha\beta}^{k\tau_z s} \tag{7.167}$$

Nucleus $\boldsymbol{K}_{\mathcal{D}u}^{k\tau s}$ of (1 × 3) dimension is:

$$\begin{aligned} K_{\mathcal{D}u_{11}}^{k\tau s} &= -\bar{\alpha} J_{\beta}^{k\tau s}\tilde{C}_{\mathcal{E}_n\epsilon_p 11}^{k}, \quad K_{\mathcal{D}u_{12}}^{k\tau s} = -\bar{\beta} J_{\alpha}^{k\tau s}\tilde{C}_{\mathcal{E}_n\epsilon_p 12}^{k} \\ K_{\mathcal{D}u_{13}}^{k\tau s} &= \frac{1}{R_{\alpha}^{k}} J_{\beta}^{k\tau s}\tilde{C}_{\mathcal{E}_n\epsilon_p 11}^{k} + \frac{1}{R_{\beta}^{k}} J_{\alpha}^{k\tau s}\tilde{C}_{\mathcal{E}_n\epsilon_p 12}^{k} \end{aligned} \tag{7.168}$$

Nucleus $\boldsymbol{K}_{\mathcal{D}\sigma}^{k\tau s}$ of (1×3) dimension is:

$$K_{\mathcal{D}\sigma_{11}}^{k\tau s} = K_{\mathcal{D}\sigma_{12}}^{k\tau s} = 0, \quad K_{\mathcal{D}\sigma_{13}}^{k\tau s} = J_{\alpha\beta}^{k\tau s} \tilde{C}_{\mathcal{E}_n\sigma_n 13}^{k} \tag{7.169}$$

Nucleus $\boldsymbol{K}_{\mathcal{D}\Phi}^{k\tau s}$ of (1×1) dimension is:

$$K_{\mathcal{D}\Phi_{11}}^{k\tau s} = J_{\alpha\beta}^{k\tau_z s} \tag{7.170}$$

Nucleus $\boldsymbol{K}_{\mathcal{D}\mathcal{D}}^{k\tau s}$ of (1×1) dimension is:

$$K_{\mathcal{D}\mathcal{D}_{11}}^{k\tau s} = J_{\alpha\beta}^{k\tau s} \tilde{C}_{\mathcal{E}_n\mathcal{D}_n 11}^{k} \tag{7.171}$$

The fundamental nucleus $\boldsymbol{M}_{uu}^{k\tau s}$ is of (3×3) dimension with the diagonal elements as already given in Equation (7.46).

$\bar{\alpha} = m\pi/a$ and $\bar{\beta} = n\pi/b$, where m and n are the wave numbers in the in-plane directions and a and b are the shell dimensions in the α and β directions, respectively. The explicit form of coefficients $\tilde{\boldsymbol{C}}$ and their components are given in Equations (2.88) and (6.207)–(6.222), respectively.

A Navier-type closed-form solution is obtained via substitution of the harmonic expressions for the displacements, transverse stresses, electric potential, and transverse normal electric displacement and by considering the following material coefficients to be equal to zero: $Q_{16} = Q_{26} = Q_{36} = Q_{45} = 0$ and $e_{25} = e_{14} = e_{36} = \varepsilon_{12} = 0$. The harmonic assumptions used for the displacements, the electric potential, and the transverse normal electric displacement are:

$$(u_\tau^k, \sigma_{\alpha z\tau}^k) = \sum_{m,n} (\hat{U}_\tau^k, \hat{\sigma}_{\alpha z\tau}^k) \cos\left(\frac{m\pi\alpha}{a}\right) \sin\left(\frac{n\pi\beta}{b}\right), \quad k = 1, N_l \tag{7.172}$$

$$(v_\tau^k, \sigma_{\beta z\tau}^k) = \sum_{m,n} (\hat{V}_\tau^k, \hat{\sigma}_{\beta z\tau}^k) \sin\left(\frac{m\pi\alpha}{a}\right) \cos\left(\frac{n\pi\beta}{b}\right), \quad \tau = t, b, r \tag{7.173}$$

$$(w_\tau^k, \Phi_\tau^k, \mathcal{D}_{z\tau}^k, \sigma_{zz\tau}^k) = \sum_{m,n} (\hat{W}_\tau^k, \hat{\Phi}_\tau^k, \hat{\mathcal{D}}_{z\tau}^k, \hat{\sigma}_{zz\tau}^k) \sin\left(\frac{m\pi\alpha}{a}\right) \sin\left(\frac{n\pi\beta}{b}\right), \quad r = 2, N \tag{7.174}$$

where $\hat{U}_\tau^k$, $\hat{V}_\tau^k$, $\hat{W}_\tau^k$ are the displacement amplitudes, $\hat{\sigma}_{\alpha z\tau}^k$, $\hat{\sigma}_{\beta z\tau}^k$, $\hat{\sigma}_{zz\tau}^k$ are the transverse stress amplitudes, $\hat{\Phi}_\tau^k$ is the electric potential amplitude, and $\hat{\mathcal{D}}_{z\tau}^k$ is the transverse normal electric displacement amplitude; k indicates the layer and N_l is the total number of layers. τ is the index for the order of expansion, where t and b indicate the top and bottom of the layer, respectively, while r indicates the higher orders of expansion until $N = 4$. Details on the assembly procedure of the fundamental nuclei and on the acronyms are given in Sections 7.7 and 7.8, respectively, as already discussed for the plate case in Sections 6.7 and 6.8.

7.7 Assembly procedure for fundamental nuclei

The models proposed in Sections 7.3–7.6 are the same ones that were given in the previous chapters for the plate case. The PVD for electromechanical problems, PVD($\boldsymbol{u}$, Φ), given in Section 6.3 for a plate geometry, is extended in Section 7.3 to shells. It has two primary variables in the governing equations: the displacement vector $\boldsymbol{u}^k = (u^k, v^k, w^k)$ and the electric potential Φ^k. Three extensions of the RMVT to electro-mechanical problems are possible. RMVT($\boldsymbol{u}$, Φ, $\boldsymbol{\sigma}_n$), shown in Section 6.4 for plates, is extended to a shell geometry in Section 7.4, and three primary variables are considered in the governing equations: the displacement vector $\boldsymbol{u}^k = (u^k, v^k, w^k)$, the electric potential Φ^k, and the transverse stress components vector $\boldsymbol{\sigma}_n^k = (\sigma_{\alpha z}^k, \sigma_{\beta z}^k, \sigma_{zz}^k)$. RMVT($\boldsymbol{u}$, Φ, $\boldsymbol{\mathcal{D}}_n$), developed in Section 6.5 for a plate geometry, is extended to shells in Section 7.5 and has three primary variables: the displacement vector $\boldsymbol{u}^k = (u^k, v^k, w^k)$, the electric potential Φ^k, and the transverse normal electric displacement $\boldsymbol{\mathcal{D}}_n^k = (\mathcal{D}_z^k)$. Finally, RMVT($\boldsymbol{u}$, Φ, $\boldsymbol{\sigma}_n$, $\boldsymbol{\mathcal{D}}_n$), shown in Section 6.6 for plates, is given in Section 7.6 for a shell geometry; the four primary variables are the displacement vector, the electric potential, the transverse normal electric displacement, and the transverse stress components.

In analogy with the plate case, the choice made in this chapter for a shell geometry is that the displacement $\boldsymbol{u}^k = (u^k, v^k, w^k)$ can be modeled in both ESL and LW form; the other three variables are always modeled in LW form, which means that an electromechanical model is defined as ESL or LW, according to the choice made for the displacement unknowns. Each modeled variable, regardless of which multilayer assembly procedure is considered (ESL or LW), has the same order of expansion in the thickness direction (from linear $N = 1$ to fourth order $N = 4$). A typical Taylor expansion is used in the case of an ESL assembly procedure, while a combination of Legendre polynomials is used as thickness functions in the case of an LW assembly procedure. In the ESL approach, the multilayered shell is considered as one equivalent shell and the stiffnesses of each embedded layer are simply summed, while in the LW approach, each embedded layer is considered as an independent shell and in the global stiffness matrix each contribution is partially summed considering the compatibility and/or equilibrium conditions at each layer interface.

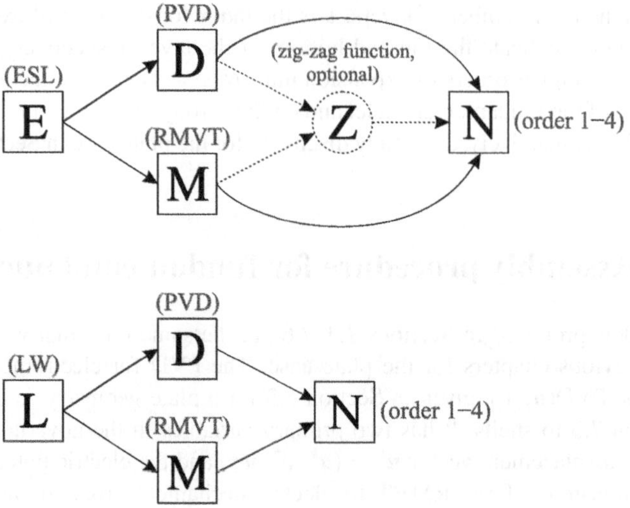

Figure 7.10 Acronym scheme for refined and advanced shell models.

Fundamental nuclei $\boldsymbol{K}_{uu}^{k\tau s}$ can be assembled in ESL or LW form for an example of a three-layered shell, as indicated in Figure 6.10 for the plate case: the introduction of the curvature does not add any other effects. The stiffness is first obtained for each layer by expansion via the indexes τ and s, which consider the order of expansion in the thickness direction, then the three obtained stiffnesses of each layer can be assembled at a multilayer level in ESL form (on the left) by means of a simple summation, or assembled at a multilayer level in LW form (on the right) by considering the compatibility conditions for the displacement components at each layer interface.

Fundamental nuclei $\boldsymbol{K}_{\sigma\sigma}^{k\tau s}$, $\boldsymbol{K}_{\sigma\Phi}^{k\tau s}$, $\boldsymbol{K}_{\Phi\sigma}^{k\tau s}$, $\boldsymbol{K}_{\Phi\Phi}^{k\tau s}$, $\boldsymbol{K}_{\Phi\mathcal{D}}^{k\tau s}$, $\boldsymbol{K}_{\mathcal{D}\Phi}^{k\tau s}$, $\boldsymbol{K}_{\mathcal{D}\mathcal{D}}^{k\tau s}$, $\boldsymbol{K}_{\mathcal{D}\sigma}^{k\tau s}$ and $\boldsymbol{K}_{\sigma\mathcal{D}}^{k\tau s}$ are always assembled in LW form, as indicated on the right of Figure 6.10 for the plate case; in these cases, the partial summation of the stiffness matrices of each layer, for the multilayer assembly procedure, is done by imposing continuity of the electric potential and/or the transverse stresses and/or the transverse normal electric displacement at each layer interface. The introduction of curvatures in the shell geometry does not involve any other difficulties in imposing the compatibility and equilibrium conditions at each shell layer interface.

Fundamental nuclei $\boldsymbol{K}_{u\sigma}^{k\tau s}$, $\boldsymbol{K}_{u\Phi}^{k\tau s}$, and $\boldsymbol{K}_{u\mathcal{D}}^{k\tau s}$ can be assembled in LW form, as indicated on the right of Figure 6.10, by imposing the continuity of the displacements and/or transverse stresses and/or electric potential and/or transverse normal electric displacement at each layer interface, as already

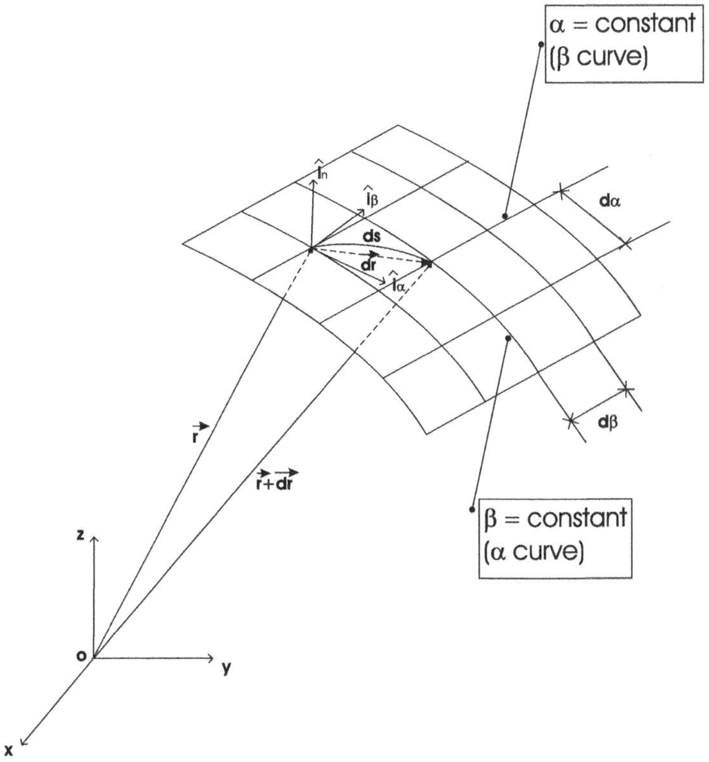

Figure 7.11 Middle surface coordinates.

seen for the plate case. They can also be assembled in a partial ESL form, as indicated on the left of Figure 6.11 for the plate, where the global stiffness matrix at a multilayer level is obtained by assembling the displacements in ESL form (see the rows) and the other variables in LW form (see the columns). This assembly procedure by rows and columns is the same as that of the plate case, since the curvature effect is already considered inside the fundamental nuclei.

Fundamental nuclei $\boldsymbol{K}_{\sigma u}^{k\tau s}$, $\boldsymbol{K}_{\Phi u}^{k\tau s}$, and $\boldsymbol{K}_{\mathcal{D} u}^{k\tau s}$ can be assembled in LW form, as indicated on the right of Figure 6.10 for plates, by imposing the continuity of the displacements and/or transverse stresses and/or electric potential and/or transverse normal electric displacement at each layer interface. They can also be assembled in partial ESL form, as indicated on the right of Figure 6.11 for plates, where the global stiffness matrix at a multilayer level is obtained by assembling the other variables in LW form (see the rows) and the displacements in ESL form (see the columns). As for the other cases, the assembly procedure by rows and columns is the same as that of the plate case, since the curvature effect is already considered inside the fundamental nuclei.

Independently of the curvature effect, the general fundamental nucleus $\boldsymbol{K}^{k\tau s}$ has index k for the kth layer, which permits the multilayer assembly procedure (both ESL and LW approaches), while the indexes τ and s permit expansion in the thickness direction until the considered order N.

7.8 Acronyms for refined and advanced models

The system of acronyms employed for the shell geometry is the same as that mentioned in Section 6.8 for the plate case, and it is summarized in Figure 7.10. The refined and advanced electromechanical models obtained in this chapter, by means of the PVD, and the three extensions of the RMVT can be defined by means of a system of acronyms that explains the multilayer approach (ESL or LW) for the displacements (the other variables are always LW), the employed variational statement (PVD or one of the three possible extensions of RMVT), and the order of expansion in the thickness direction, which is the same for all the variables (from linear to fourth order).

The modeled a priori variables are given in parentheses at the end of the acronym. We have $(\boldsymbol{u}, \Phi)$ for the PVD in Section 7.3, while the variables are $(\boldsymbol{u}, \Phi, \boldsymbol{\sigma}_n)$ for the RMVT in Section 7.4. We add $(\boldsymbol{u}, \Phi, \boldsymbol{\mathcal{D}}_n)$ for the RMVT in Section 7.5 and, finally, the modeled variables are $(\boldsymbol{u}, \Phi, \boldsymbol{\sigma}_n, \boldsymbol{\mathcal{D}}_n)$ for the RMVT in Section 7.6.

For example, an ESL model based on the PVD in Section 7.3 with a second order of expansion in the thickness direction for the modeled variables has the acronym ED2$(\boldsymbol{u}, \Phi)$. A LW model, based on the RMVT in Section 7.5, with a third order of expansion in the thickness direction for the modeled variables has the acronym LM3$(\boldsymbol{u}, \Phi, \boldsymbol{\mathcal{D}}_n)$.

7.9 Pure mechanical problems as particular cases, PVD($\boldsymbol{u}$) and RMVT($\boldsymbol{u}, \boldsymbol{\sigma}_n$)

Pure mechanical models for shells can be considered as particular cases of the electro-mechanical models proposed in this chapter. Pure mechanical refined models are obtained from the variational statement PVD$(\boldsymbol{u})$, which is a particular case of the electromechanical PVD$(\boldsymbol{u}, \Phi)$ given in Section 7.3 (Carrera 2002; Carrera *et al.* 2008). Pure mechanical advanced mixed models are obtained from the variational statement RMVT$(\boldsymbol{u}, \boldsymbol{\sigma}_n)$, which is a particular case of the electromechanical RMVT$(\boldsymbol{u}, \Phi, \boldsymbol{\sigma}_n)$ shown in Section 7.4 (Brischetto and Carrera 2010; Carrera and Boscolo 2007; Carrera *et al.* 2008).

PVD($\boldsymbol{u}$) is obtained from Equation (7.22) simply by discarding the internal electrical work $\delta\boldsymbol{\mathcal{E}}_G^{k\,T}\boldsymbol{\mathcal{D}}_C^k$ and has the same form already given in Equation (6.223). The relative constitutive equations are obtained from Equations (7.23)–(7.25) simply by discarding the electrical contributions and the electromechanical coupling, and they have the same form as those given in Equations (6.224) and (6.225). The only difference with respect to the plate case described in Section 6.9 are the geometrical relations, which, in the case of shells, are given in Equations (2.31)–(2.37). As already illustrated in Section 7.3, by substituting Equations (6.224)–(6.225) in the variational statement of Equation (6.223) and referring to the CUF for the 2D approximation (Carrera 1995), the governing equations and the relative fundamental nuclei can be obtained after integration by parts. However, the governing equations can be obtained in a simpler way from Equations (7.58) simply by discarding the second row and column:

$$\delta\boldsymbol{u}_s^k : \quad \boldsymbol{K}_{uu}^{k\tau s}\,\boldsymbol{u}_\tau^k = \boldsymbol{p}_{us}^k - \boldsymbol{M}_{uu}^{k\tau s}\,\ddot{\boldsymbol{u}}_\tau^k \tag{7.175}$$

The fundamental nuclei $\boldsymbol{K}_{uu}^{k\tau s}$ and $\boldsymbol{M}_{uu}^{k\tau s}$ are the ones that were given in Equations (7.32) and (7.36), respectively.

RMVT($\boldsymbol{u}, \boldsymbol{\sigma}_n$) is obtained from Equation (7.52) simply by discarding the internal electrical work $\delta\boldsymbol{\mathcal{E}}_G^{k\,T}\boldsymbol{\mathcal{D}}_C^k$ and has the same form as that given in Equation (6.227). The relative constitutive equations are obtained from Equations (7.53)–(7.56) simply by discarding the electrical contributions and the electromechanical coupling, and they have the same form as those given in Equations (6.228) and (6.229). The only difference, with respect to the plate case described in Section 6.9, are the geometrical relations, which, in the case of shells, are given in Equations (2.31)–(2.37). As already illustrated in Section 7.5, by substituting Equations (6.228)–(6.229) in the variational statement of Equation (6.227) and referring to the CUF for the 2D approximation (Carrera 1995), the governing equations and the relative fundamental nuclei can be obtained after integration by parts. However, the governing equations can be obtained in a simpler way from Equations (7.58) simply by discarding the third row and column:

$$\begin{aligned}
\delta\boldsymbol{u}_s^k &: \boldsymbol{K}_{uu}^{k\tau s}\,\boldsymbol{u}_\tau^k + \boldsymbol{K}_{u\sigma}^{k\tau s}\,\boldsymbol{\sigma}_{nM\tau}^k = \boldsymbol{p}_{us}^k - \boldsymbol{M}_{uu}^{k\tau s}\,\ddot{\boldsymbol{u}}_\tau^k \\
\delta\boldsymbol{\sigma}_{ns}^k &: \boldsymbol{K}_{\sigma u}^{k\tau s}\,\boldsymbol{u}_\tau^k + \boldsymbol{K}_{\sigma\sigma}^{k\tau s}\,\boldsymbol{\sigma}_{nM\tau}^k = 0
\end{aligned} \tag{7.176}$$

The fundamental nuclei $\boldsymbol{K}_{uu}^{k\tau s}$, $\boldsymbol{K}_{u\sigma}^{k\tau s}$, $\boldsymbol{K}_{\sigma u}^{k\tau s}$, $\boldsymbol{K}_{\sigma\sigma}^{k\tau s}$, and $\boldsymbol{M}_{uu}^{k\tau s}$ are those already given in Equations (7.60), (7.61), (7.63), (7.64), and (7.36), respectively.

7.10 Classical shell theories as particular cases of unified formulation

Classical shell theories, such as CLT and FSDT, can be obtained as particular cases of CUF theory, by means of typical penalty techniques, without the complications shown in Section 3.5.

The starting point is the ED1($\boldsymbol{u}$) theory, which is an ESL model where the three displacement components are linear through the thickness direction z. The FSDT (see Equations (3.58)–(3.66)) can be obtained by means of a typical penalty technique, applied to the global stiffness matrix, which allows one to discard the linear term in the displacement component w. FSDT has the curvature information inside the relative stiffness matrix. Both ED1($\boldsymbol{u}$) and FSDT($\boldsymbol{u}$) have Poisson locking phenomena which can be overcome by means of reduced elastic coefficients in the constitutive equations, imposing the condition $\sigma_{zz} = 0$. Further details about these phenomena can be found in Carrera and Brischetto (2008a,b).

CLT (see Equations (3.58)–(3.68)) can be considered a particular case of the above FSDT($\boldsymbol{u}$) model. In CLT($\boldsymbol{u}$) the transverse shear strains $\gamma_{\alpha z}$ and $\gamma_{\beta z}$ are zero, therefore in the FSDT($\boldsymbol{u}$) model, we can penalize the coefficients Q_{55} and Q_{44} in Equations (2.22) and (2.26) in order to impose $\gamma_{\alpha z} = \gamma_{\beta z} = 0$. Poisson locking also appears in CLT($\boldsymbol{u}$) as for the FSDT($\boldsymbol{u}$) and ED1($\boldsymbol{u}$) cases, and it can be corrected in the same way (Carrera and Brischetto, 2008a,b).

CLT and FSDT, in the case of electromechanical problems for shells, extended to smart structures, have been discussed in this book in Section 3.5. Kinematic models have been introduced by considering a LW linear through-the-thickness electric potential. CLT($\boldsymbol{u}$, Φ) and FSDT($\boldsymbol{u}$, Φ) can also be obtained from the CUF by considering the variational statement PVD($\boldsymbol{u}$, Φ) in Equation (7.22) and the relative governing relations in Equation (7.30). They can be considered as particular cases of the ED1($\boldsymbol{u}$, Φ) model, in which the mechanical part is penalized in the same way as described for the pure mechanical case and the electric potential remains LW and linearly expanded through the thickness-layer direction. The curvatures are introduced directly into the stiffness matrices by considering the geometrical relations for shells (see Equations (2.31)–(2.37)). No Poisson locking corrections are considered for the CLT($\boldsymbol{u}$, Φ) or FSDT($\boldsymbol{u}$, Φ) employed in this book.

7.11 Geometry of shells

The main features of the shell geometry are explained in this section in order to gain a better understanding of the meaning of the metric coefficients H_α^k and

H_{β}^{k}, and to justify the degeneration of shell models into plate models as will be described in Section 7.12.

A thin shell is defined as a 3D body bounded by two closely spaced curved surfaces, where the distance between the two surfaces must be small in comparison to the other dimensions. The middle surface of the shell is the locus of the points which lie midway between these surfaces. The distance between the surfaces measured along the normal to the middle surface is the *thickness* of the shell at that point (Leissa 1973). Shells may be seen as generalizations of a flat plate (Leissa 1969); conversely, a flat plate is a special case of a shell with no curvature. The fundamental equations of the thin shell theory are presented in this section in order to obtain geometrical relations for multifield problems. Geometrical relations for plates can be seen as particular cases of those for shells.

The material is assumed to be linearly elastic and homogeneous, displacements are assumed to be small, and thereby yield linear equations; shear deformation and rotary inertia effects are neglected, and the thickness is taken to be small.

The deformation of a thin shell is determined completely by the displacements of its middle surface (Leissa 1973). The equation of the undeformed middle surface is given, in terms of two independent parameters, α and β, by the radius vector:

$$\vec{r} = \vec{r}(\alpha, \beta) \tag{7.177}$$

Equation (7.177) determines a space curve on the surface. Such curves are called β curves and α curves, see Figure 7.11. We can assume that the parameters α and β always vary within a definite region, and that a one-to-one correspondence exists between the points in this region and the points on the portion of the surface of interest:

$$\vec{r}_{,\alpha} = \frac{\partial \vec{r}}{\partial \alpha}, \quad \vec{r}_{,\beta} = \frac{\partial \vec{r}}{\partial \beta} \tag{7.178}$$

The vectors $\vec{r}_{,\alpha}$ and $\vec{r}_{,\beta}$ are tangent to the α and β curves, respectively. Their length is:

$$|\vec{r}_{,\alpha}| = A, \quad |\vec{r}_{,\beta}| = B \tag{7.179}$$

Consequently, $\vec{r}_{,\alpha}/A$ and $\vec{r}_{,\beta}/B$ are unit vectors that are tangent to the coordinate curves.

The angle between the coordinate curves is χ:

$$\frac{\vec{r}_{,\alpha}}{A} \cdot \frac{\vec{r}_{,\beta}}{B} = \cos\chi \tag{7.180}$$

where

$$\frac{\vec{r}_{,\alpha}}{A} = \hat{\imath}_\alpha, \quad \frac{\vec{r}_{,\beta}}{B} = \hat{\imath}_\beta, \quad \hat{\imath}_n = \frac{\hat{\imath}_\alpha \times \hat{\imath}_\beta}{\sin\chi} \tag{7.181}$$

$\hat{\imath}_n$ is the unit vector of the normal to the surface and it is orthogonal to the vectors $\hat{\imath}_\alpha$ and $\hat{\imath}_\beta$. The unit vectors $\hat{\imath}_\alpha$, $\hat{\imath}_\beta$, and $\hat{\imath}_n$ are usually called the basic vectors of the surface (Leissa 1973).

7.11.1 First quadratic form

If we consider two points (α, β) and $(\alpha + d\alpha, \beta + d\beta)$ arbitrarily near to each other and both lying on the surface, the increment in the vector $\vec{r}$, when moving from the first point to the second one, is:

$$d\vec{r} = \vec{r}_{,\alpha} d\alpha + \vec{r}_{,\beta} d\beta \tag{7.182}$$

By considering Equations (7.179), (7.180), (7.181), and (7.182), we can obtain the square of the differential of the arc length on the surface:

$$d\vec{r} \cdot d\vec{r} = ds^2 = A^2 d\alpha^2 + 2AB\cos\chi \; d\alpha \; d\beta + B^2 d\beta^2 \tag{7.183}$$

The right-hand side of Equation (7.183) is the *first quadratic form of the surface*. This form determines the infinitesimal lengths, the angle between the curves, and the area on the surface: the intrinsic geometry of the surface. However, it does not determine a surface by itself. The terms A^2, $AB\cos\chi$, and B^2 are called *first fundamental quantities*.

7.11.2 Second quadratic form

The problem of finding the curvature of a curve which lies on the surface can be solved by considering the *second quadratic form of the surface*. $\vec{r} = \vec{r}(s)$ is the vectorial equation of a curve on the surface (s is the arc length from a certain origin). $\hat{t}$ is the unit vector along the tangent to the curve:

$$\hat{t} = \frac{d\vec{r}}{ds} = \vec{r}_{,\alpha} \frac{d\alpha}{ds} + \vec{r}_{,\beta} \frac{d\beta}{ds} \tag{7.184}$$

According to the Frenet formula (Kreyszig 1966), the derivative of this vector is:

$$\frac{d\hat{\tau}}{ds} = \frac{\hat{N}}{\rho} \tag{7.185}$$

where $1/\rho$ is the curvature of the curve, and $\hat{N}$ is the unit vector of the principal normal to the curve.

By omitting the middle passages, described in detail in Leissa (1973), it is possible to obtain the expression for the *second quadratic form*:

$$L d\alpha^2 + 2M d\alpha d\beta + N d\beta^2 \tag{7.186}$$

where L, M, and N are the coefficients of the form. The second quadratic form is thus related to the curvatures of the curves on the surface. The curvatures of the α curves and the β curves take β = constant and α = constant, respectively:

$$\frac{1}{R_\alpha} = -\frac{L}{A^2}, \quad \frac{1}{R_\beta} = -\frac{N}{B^2} \tag{7.187}$$

When A, B, R_α, and R_β are given, they uniquely determine a surface, except for the position and orientation in space (Leissa 1973). R_α and R_β are the radii of curvature.

7.11.3 Strain–displacement equations

In order to describe the location of an arbitrary point in the space occupied by a thin shell, the position vector is defined as:

$$\vec{R}(\alpha, \beta, z) = \vec{r}(\alpha, \beta) + z\hat{\imath}_n \tag{7.188}$$

where z measures the distance of the point from the corresponding point on the middle surface along $\hat{\imath}_n$ and varies over the thickness $-h/2 \leq z \leq h/2$. The magnitude of an arbitrary infinitesimal change in the vector $\vec{R}(\alpha, \beta, z)$ is determined by:

$$(ds)^2 = d\vec{R} \cdot d\vec{R} = (d\vec{r} + z d\hat{\imath}_n + \hat{\imath}_n dz)(d\vec{r} + z d\hat{\imath}_n + \hat{\imath}_n dz) \tag{7.189}$$

Recalling the orthogonality of the coordinate system, and the chain rule:

$$d\hat{\imath}_n = \frac{\partial \hat{\imath}_n}{\partial \alpha} d\alpha + \frac{\partial \hat{\imath}_n}{\partial \beta} d\beta \tag{7.190}$$

one obtains:

$$(ds)^2 = g_1 d\alpha^2 + g_2 d\beta^2 + g_3 dz^2 \tag{7.191}$$

where

$$g_1 = [A(1 + z/R_\alpha)]^2, \quad g_2 = [B(1 + z/R_\beta)]^2, \quad g_3 = 1 \tag{7.192}$$

The quantities g_1, g_2, g_3, A, B, R_α, R_β are connected by the Lamb equations (Vlasov 1951), since the 3D space (the space in which the three independent variables α, β, z vary) is a Euclidean space.

The fundamental shell element is a differential element bounded by two surfaces dz apart by a distance z from the middle surface, and four ruled surfaces whose generators are the normals to the middle surface along the parametric curves $\alpha = \alpha_0$, $\alpha = \alpha_0 + d\alpha$, $\beta = \beta_0$, and $\beta = \beta_0 + d\beta$ (Leissa 1973). The lengths of the edges of this fundamental element are (see Figure 7.12):

$$\begin{aligned} ds_\alpha^{(z)} &= A(1 + z/R_\alpha)d\alpha \\ ds_\beta^{(z)} &= B(1 + z/R_\beta)d\beta \end{aligned} \tag{7.193}$$

The differential areas of the edge faces of the fundamental element are (see Figure 7.12):

$$\begin{aligned} dA_\alpha^{(z)} &= A(1 + z/R_\alpha)d\alpha dz \\ dA_\beta^{(z)} &= B(1 + z/R_\beta)d\beta dz \end{aligned} \tag{7.194}$$

while the volume of the fundamental element is:

$$dV^{(z)} = [A(1 + z/R_\alpha)][B(1 + z/R_\beta)]d\alpha d\beta dz \tag{7.195}$$

The well-known strain–displacement equations of the 3D theory of elasticity in orthogonal curvilinear coordinates were obtained by Sokolnikoff (1956):

$$e_i = \frac{\partial}{\partial \alpha_i}\left(\frac{U_i}{\sqrt{g_i}}\right) + \frac{1}{2g_i}\sum_{k=1}^{3}\frac{\partial g_i}{\partial \alpha_k}\frac{U_k}{\sqrt{g_k}}, \qquad i = 1, 2, 3 \tag{7.196}$$

$$\gamma_{ij} = \frac{1}{\sqrt{g_i g_j}}\left[g_i \frac{\partial}{\partial \alpha_j}\left(\frac{U_i}{\sqrt{g_i}}\right) + g_j \frac{\partial}{\partial \alpha_i}\left(\frac{U_j}{\sqrt{g_j}}\right)\right],$$
$$i, j = 1, 2, 3, \quad i \neq j \tag{7.197}$$

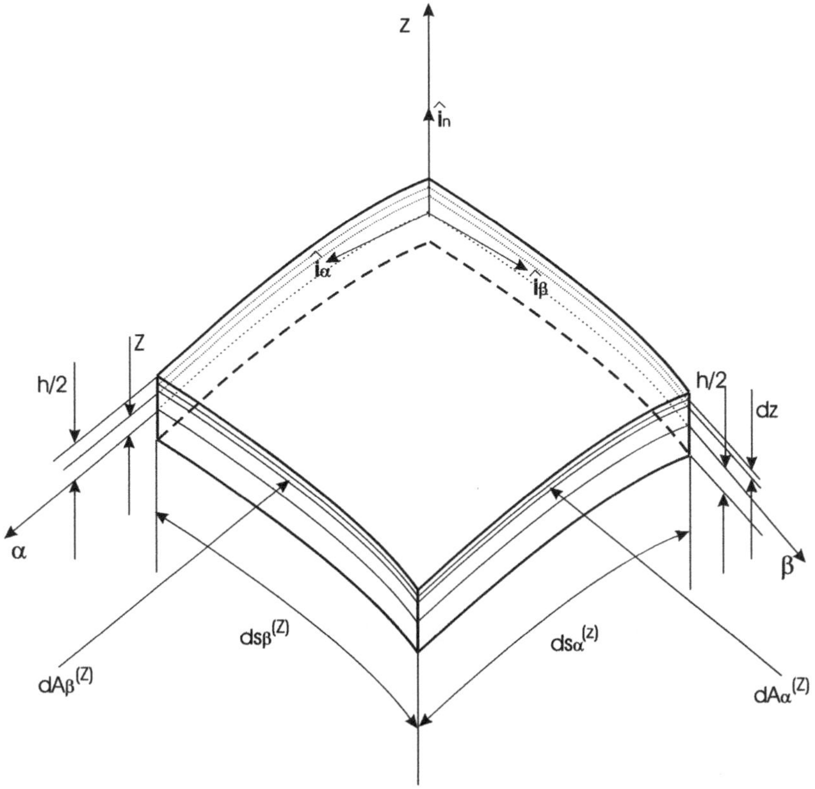

Figure 7.12 Notation and positive directions in shell coordinates.

where the e_i, γ_{ij}, and U_i are the normal strains, shear strains, and displacement components, respectively, at an arbitrary point. In the shell coordinates, the indexes 1, 2, and 3 are replaced by α, β, and z, respectively, while the displacements U_1, U_2, and U_3 are replaced by u, v, and w, respectively. The coefficients of the metric tensor are given by Equation (7.192), and thus yield:

$$e_\alpha = \frac{1}{(1 + z/R_\alpha)}\left(\frac{1}{A}\frac{\partial u}{\partial \alpha} + \frac{v}{AB}\frac{\partial A}{\partial \beta} + \frac{w}{R_\alpha}\right) \tag{7.198}$$

$$e_\beta = \frac{1}{(1 + z/R_\beta)}\left(\frac{u}{AB}\frac{\partial B}{\partial \alpha} + \frac{1}{B}\frac{\partial v}{\partial \beta} + \frac{w}{R_\beta}\right) \tag{7.199}$$

$$e_z = \frac{\partial w}{\partial z} \tag{7.200}$$

$$\gamma_{\alpha\beta} = \frac{A(1+z/R_\alpha)}{B(1+z/R_\beta)}\frac{\partial}{\partial\beta}\left[\frac{u}{A(1+z/R_\alpha)}\right] + \frac{B(1+z/R_\beta)}{A(1+z/R_\alpha)}\frac{\partial}{\partial\alpha}\left[\frac{v}{B(1+z/R_\beta)}\right] \tag{7.201}$$

$$\gamma_{\alpha z} = \frac{1}{A(1+z/R_\alpha)}\frac{\partial w}{\partial\alpha} + A(1+z/R_\alpha)\frac{\partial}{\partial z}\left[\frac{u}{A(1+z/R_\alpha)}\right] \tag{7.202}$$

$$\gamma_{\beta z} = \frac{1}{B(1+z/R_\beta)}\frac{\partial w}{\partial\beta} + B(1+z/R_\beta)\frac{\partial}{\partial z}\left[\frac{v}{B(1+z/R_\beta)}\right] \tag{7.203}$$

The parametric coefficients are given as $H_\alpha = \sqrt{g_1} = 1 + z/R_\alpha$, $H_\beta = \sqrt{g_2} = 1 + z/R_\beta$ and $H_z = \sqrt{g_3} = 1$ (see Equation (7.192)). Equations (7.191)–(7.195) explain the meaning of $d\Omega_k$ in Equations (7.27), (7.57), (7.93), and (7.129). The geometrical relations in Equations (2.31)–(2.42) for plates and shells, in the case of multifield problems, are a consequence of Equations (7.198)–(7.203).

7.12 Plate models as particular cases of shell models

The electro-mechanical models proposed in Chapter 6 for the plate case can also be seen as particular cases of the electro-mechanical shell models given in the present chapter. The decision to split the plate and shell models into two different chapters was only taken to facilitate the understanding of the book for those readers who may be interested in one of the two geometries. However, it is important to explain that the governing equations are very general, and the models obtained for the shell geometries contain the plate geometry as a particular case. PVD($\boldsymbol{u}$, Φ) for shells in Section 7.3 degenerates into PVD($\boldsymbol{u}$, Φ) for plates in Section 6.3, simply by considering the infinite radii of curvature R_α and R_β and the parametric coefficients H_α and H_β equal to one. The same thing happens for the other three RMVT applications. RMVT($\boldsymbol{u}$, Φ, $\boldsymbol{\sigma}_n$) shown in Section 7.4 for shells degenerates into that for plates in Section 6.4. RMVT($\boldsymbol{u}$, Φ, $\mathcal{D}_n$) in Section 7.5, written for a shell geometry, degenerates into that of Section 6.5 for the plate case, and, finally, RMVT($\boldsymbol{u}$, Φ, $\boldsymbol{\sigma}_n$, $\mathcal{D}_n$) in Section 7.6 for a shell degenerates into that written in Section 6.6 for a plate case.

An example is given here in order to clarify this aspect. Let us consider the fundamental nucleus $\boldsymbol{K}_{uu}^{k\tau s}$ of the PVD($\boldsymbol{u}$, Φ) case, although the concept is valid for each fundamental nucleus and each variational statement (PVD and RMVT).

The first three components of the fundamental nucleus $\boldsymbol{K}_{uu}^{k\tau s}$ in Equation (7.42) written for a spherical shell in the case of PVD($\boldsymbol{u}$, Φ) are:

$$
\begin{aligned}
K_{uu_{11}} &= Q_{55}^k J_{\alpha\beta}^{k\tau_z s_z} + \frac{1}{R_\alpha^k} Q_{55}^k \left(-J_\beta^{k\tau_z s} - J_\beta^{k\tau s_z} + \frac{1}{R_\alpha^k} J_{\beta/\alpha}^{k\tau s}\right) \\
&\quad + Q_{11}^k J_{\beta/\alpha}^{k\tau s} \bar{\alpha}^2 + Q_{66}^k J_{\alpha/\beta}^{k\tau s} \bar{\beta}^2 \\
K_{uu_{12}} &= J^{k\tau s} \bar{\alpha}\bar{\beta}(Q_{12}^k + Q_{66}^k) \\
K_{uu_{13}} &= Q_{55}^k \left(J_\beta^{k\tau_z s} \bar{\alpha} - \frac{1}{R_\alpha^k} J_{\beta/\alpha}^{k\tau s} \bar{\alpha}\right) - Q_{13}^k J_\beta^{k\tau s_z} \bar{\alpha} \\
&\quad - \frac{1}{R_\alpha^k} Q_{11}^k J_{\beta/\alpha}^{k\tau s} \bar{\alpha} - Q_{12}^k J^{k\tau s} \bar{\alpha} \frac{1}{R_\beta^k}
\end{aligned}
\tag{7.204}
$$

If we consider a cylindrical shell panel with radius of curvature $R_\alpha^k = \infty$ (which means $1/R_\alpha^k = 0$), the components in Equation (7.204) are simplified because some terms disappear and some integrals in the thickness direction z contain the parametric coefficient $H_\alpha^k = 1$:

$$
\begin{aligned}
K_{uu_{11}} &= Q_{55}^k J_\beta^{k\tau_z s_z} + Q_{11}^k J_\beta^{k\tau s} \bar{\alpha}^2 + Q_{66}^k J_{1/\beta}^{k\tau s} \bar{\beta}^2 \\
K_{uu_{12}} &= J^{k\tau s} \bar{\alpha}\bar{\beta}(Q_{12}^k + Q_{66}^k) \\
K_{uu_{13}} &= Q_{55}^k J_\beta^{k\tau_z s} \bar{\alpha} - Q_{13}^k J_\beta^{k\tau s_z} \bar{\alpha} - Q_{12}^k J^{k\tau s} \bar{\alpha} \frac{1}{R_\beta^k}
\end{aligned}
\tag{7.205}
$$

In the case of a plate geometry, the radius of curvature R_β^k is also infinite (which also means $1/R_\beta^k = 0$), and both the parametric coefficients H_α^k and H_β^k are equal to one. In this way, the components in Equation (7.205) are further simplified:

$$
\begin{aligned}
K_{uu_{11}} &= Q_{55}^k J^{k\tau_z s_z} + Q_{11}^k J^{k\tau s} \bar{\alpha}^2 + Q_{66}^k J^{k\tau s} \bar{\beta}^2 \\
K_{uu_{12}} &= J^{k\tau s} \bar{\alpha}\bar{\beta}(Q_{12}^k + Q_{66}^k) \\
K_{uu_{13}} &= Q_{55}^k J^{k\tau_z s} \bar{\alpha} - Q_{13}^k J^{k\tau s_z} \bar{\alpha}
\end{aligned}
\tag{7.206}
$$

The components in Equations (7.206) (obtained from the degeneration of the shell case in Equations (7.204)) coincide with those obtained in Equations (6.42) for the plate case procedure.

References

Ballhause D, D'Ottavio M, Kröplin B, and Carrera E 2005 A unified formulation to assess multilayered theories for piezoelectric plates. *Comput. Struct.* **83**, 1217–1235.

Brischetto S 2009 Classical and mixed advanced models for sandwich plates embedding functionally graded cores. *J. Mech. Mater. Struct.* **4**, 13–33.

Brischetto S and Carrera E 2009 Refined 2D models for the analysis of functionally graded piezoelectric plates. *J. Intell. Mater. Syst. Struct.* **20**, 1783–1797.

Brischetto S and Carrera E 2010 Advanced mixed theories for bending analysis of functionally graded plates. *Comp. Struct.* **88**, 1474–1483.

Brischetto S, Carrera E, and Demasi L 2009a Improved bending analysis of sandwich plates using a zig-zag function. *Comp. Struct.* **89**, 408–415.

Brischetto S, Carrera E, and Demasi L 2009b Free vibration of sandwich plates and shells by using zig-zag function. *Shock Vib.* **16**, 495–503.

Brischetto S, Carrera E, and Demasi L 2009c Improved response of unsymmetrically laminated sandwich plates by using zig-zag functions. *J. Sandwich Struct. Mater.* **11**, 257–267.

Carrera E 1995 A class of two-dimensional theories for anisotropic multilayered plates analysis. *Accad Sci. Torino, Mem. Sci. Fis.* **19–20**, 1–39.

Carrera E 1999a Multilayered shell theories that account for a layer-wise mixed description. Part I. Governing equations. *AIAA J.* **37**, 1107–1116.

Carrera E 1999b Multilayered shell theories that account for a layer-wise mixed description. Part II. Numerical evaluations. *AIAA J.* **37**, 1117–1124.

Carrera E 2002 Theories and finite elements for multilayered anisotropic, composite plates and shells. *Arch. Comput. Methods. Eng.* **9**, 87–140.

Carrera E 2003 Historical review of zig-zag theories for multilayered plates and shells. *Appl. Mech. Rev.* **56**, 287–309.

Carrera E and Boscolo M 2007 Classical and mixed finite elements for static and dynamic analysis of piezoelectric plates. *Int. J. Numer. Methods Eng.* **70**, 1135–1181.

Carrera E, Boscolo M, and Robaldo A 2007 Hierarchic multilayered plate elements for coupled multifield problems of piezoelectric adaptive structures: formulation and numerical assessment. *Arch. Comput. Methods. Eng.* **14**, 383–430.

Carrera E, Brischetto S, and Cinefra M 2010a Variable kinematics and advanced variational statements for free vibrations analysis of piezoelectric plates and shells. *Comput. Model. Eng. Sc.* **65**, 259–341.

Carrera E, Brischetto S, and Nali P 2008 Variational statements and computational models for multifield problems and multilayered structures. *Mech. Adv. Mater. Struct.* **15**, 182–198.

Carrera E and Brischetto S 2007a Piezoelectric shell theories with "a priori" continuous transverse electro-mechanical variables. *J. Mech. Mater. Struct.* **2**, 377–398.

Carrera E and Brischetto S 2007b Reissner mixed theorem applied to static analysis of piezoelectric shells. *J. Int. Mater. Syst. Struct.* **18**, 1083–1107.

Carrera E and Brischetto S 2008a Analysis of thickness locking in classical, refined and mixed multilayered plate theories. *Comp. Struct.* **82**, 549–562.

Carrera E and Brischetto S 2008b Analysis of thickness locking in classical, refined and mixed theories for layered shells. *Comp. Struct.* **85**, 83–90.

Carrera E and Brischetto S 2009a A survey with numerical assessment of classical and refined theories for the analysis of sandwich plates. *Appl. Mech. Rev.* **62**, 1–17.

Carrera E and Brischetto S 2009b A comparison of various kinematic models for sandwich shell panels with soft core. *J. Compos. Mater.* **43**, 2201–2221.

Carrera E, Nali P, Brischetto S, and Cinefra M 2010b Hierarchic plate and shell theories with direct evaluation of transverse electric displacement. In *Proceedings of 17th AIAA/ASME/AHS Adaptive Strutctures Conference.*

D'Ottavio M and Kröplin B 2006 An extension of Reissner mixed variational theorem to piezoelectric laminates. *Mech. Adv. Mater. Struct.* **13**, 139–150.

Hsu T and Wang JT 1970 A theory of laminated cylindrical shells consisting of layers of orthotropic laminae. *AIAA J.* **8**, 2141–2146.

Hsu T and Wang JT 1971 Rotationally symmetric vibrations of orthotropic layered cylindrical shells. *J. Sound Vib.* **16**, 473–487.

Ikeda T 1996 *Fundamentals of Piezoelectricity.* Oxford University Press.

Kreyszig E 1966 *Advanced Engineering Mathematics.* John Wiley & Sons, Inc., USA.

Leissa AW 1969 *Vibration of Plates.* NASA SP-160.

Leissa AW 1973 *Vibration of Shells.* NASA SP-288.

Librescu L and Schmidt R 1988 Refined theories of elastic anisotropic shells accounting for small strains and moderate rotations. *Int. J. Non-linear Mech.* **23**, 217–229.

Librescu L and Wu EM 1977 A higher-order theory of plate deformation. Part 2: laminated plates. *J. Appl. Mech.* **44**, 669–676.

Murakami H 1985 Laminated composite plate theory with improved in-plane responses. In *ASME Proceedings of Pressure Vessels & Piping Conference.*

Murakami H 1986 Laminated composite plate theory with improved in-plane responses. *J. Appl. Mech.* **53**, 661–666.

Reddy JN 2004 *Mechanics of Laminated Composite Plates and Shells: Theory and Analysis.* CRC Press.

Reissner E 1984 On a certain mixed variational theory and a proposed application. *Int. J. Numer. Methods Eng.* **20**, 1366–1368.

Robbins DH Jr. and Reddy JN 1993 Modeling of thick composites using a layer-wise theory. *Int. J. Numer. Methods Eng.* **36**, 655–677.

Sokolnikoff IS 1956 *Mathematical Theory of Elasticity.* McGraw-Hill.

Srinivas S 1973 A refined analysis of composite laminates. *J. Sound Vib.* **30**, 495–507.

Vlasov VZ Osnovnye Differentsialnye Uravnemia Obshche Teorii Uprugikh Obolochek (1951) Pinkl. Mat. Mekh. (English Translation NACA TM-1241, *Basic Differential Equations in General Theory of Elastic Shells*, 1951).

8

Refined and advanced finite elements for plates

Refined and advanced kinematic descriptions for plate modeling were discussed in Chapter 6, which was devoted to analytical solutions. In this chapter, which pertains to refined and advanced plate finite elements, some of those concepts are presented again in order to avoid dependence on Chapter 6.

In refined and advanced plate models, the electromechanical quantities along the thickness direction of the plate are assumed according to higher order expansions. These axiomatic 2D models can have an ESL or a LW kinematic description. All these kinematic descriptions are included in the framework of the CUF. According to the CUF, the obtained theories can have an order of expansion which goes from first- to higher order values, and, depending on the thickness functions used, a model can be ESL or LW. A FE version of the MUL2 in-house academic code has also been developed for the static and dynamic analysis of smart structures.

8.1 Unified formulation: refined models

We define *refined models* as those *displacement models* where higher orders of expansion in the thickness direction z are assumed for the three displacement components. These axiomatic 2D models can be seen in ESL form when the layers included in the multilayer are considered as one equivalent structure, and in

Plates and Shells for Smart Structures: Classical and Advanced Theories for Modeling and Analysis, First Edition.
Erasmo Carrera, Salvatore Brischetto and Pietro Nali.

LW form when each layer embedded in the multilayer is separately considered in order to write the expansions in z for each layer k. In the case of electromechanical problems, *refined models* are those where the extension is made by considering the electric potential in addition to the displacement vector as the primary variables. These models are obtained by using the PVD (Carrera 2002) and its extensions to multifield problems (Carrera *et al.* 2007; Ikeda 1996).

The CUF is a technique which handles a large variety of plate models in a unified manner (Carrera 1995). According to the CUF, the governing equations are written in terms of a few fundamental nuclei which do not formally depend on the order of expansion N used in the z direction, and on the description of the variables (LW or ESL) (Demasi 2008a,b). The application of a 2D method for plates allows one to express the unknown variables as a set of thickness functions that depend only on the thickness coordinate z, and the corresponding variable that depends on the in-plane coordinates x and y. The generic variable $\boldsymbol{f}(x, y, z)$, for instance a displacement, and its variation $\delta \boldsymbol{f}(x, y, z)$ are written according to the following general expansion:

$$\boldsymbol{f}(x, y, z) = F_\tau(z)\boldsymbol{f}_\tau(x, y), \quad \delta \boldsymbol{f}(x, y, z) = F_s(z)\delta \boldsymbol{f}_s(x, y),$$
$$\text{with} \quad \tau, s = 1, \ldots, N \tag{8.1}$$

where the bold letters denote arrays, (x,y) are the in-plane coordinates, and z the thickness one. The summing convention is assumed with repeated indexes τ and s. The order of expansion N goes from first- to higher order values. Depending on the employed thickness functions, a model can be: ESL, when the variable is assumed for the whole laminate, and a Taylor expansion is employed as thickness functions $F(z)$; LW, when the variable is considered independent in each layer and a combination of Legendre polynomials is used as thickness functions $F(z)$. The maximum order of expansion N along the z direction is the fourth.

According to the FE approximation, by employing shape functions $N_i(\xi, \eta)$ (see Section 4.2), which are defined in the natural reference system, the primary unknowns $\boldsymbol{f}_\tau$ and the corresponding variations $\delta \boldsymbol{f}_s$ can be expressed in terms of nodal values $\boldsymbol{Q}_{\tau i}$ and nodal virtual variations $\delta \boldsymbol{Q}_{sj}$:

$$\boldsymbol{f}_\tau(x, y) = N_i(\xi, \eta)\boldsymbol{Q}_{\tau i} \tag{8.2}$$

$$\delta \boldsymbol{f}_s(x, y) = N_j(\xi, \eta)\delta \boldsymbol{Q}_{sj} \tag{8.3}$$

Consequently, variables $\boldsymbol{f}(x, y, z)$ and their variations $\delta \boldsymbol{f}(x, y, z)$ can be expressed as follows:

$$\boldsymbol{f}(x, y, z) = F_\tau(z)N_i(\xi, \eta)\boldsymbol{Q}_{\tau i} \tag{8.4}$$

$$\delta \boldsymbol{f}(x, y, z) = F_s(z)N_j(\xi, \eta)\delta \boldsymbol{Q}_{sj} \tag{8.5}$$

where $\tau, s = 1, \ldots, N$ and $i, j = 1, \ldots, N_n$, with N_n indicating the number of nodes of the element. The employed elements can have four nodes (Q4), eight nodes (Q8), and nine nodes (Q9) (see Zienkiewicz *et al.* (2005) and Section 4.2 for further details). The unknowns in the FE are expressed in terms of their nodal values via the shape functions N_i and N_j. The latter functions assume unit value in the nodes and allow one to express the unknowns in the points that are different from the nodes as linear combinations of the 4, 8, or 9 node values. The Q4, Q8, and Q9 elements are clearly indicated in Zienkiewicz *et al.* (2005) and a natural coordinate system (ξ, η) is defined in Section 4.2, which goes from -1 to $+1$.

8.1.1 ESL theories

The displacement $\boldsymbol{u} = (u, v, w)$ is described according to the ESL description if the unknowns are the same for the whole plate (Librescu and Wu 1977; Librescu and Schmidt 1988). The z expansion is obtained via Taylor polynomials, that is:

$$\begin{aligned} u &= F_0 u_0 + F_1 u_1 + \cdots + F_N u_N = F_\tau u_\tau \\ v &= F_0 v_0 + F_1 v_1 + \cdots + F_N v_N = F_\tau v_\tau \\ w &= F_0 w_0 + F_1 w_1 + \cdots + F_N w_N = F_\tau w_\tau \end{aligned} \tag{8.6}$$

with $\tau = 0, 1, \ldots, N$, and N is the order of expansion that ranges from 1 (linear) to 4:

$$F_0 = z^0 = 1, \quad F_1 = z^1 = z, \ldots, F_N = z^N \tag{8.7}$$

Equation (8.6) can be written in vectorial form:

$$\begin{aligned} &\boldsymbol{u}(x, y, z) = F_\tau(z)\boldsymbol{u}_\tau(x, y), \quad \delta\boldsymbol{u}(x, y, z) = F_s(z)\delta\boldsymbol{u}_s(x, y), \\ &\text{with} \quad \tau, s = 1, \ldots, N \end{aligned} \tag{8.8}$$

Considering the FE discretization and introducing the nodal values, Equations (8.8) become:

$$\boldsymbol{u}(x, y, z) = F_\tau(z)N_i(\xi, \eta)\boldsymbol{Q}_{u\tau i} \tag{8.9}$$

$$\delta\boldsymbol{u}(x, y, z) = F_s(z)N_j(\xi, \eta)\delta\boldsymbol{Q}_{usj} \tag{8.10}$$

where the nodal displacements are considered in vector $\boldsymbol{Q}_u$; $\tau, s = 1, \ldots, N$ and $i, j = 1, \ldots, N_n$.

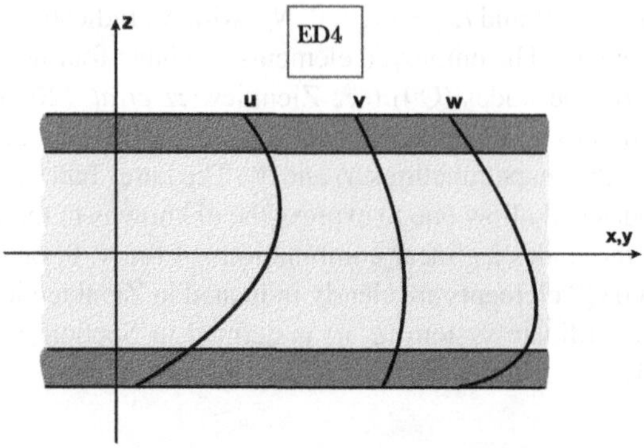

Figure 8.1 ED4: displacements u, v, and w through the thickness direction z.

The 2D models obtained from Equations (8.6)–(8.10) are denoted by the acronym EDN, where E indicates that an ESL approach has been employed, D indicates that the theory is a displacement formulation, and N indicates the order of expansion in the thickness direction. For example, an ED2 model has a quadratic expansion in z, an ED4 has a fourth order of expansion in z, and so on. A typical displacement field is indicated in Figure 8.1 for a three-layered structure for the case of an ED4 model. Figure 8.2 considers the displacement and the transverse stresses along the z direction for an ED2 model: the displacements are quadratic in z, therefore the transverse stresses are linear (no longer constant as in classical theories) but discontinuous at each interface. Simpler theories can be obtained from EDN models, such as those

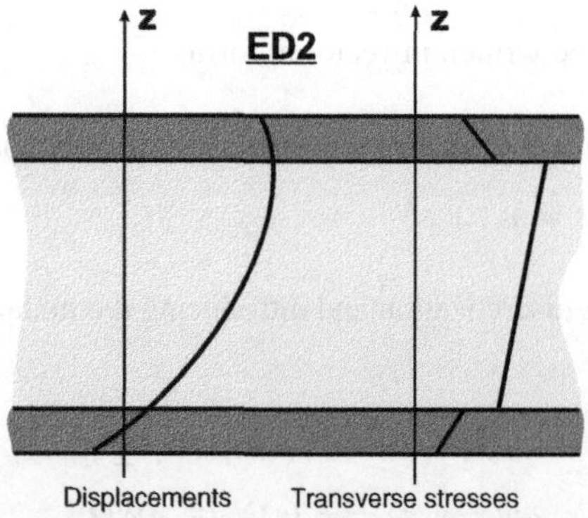

Figure 8.2 ED2: displacements and transverse shear stresses through the thickness direction z.

which discard the ϵ_{zz} effect; in this case, it is sufficient to impose that the transverse displacement w is constant in z. Such theories are denoted as EDNd. The ED1d model coincides with FSDT. CLT is obtained from FSDT via an opportune penalty technique which imposes an infinite shear correction factor. It is important to remember that all the EDNd theories which have constant transverse displacement and zero transverse normal strain ϵ_{zz}, and the ED1 model, show the Poisson locking phenomenon in the case of pure mechanical problems; this can be overcome via plane stress conditions in the constitutive equations (Carrera and Brischetto 2008a,b).

8.1.2 Murakami zigzag function

The ESL models proposed in the previous section do not consider the typical zigzag (ZZ) form of displacements in the z direction, which is typical of multilayered structures with transverse anisotropy (Carrera 2003). A remedy for this limitation is to introduce an opportune zigzag function in the ESL displacement model, in order to recover the ZZ form of displacements without the use of LW models. The latter models have intrinsic ZZ behavior, but are more computationally expansive than ESL models (Carrera and Brischetto 2009a,b). A possible choice for the zigzag function is the so-called *Murakami zigzag function* (MZZF) (Murakami 1985, 1986). MZZF can simply be added to the displacement model and leads to remarkable improvements in the solution as it satisfies the typical ZZ form of displacements in multi-layered structures.

The MZZF $Z(z)$ is defined according to:

$$F_Z = Z(z) = (-1)^k \zeta_k \tag{8.11}$$

with the non-dimensioned layer coordinate $\zeta_k = (2z_k)/h_k$, where z_k is the transverse thickness coordinate and h_k is the thickness of the k layer, therefore $-1 \leq \zeta_k \leq 1$. $Z(z)$ has the following properties: it is a piecewise linear function of the layer coordinates z_k; $Z(z)$ has unit amplitude for the whole layer; and the slope $Z'(z) = dZ/dz$ assumes an opposite sign between two adjacent layers. Its amplitude is layer thickness independent (Murakami 1986). The displacement model including MZZF is:

$$\begin{aligned}
u &= F_0 u_0 + F_1 u_1 + \cdots + F_N u_N + F_Z u_Z = F_\tau u_\tau \\
v &= F_0 v_0 + F_1 v_1 + \cdots + F_N v_N + F_Z v_Z = F_\tau v_\tau \\
w &= F_0 w_0 + F_1 w_1 + \cdots + F_N w_N + F_Z w_Z = F_\tau w_\tau
\end{aligned} \tag{8.12}$$

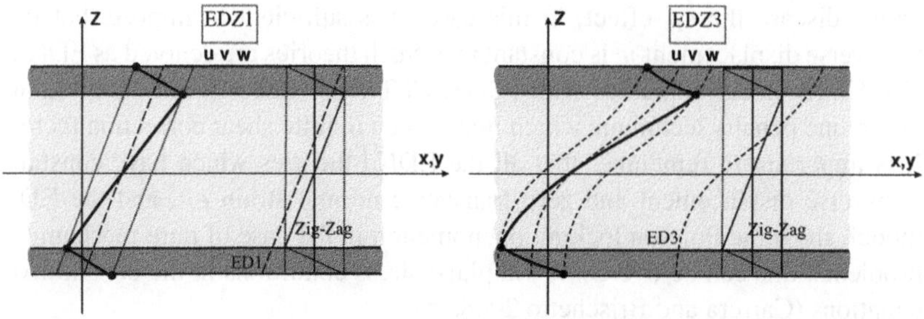

Figure 8.3 Displacement model in the EDZ1 and EDZ3 theories. Inclusion of MZZF in an ESL model.

with $\tau = 0, 1, \ldots, (N+1)$, and N the order of expansion, which ranges from 1 (linear) to 4:

$$F_0 = z^0 = 1, \quad F_1 = z^1 = z, \ldots, F_N = z^N, \quad F_{N+1} = F_Z = (-1)^k \zeta_k \tag{8.13}$$

The acronym to indicate such models is EDZN, where E stands for the ESL approach, D for displacements formulation, and N is the order of expansion in the z direction. Z indicates that MZZF has been added (Brischetto *et al.* 2009a,b). The following remarks can be made: the additional degree of freedom $\boldsymbol{u}_Z$ has a displacement meaning; the amplitude $\boldsymbol{u}_Z$ is layer independent since $\boldsymbol{u}_Z$ has an intrinsic ESL description; and MZZF can be used for both in-plane and out-of-plane displacement components (Brischetto *et al.* 2009b,c). Figure 8.3 clearly shows the meaning of MZZF and how to add it to displacement components. The vectorial form in Equation (8.12) can be written by considering the MZZF $F_Z = Z(z) = (-1)^k \zeta_k$ as the $(N+1)$th thickness function and by expressing the displacements through the nodal values:

$$\boldsymbol{u}(x, y, z) = F_\tau(z) N_i(\xi, \eta) \boldsymbol{Q}_{u\tau i} \tag{8.14}$$

$$\delta \boldsymbol{u}(x, y, z) = F_s(z) N_j(\xi, \eta) \boldsymbol{Q}_{usj} \tag{8.15}$$

where the nodal displacements are considered in vector $\boldsymbol{Q}_u$; $\tau, s = 1, \ldots, (N+1)$ and $i, j = 1, \ldots, N_n$.

Some typical displacements and transverse shear stresses along thickness z are shown in Figure 8.4 for an EDZ1 model: the inclusion of MZZF allows one to recover the typical ZZ form of the displacement vector for the case of multilayered transverse-anisotropy structures. In analogy with the EDN models,

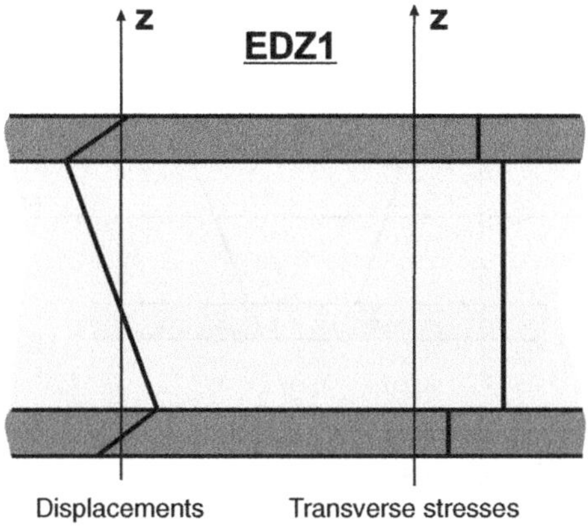

Figure 8.4 EDZ1: displacements and transverse shear stresses through the thickness direction z.

it is possible to impose constant transverse displacements w. Such models are denoted as EDZNd models. It is necessary to correct the Poisson locking phenomenon in EDZNd models, in the case of pure mechanical problems, as indicated in Carrera and Brischetto (2008a,b).

8.1.3 LW theories

When each layer of a multilayered structure is described as an independent plate, a LW approach is considered (Reddy 2004). The displacement $\boldsymbol{u}^k = (u, v, w)^k$ is described for each k layer. In this way, the ZZ form of the displacement in multilayered transverse-anisotropy structures is easily obtained (Hsu and Wang 1970, 1971; Robbins and Reddy 1993; Srinivas 1973). The recovering of the ZZ effect via LW models is explained in detail in Carrera and Brischetto (2009a,b) and shown in Figure 8.5. The z expansion for the displacement components is made for each k layer:

$$\begin{aligned} u^k &= F_0 u_0^k + F_1 u_1^k + \cdots + F_N u_N^k = F_\tau u_\tau^k \\ v^k &= F_0 v_0^k + F_1 v_1^k + \cdots + F_N v_N^k = F_\tau v_\tau^k \\ w^k &= F_0 w_0^k + F_1 w_1^k + \cdots + F_N w_N^k = F_\tau w_\tau^k \end{aligned} \tag{8.16}$$

with $\tau = 0, 1, \ldots, N$, and N the order of expansion, which ranges from 1 (linear) to 4; $k = 1, \ldots, N_l$ where N_l indicates the number of layers.

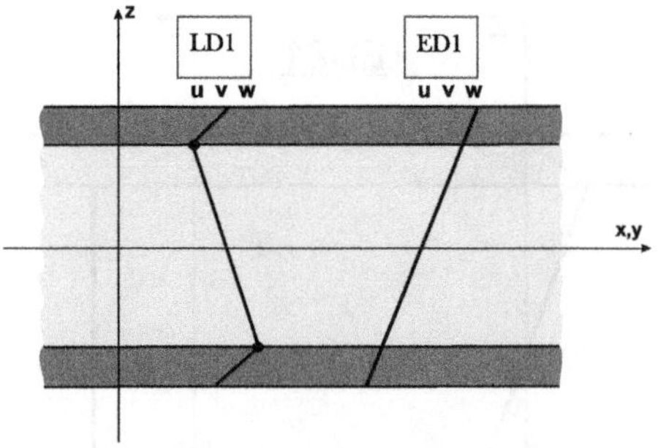

Figure 8.5 Linear expansion in the z direction for displacement components: LW approach vs. ESL approach.

Equation (8.16) is written in vectorial form as:

$$\boldsymbol{u}^k(x,y,z) = F_\tau(z)N_i(\xi,\eta)\boldsymbol{Q}^k_{u\tau i} \tag{8.17}$$

$$\delta\boldsymbol{u}^k(x,y,z) = F_s(z)N_j(\xi,\eta)\delta\boldsymbol{Q}^k_{usj} \tag{8.18}$$

where the nodal displacements are considered in vector $\boldsymbol{Q}^k_u$; $\tau, s = t, b, r$, $k = 1,\ldots,N_l$, and $i, j = 1,\ldots,N_n$. Moreover, t and b indicate the top and bottom of each layer k, respectively, N_l is the number of total layers, and r indicates the higher orders of expansion in the thickness direction: $r = 2,\ldots,N$. The thickness functions $F_\tau(\zeta_k)$ and $F_s(\zeta_k)$ have now been defined at the k-layer level, and are a linear combination of the Legendre polynomials $P_j = P_j(\zeta_k)$ of the jth order defined in the ζ_k domain ($\zeta_k = 2z_k/h_k$ where z_k is the local coordinate and h_k the thickness, both of which refer to the kth layer, therefore $-1 \leq \zeta_k \leq 1$). The first five Legendre polynomials are:

$$P_0 = 1, \quad P_1 = \zeta_k, \quad P_2 = \frac{(3\zeta_k^{\,2}-1)}{2}, \quad P_3 = \frac{5\zeta_k^{\,3}}{2} - \frac{3\zeta_k}{2},$$
$$P_4 = \frac{35\zeta_k^{\,4}}{8} - \frac{15\zeta_k^{\,2}}{4} + \frac{3}{8} \tag{8.19}$$

Their combinations for the thickness functions are:

$$F_t = F_0 = \frac{P_0 + P_1}{2}, \quad F_b = F_1 = \frac{P_0 - P_1}{2}, \quad F_r = P_r - P_{r-2}$$
$$\text{with} \quad r = 2,\ldots,N \tag{8.20}$$

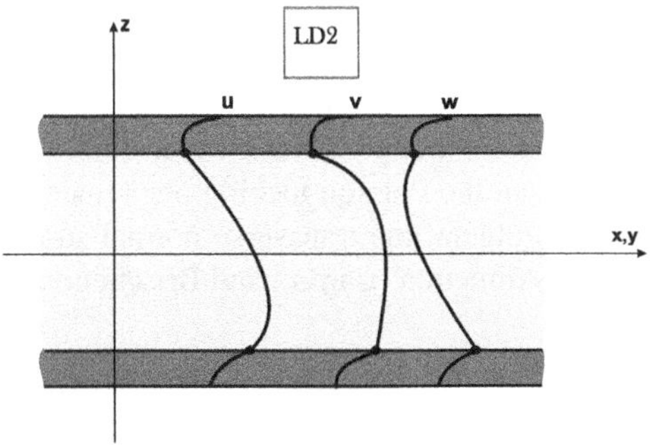

Figure 8.6 LD2: displacements u, v, and w through the thickness direction z.

The chosen functions have the following interesting properties:

$$\zeta_k = 1 : F_t = 1; \quad F_b = 0; \quad F_r = 0 \quad \text{at the top} \tag{8.21}$$

$$\zeta_k = -1 : F_t = 0; \quad F_b = 1; \quad F_r = 0 \quad \text{at the bottom} \tag{8.22}$$

In other words, the interface values of the variables are considered as variable unknowns. This fact allows one to easily impose the compatibility conditions for the displacements at each layer interface. The acronym to indicate such theories is LDN, where L stands for the LW approach, D indicates the displacement formulation, and N is the order of expansion in each k layer. Typical displacement behavior for a three-layered structure is shown in Figure 8.6 for a LD2 model. Figure 8.7 indicates the displacements and transverse shear

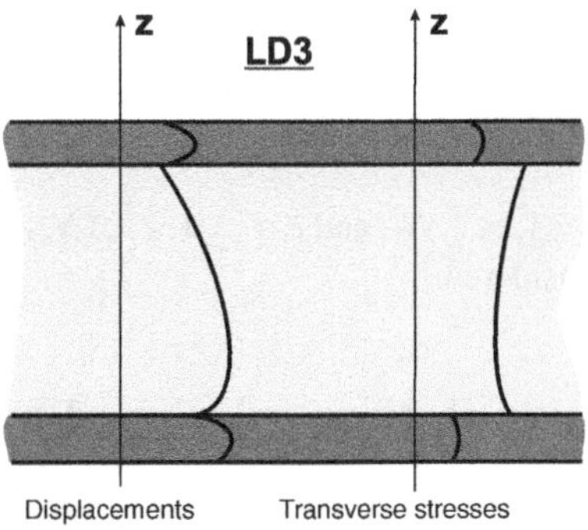

Figure 8.7 LD3: displacements and transverse shear stresses through the thickness direction z.

stresses of a LD3 model. The transverse shear/normal stresses are obtained via the constitutive equations but this does not ensure interlaminar continuity (IC). IC could be enforced by a priori modeling of the transverse shear/normal stresses. In LW models, even though a linear expansion in z is considered for transverse displacement w, the Poisson locking phenomenon does not appear for a pure mechanical problem: the transverse normal strain ϵ_{zz} is piecewise constant in the thickness direction (Carrera and Brischetto 2008a,b).

8.1.4 Refined models for the electromechanical case

The primary variables, in the case of electromechanical problems, are the displacement vector $\boldsymbol{u} = (u, v, w)$ and the scalar electric potential Φ. By considering the higher spatial gradient of the electric potential, the variable Φ^k is always modeled as LW (Brischetto and Carrera 2009; Carrera and Brischetto 2007a,b):

$$\Phi^k(x, y, z) = F_\tau(z)\Phi^k_\tau(x, y), \quad \delta\Phi^k(x, y, z) = F_s(z)\delta\Phi^k_s(x, y),$$
$$\text{with} \quad \tau, s = t, b, r \quad \text{and} \quad k = 1, \ldots, N_l \tag{8.23}$$

where t and b indicate the top and bottom of each k layer, respectively. N_l is the total number of layers; r indicates the higher orders of expansion in the thickness direction: $r = 2, \ldots, N$. The thickness functions are a combination of Legendre polynomials, as indicated in the previous section. A 2D model for electromechanical problems is defined as ESL, ESL+MZZF, or LW, depending on the choice made for the displacement vector: the electric potential is always considered LW (Ballhause *et al.* 2005; Carrera and Boscolo 2007).

By considering shape functions and defining $\boldsymbol{Q}^k_{\Phi\tau i}$ and $\delta\boldsymbol{Q}^k_{\Phi sj}$, the electric potential nodal values and the corresponding virtual variations, Equations (8.23) are rewritten as:

$$\Phi^k(x, y, z) = F_\tau(z)N_i(\xi, \eta)\boldsymbol{Q}^k_{\Phi\tau i} \tag{8.24}$$

$$\delta\Phi^k(x, y, z) = F_s(z)N_j(\xi, \eta)\delta\boldsymbol{Q}^k_{\Phi sj} \tag{8.25}$$

with $\tau, s = t, b, r$; $k = 1, \ldots, N_l$; and $i, j = 1, \ldots, N_n$. N_n is the total number of nodes of the FE considered.

8.2 Unified formulation: advanced mixed models

In the case of electromechanical problems, we define *advanced mixed models* as those 2D models that are obtained by employing the RMVT (Reissner 1984) and its extensions to electromechanical coupling (Carrera *et al.* 2008). These extensions allow one to a priori model some transverse quantities that, in PVD applications, are obtained via post-processing. Transverse shear/normal stresses $\boldsymbol{\sigma}_n = (\sigma_{xz}, \sigma_{yz}, \sigma_{zz})$ and/or transverse normal electric displacement $\boldsymbol{\mathcal{D}}_n = (\mathcal{D}_z)$ are a priori modeled and considered in LW form. The main advantage of obtaining these variables directly from the governing equations is the fulfillment of IC (Brischetto 2009; Brischetto and Carrera 2010). These advanced models are obtained by means of the CUF (Carrera 2002), which has been dealt with in previous sections.

8.2.1 Transverse shear/normal stress modeling

An *advanced mixed model* for a pure mechanical problem considers both the displacements $\boldsymbol{u} = (u, v, w)$ and the transverse shear/normal stresses $\boldsymbol{\sigma}_n = (\sigma_{xz}, \sigma_{yz}, \sigma_{zz})$ as the primary variables (Brischetto and Carrera 2010). The displacements can be modeled as ESL (Section 8.1.1), ESL+MZZF (Section 8.1.2), or LW (Section 8.1.3), and this choice allows one to define the considered advanced model as ESL, ESL+MZZF, or LW, respectively: the transverse shear/normal stresses $\boldsymbol{\sigma}_{nM}$ are always LW (the subscript M means that the stresses are modeled and not obtained from the constitutive equations). The LW model for stresses is:

$$\begin{aligned}
\sigma_{xz}^k &= F_0\sigma_{xz0}^k + F_1\sigma_{xz1}^k + \cdots + F_N\sigma_{xzN}^k = F_\tau\sigma_{xz\tau}^k \\
\sigma_{yz}^k &= F_0\sigma_{yz0}^k + F_1\sigma_{yz1}^k + \cdots + F_N\sigma_{yzN}^k = F_\tau\sigma_{yz\tau}^k \\
\sigma_{zz}^k &= F_0\sigma_{zz0}^k + F_1\sigma_{zz1}^k + \cdots + F_N\sigma_{zzN}^k = F_\tau\sigma_{zz\tau}^k
\end{aligned} \tag{8.26}$$

with $\tau = 0, 1, \ldots, N$, and N the order of expansion, which ranges from 1 (linear) to 4; $k = 1, \ldots, N_l$ where N_l indicates the number of layers. Equation (8.26), written in vectorial form, is:

$$\begin{aligned}
&\boldsymbol{\sigma}_{nM}^k(x, y, z) = F_\tau(z)\boldsymbol{\sigma}_{nM\tau}^k(x, y), \quad \delta\boldsymbol{\sigma}_{nM}^k(x, y, z) = F_s(z)\delta\boldsymbol{\sigma}_{nMs}^k(x, y), \\
&\text{with} \quad \tau, s = t, b, r \quad \text{and} \quad k = 1, \ldots, N_l
\end{aligned} \tag{8.27}$$

where t and b indicate the top and bottom of each k layer, respectively; r indicates the higher orders of expansion in the thickness direction: $r = 2, \ldots, N$. The thickness functions $F_\tau(\zeta_k)$ and $F_s(\zeta_k)$ have now been defined at the k-layer level, and are a linear combination of the Legendre polynomials. If the nodal transverse stresses and the corresponding virtual variations are collected in vectors $\boldsymbol{Q}_{\sigma_n\tau i}^k$ and $\delta\boldsymbol{Q}_{\sigma_n sj}^k$, respectively, Equations (8.27) become:

$$\boldsymbol{\sigma}_{nM}^k(x, y, z) = F_\tau(z)N_i(\xi, \eta)\boldsymbol{Q}_{\sigma_n\tau i}^k \tag{8.28}$$

$$\delta\boldsymbol{\sigma}_{nM}^k(x, y, z) = F_s(z)N_j(\xi, \eta)\delta\boldsymbol{Q}_{\sigma_n sj}^k \tag{8.29}$$

with $\tau, s = t, b, r; k = 1, \ldots, N_l$; and $i, j = 1, \ldots, N_n$. The use of such thickness functions allows one to easily write the IC for the transverse stresses:

$$\sigma_{nt}^{k} = \sigma_{nb}^{k+1} \quad \text{with} \quad k = 1, \ldots, (N_l - 1) \tag{8.30}$$

which means the top value of the k layer in each interface is equal to the bottom value of the $(k+1)$ layer. The same property can be used for the displacements in LW form, in order to impose the compatibility conditions:

$$\boldsymbol{u}_t^{k} = \boldsymbol{u}_b^{k+1} \quad \text{with} \quad k = 1, \ldots, (N_l - 1) \tag{8.31}$$

We define EMN as those models which have the displacements in ESL form (E) and the transverse stresses in LW form, where M means mixed formulation (use of RMVT), and N is the order of expansion, which is the same for both variables. EMZN models consider the displacements modeled in ESL form with the inclusion of MZZF. LMN models consider both the displacements and the transverse stresses in LW form. Figure 8.8 gives the displacements and transverse stresses for an EM2 model. The displacements are considered ESL, while the transverse stresses are a priori modeled and obtained directly from the governing equations: they are considered in LW form, and this allows one to satisfy both the ZZ form and IC. If transverse stresses are obtained from the constitutive equations via post-processing, IC might not be ensured. Figure 8.9 shows the displacements and stresses for a LM2 model; in this case, the displacements are also LW, and the ZZ form and IC are ensured for both the displacement and transverse stress components. The transverse stresses obtained from the constitutive equations might not satisfy IC (Brischetto 2009).

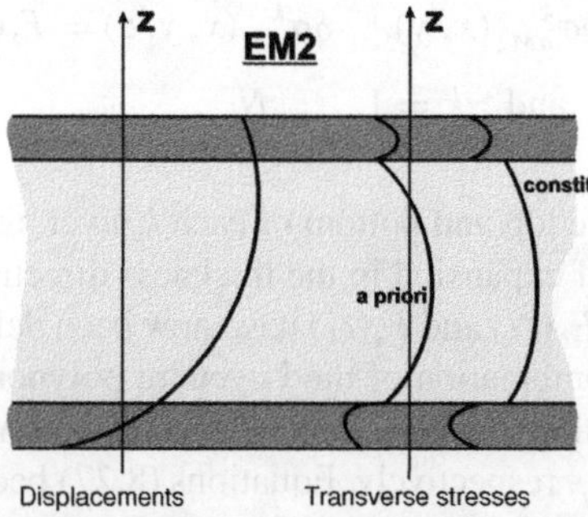

Figure 8.8 EM2: displacements and transverse shear stresses through the thickness direction z.

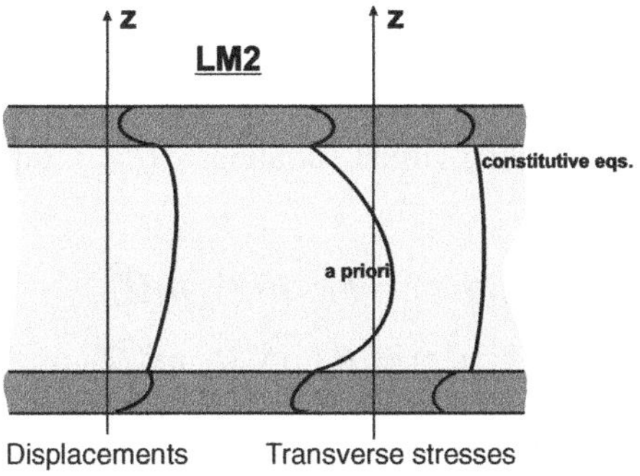

Figure 8.9 LM2: displacements and transverse shear stresses through the thickness direction z.

8.2.2 Advanced mixed models for the electromechanical case

In the case of electromechanical problems, several extensions of RMVT can be considered (Reissner 1984; Carrera *et al.* 2008). In such models, displacements $\boldsymbol{u}$ and electric potential Φ are always considered in the governing equations, the electric potential Φ is always modeled in LW form, as discussed for the PVD case in the previous sections, the displacement components $\boldsymbol{u}$ are modeled as ESL, ESL+MZZF, or LW, and this choice defines the considered advanced models as ESL, ESL+MZZF, or LW.

Three different extensions of RMVT to electromechanical problems can be considered. In addition to displacements $\boldsymbol{u}$ and electric potential Φ, the other modeled variables are:

1. Using only one Lagrange multiplier (Reissner 1984), the transverse stresses $\boldsymbol{\sigma}_{nM}$ are a priori modeled (LW form as described in the previous sections) (Carrera and Brischetto 2007b).
2. Using only one Lagrange multiplier, the transverse normal electric displacement $\mathcal{D}_{nM} = \mathcal{D}_z$ is a priori obtained in LW form (Carrera and Brischetto 2007a).
3. Considering two Lagrange multipliers, both the transverse stresses and the transverse normal electric displacement are a priori modeled in LW form (Carrera and Brischetto 2007a).

The LW expansion for the transverse normal electric displacement $\mathcal{D}_{nM} = \mathcal{D}_z$ is:

$$\mathcal{D}_z^k(x, y, z) = F_\tau(z)\mathcal{D}_{z\tau}^k(x, y), \quad \delta\mathcal{D}_z^k(x, y, z) = F_s(z)\delta\mathcal{D}_{zs}^k(x, y),$$
$$\text{with} \quad \tau, s = t, b, r \quad \text{and} \quad k = 1, \ldots, N_l \tag{8.32}$$

where t and b indicate the top and bottom of each k layer, respectively; r indicates the higher orders of expansion in the thickness direction: $r = 2, \ldots, N$.

By introducing nodal values of the transverse normal electric displacement $\boldsymbol{Q}^k_{\mathcal{D}_z\tau i}$ and the corresponding virtual variations $\delta \boldsymbol{Q}^k_{\mathcal{D}_z sj}$, Equations (8.32) can be rewritten as:

$$\mathcal{D}^k_z(x, y, z) = F_\tau(z) N_i(\xi, \eta) \boldsymbol{Q}^k_{\mathcal{D}_z\tau i} \tag{8.33}$$

$$\delta \mathcal{D}^k_z(x, y, z) = F_s(z) N_j(\xi, \eta) \delta \boldsymbol{Q}^k_{\mathcal{D}_z sj} \tag{8.34}$$

with $\tau, s = t, b, r$; $k = 1, \ldots, N_l$; and $i, j = 1, \ldots, N_n$.

The modeled variables for these three advanced models are:

1. displacements $\boldsymbol{u}$, transverse stresses $\boldsymbol{\sigma}_{nM}$, and electric potential Φ for case 1;
2. displacements $\boldsymbol{u}$, electric potential Φ, and transverse normal electric displacement $\mathcal{D}_{nM} = \mathcal{D}_z$ for case 2;
3. displacements $\boldsymbol{u}$, electric potential Φ, transverse stresses $\boldsymbol{\sigma}_{nM}$, and transverse normal electric displacement $\mathcal{D}_{nM} = \mathcal{D}_z$ for case 3.

The acronyms for such advanced mixed models were explained in previous chapters.

8.3 PVD(u, Φ) for the electromechanical plate case

The PVD extended to smart structures, as in Equations (2.53) and (6.21), is:

$$\int_V \left(\delta \boldsymbol{\epsilon}^T_{pG} \boldsymbol{\sigma}_{pC} + \delta \boldsymbol{\epsilon}^T_{nG} \boldsymbol{\sigma}_{nC} - \delta \boldsymbol{\mathcal{E}}^T_G \boldsymbol{\mathcal{D}}_C \right) dV = \delta L_e - \delta L_{in} \tag{8.35}$$

The variational statement in Equation (8.35) was applied in the previous chapters for plate and shell geometries in the case of analytical closed-form solutions. In the present chapter, an alternative procedure, with a more compact notation, will be used to obtain the governing equations for the FE analysis.

Equation (8.35) can be rewritten as:

$$\int_V \left(\delta \bar{\boldsymbol{\epsilon}}^T_G \, \bar{\boldsymbol{\sigma}}_C \right) dV = \delta \bar{L}_e - \delta L_{in} \tag{8.36}$$

where the overbar indicates that the same array could contain both mechanical and electrical variables. Generalized electromechanical intensive and extensive arrays are introduced into Equation (8.36). Their explicit form is:

$$\bar{\boldsymbol{\epsilon}}^T = \left\{\epsilon_{xx} \quad \epsilon_{yy} \quad \gamma_{xy} \quad \mathcal{E}_x \quad \mathcal{E}_y \quad \epsilon_{zz} \quad \gamma_{xz} \quad \gamma_{yz} \quad \mathcal{E}_z\right\} \tag{8.37}$$

$$\bar{\boldsymbol{\sigma}}^T = \left\{\sigma_{xx} \quad \sigma_{yy} \quad \sigma_{xy} \quad -\mathcal{D}_x \quad -\mathcal{D}_y \quad \sigma_{zz} \quad \sigma_{xz} \quad \sigma_{yz} \quad -\mathcal{D}_z\right\} \tag{8.38}$$

where the subscripts G and C mean substitution of the geometrical and constitutive equations, respectively. The electromechanical physical constitutive coefficients can be grouped in the matrix $\bar{\boldsymbol{H}}$ (see Equations (2.12)–(2.18)):

$$\bar{\boldsymbol{H}} = \begin{bmatrix} Q_{11}^k & Q_{12}^k & Q_{16}^k & 0 & 0 & Q_{13}^k & 0 & 0 & -e_{31}^k \\ Q_{12}^k & Q_{22}^k & Q_{26}^k & 0 & 0 & Q_{23}^k & 0 & 0 & -e_{32}^k \\ Q_{16}^k & Q_{26}^k & Q_{66}^k & 0 & 0 & Q_{36}^k & 0 & 0 & -e_{36}^k \\ 0 & 0 & 0 & -\varepsilon_{11}^k & -\varepsilon_{12}^k & 0 & -e_{15}^k & -e_{14}^k & 0 \\ 0 & 0 & 0 & -\varepsilon_{12}^k & -\varepsilon_{22}^k & 0 & -e_{25}^k & -e_{24}^k & 0 \\ Q_{13}^k & Q_{23}^k & Q_{36}^k & 0 & 0 & Q_{33}^k & 0 & 0 & -e_{33}^k \\ 0 & 0 & 0 & -e_{15}^k & -e_{25}^k & 0 & Q_{55}^k & Q_{45}^k & 0 \\ 0 & 0 & 0 & -e_{14}^k & -e_{24}^k & 0 & Q_{45}^k & Q_{44}^k & 0 \\ -e_{31}^k & -e_{32}^k & -e_{36}^k & 0 & 0 & -e_{33}^k & 0 & 0 & -\varepsilon_{33}^k \end{bmatrix} \tag{8.39}$$

The constitutive equations, already discussed in Equations (6.23)–(6.25), are written in compact form as:

$$\bar{\boldsymbol{\sigma}}_C = \bar{\boldsymbol{H}}\bar{\boldsymbol{\epsilon}}_G \tag{8.40}$$

If the electromechanical primary unknowns are collected in the vector $\bar{\boldsymbol{U}}^T = \{u \ v \ w \ \Phi\}$, the geometrical relations, as in Equations (2.38)–(2.42), can be written in compact form:

$$\bar{\boldsymbol{\epsilon}}_G = \boldsymbol{D}\bar{\boldsymbol{U}} \tag{8.41}$$

where $\boldsymbol{D}$ is the following differential operator:

$$\boldsymbol{D} = \begin{bmatrix} \partial_x & 0 & 0 & 0 \\ 0 & \partial_y & 0 & 0 \\ \partial_y & \partial_x & 0 & 0 \\ 0 & 0 & 0 & -\partial_x \\ 0 & 0 & 0 & -\partial_y \\ 0 & 0 & \partial_z & 0 \\ \partial_z & 0 & \partial_x & 0 \\ 0 & \partial_z & \partial_y & 0 \\ 0 & 0 & 0 & -\partial_z \end{bmatrix} \tag{8.42}$$

In the framework of the CUF (Carrera 1995), $\bar{U}$ can be expressed through the thickness functions, according to a generic kinematic description:

$$\bar{U}(x, y, z) = F_\tau(z)\,\bar{U}_\tau(x, y) \tag{8.43}$$

Similarly for the virtual variations:

$$\delta\bar{U}(x, y, z) = F_s(z)\,\delta\bar{U}_s(x, y) \tag{8.44}$$

If the FE approach is addressed, $\bar{U}_\tau$ and $\delta\bar{U}_s$ can be expressed in terms of the nodal values (array $\bar{Q}^T = \{Q_u\ Q_\Phi\}^T$) through the shape functions, which are defined in the natural coordinate system of the element (see Chapter 4):

$$\bar{U}_\tau^{(e)}(x, y) = N_i^{(e)}(\xi, \eta)\,\bar{Q}_{\tau i}^{(e)} \tag{8.45}$$

$$\delta\bar{U}_s^{(e)}(x, y) = N_j^{(e)}(\xi, \eta)\,\delta\bar{Q}_{sj}^{(e)} \tag{8.46}$$

where the superscript (e) indicates the FE.

Substituting Equations (8.40)–(8.46) in the variational statement of Equation (8.36) and considering a multilayered plate composed of N_l layers, a set of equilibrium equations is obtained, which can be given formally in the following compact form:

$$\delta\bar{Q}_{sj}^{(e)k} : \boldsymbol{M}^{(e)k\tau sij}\,\ddot{\bar{Q}}_{\tau i}^{(e)k} + \bar{\boldsymbol{K}}^{(e)k\tau sij}\,\bar{Q}_{\tau i}^{(e)k} = \bar{\boldsymbol{F}}_{sj}^{(e)k} \tag{8.47}$$

where $\bar{F}$ is the vector of the nodal loads, while the boundary conditions are $\bar{\bar{Q}}$.

In Equation (8.47), τ and s vary from 0 to N (order of expansion), i and j vary from 1 to the element node number N_n, and k ranges from 1 to N_l.

Matrix $\bar{\boldsymbol{K}}^{(e)k\tau sij}$ is the fundamental stiffness nucleus of the FE (e) and can be calculated through numerical integration of the shape and thickness functions, according to the following product:

$$\bar{\boldsymbol{K}}^{(e)k\tau sij} = \int_{V^{(e)}} \left((N_j \boldsymbol{I} F_s)\boldsymbol{D}^T\,\bar{\boldsymbol{H}}\,\boldsymbol{D}(F_\tau \boldsymbol{I} N_i)\right) dV^{(e)} \tag{8.48}$$

where $\boldsymbol{I}$ is the 4×4 identity matrix and $V^{(e)}$ is the finite volume element.

The following symbols are introduced in order to simplify the notation, which indicates the in-plane integration and the through-the-thickness integration at the element level:

$$\triangleleft \cdots \triangleright_{\Omega} = \int_{\Omega_0^{(e)}} (\cdots) \, d\Omega^{(e)} \tag{8.49}$$

$$\triangleleft \cdots \triangleright_{A} = \int_{A^{(e)k}} (\cdots) \, dz \tag{8.50}$$

where $\Omega_0^{(e)}$ is the midsurface element. The explicit form of $\bar{\boldsymbol{K}}^{(e)k\tau sij}$ for PVD($\boldsymbol{u}$, Φ) is given below and has a (4×4) dimension because there are four variables in the PVD($\boldsymbol{u}$, Φ) (three displacement components and the scalar electric potential):

$$\bar{\boldsymbol{K}}^{(e)k\tau sij} = \begin{bmatrix} K_{11} & K_{12} & K_{13} & K_{14} \\ K_{21} & K_{22} & K_{23} & K_{24} \\ K_{31} & K_{32} & K_{33} & K_{34} \\ K_{41} & K_{42} & K_{43} & K_{44} \end{bmatrix}^{k\tau sij} \tag{8.51}$$

where each component is:

$$\begin{aligned}
K_{11}^{k\tau sij} &= Q_{55}^k \triangleleft N_i N_j \triangleright_{\Omega} \triangleleft F_{\tau,z} F_{s,z} \triangleright_A + Q_{11}^k \triangleleft N_{i,x} N_{j,x} \triangleright_{\Omega} \triangleleft F_{\tau} F_{s} \triangleright_A \\
&+ Q_{16}^k \triangleleft N_{i,y} N_{j,x} \triangleright_{\Omega} \triangleleft F_{\tau} F_{s} \triangleright_A + Q_{16}^k \triangleleft N_{i,x} N_{j,y} \triangleright_{\Omega} \triangleleft F_{\tau} F_{s} \triangleright_A \\
&+ Q_{66}^k \triangleleft N_{i,y} N_{j,y} \triangleright_{\Omega} \triangleleft F_{\tau} F_{s} \triangleright_A \\
K_{21}^{k\tau sij} &= Q_{45}^k \triangleleft N_i N_j \triangleright_{\Omega} \triangleleft F_{\tau,z} F_{s,z} \triangleright_A + Q_{16}^k \triangleleft N_{i,x} N_{j,x} \triangleright_{\Omega} \triangleleft F_{\tau} F_{s} \triangleright_A \\
&+ Q_{12}^k \triangleleft N_{i,y} N_{j,x} \triangleright_{\Omega} \triangleleft F_{\tau} F_{s} \triangleright_A + Q_{66}^k \triangleleft N_{i,x} N_{j,y} \triangleright_{\Omega} \triangleleft F_{\tau} F_{s} \triangleright_A \\
&+ Q_{26}^k \triangleleft N_{i,y} N_{j,y} \triangleright_{\Omega} \triangleleft F_{\tau} F_{s} \triangleright_A \\
K_{31}^{k\tau sij} &= Q_{55}^k \triangleleft N_{i,x} N_j \triangleright_{\Omega} \triangleleft F_{\tau} F_{s,z} \triangleright_A + Q_{45}^k \triangleleft N_{i,y} N_j \triangleright_{\Omega} \triangleleft F_{\tau} F_{s,z} \triangleright_A \\
&+ Q_{13}^k \triangleleft N_i N_{j,x} \triangleright_{\Omega} \triangleleft F_{\tau,z} F_{s} \triangleright_A + Q_{36}^k \triangleleft N_i N_{j,y} \triangleright_{\Omega} \triangleleft F_{\tau,z} F_{s} \triangleright_A \\
K_{41}^{k\tau sij} &= e_{31}^k \triangleleft N_i N_{j,x} \triangleright_{\Omega} \triangleleft F_{\tau,z} F_{s} \triangleright_A + e_{36}^k \triangleleft N_i N_{j,y} \triangleright_{\Omega} \triangleleft F_{\tau,z} F_{s} \triangleright_A \\
&+ e_{15}^k \triangleleft N_{i,x} N_j \triangleright_{\Omega} \triangleleft F_{\tau} F_{s,z} \triangleright_A + e_{25}^k \triangleleft N_{i,y} N_j \triangleright_{\Omega} \triangleleft F_{\tau} F_{s,z} \triangleright_A \\
K_{12}^{k\tau sij} &= Q_{45}^k \triangleleft N_i N_j \triangleright_{\Omega} \triangleleft F_{\tau,z} F_{s,x} \triangleright_A + Q_{16}^k \triangleleft N_{i,x} N_{j,x} \triangleright_{\Omega} \triangleleft F_{\tau} F_{s} \triangleright_A \\
&+ Q_{66}^k \triangleleft N_{i,y} N_{j,x} \triangleright_{\Omega} \triangleleft F_{\tau} F_{s} \triangleright_A + Q_{12}^k \triangleleft N_{i,x} N_{j,y} \triangleright_{\Omega} \triangleleft F_{\tau} F_{s} \triangleright_A \\
&+ Q_{26}^k \triangleleft N_{i,y} N_{j,y} \triangleright_{\Omega} \triangleleft F_{\tau} F_{s} \triangleright_A \\
K_{22}^{k\tau sij} &= Q_{44}^k \triangleleft N_i N_j \triangleright_{\Omega} \triangleleft F_{\tau,z} F_{s,z} \triangleright_A + Q_{66}^k \triangleleft N_{i,x} N_{j,x} \triangleright_{\Omega} \triangleleft F_{\tau} F_{s} \triangleright_A \\
&+ Q_{26}^k \triangleleft N_{i,y} N_{j,x} \triangleright_{\Omega} \triangleleft F_{\tau} F_{s} \triangleright_A + Q_{26}^k \triangleleft N_{i,x} N_{j,y} \triangleright_{\Omega} \triangleleft F_{\tau} F_{s} \triangleright_A \\
&+ Q_{22}^k \triangleleft N_{i,y} N_{j,y} \triangleright_{\Omega} \triangleleft F_{\tau} F_{s} \triangleright_A
\end{aligned}$$

$$
\begin{aligned}
K_{32}^{k\tau sij} &= Q_{45}^k \triangleleft N_{i,x} N_j \triangleright_\Omega \triangleleft F_\tau F_{s,z} \triangleright_A + Q_{44}^k \triangleleft N_{i,y} N_j \triangleright_\Omega \triangleleft F_\tau F_{s,z} \triangleright_A \\
&\quad + Q_{36}^k \triangleleft N_i N_{j,x} \triangleright_\Omega \triangleleft F_{\tau,z} F_s \triangleright_A + Q_{23}^k \triangleleft N_i N_{j,y} \triangleright_\Omega \triangleleft F_{\tau,z} F_s \triangleright_A \\
K_{42}^{k\tau sij} &= e_{32}^k \triangleleft N_i N_{j,y} \triangleright_\Omega \triangleleft F_{\tau,z} F_s \triangleright_A + e_{36}^k \triangleleft N_i N_{j,x} \triangleright_\Omega \triangleleft F_{\tau,z} F_s \triangleright_A \\
&\quad + e_{14}^k \triangleleft N_{i,x} N_j \triangleright_\Omega \triangleleft F_\tau F_{s,z} \triangleright_A + e_{24}^k \triangleleft N_{i,y} N_j \triangleright_\Omega \triangleleft F_\tau F_{s,z} \triangleright_A \qquad (8.52) \\
K_{13}^{k\tau sij} &= Q_{13}^k \triangleleft N_{i,x} N_j \triangleright_\Omega \triangleleft F_\tau F_{s,z} \triangleright_A + Q_{36}^k \triangleleft N_{i,y} N_j \triangleright_\Omega \triangleleft F_\tau F_{s,z} \triangleright_A \\
&\quad + Q_{55}^k \triangleleft N_i N_{j,x} \triangleright_\Omega \triangleleft F_{\tau,z} F_s \triangleright_A + Q_{45}^k \triangleleft N_i N_{j,y} \triangleright_\Omega \triangleleft F_{\tau,z} F_s \triangleright_A \\
K_{23}^{k\tau sij} &= Q_{36}^k \triangleleft N_{i,x} N_j \triangleright_\Omega \triangleleft F_\tau F_{s,z} \triangleright_A + Q_{23}^k \triangleleft N_{i,y} N_j \triangleright_\Omega \triangleleft F_\tau F_{s,z} \triangleright_A \\
&\quad + Q_{45}^k \triangleleft N_i N_{j,x} \triangleright_\Omega \triangleleft F_{\tau,z} F_s \triangleright_A + Q_{44}^k \triangleleft N_i N_{j,y} \triangleright_\Omega \triangleleft F_{\tau,z} F_s \triangleright_A \\
K_{33}^{k\tau sij} &= Q_{33}^k \triangleleft N_i N_j \triangleright_\Omega \triangleleft F_{\tau,z} F_{s,z} \triangleright_A + Q_{55}^k \triangleleft N_{i,x} N_{j,x} \triangleright_\Omega \triangleleft F_\tau F_s \triangleright_A \\
&\quad + Q_{45}^k \triangleleft N_{i,y} N_{j,x} \triangleright_\Omega \triangleleft F_\tau F_s \triangleright_A + Q_{45}^k \triangleleft N_{i,x} N_{j,y} \triangleright_\Omega \triangleleft F_\tau F_s \triangleright_A \\
&\quad + Q_{44}^k \triangleleft N_{i,y} N_{j,y} \triangleright_\Omega \triangleleft F_\tau F_s \triangleright_A \\
K_{43}^{k\tau sij} &= e_{33}^k \triangleleft N_i N_j \triangleright_\Omega \triangleleft F_{\tau,z} F_{s,z} \triangleright_A + e_{14}^k \triangleleft N_{i,x} N_{j,y} \triangleright_\Omega \triangleleft F_\tau F_s \triangleright_A \\
&\quad + e_{24}^k \triangleleft N_{i,y} N_{j,y} \triangleright_\Omega \triangleleft F_\tau F_s \triangleright_A + e_{15}^k \triangleleft N_{i,x} N_{j,x} \triangleright_\Omega \triangleleft F_\tau F_s \triangleright_A \\
&\quad + e_{25}^k \triangleleft N_{i,y} N_{j,x} \triangleright_\Omega \triangleleft F_\tau F_s \triangleright_A \\
K_{14}^{k\tau sij} &= e_{31}^k \triangleleft N_{i,x} N_j \triangleright_\Omega \triangleleft F_\tau F_{s,z} \triangleright_A + e_{36}^k \triangleleft N_{i,y} N_j \triangleright_\Omega \triangleleft F_\tau F_{s,z} \triangleright_A \\
&\quad + e_{15}^k \triangleleft N_i N_{j,x} \triangleright_\Omega \triangleleft F_{\tau,z} F_s \triangleright_A + e_{25}^k \triangleleft N_i N_{j,y} \triangleright_\Omega \triangleleft F_{\tau,z} F_s \triangleright_A \\
K_{24}^{k\tau sij} &= e_{32}^k \triangleleft N_{i,y} N_j \triangleright_\Omega \triangleleft F_\tau F_{s,z} \triangleright_A + e_{36}^k \triangleleft N_{i,x} N_j \triangleright_\Omega \triangleleft F_\tau F_{s,z} \triangleright_A \\
&\quad + e_{14}^k \triangleleft N_i N_{j,x} \triangleright_\Omega \triangleleft F_{\tau,z} F_s \triangleright_A + e_{24}^k \triangleleft N_i N_{j,y} \triangleright_\Omega \triangleleft F_{\tau,z} F_s \triangleright_A \\
K_{34}^{k\tau sij} &= e_{33}^k \triangleleft N_i N_j \triangleright_\Omega \triangleleft F_{\tau,z} F_{s,z} \triangleright_A + e_{14}^k \triangleleft N_{i,y} N_{j,x} \triangleright_\Omega \triangleleft F_\tau F_s \triangleright_A \\
&\quad + e_{24}^k \triangleleft N_{i,y} N_{j,y} \triangleright_\Omega \triangleleft F_\tau F_s \triangleright_A + e_{15}^k \triangleleft N_{i,x} N_{j,x} \triangleright_\Omega \triangleleft F_\tau F_s \triangleright_A \\
&\quad + e_{25}^k \triangleleft N_{i,x} N_{j,y} \triangleright_\Omega \triangleleft F_\tau F_s \triangleright_A \\
K_{44}^{k\tau sij} &= -\varepsilon_{33}^k \triangleleft N_i N_j \triangleright_\Omega \triangleleft F_{\tau,z} F_{s,z} \triangleright_A - \varepsilon_{11}^k \triangleleft N_{i,x} N_{j,x} \triangleright_\Omega \triangleleft F_\tau F_s \triangleright_A \\
&\quad - \varepsilon_{12}^k \triangleleft N_{i,x} N_{j,y} \triangleright_\Omega \triangleleft F_\tau F_s \triangleright_A - \varepsilon_{12}^k \triangleleft N_{i,y} N_{j,x} \triangleright_\Omega \triangleleft F_\tau F_s \triangleright_A \\
&\quad - \varepsilon_{22}^k \triangleleft N_{i,y} N_{j,y} \triangleright_\Omega \triangleleft F_\tau F_s \triangleright_A
\end{aligned}
$$

Matrix $\boldsymbol{M}^{(e)k\tau sij}$ is the fundamental mass nucleus of the element and is representative of inertial effects. $\boldsymbol{M}^{(e)k\tau sij}$ can be calculated through the numerical integration of the shape and thickness functions, according to the following product:

$$\boldsymbol{M}^{(e)k\tau sij} = \int_{V^{(e)}} \left((N_j \boldsymbol{I} F_s)\, \rho^k (F_\tau \boldsymbol{I} N_i) \right) dV^{(e)} \qquad (8.53)$$

where ρ^k is the mass density of each layer. The non-zero elements of the fundamental mass nucleus are:

$$M_{11}^{k\tau sij} = M_{22}^{k\tau sij} = M_{33}^{k\tau sij} = \rho^k \triangleleft N_i N_j \triangleright_\Omega \triangleleft F_\tau F_s \triangleright_A \tag{8.54}$$

The pure mechanical problem corresponding to PVD($\boldsymbol{u}$) can be obtained as a particular case of PVD($\boldsymbol{u}$, Φ), coherently with what has already been explained in Section 6.9 for the analytical solution. The assembly procedure for the FE method has two additional indexes (i, j) for the nodes (with respect to the analytical closed-form solutions). Details on the FE assembly procedure are given at the end of this chapter.

8.4 RMVT(u, Φ, σ_n) for the electromechanical plate case

A possible extension of the RMVT (Reissner 1984) has been indicated in Equations (2.60) and (6.51), where the internal electrical work has been added and the Lagrange multiplier is employed in order to increase the accuracy of the computed transverse stress components (Carrera and Brischetto 2007a; Carrera *et al.* 2008):

$$\int_V \left(\delta\boldsymbol{\epsilon}_{pG}^T \boldsymbol{\sigma}_{pC} + \delta\boldsymbol{\epsilon}_{nG}^T \boldsymbol{\sigma}_{nM} - \delta\boldsymbol{\mathcal{E}}_{pG}^T \boldsymbol{\mathcal{D}}_{pC} - \delta\boldsymbol{\mathcal{E}}_{nG}^T \boldsymbol{\mathcal{D}}_{nC} + \delta\boldsymbol{\sigma}_{nM}^T(\boldsymbol{\epsilon}_{nG} - \boldsymbol{\epsilon}_{nC})\right) dV$$
$$= \delta L_e - \delta L_{in} \tag{8.55}$$

According to the condensed notation, Equation (8.55) can be rewritten as:

$$\int_V \left(\delta\bar{\boldsymbol{\epsilon}}_{aG}^T\, \bar{\boldsymbol{\sigma}}_{aC} + \delta\bar{\boldsymbol{\epsilon}}_{bG}^T\, \bar{\boldsymbol{\sigma}}_b + \delta\bar{\boldsymbol{\sigma}}_b^T(\bar{\boldsymbol{\epsilon}}_{bG} - \bar{\boldsymbol{\epsilon}}_{bC})\right) dV = \delta\bar{L}_e - \delta L_{in} \tag{8.56}$$

where the adopted arrays mean:

$\bar{\boldsymbol{\sigma}}_{aC}^T = \{\sigma_{xx} \quad \sigma_{yy} \quad \sigma_{xy} \quad -\mathcal{D}_x \quad -\mathcal{D}_y \quad -\mathcal{D}_z\}_C$ is the vector of the extensive non-modeled variables, which are calculated by means of the constitutive relations (subscript C);

$\bar{\boldsymbol{\sigma}}_b^T = \{\sigma_{zz} \quad \sigma_{xz} \quad \sigma_{yz}\}$ is the vector of the extensive modeled variables;

$\bar{\boldsymbol{\epsilon}}_{aG}^T = \{\epsilon_{xx} \quad \epsilon_{yy} \quad \gamma_{xy} \quad \mathcal{E}_x \quad \mathcal{E}_y \quad \mathcal{E}_z\}_G$ is the vector of the intensive variables associated to $\bar{\boldsymbol{\sigma}}_a$, which is calculated by means of the geometrical relations (subscript G);

$\bar{\boldsymbol{\epsilon}}_{bG}^T = \{\epsilon_{zz} \quad \gamma_{xz} \quad \gamma_{yz}\}_G$ is the vector of the intensive variables associated to $\bar{\boldsymbol{\sigma}}_b$, which is calculated by means of the geometrical relations (subscript G);

$\bar{\boldsymbol{\epsilon}}_{bC}^T = \{\epsilon_{zz} \quad \gamma_{xz} \quad \gamma_{yz}\}_C$ is the vector of intensive variables associated to $\bar{\boldsymbol{\sigma}}_b$, which is calculated by means of the constitutive relations (subscript C).

It is convenient to specify that the primary unknown variables are collected in the vector:

$$\bar{\boldsymbol{U}}^T = \{u \quad v \quad w \quad \Phi \quad \sigma_{zz} \quad \sigma_{xz} \quad \sigma_{yz}\} \tag{8.57}$$

The geometrical relations can be written, as in Equations (2.38)–(2.42), as:

$$\bar{\boldsymbol{\epsilon}}_{aG} = \boldsymbol{D}_a \bar{\boldsymbol{U}} \tag{8.58}$$

$$\bar{\boldsymbol{\epsilon}}_{bG} = \boldsymbol{D}_b \bar{\boldsymbol{U}} \tag{8.59}$$

$$\bar{\boldsymbol{\sigma}}_b = \bar{\boldsymbol{\sigma}}_{bG} = \boldsymbol{D}_{b'} \bar{\boldsymbol{U}} \tag{8.60}$$

where the differential matrices in explicit form read:

$$\boldsymbol{D}_a = \begin{bmatrix} \partial_x & 0 & 0 & 0 & 0 & 0 & 0 \\ 0 & \partial_y & 0 & 0 & 0 & 0 & 0 \\ \partial_y & \partial_x & 0 & 0 & 0 & 0 & 0 \\ 0 & 0 & 0 & -\partial_x & 0 & 0 & 0 \\ 0 & 0 & 0 & -\partial_y & 0 & 0 & 0 \\ 0 & 0 & 0 & -\partial_z & 0 & 0 & 0 \end{bmatrix}, \quad \boldsymbol{D}_b = \begin{bmatrix} 0 & 0 & \partial_z & 0 & 0 & 0 & 0 \\ \partial_z & 0 & \partial_x & 0 & 0 & 0 & 0 \\ 0 & \partial_z & \partial_y & 0 & 0 & 0 & 0 \end{bmatrix}$$

$$\boldsymbol{D}_{b'} = \begin{bmatrix} 0 & 0 & 0 & 0 & 1 & 0 & 0 \\ 0 & 0 & 0 & 0 & 0 & 1 & 0 \\ 0 & 0 & 0 & 0 & 0 & 0 & 1 \end{bmatrix} \tag{8.61}$$

The constitutive relations for RMVT($\boldsymbol{u}$, Φ, $\boldsymbol{\sigma}_n$), as in Equations (2.61)–(2.64), can be summarized as:

$$\bar{\boldsymbol{\sigma}}_C = \hat{\boldsymbol{C}}\, \bar{\boldsymbol{\epsilon}}_G \tag{8.62}$$

where $\bar{\boldsymbol{\sigma}}_C$ is composed of the vector of the extensive non-modeled variables $\bar{\boldsymbol{\sigma}}_{aC}$ and the vector of the intensive variables $\bar{\boldsymbol{\epsilon}}_{bC}$ (which is associated to $\bar{\boldsymbol{\sigma}}_b$); $\bar{\boldsymbol{\epsilon}}_G$ is composed of the vector of the intensive variables $\bar{\boldsymbol{\epsilon}}_{aG}$ (which is associated to $\bar{\boldsymbol{\sigma}}_a$) and the vector of the extensive modeled variables $\bar{\boldsymbol{\sigma}}_b$:

$$\begin{aligned} \bar{\boldsymbol{\sigma}}_C^T &= \{\bar{\boldsymbol{\sigma}}_{aC}^T \bar{\boldsymbol{\epsilon}}_{bC}^T\} \\ \bar{\boldsymbol{\epsilon}}_G^T &= \{\bar{\boldsymbol{\epsilon}}_{aG}^T \bar{\boldsymbol{\sigma}}_{bG}^T\} \end{aligned} \tag{8.63}$$

The physical constitutive matrix $\bar{\boldsymbol{H}}$ can be partitioned by dividing cells into modeled and not modeled quantities:

$$\bar{\boldsymbol{H}} = \begin{Bmatrix} \bar{\boldsymbol{H}}_{aa} & \bar{\boldsymbol{H}}_{ab} \\ \bar{\boldsymbol{H}}_{ba} & \bar{\boldsymbol{H}}_{bb} \end{Bmatrix} \tag{8.64}$$

where $\bar{\boldsymbol{H}}_{ab} = \bar{\boldsymbol{H}}_{ba}^T$.

In explicit form (see also Equation (8.39)) for each k layer:

$$\bar{\boldsymbol{H}}_{aa} = \begin{bmatrix} Q_{11} & Q_{12} & Q_{16} & 0 & 0 & -e_{31} \\ Q_{12} & Q_{22} & Q_{26} & 0 & 0 & -e_{32} \\ Q_{16} & Q_{26} & Q_{66} & 0 & 0 & -e_{36} \\ 0 & 0 & 0 & -\varepsilon_{11} & -\varepsilon_{12} & 0 \\ 0 & 0 & 0 & -\varepsilon_{12} & -\varepsilon_{22} & 0 \\ -e_{31} & -e_{32} & -e_{36} & 0 & 0 & -\varepsilon_{33} \end{bmatrix}$$

$$\bar{\boldsymbol{H}}_{ab} = \begin{bmatrix} Q_{13} & 0 & 0 \\ Q_{23} & 0 & 0 \\ Q_{36} & 0 & 0 \\ 0 & -e_{15} & -e_{14} \\ 0 & -e_{25} & -e_{24} \\ -e_{33} & 0 & 0 \end{bmatrix} \tag{8.65}$$

$$\bar{\boldsymbol{H}}_{ba} = \begin{bmatrix} Q_{13} & Q_{23} & Q_{36} & 0 & 0 & -e_{33} \\ 0 & 0 & 0 & -e_{15} & -e_{25} & 0 \\ 0 & 0 & 0 & -e_{14} & -e_{24} & 0 \end{bmatrix}, \quad \bar{\boldsymbol{H}}_{bb} = \begin{bmatrix} Q_{33} & 0 & 0 \\ 0 & Q_{55} & Q_{45} \\ 0 & Q_{45} & Q_{44} \end{bmatrix}$$

The physical constitutive relations can be arranged according to the above partitioning (see also Equations (8.40) and (2.61)–(2.64)):

$$\bar{\boldsymbol{\sigma}}_{aC} = \bar{\boldsymbol{H}}_{aa}\bar{\boldsymbol{\epsilon}}_{aG} + \bar{\boldsymbol{H}}_{ab}\bar{\boldsymbol{\epsilon}}_{bG}, \quad \bar{\boldsymbol{\sigma}}_{bC} = \bar{\boldsymbol{H}}_{ba}\bar{\boldsymbol{\epsilon}}_{aG} + \bar{\boldsymbol{H}}_{bb}\bar{\boldsymbol{\epsilon}}_{bG} \tag{8.66}$$

From Equations (8.66), considering vectors $\bar{\boldsymbol{\sigma}}_C$ and $\bar{\boldsymbol{\epsilon}}_G$ in Equations (8.63), it is possible to write:

$$\bar{\boldsymbol{\sigma}}_{aC} = \hat{\boldsymbol{C}}_{aa}\bar{\boldsymbol{\epsilon}}_{aG} + \hat{\boldsymbol{C}}_{ab}\bar{\boldsymbol{\sigma}}_{bG}, \quad \bar{\boldsymbol{\epsilon}}_{bC} = \hat{\boldsymbol{C}}_{ba}\bar{\boldsymbol{\epsilon}}_{aG} + \hat{\boldsymbol{C}}_{bb}\bar{\boldsymbol{\sigma}}_{bG} \tag{8.67}$$

with:

$$\begin{aligned} \hat{\boldsymbol{C}}_{aa} &= \bar{\boldsymbol{H}}_{aa} - \bar{\boldsymbol{H}}_{ab}(\bar{\boldsymbol{H}}_{bb})^{-1}\bar{\boldsymbol{H}}_{ba}, \quad \hat{\boldsymbol{C}}_{ab} = \bar{\boldsymbol{H}}_{ab}(\bar{\boldsymbol{H}}_{bb})^{-1} \\ \hat{\boldsymbol{C}}_{ba} &= -(\bar{\boldsymbol{H}}_{bb})^{-1}\bar{\boldsymbol{H}}_{ba}, \quad \hat{\boldsymbol{C}}_{bb} = (\bar{\boldsymbol{H}}_{bb})^{-1} \end{aligned} \tag{8.68}$$

Matrix $\hat{C}$ in Equation (8.62) can be written as:

$$\hat{C} = \begin{Bmatrix} \hat{C}_{aa} & \hat{C}_{ab} \\ \hat{C}_{ba} & \hat{C}_{bb} \end{Bmatrix} \tag{8.69}$$

The explicit form of matrices $\hat{C}_{aa}$, $\hat{C}_{ba}$, $\hat{C}_{ab}$, and $\hat{C}_{bb}$ is given below for each k layer:

$$\hat{C}_{aa11} = Q_{11} - \frac{Q_{13}^2}{Q_{33}}, \quad \hat{C}_{aa12} = Q_{12} - \frac{Q_{13}Q_{23}}{Q_{33}}, \quad \hat{C}_{aa13} = Q_{16} - \frac{Q_{13}Q_{36}}{Q_{33}}$$

$$\hat{C}_{aa14} = 0, \quad \hat{C}_{aa15} = 0, \quad \hat{C}_{aa16} = -e_{31} + \frac{Q_{13}e_{33}}{Q_{33}}$$

$$\hat{C}_{aa21} = Q_{12} - \frac{Q_{13}e_{23}}{Q_{33}}, \quad \hat{C}_{aa22} = Q_{22} - \frac{Q_{23}^2}{Q_{33}}, \quad \hat{C}_{aa23} = Q_{26} - \frac{Q_{23}Q_{36}}{Q_{33}}$$

$$\hat{C}_{aa24} = 0, \quad \hat{C}_{aa25} = 0, \quad \hat{C}_{aa26} = -e_{32} + \frac{Q_{23}e_{33}}{Q_{33}}$$

$$\hat{C}_{aa31} = Q_{16} - \frac{Q_{13}Q_{36}}{Q_{33}}, \quad \hat{C}_{aa32} = Q_{26} - \frac{Q_{23}Q_{36}}{Q_{33}}, \quad \hat{C}_{aa33} = Q_{66} - \frac{Q_{36}^2}{Q_{33}}$$

$$\hat{C}_{aa34} = 0, \quad \hat{C}_{aa35} = 0, \quad \hat{C}_{aa36} = \frac{Q_{36}e_{33}}{Q_{33}} - e_{36}, \quad \hat{C}_{aa41} = 0$$

$$\hat{C}_{aa42} = 0, \quad \hat{C}_{aa43} = 0, \quad \hat{C}_{aa44} = \frac{Q_{55}e_{14}^2 - 2Q_{45}e_{14}e_{15} + Q_{44}e_{15}^2}{Q_{45}^2 - Q_{44}Q_{55}} - \varepsilon_{11}$$

$$\hat{C}_{aa45} = \frac{Q_{55}e_{14}e_{24} + Q_{44}e_{15}e_{25} - Q_{45}(e_{15}e_{24} + e_{14}e_{25})}{Q_{45}^2 - Q_{44}Q_{55}} - \varepsilon_{12}, \quad \hat{C}_{aa46} = 0$$

$$\hat{C}_{aa51} = 0, \quad \hat{C}_{aa52} = 0, \quad \hat{C}_{aa53} = 0$$

$$\hat{C}_{aa54} = \frac{Q_{55}e_{14}e_{24} + Q_{44}e_{15}e_{25} - Q_{45}(e_{15}e_{24} + e_{14}e_{25})}{Q_{45}^2 - Q_{44}Q_{55}} - \varepsilon_{12}$$

$$\hat{C}_{aa55} = \frac{Q_{55}e_{24}^2 - 2Q_{45}e_{24}e_{25} + Q_{44}e_{25}^2}{Q_{45}^2 - Q_{44}Q_{55}} - \varepsilon_{22}, \quad \hat{C}_{aa56} = 0$$

$$\hat{C}_{aa61} = -e_{31} + \frac{Q_{13}e_{33}}{Q_{33}}, \quad \hat{C}_{aa62} = -e_{32} + \frac{Q_{23}e_{33}}{Q_{33}}, \quad \hat{C}_{aa63} = \frac{Q_{36}e_{33}}{Q_{33}} - e_{36}$$

$$\hat{C}_{aa64} = 0, \quad \hat{C}_{aa65} = 0, \quad \hat{C}_{aa66} = -\frac{e_{33}^2}{Q_{33}} - \varepsilon_{33} \tag{8.70}$$

$$\hat{\boldsymbol{C}}_{ab} = \begin{bmatrix} \dfrac{Q_{13}}{Q_{33}} & 0 & 0 \\ \dfrac{Q_{23}}{Q_{33}} & 0 & 0 \\ \dfrac{Q_{36}}{Q_{33}} & 0 & 0 \\ 0 & \dfrac{-Q_{45}e_{14} - Q_{44}e_{15}}{Q_{45}^2 - Q_{44}Q_{55}} & \dfrac{-Q_{55}e_{14} - Q_{45}e_{15}}{Q_{15}^2 - Q_{45}Q_{55}} \\ 0 & \dfrac{-Q_{45}e_{24} - Q_{44}e_{25}}{Q_{45}^2 - Q_{44}Q_{55}} & \dfrac{-Q_{55}e_{24} - Q_{45}e_{25}}{Q_{45}^2 - Q_{44}Q_{55}} \\ -\dfrac{e_{33}}{Q_{33}} & 0 & 0 \end{bmatrix} \tag{8.71}$$

$$\hat{\boldsymbol{C}}_{ba} = \begin{bmatrix} \dfrac{Q_{13}}{Q_{33}} & \dfrac{Q_{23}}{Q_{33}} & \dfrac{Q_{36}}{Q_{33}} & 0 & 0 & -\dfrac{e_{33}}{Q_{33}} \\ 0 & 0 & 0 & \dfrac{-Q_{45}e_{14} - Q_{44}e_{15}}{Q_{45}^2 - Q_{44}Q_{55}} & \dfrac{-Q_{45}e_{24} - Q_{44}e_{25}}{Q_{44}^2 - Q_{44}Q_{55}} & 0 \\ 0 & 0 & 0 & \dfrac{-Q_{55}e_{14} - Q_{45}e_{15}}{Q_{15}^2 - Q_{44}Q_{55}} & \dfrac{-Q_{55}e_{24} - Q_{45}e_{25}}{Q_{45}^2 - Q_{44}Q_{55}} & 0 \end{bmatrix} \tag{8.72}$$

$$\hat{\boldsymbol{C}}_{bb} = \begin{bmatrix} \dfrac{1}{Q_{33}} & 0 & 0 \\ 0 & \dfrac{1}{-Q_{44}Q_{45}^2 + Q_{55}} & \dfrac{1}{Q_{45}^2 - Q_{44}Q_{55}} \\ 0 & \dfrac{1}{Q_{45}^2 - Q_{44}Q_{55}} & \dfrac{1}{Q_{44} - \dfrac{Q_{45}^2}{Q_{55}}} \end{bmatrix} \tag{8.73}$$

In analogy with PVD($\boldsymbol{u}$,Φ) in Equation (8.36), by substituting the constitutive and geometrical relations in the RMVT($\boldsymbol{u}$,Φ,$\boldsymbol{\sigma}_n$) variational statement in Equation (8.56), and referring to a multilayered structure by employing thickness functions for the kinematic description and shape functions for FE discretization, we obtain the following fundamental stiffness nucleus $\bar{\boldsymbol{K}}^{(e)k\tau sij}$ at the element level. The FE approximation for the primary unknowns $\bar{\boldsymbol{U}}^T = \{u\ v\ w\ \Phi\ \sigma_{zz}\ \sigma_{xz}\ \sigma_{yz}\}$ and their virtual variation $\delta\bar{\boldsymbol{U}}$ is:

$$\bar{\boldsymbol{U}}_\tau^{(e)}(x, y) = N_i^{(e)}(\xi, \eta)\, \bar{\boldsymbol{Q}}_{\tau i}^{(e)} \tag{8.74}$$

$$\delta\bar{\boldsymbol{U}}_s^{(e)}(x, y) = N_j^{(e)}(\xi, \eta)\, \delta\bar{\boldsymbol{Q}}_{sj}^{(e)} \tag{8.75}$$

where the nodal values are $\bar{Q}^T = \{Q_u \; Q_\Phi \; Q_{\sigma_n}\}$. The explicit form of the fundamental stiffness nucleus at the element level is:

$$\bar{K}^{(e)k\tau sij} = \begin{bmatrix} K_{11} & K_{12} & K_{13} & K_{14} & K_{15} & K_{16} & K_{17} \\ K_{21} & K_{22} & K_{23} & K_{24} & K_{25} & K_{26} & K_{27} \\ K_{31} & K_{32} & K_{33} & K_{34} & K_{35} & K_{36} & K_{37} \\ K_{41} & K_{42} & K_{43} & K_{44} & K_{45} & K_{46} & K_{47} \\ K_{51} & K_{52} & K_{53} & K_{54} & K_{55} & K_{56} & K_{57} \\ K_{61} & K_{62} & K_{63} & K_{64} & K_{65} & K_{66} & K_{67} \\ K_{71} & K_{72} & K_{73} & K_{74} & K_{75} & K_{76} & K_{77} \end{bmatrix}^{k\tau sij} \tag{8.76}$$

with:

$$\begin{aligned}
K_{11}^{k\tau sij} &= \triangleleft F_s F_\tau \triangleright_A \hat{C}_{aa11} \triangleleft N_{i,x} N_{j,x} \triangleright_\Omega + \triangleleft F_s F_\tau \triangleright_A \hat{C}_{aa31} \triangleleft N_{i,y} N_{j,x} \triangleright_\Omega \\
&\quad + \triangleleft F_s F_\tau \triangleright_A \hat{C}_{aa13} \triangleleft N_{i,x} N_{j,y} \triangleright_\Omega + \triangleleft F_s F_\tau \triangleright_A \hat{C}_{aa33} \triangleleft N_{i,y} N_{j,y} \triangleright_\Omega \\
K_{21}^{k\tau sij} &= \triangleleft F_s F_\tau \triangleright_A \hat{C}_{aa31} \triangleleft N_{i,x} N_{j,x} \triangleright_\Omega + \triangleleft F_s F_\tau \triangleright_A \hat{C}_{aa21} \triangleleft N_{i,y} N_{j,x} \triangleright_\Omega \\
&\quad + \triangleleft F_s F_\tau \triangleright_A \hat{C}_{aa33} \triangleleft N_{i,x} N_{j,y} \triangleright_\Omega + \triangleleft F_s F_\tau \triangleright_A \hat{C}_{aa23} \triangleleft N_{i,y} N_{j,y} \triangleright_\Omega \\
K_{31}^{k\tau sij} &= 0 \\
K_{41}^{k\tau sij} &= -\triangleleft F_s F_{\tau,z} \triangleright_A \hat{C}_{aa61} \triangleleft N_i N_{j,x} \triangleright_\Omega - \triangleleft F_s F_{\tau,z} \triangleright_A \hat{C}_{aa63} \triangleleft N_i N_{j,y} \triangleright_\Omega \\
K_{51}^{k\tau sij} &= -\triangleleft F_s F_\tau \triangleright_A \hat{C}_{ba11} \triangleleft N_i N_{j,x} \triangleright_\Omega - \triangleleft F_s F_\tau \triangleright_A \hat{C}_{ba13} \triangleleft N_i N_{j,y} \triangleright_\Omega \\
K_{61}^{k\tau sij} &= \triangleleft F_{s,z} F_\tau \triangleright_A \triangleleft N_i N_j \triangleright_\Omega, \quad K_{71}^{k\tau sij} = 0 \\
K_{12}^{k\tau sij} &= \triangleleft F_s F_\tau \triangleright_A \hat{C}_{aa13} \triangleleft N_{i,x} N_{j,x} \triangleright_\Omega + \triangleleft F_s F_\tau \triangleright_A \hat{C}_{aa33} \triangleleft N_{i,y} N_{j,x} \triangleright_\Omega \\
&\quad + \triangleleft F_s F_\tau \triangleright_A \hat{C}_{aa12} \triangleleft N_{i,x} N_{j,y} \triangleright_\Omega + \triangleleft F_s F_\tau \triangleright_A \hat{C}_{aa32} \triangleleft N_{i,y} N_{j,y} \triangleright_\Omega \\
K_{22}^{k\tau sij} &= \triangleleft F_s F_\tau \triangleright_A \hat{C}_{aa33} \triangleleft N_{i,x} N_{j,x} \triangleright_\Omega + \triangleleft F_s F_\tau \triangleright_A \hat{C}_{aa23} \triangleleft N_{i,y} N_{j,x} \triangleright_\Omega \\
&\quad + \triangleleft F_s F_\tau \triangleright_A \hat{C}_{aa32} \triangleleft N_{i,x} N_{j,y} \triangleright_\Omega + \triangleleft F_s F_\tau \triangleright_A \hat{C}_{aa22} \triangleleft N_{i,y} N_{j,y} \triangleright_\Omega \\
K_{32}^{k\tau sij} &= 0 \\
K_{42}^{k\tau sij} &= -\triangleleft F_s F_{\tau,z} \triangleright_A \hat{C}_{aa63} \triangleleft N_i N_{j,x} \triangleright_\Omega - \triangleleft F_s F_{\tau,z} \triangleright_A \hat{C}_{aa62} \triangleleft N_i N_{j,y} \triangleright_\Omega \\
K_{52}^{k\tau sij} &= -\triangleleft F_s F_\tau \triangleright_A \hat{C}_{ba13} \triangleleft N_i N_{j,x} \triangleright_\Omega - \triangleleft F_s F_\tau \triangleright_A \hat{C}_{ba12} \triangleleft N_i N_{j,y} \triangleright_\Omega \\
K_{62}^{k\tau sij} &= 0, \quad K_{72}^{k\tau sij} = \triangleleft F_{s,z} F_\tau \triangleright_A \triangleleft N_i N_j \triangleright_\Omega, \quad K_{13}^{k\tau sij} = 0 \\
K_{23}^{k\tau sij} &= 0, \quad K_{33}^{k\tau sij} = 0, \quad K_{43}^{k\tau sij} = 0, \quad K_{53}^{k\tau sij} = \triangleleft F_{s,z} F_\tau \triangleright_A \triangleleft N_i N_j \triangleright_\Omega \\
K_{63}^{k\tau sij} &= \triangleleft F_s F_\tau \triangleright_A \triangleleft N_i N_{j,x} \triangleright_\Omega, \quad K_{73}^{k\tau sij} = \triangleleft F_s F_\tau \triangleright_A \triangleleft N_i N_{j,y} \triangleright_\Omega
\end{aligned}$$

$$
\begin{aligned}
K_{14}^{k\tau sij} &= -\triangleleft F_{s,z}F_{\tau}\triangleright_A \hat{C}_{aa16}\triangleleft N_{i,x}N_j\triangleright_\Omega - \triangleleft F_{s,z}F_{\tau}\triangleright_A \hat{C}_{aa36}\triangleleft N_{i,y}N_j\triangleright_\Omega \\
K_{24}^{k\tau sij} &= -\triangleleft F_{s,z}F_{\tau}\triangleright_A \hat{C}_{aa36}\triangleleft N_{i,x}N_j\triangleright_\Omega - \triangleleft F_{s,z}F_{\tau}\triangleright_A \hat{C}_{aa26}\triangleleft N_{i,y}N_j\triangleright_\Omega \\
K_{34}^{k\tau sij} &= 0 \qquad (8.77)\\
K_{44}^{k\tau sij} &= \triangleleft F_{s,z}F_{\tau,z}\triangleright_A \hat{C}_{aa66}\triangleleft N_iN_j\triangleright_\Omega + \triangleleft F_{s}F_{\tau}\triangleright_A \hat{C}_{aa44}\triangleleft N_{i,x}N_{j,x}\triangleright_\Omega \\
&\quad + \triangleleft F_{s}F_{\tau}\triangleright_A \hat{C}_{aa54}\triangleleft N_{i,y}N_{j,x}\triangleright_\Omega + \triangleleft F_{s}F_{\tau}\triangleright_A \hat{C}_{aa45}\triangleleft N_{i,x}N_{j,y}\triangleright_\Omega \\
&\quad + \triangleleft F_{s}F_{\tau}\triangleright_A \hat{C}_{aa55}\triangleleft N_{i,y}N_{j,y}\triangleright_\Omega \\
K_{54}^{k\tau sij} &= \triangleleft F_{s,z}F_{\tau}\triangleright_A \hat{C}_{ba16}\triangleleft N_iN_j\triangleright_\Omega \\
K_{64}^{k\tau sij} &= \triangleleft F_{s}F_{\tau}\triangleright_A \hat{C}_{ba24}\triangleleft N_iN_{j,x}\triangleright_\Omega + \triangleleft F_{s}F_{\tau}\triangleright_A \hat{C}_{ba25}\triangleleft N_iN_{j,y}\triangleright_\Omega \\
K_{74}^{k\tau sij} &= \triangleleft F_{s}F_{\tau}\triangleright_A \hat{C}_{ba34}\triangleleft N_iN_{j,x}\triangleright_\Omega + \triangleleft F_{s}F_{\tau}\triangleright_A \hat{C}_{ba35}\triangleleft N_iN_{j,y}\triangleright_\Omega \\
K_{15}^{k\tau sij} &= \triangleleft F_{s}F_{\tau}\triangleright_A \hat{C}_{\mathrm{ab}11}\triangleleft N_{i,x}N_j\triangleright_\Omega + \triangleleft F_{s}F_{\tau}\triangleright_A \hat{C}_{\mathrm{ab}31}\triangleleft N_{i,y}N_j\triangleright_\Omega \\
K_{25}^{k\tau sij} &= \triangleleft F_{s}F_{\tau}\triangleright_A \hat{C}_{\mathrm{ab}31}\triangleleft N_{i,x}N_j\triangleright_\Omega + \triangleleft F_{s}F_{\tau}\triangleright_A \hat{C}_{\mathrm{ab}21}\triangleleft N_{i,y}N_j\triangleright_\Omega \\
K_{35}^{k\tau sij} &= \triangleleft F_{s}F_{\tau,z}\triangleright_A \triangleleft N_iN_j\triangleright_\Omega, \quad K_{45}^{k\tau sij} = -\triangleleft F_{s}F_{\tau,z}\triangleright_A \hat{C}_{\mathrm{ab}61}\triangleleft N_iN_j\triangleright_\Omega \\
K_{55}^{k\tau sij} &= -\triangleleft F_{s}F_{\tau}\triangleright_A \hat{C}_{\mathrm{bb}11}\triangleleft N_iN_j\triangleright_\Omega \\
K_{65}^{k\tau sij} &= 0, \quad K_{75}^{k\tau sij} = 0, \quad K_{16}^{k\tau sij} = \triangleleft F_{s}F_{\tau,z}\triangleright_A \triangleleft N_iN_j\triangleright_\Omega \\
K_{26}^{k\tau sij} &= 0, \quad K_{36}^{k\tau sij} = \triangleleft F_{s}F_{\tau}\triangleright_A \triangleleft N_{i,x}N_j\triangleright_\Omega \\
K_{46}^{k\tau sij} &= -\triangleleft F_{s}F_{\tau}\triangleright_A \hat{C}_{\mathrm{ab}42}\triangleleft N_{i,x}N_j\triangleright_\Omega - \triangleleft F_{s}F_{\tau}\triangleright_A \hat{C}_{\mathrm{ab}52}\triangleleft N_{i,y}N_j\triangleright_\Omega \\
K_{56}^{k\tau sij} &= 0, \quad K_{66}^{k\tau sij} = -\triangleleft F_{s}F_{\tau}\triangleright_A \hat{C}_{\mathrm{bb}22}\triangleleft N_iN_j\triangleright_\Omega \\
K_{76}^{k\tau sij} &= -\triangleleft F_{s}F_{\tau}\triangleright_A \hat{C}_{\mathrm{bb}32}\triangleleft N_iN_j\triangleright_\Omega, \quad K_{17}^{k\tau sij} = 0 \\
K_{27}^{k\tau sij} &= \triangleleft F_{s}F_{\tau,z}\triangleright_A \triangleleft N_iN_j\triangleright_\Omega, \quad K_{37}^{k\tau sij} = \triangleleft F_{s}F_{\tau}\triangleright_A \triangleleft N_{i,y}N_j\triangleright_\Omega \\
K_{47}^{k\tau sij} &= -\triangleleft F_{s}F_{\tau}\triangleright_A \hat{C}_{\mathrm{ab}43}\triangleleft N_{i,x}N_j\triangleright_\Omega - \triangleleft F_{s}F_{\tau}\triangleright_A \hat{C}_{\mathrm{ab}53}\triangleleft N_{i,y}N_j\triangleright_\Omega \\
K_{57}^{k\tau sij} &= 0, \quad K_{67}^{k\tau sij} = -\triangleleft F_{s}F_{\tau}\triangleright_A \hat{C}_{\mathrm{bb}23}\triangleleft N_iN_j\triangleright_\Omega \\
K_{77}^{k\tau sij} &= -\triangleleft F_{s}F_{\tau}\triangleright_A \hat{C}_{\mathrm{bb}33}\triangleleft N_iN_j\triangleright_\Omega
\end{aligned}
$$

The subscripts after the commas indicate derivatives. The following integrals are also defined in the in-plane and thickness directions, respectively:

$$
\triangleleft(\cdots)\triangleright_\Omega = \int_{\Omega_0^{(e)}} (\cdots)\, d\Omega^{(e)} \quad \text{and} \quad \triangleleft(\cdots)\triangleright_A = \int_{A^{(e)k}} (\cdots)\, dz \qquad (8.78)
$$

where $\Omega_0^{(e)}$ is the midsurface element. The non-zero elements of the fundamental mass nucleus $\boldsymbol{M}^{(e)k\tau sij}$ are (in analogy with the previous sections):

$$M_{11}^{k\tau sij} = M_{22}^{k\tau sij} = M_{33}^{k\tau sij} = \rho^k \triangleleft N_i N_j \triangleright_\Omega \triangleleft F_\tau F_s \triangleright_A \tag{8.79}$$

The obtained fundamental nuclei are used in the governing equations, which do not formally change with respect to the PVD($\boldsymbol{u}$,Φ) case (see Equation (8.47)). The pure mechanical problem, corresponding to RMVT($\boldsymbol{u}$, $\boldsymbol{\sigma}_n$), can be obtained as a particular case of RMVT($\boldsymbol{u}$, Φ, $\boldsymbol{\sigma}_n$), coherently with what has already been explained in Section 6.9 for the analytical solution.

8.5 RMVT($\boldsymbol{u}$, $\boldsymbol{\Phi}$, $\boldsymbol{\mathcal{D}}_n$) for the electromechanical plate case

Another possible extension of the RMVT (Reissner 1984) has been indicated in Equations (2.74) and (6.103), where the internal electrical work has been added and the Lagrange multiplier is employed in order to increase the accuracy of the computed transverse normal electric displacement (Carrera and Brischetto 2007a; Carrera *et al.* 2008):

$$\int_V \Big(\delta\boldsymbol{\epsilon}_{pG}^T \boldsymbol{\sigma}_{pC} + \delta\boldsymbol{\epsilon}_{nG}^T \boldsymbol{\sigma}_{nC} - \delta\boldsymbol{\mathcal{E}}_{pG}^T \boldsymbol{\mathcal{D}}_{pC} - \delta\boldsymbol{\mathcal{E}}_{nG}^T \boldsymbol{\mathcal{D}}_{nM}$$
$$- \delta\boldsymbol{\mathcal{D}}_{nM}^T(\boldsymbol{\mathcal{E}}_{nG} - \boldsymbol{\mathcal{E}}_{nC})\Big)dV = \delta L_e - \delta L_{in} \tag{8.80}$$

On the basis of a condensed notation, Equation (8.80) can be rewritten as:

$$\int_V \left(\delta\bar{\boldsymbol{\epsilon}}_{aG}^T\, \bar{\boldsymbol{\sigma}}_{aC} + \delta\bar{\boldsymbol{\epsilon}}_{bG}^T\, \bar{\boldsymbol{\sigma}}_b + \delta\bar{\boldsymbol{\sigma}}_b^T(\bar{\boldsymbol{\epsilon}}_{bG} - \bar{\boldsymbol{\epsilon}}_{bC})\right) dV = \delta\bar{L}_e - \delta L_{in} \tag{8.81}$$

Equation (8.81) is formally identical to Equation (8.56), but the meaning of the arrays changes, therefore the following arrays can be introduced:

$\bar{\boldsymbol{\sigma}}_{aC}^T = \{\sigma_{xx} \quad \sigma_{yy} \quad \sigma_{xy} \quad -\mathcal{D}_x \quad -\mathcal{D}_y \quad \sigma_{zz} \quad \sigma_{xz} \quad \sigma_{yz}\}_C$ is the vector of the extensive non-modeled variables, which are calculated by means of the constitutive relations (subscript C);

$\bar{\boldsymbol{\sigma}}_b^T = \{-\mathcal{D}_z\}$ is the vector of extensive modeled variables;

$\bar{\boldsymbol{\epsilon}}_{aG}^T = \{\epsilon_{xx} \quad \epsilon_{yy} \quad \gamma_{xy} \quad \mathcal{E}_x \quad \mathcal{E}_y \quad \epsilon_{zz} \quad \gamma_{xz} \quad \gamma_{yz}\}_G$ is the vector of the intensive variables associated to $\bar{\boldsymbol{\sigma}}_a$, which is calculated by means of the geometrical relations (subscript G);

$\bar{\boldsymbol{\epsilon}}_{bG}^T = \{\mathcal{E}_z\}_G$ is the vector of the intensive variables associated to $\bar{\boldsymbol{\sigma}}_b$, which is calculated by means of the geometrical relations (subscript G);

$\bar{\boldsymbol{\epsilon}}_{bC}^T = \{\mathcal{E}_z\}_C$ is the vector of the intensive variables associated to $\bar{\boldsymbol{\sigma}}_b$, which is calculated by means of the constitutive relations (subscript C).

It is convenient to specify that the primary unknown variables are collected in the vector:

$$\bar{\boldsymbol{U}}^T = \{u \quad v \quad w \quad \Phi \quad \mathcal{D}_z\} \tag{8.82}$$

The geometrical relations, can be written, as in Equations (2.38)–(2.42), as:

$$\bar{\boldsymbol{\epsilon}}_{aG} = \boldsymbol{D}_a \bar{\boldsymbol{U}} \tag{8.83}$$

$$\bar{\boldsymbol{\epsilon}}_{bG} = \boldsymbol{D}_b \bar{\boldsymbol{U}} \tag{8.84}$$

$$\bar{\boldsymbol{\sigma}}_b = \bar{\boldsymbol{\sigma}}_{bG} = \boldsymbol{D}_{b'} \bar{\boldsymbol{U}} \tag{8.85}$$

which are formally identical to Equations (8.58)–(8.60), while the differential matrices in explicit form read:

$$\boldsymbol{D}_a = \begin{bmatrix} \partial_x & 0 & 0 & 0 & 0 \\ 0 & \partial_y & 0 & 0 & 0 \\ \partial_y & \partial_x & 0 & 0 & 0 \\ 0 & 0 & 0 & -\partial_x & 0 \\ 0 & 0 & 0 & -\partial_y & 0 \\ 0 & 0 & \partial_z & 0 & 0 \\ \partial_z & 0 & \partial_x & 0 & 0 \\ 0 & \partial_z & \partial_y & 0 & 0 \end{bmatrix}, \quad \boldsymbol{D}_b = (0 \quad 0 \quad 0 \quad -\partial_z \quad 0)$$

$$\boldsymbol{D}_{b'} = (0 \quad 0 \quad 0 \quad 0 \quad -1) \tag{8.86}$$

The constitutive relations for RMVT($\boldsymbol{u}$, Φ, $\mathcal{D}_n$) can be summarized, as in Equations (2.75)–(2.78), as:

$$\bar{\boldsymbol{\sigma}}_C = \bar{\boldsymbol{C}}\,\bar{\boldsymbol{\epsilon}}_G \tag{8.87}$$

where $\bar{\boldsymbol{\sigma}}_C$ is composed of the vector of extensive non-modeled variables $\bar{\boldsymbol{\sigma}}_{aC}$ and the vector of the intensive variables $\bar{\boldsymbol{\epsilon}}_{bC}$ (which is associated to $\bar{\boldsymbol{\sigma}}_b$); $\bar{\boldsymbol{\epsilon}}_G$ is composed of the vector of the intensive variables $\bar{\boldsymbol{\epsilon}}_{aG}$ (which is associated to $\bar{\boldsymbol{\sigma}}_a$) and the vector of the modeled extensive variables $\bar{\boldsymbol{\sigma}}_{bG}$:

$$\begin{aligned} \bar{\boldsymbol{\sigma}}_C^T &= \left\{\bar{\boldsymbol{\sigma}}_{aC}^T \bar{\boldsymbol{\epsilon}}_{bC}^T\right\} \\ \bar{\boldsymbol{\epsilon}}_G^T &= \left\{\bar{\boldsymbol{\epsilon}}_{aG}^T \bar{\boldsymbol{\sigma}}_{bG}^T\right\} \end{aligned} \tag{8.88}$$

The physical constitutive matrix $\bar{\boldsymbol{H}}$ can be partitioned by dividing the cells into modeled and non-modeled quantities:

$$\bar{\boldsymbol{H}} = \begin{Bmatrix} \bar{\boldsymbol{H}}_{aa} & \bar{\boldsymbol{H}}_{ab} \\ \bar{\boldsymbol{H}}_{ba} & \bar{\boldsymbol{H}}_{bb} \end{Bmatrix} \tag{8.89}$$

where $\bar{\boldsymbol{H}}_{ab} = \bar{\boldsymbol{H}}_{ba}^T$.

In explicit form (see also Equation (8.39)):

$$\bar{\boldsymbol{H}}_{aa} = \begin{bmatrix} Q_{11}^k & Q_{12}^k & Q_{16}^k & 0 & 0 & Q_{13}^k & 0 & 0 \\ Q_{12}^k & Q_{22}^k & Q_{26}^k & 0 & 0 & Q_{23}^k & 0 & 0 \\ Q_{16}^k & Q_{26}^k & Q_{66}^k & 0 & 0 & Q_{36}^k & 0 & 0 \\ 0 & 0 & 0 & -\varepsilon_{11}^k & -\varepsilon_{12}^k & 0 & -e_{15}^k & -e_{14}^k \\ 0 & 0 & 0 & -\varepsilon_{12}^k & -\varepsilon_{22}^k & 0 & -e_{25}^k & -e_{24}^k \\ Q_{13}^k & Q_{23}^k & Q_{36}^k & 0 & 0 & Q_{33}^k & 0 & 0 \\ 0 & 0 & 0 & -e_{15}^k & -e_{25}^k & 0 & Q_{55}^k & Q_{45}^k \\ 0 & 0 & 0 & -e_{14}^k & -e_{24}^k & 0 & Q_{45}^k & Q_{44}^k \end{bmatrix}$$

$$\bar{\boldsymbol{H}}_{ab} = \begin{bmatrix} -e_{31}^k \\ -e_{32}^k \\ -e_{36}^k \\ 0 \\ 0 \\ -e_{33}^k \\ 0 \\ 0 \end{bmatrix} \tag{8.90}$$

$$\bar{\boldsymbol{H}}_{ba} = \begin{pmatrix} -e_{31}^k & -e_{32}^k & -e_{36}^k & 0 & 0 & -e_{33}^k & 0 & 0 \end{pmatrix}, \quad \bar{\boldsymbol{H}}_{bb} = \left(-\varepsilon_{33}^k\right)$$

The physical constitutive relations can be arranged according to the above partitioning (see also Equations (8.40) and (2.75)–(2.78)):

$$\bar{\boldsymbol{\sigma}}_{aC} = \bar{\boldsymbol{H}}_{aa}\bar{\boldsymbol{\epsilon}}_{aG} + \bar{\boldsymbol{H}}_{ab}\bar{\boldsymbol{\epsilon}}_{bG}, \quad \bar{\boldsymbol{\sigma}}_{bC} = \bar{\boldsymbol{H}}_{ba}\bar{\boldsymbol{\epsilon}}_{aG} + \bar{\boldsymbol{H}}_{bb}\bar{\boldsymbol{\epsilon}}_{bG} \tag{8.91}$$

From Equations (8.91), considering vectors $\bar{\boldsymbol{\sigma}}_C$ and $\bar{\boldsymbol{\epsilon}}_G$ of Equations (8.88), it is possible to write:

$$\bar{\boldsymbol{\sigma}}_{aC} = \bar{\boldsymbol{C}}_{aa}\bar{\boldsymbol{\epsilon}}_{aG} + \bar{\boldsymbol{C}}_{ab}\bar{\boldsymbol{\sigma}}_{bG}, \quad \bar{\boldsymbol{\epsilon}}_{bC} = \bar{\boldsymbol{C}}_{ba}\bar{\boldsymbol{\epsilon}}_{aG} + \bar{\boldsymbol{C}}_{bb}\bar{\boldsymbol{\sigma}}_{bG} \tag{8.92}$$

with:

$$\begin{aligned} \bar{\boldsymbol{C}}_{aa} &= \bar{\boldsymbol{H}}_{aa} - \bar{\boldsymbol{H}}_{ab}(\bar{\boldsymbol{H}}_{bb})^{-1}\bar{\boldsymbol{H}}_{ba}, \quad \bar{\boldsymbol{C}}_{ab} = \bar{\boldsymbol{H}}_{ab}(\bar{\boldsymbol{H}}_{bb})^{-1} \\ \bar{\boldsymbol{C}}_{ba} &= -(\bar{\boldsymbol{H}}_{bb})^{-1}\bar{\boldsymbol{H}}_{ba}, \quad \bar{\boldsymbol{C}}_{bb} = (\bar{\boldsymbol{H}}_{bb})^{-1} \end{aligned} \tag{8.93}$$

Matrix $\bar{\boldsymbol{C}}$ in Equation (8.87) can be written as:

$$\bar{\boldsymbol{C}} = \begin{Bmatrix} \bar{\boldsymbol{C}}_{aa} & \bar{\boldsymbol{C}}_{ab} \\ \bar{\boldsymbol{C}}_{ba} & \bar{\boldsymbol{C}}_{bb} \end{Bmatrix} \tag{8.94}$$

The explicit form of matrices $\bar{\boldsymbol{C}}_{aa}$, $\bar{\boldsymbol{C}}_{ba}$, $\bar{\boldsymbol{C}}_{ab}$, and $\bar{\boldsymbol{C}}_{bb}$ is given below for each k layer:

$$\bar{\boldsymbol{C}}_{aa} = \left[\begin{matrix} Q_{11}+e_{31}^2/\varepsilon_{33} & Q_{12}+e_{31}e_{32}/\varepsilon_{33} & Q_{16}+e_{31}e_{36}/\varepsilon_{33} & 0 \\ Q_{12}+e_{31}e_{32}/\varepsilon_{33} & Q_{22}+e_{32}^2/\varepsilon_{33} & Q_{26}+e_{32}e_{36}/\varepsilon_{33} & 0 \\ Q_{16}+e_{31}e_{36}/\varepsilon_{33} & Q_{26}+e_{32}e_{36}/\varepsilon_{33} & Q_{66}+e_{36}^2/\varepsilon_{33} & 0 \\ 0 & 0 & 0 & -\varepsilon_{11} \\ 0 & 0 & 0 & -\varepsilon_{12} \\ Q_{13}+e_{31}e_{33}/\varepsilon_{33} & Q_{23}+e_{32}e_{33}/\varepsilon_{33} & Q_{36}+e_{33}e_{36}/\varepsilon_{33} & 0 \\ 0 & 0 & 0 & -e_{15} \\ 0 & 0 & 0 & -e_{14} \end{matrix}\right.$$
$$\left.\begin{matrix} 0 & Q_{13}+e_{31}e_{33}/\varepsilon_{33} & 0 & 0 \\ 0 & Q_{23}+e_{32}e_{33}/\varepsilon_{33} & 0 & 0 \\ 0 & Q_{36}+e_{33}e_{36}/\varepsilon_{33} & 0 & 0 \\ -\varepsilon_{12} & 0 & -e_{15} & -e_{14} \\ -\varepsilon_{22} & 0 & -e_{25} & -e_{24} \\ 0 & Q_{33}+e_{33}^2/\varepsilon_{33} & 0 & 0 \\ -e_{25} & 0 & Q_{55} & Q_{45} \\ -e_{24} & 0 & Q_{45} & Q_{44} \end{matrix}\right] \tag{8.95}$$

$$\bar{\boldsymbol{C}}_{ab}^T = [e_{31}/\varepsilon_{33} \quad e_{32}/\varepsilon_{33} \quad e_{36}/\varepsilon_{33} \quad 0 \quad 0 \quad e_{33}/\varepsilon_{33} \quad 0 \quad 0] \tag{8.96}$$

$$\bar{\boldsymbol{C}}_{ba} = [-e_{31}/\varepsilon_{33} \quad -e_{32}/\varepsilon_{33} \quad -e_{36}/\varepsilon_{33} \quad 0 \quad 0 \quad -e_{33}/\varepsilon_{33} \quad 0 \quad 0] \tag{8.97}$$

$$\bar{\boldsymbol{C}}_{bb} = [-1/\varepsilon_{33}] \tag{8.98}$$

In analogy with PVD($\boldsymbol{u}$, Φ) in Equation (8.36), by substituting the constitutive and geometrical relations in the RMVT($\boldsymbol{u}$, Φ, $\mathcal{D}_n$) variational statement of Equation (8.81), and referring to a multilayered structure by employing thickness functions for the kinematic description and shape functions for FE discretization, we obtain the following stiffness fundamental nucleus $\bar{\boldsymbol{K}}^{(e)k\tau sij}$ at the element level; its form does not change with respect to the previous section. The FE approximation for the primary unknowns $\bar{\boldsymbol{U}}^T = \{u \; v \; w \; \Phi \; \mathcal{D}_z\}$ and their virtual variation $\delta\bar{\boldsymbol{U}}$ is:

$$\bar{\boldsymbol{U}}_\tau^{(e)}(x,y) = N_i^{(e)}(\xi,\eta)\,\bar{\boldsymbol{Q}}_{\tau i}^{(e)} \tag{8.99}$$

$$\delta\bar{\boldsymbol{U}}_s^{(e)}(x,y) = N_j^{(e)}(\xi,\eta)\,\delta\bar{\boldsymbol{Q}}_{sj}^{(e)} \tag{8.100}$$

where the nodal values are $\bar{Q}^T = \{Q_u \ Q_\Phi \ Q_{\mathcal{D}_z}\}$. The explicit form of the fundamental stiffness nucleus at the element level is:

$$\bar{K}^{(e)k\tau sij} = \begin{bmatrix} K_{11} & K_{12} & K_{13} & K_{14} & K_{15} \\ K_{21} & K_{22} & K_{23} & K_{24} & K_{25} \\ K_{31} & K_{32} & K_{33} & K_{34} & K_{35} \\ K_{41} & K_{42} & K_{43} & K_{44} & K_{45} \\ K_{51} & K_{52} & K_{53} & K_{54} & K_{55} \end{bmatrix}^{k\tau sij} \tag{8.101}$$

with:

$$\begin{aligned}
K_{11}^{k\tau sij} &= \bar{C}_{aa77} \triangleleft N_i N_j \triangleright_\Omega \triangleleft F_{\tau,z} F_{s,z} \triangleright_A + \bar{C}_{aa11} \triangleleft N_{i,x} N_{j,x} \triangleright_\Omega \triangleleft F_\tau F_s \triangleright_A \\
&+ \bar{C}_{aa31} \triangleleft N_{i,y} N_{j,x} \triangleright_\Omega \triangleleft F_\tau F_s \triangleright_A + \bar{C}_{aa13} \triangleleft N_{i,x} N_{j,y} \triangleright_\Omega \triangleleft F_\tau F_s \triangleright_A \\
&+ \bar{C}_{aa33} \triangleleft N_{i,y} N_{j,y} \triangleright_\Omega \triangleleft F_\tau F_s \triangleright_A \\
K_{21}^{k\tau sij} &= \bar{C}_{aa87} \triangleleft N_i N_j \triangleright_\Omega \triangleleft F_{\tau,z} F_{s,z} \triangleright_A + \bar{C}_{aa31} \triangleleft N_{i,x} N_{j,x} \triangleright_\Omega \triangleleft F_\tau F_s \triangleright_A \\
&+ \bar{C}_{aa21} \triangleleft N_{i,y} N_{j,x} \triangleright_\Omega \triangleleft F_\tau F_s \triangleright_A + \bar{C}_{aa33} \triangleleft N_{i,x} N_{j,y} \triangleright_\Omega \triangleleft F_\tau F_s \triangleright_A \\
&+ \bar{C}_{aa23} \triangleleft N_{i,y} N_{j,y} \triangleright_\Omega \triangleleft F_\tau F_s \triangleright_A \\
K_{31}^{k\tau sij} &= \bar{C}_{aa77} \triangleleft N_{i,x} N_j \triangleright_\Omega \triangleleft F_\tau F_{s,z} \triangleright_A + \bar{C}_{aa87} \triangleleft N_{i,y} N_j \triangleright_\Omega \triangleleft F_\tau F_{s,z} \triangleright_A \\
&+ \bar{C}_{aa61} \triangleleft N_i N_{j,x} \triangleright_\Omega \triangleleft F_{\tau,z} F_s \triangleright_A + \bar{C}_{aa63} \triangleleft N_i N_{j,y} \triangleright_\Omega \triangleleft F_{\tau,z} F_s \triangleright_A \\
K_{41}^{k\tau sij} &= -\bar{C}_{aa47} \triangleleft N_{i,x} N_j \triangleright_\Omega \triangleleft F_\tau F_{s,z} \triangleright_A - \bar{C}_{aa57} \triangleleft N_{i,y} N_j \triangleright_\Omega \triangleleft F_\tau F_{s,z} \triangleright_A \\
K_{51}^{k\tau sij} &= \bar{C}_{ba11} \triangleleft N_i N_{j,x} \triangleright_\Omega \triangleleft F_\tau F_s \triangleright_A + \bar{C}_{ba13} \triangleleft N_i N_{j,y} \triangleright_\Omega \triangleleft F_\tau F_s \triangleright_A \\
K_{12}^{k\tau sij} &= \bar{C}_{aa78} \triangleleft N_i N_j \triangleright_\Omega \triangleleft F_{\tau,z} F_{s,z} \triangleright_A + \bar{C}_{aa13} \triangleleft N_{i,x} N_{j,x} \triangleright_\Omega \triangleleft F_\tau F_s \triangleright_A \\
&+ \bar{C}_{aa33} \triangleleft N_{i,y} N_{j,x} \triangleright_\Omega \triangleleft F_\tau F_s \triangleright_A + \bar{C}_{aa12} \triangleleft N_{i,x} N_{j,y} \triangleright_\Omega \triangleleft F_\tau F_s \triangleright_A \\
&+ \bar{C}_{aa32} \triangleleft N_{i,y} N_{j,y} \triangleright_\Omega \triangleleft F_\tau F_s \triangleright_A \\
K_{22}^{k\tau sij} &= \bar{C}_{aa88} \triangleleft N_i N_j \triangleright_\Omega \triangleleft F_{\tau,z} F_{s,z} \triangleright_A + \bar{C}_{aa33} \triangleleft N_{i,x} N_{j,x} \triangleright_\Omega \triangleleft F_\tau F_s \triangleright_A \\
&+ \bar{C}_{aa23} \triangleleft N_{i,y} N_{j,x} \triangleright_\Omega \triangleleft F_\tau F_s \triangleright_A + \bar{C}_{aa32} \triangleleft N_{i,x} N_{j,y} \triangleright_\Omega \triangleleft F_\tau F_s \triangleright_A \\
&+ \bar{C}_{aa22} \triangleleft N_{i,y} N_{j,y} \triangleright_\Omega \triangleleft F_\tau F_s \triangleright_A \\
K_{32}^{k\tau sij} &= \bar{C}_{aa78} \triangleleft N_{i,x} N_j \triangleright_\Omega \triangleleft F_\tau F_{s,z} \triangleright_A + \bar{C}_{aa88} \triangleleft N_{i,y} N_j \triangleright_\Omega \triangleleft F_\tau F_{s,z} \triangleright_A \\
&+ \bar{C}_{aa63} \triangleleft N_i N_{j,x} \triangleright_\Omega \triangleleft F_{\tau,z} F_s \triangleright_A + \bar{C}_{aa62} \triangleleft N_i N_{j,y} \triangleright_\Omega \triangleleft F_{\tau,z} F_s \triangleright_A \\
K_{42}^{k\tau sij} &= -\bar{C}_{aa48} \triangleleft N_{i,x} N_j \triangleright_\Omega \triangleleft F_\tau F_{s,z} \triangleright_A - \bar{C}_{aa58} \triangleleft N_{i,y} N_j \triangleright_\Omega \triangleleft F_\tau F_{s,z} \triangleright_A \\
K_{52}^{k\tau sij} &= \bar{C}_{ba13} \triangleleft N_i N_{j,x} \triangleright_\Omega \triangleleft F_\tau F_s \triangleright_A + \bar{C}_{ba12} \triangleleft N_i N_{j,y} \triangleright_\Omega \triangleleft F_\tau F_s \triangleright_A
\end{aligned}$$

$$
\begin{aligned}
K_{13}^{k\tau sij} &= \bar{C}_{aa16}\triangleleft N_{i,x}N_j\triangleright_\Omega\triangleleft F_\tau F_{s,z}\triangleright_A + \bar{C}_{aa36}\triangleleft N_{i,y}N_j\triangleright_\Omega\triangleleft F_\tau F_{s,z}\triangleright_A \\
&\quad + \bar{C}_{aa77}\triangleleft N_i N_{j,x}\triangleright_\Omega\triangleleft F_{\tau,z} F_s\triangleright_A + \bar{C}_{aa78}\triangleleft N_i N_{j,y}\triangleright_\Omega\triangleleft F_{\tau,z} F_s\triangleright_A \\
K_{23}^{k\tau sij} &= \bar{C}_{aa36}\triangleleft N_{i,x}N_j\triangleright_\Omega\triangleleft F_\tau F_{s,z}\triangleright_A + \bar{C}_{aa26}\triangleleft N_{i,y}N_j\triangleright_\Omega\triangleleft F_\tau F_{s,z}\triangleright_A \\
&\quad + \bar{C}_{aa87}\triangleleft N_i N_{j,x}\triangleright_\Omega\triangleleft F_{\tau,z} F_s\triangleright_A + \bar{C}_{aa88}\triangleleft N_i N_{j,y}\triangleright_\Omega\triangleleft F_{\tau,z} F_s\triangleright_A \\
K_{33}^{k\tau sij} &= \bar{C}_{aa66}\triangleleft N_i N_j\triangleright_\Omega\triangleleft F_{\tau,z} F_{s,z}\triangleright_A + \bar{C}_{aa77}\triangleleft N_{i,x}N_{j,x}\triangleright_\Omega\triangleleft F_\tau F_s\triangleright_A \\
&\quad + \bar{C}_{aa87}\triangleleft N_{i,y}N_{j,x}\triangleright_\Omega F_\tau F_s + \bar{C}_{aa78}\triangleleft N_{i,x}N_{j,y}\triangleright_\Omega\triangleleft F_\tau F_s\triangleright_A \\
&\quad + \bar{C}_{aa78}\triangleleft N_{i,y}N_{j,y}\triangleright_\Omega\triangleleft F_\tau F_s\triangleright_A \\
K_{43}^{k\tau sij} &= -\bar{C}_{aa47}\triangleleft N_{i,x}N_{j,x}\triangleright_\Omega\triangleleft F_\tau F_s\triangleright_A - \bar{C}_{aa57}\triangleleft N_{i,y}N_{j,x}\triangleright_\Omega\triangleleft F_\tau F_s\triangleright_A \\
&\quad - \bar{C}_{aa48}\triangleleft N_{i,x}N_{j,y}\triangleright_\Omega\triangleleft F_\tau F_s\triangleright_A - \bar{C}_{aa58}\triangleleft N_{i,y}N_{j,y}\triangleright_\Omega\triangleleft F_\tau F_s\triangleright_A \\
K_{53}^{k\tau sij} &= \bar{C}_{ba16}\triangleleft N_i N_j\triangleright_\Omega\triangleleft F_\tau F_{s,z}\triangleright_A \\
K_{14}^{k\tau sij} &= -\bar{C}_{aa74}\triangleleft N_i N_{j,x}\triangleright_\Omega\triangleleft F_{\tau,z} F_s\triangleright_A - \bar{C}_{aa75}\triangleleft N_i N_{j,y}\triangleright_\Omega\triangleleft F_{\tau,z} F_s\triangleright_A \\
K_{24}^{k\tau sij} &= -\bar{C}_{aa84}\triangleleft N_i N_{j,x}\triangleright_\Omega\triangleleft F_{\tau,z} F_s\triangleright_A - \bar{C}_{aa85}\triangleleft N_i N_{j,y}\triangleright_\Omega\triangleleft F_{\tau,z} F_s\triangleright_A \\
K_{34}^{k\tau sij} &= -\bar{C}_{aa74}\triangleleft N_{i,x}N_{j,x}\triangleright_\Omega\triangleleft F_\tau F_s\triangleright_A - \bar{C}_{aa84}\triangleleft N_{i,y}N_{j,x}\triangleright_\Omega\triangleleft F_\tau F_s\triangleright_A \\
&\quad - \bar{C}_{aa75}\triangleleft N_{i,x}N_{j,y}\triangleright_\Omega\triangleleft F_\tau F_s\triangleright_A - \bar{C}_{aa85}\triangleleft N_{i,y}N_{j,y}\triangleright_\Omega\triangleleft F_\tau F_s\triangleright_A \\
K_{44}^{k\tau sij} &= \bar{C}_{aa44}\triangleleft N_{i,x}N_{j,x}\triangleright_\Omega\triangleleft F_\tau F_s\triangleright_A + \bar{C}_{aa54}\triangleleft N_{i,y}N_{j,x}\triangleright_\Omega\triangleleft F_\tau F_s\triangleright_A \\
&\quad + \bar{C}_{aa45}\triangleleft N_{i,x}N_{j,y}\triangleright_\Omega\triangleleft F_\tau F_s\triangleright_A + \bar{C}_{aa55}\triangleleft N_{i,y}N_{j,y}\triangleright_\Omega\triangleleft F_\tau F_s\triangleright_A \\
K_{54}^{k\tau sij} &= \triangleleft N_i N_j\triangleright_\Omega\triangleleft F_\tau F_{s,z}\triangleright_A \\
K_{15}^{k\tau sij} &= -\bar{C}_{ab11}\triangleleft N_{i,x}N_j\triangleright_\Omega\triangleleft F_\tau F_s\triangleright_A - \bar{C}_{\mathrm{ab}31}\triangleleft N_{i,y}N_j\triangleright_\Omega\triangleleft F_\tau F_s\triangleright_A \\
K_{25}^{k\tau sij} &= -\bar{C}_{ab31}\triangleleft N_{i,x}N_j\triangleright_\Omega\triangleleft F_\tau F_s\triangleright_A - \bar{C}_{\mathrm{ab}21}\triangleleft N_{i,y}N_j\triangleright_\Omega\triangleleft F_\tau F_s\triangleright_A \\
K_{35}^{k\tau sij} &= -\bar{C}_{ab61}\triangleleft N_i N_j\triangleright_\Omega\triangleleft F_{\tau,z} F_s\triangleright_A \\
K_{45}^{k\tau sij} &= \triangleleft N_i N_j\triangleright_\Omega\triangleleft F_{\tau,z} F_s\triangleright_A, \quad K_{55}^{k\tau sij} = -\bar{C}_{bb}(1,1)\triangleleft N_i N_j\triangleright_\Omega\triangleleft F_\tau F_s\triangleright_A
\end{aligned}
\tag{8.102}
$$

The in-plane and through-the-thickness integrals are defined as in the previous sections. The non-zero elements of the fundamental mass nucleus $\boldsymbol{M}^{(e)k\tau sij}$ are (in analogy with the previous sections):

$$
M_{11}^{k\tau sij} = M_{22}^{k\tau sij} = M_{33}^{k\tau sij} = \rho^k\triangleleft N_i N_j\triangleright_\Omega\triangleleft F_\tau F_s\triangleright_A \tag{8.103}
$$

The obtained fundamental nuclei are used in the governing equations, which formally do not change with respect to the PVD($\boldsymbol{u}$,Φ) case (see Equation (8.47)).

8.6 RMVT($u, \Phi, \sigma_n, \mathcal{D}_n$) for the electromechanical plate case

Another possible extension of the RMVT (Reissner 1984) has been indicated in Equations (2.83) and (6.155), where the internal electrical work has been added and two Lagrange multipliers are employed in order to increase the accuracy of the computed transverse stress components and the transverse normal electric displacement (Carrera and Brischetto 2007a; Carrera *et al.* 2008):

$$\int_V \Big(\delta\boldsymbol{\epsilon}_{pG}^T\boldsymbol{\sigma}_{pC} + \delta\boldsymbol{\epsilon}_{nG}^T\boldsymbol{\sigma}_{nM} + \delta\boldsymbol{\sigma}_{nM}^T(\boldsymbol{\epsilon}_{nG} - \boldsymbol{\epsilon}_{nC}) - \delta\boldsymbol{\mathcal{E}}_{pG}^T\boldsymbol{\mathcal{D}}_{pC} - \delta\boldsymbol{\mathcal{E}}_{nG}^T\boldsymbol{\mathcal{D}}_{nM}$$

$$- \delta\boldsymbol{\mathcal{D}}_{nM}^T(\boldsymbol{\mathcal{E}}_{nG} - \boldsymbol{\mathcal{E}}_{nC})\Big)dV = \delta L_e - \delta L_{in} \tag{8.104}$$

On the basis of the condensed notation, Equation (8.104) can be rewritten as:

$$\int_V \left(\delta\bar{\boldsymbol{\epsilon}}_{aG}^T\,\bar{\boldsymbol{\sigma}}_{aC} + \delta\bar{\boldsymbol{\epsilon}}_{bG}^T\,\bar{\boldsymbol{\sigma}}_b + \delta\bar{\boldsymbol{\sigma}}_b^T(\bar{\boldsymbol{\epsilon}}_{bG} - \bar{\boldsymbol{\epsilon}}_{bC})\right) dV = \delta\bar{L}_e - \delta L_{in} \tag{8.105}$$

Equation (8.105) is formally identical to Equation (8.56), but the meaning of the arrays changes, therefore the following arrays can be introduced:

$\bar{\boldsymbol{\sigma}}_{aC}^T = \{\sigma_{xx} \quad \sigma_{yy} \quad \sigma_{xy} \quad -\mathcal{D}_x \quad -\mathcal{D}_y\}_C$ is the vector of the extensive non-modeled variables, which are calculated by means of the constitutive relations (subscript C);

$\bar{\boldsymbol{\sigma}}_b^T = \{\sigma_{zz} \quad \sigma_{xz} \quad \sigma_{yz} \quad -\mathcal{D}_z\}$ is the vector of the extensive modeled variables;

$\bar{\boldsymbol{\epsilon}}_{aG}^T = \{\epsilon_{xx} \quad \epsilon_{yy} \quad \gamma_{xy} \quad \mathcal{E}_x \quad \mathcal{E}_y\}_G$ is the vector of the intensive variables associated to $\bar{\boldsymbol{\sigma}}_a$, which is calculated by means of the geometrical relations (subscript G);

$\bar{\boldsymbol{\epsilon}}_{bG}^T = \{\epsilon_{zz} \quad \gamma_{xz} \quad \gamma_{yz} \quad \mathcal{E}_z\}_G$ is the vector of the intensive variables associated to $\bar{\boldsymbol{\sigma}}_b$, which is calculated by means of the geometrical relations (subscript G);

$\bar{\boldsymbol{\epsilon}}_{bC}^T = \{\epsilon_{zz} \quad \gamma_{xz} \quad \gamma_{yz} \quad \mathcal{E}_z\}_C$ is the vector of the intensive variables associated to $\bar{\boldsymbol{\sigma}}_b$, which is calculated by means of the constitutive relations (subscript C).

It is important to specify that the primary unknown variables are collected in the vector:

$$\bar{\boldsymbol{U}}^T = \{u \quad v \quad w \quad \Phi \quad \sigma_{zz} \quad \sigma_{xz} \quad \sigma_{yz} \quad \mathcal{D}_z\} \tag{8.106}$$

The geometrical relations can be written, as in Equations (2.38)–(2.42), as:

$$\bar{\boldsymbol{\epsilon}}_{aG} = \boldsymbol{D}_a \bar{\boldsymbol{U}} \tag{8.107}$$

$$\bar{\boldsymbol{\epsilon}}_{bG} = \boldsymbol{D}_b \bar{\boldsymbol{U}} \tag{8.108}$$

$$\bar{\boldsymbol{\sigma}}_b = \bar{\boldsymbol{\sigma}}_{bG} = \boldsymbol{D}_{b'} \bar{\boldsymbol{U}} \tag{8.109}$$

which are formally identical to Equations (8.58)–(8.60), but the differential matrices in explicit form read as:

$$\boldsymbol{D}_a = \begin{bmatrix} \partial_x & 0 & 0 & 0 & 0 & 0 & 0 & 0 \\ 0 & \partial_y & 0 & 0 & 0 & 0 & 0 & 0 \\ \partial_y & \partial_x & 0 & 0 & 0 & 0 & 0 & 0 \\ 0 & 0 & 0 & -\partial_x & 0 & 0 & 0 & 0 \\ 0 & 0 & 0 & -\partial_y & 0 & 0 & 0 & 0 \end{bmatrix}$$

$$\boldsymbol{D}_b = \begin{bmatrix} 0 & 0 & \partial_z & 0 & 0 & 0 & 0 & 0 \\ \partial_z & 0 & \partial_x & 0 & 0 & 0 & 0 & 0 \\ 0 & \partial_z & \partial_y & 0 & 0 & 0 & 0 & 0 \\ 0 & 0 & 0 & -\partial_z & 0 & 0 & 0 & 0 \end{bmatrix} \quad \boldsymbol{D}_{b'} = \begin{bmatrix} 0 & 0 & 0 & 0 & 1 & 0 & 0 & 0 \\ 0 & 0 & 0 & 0 & 0 & 1 & 0 & 0 \\ 0 & 0 & 0 & 0 & 0 & 0 & 1 & 0 \\ 0 & 0 & 0 & 0 & 0 & 0 & 0 & -1 \end{bmatrix} \tag{8.110}$$

The constitutive relations for RMVT($\boldsymbol{u}$, Φ, $\boldsymbol{\sigma}_n$, $\boldsymbol{\mathcal{D}}_n$) can be summarized, as in Equations (2.84)–(2.87), as:

$$\bar{\boldsymbol{\sigma}}_C = \tilde{\boldsymbol{C}}\, \bar{\boldsymbol{\epsilon}}_G \tag{8.111}$$

where $\bar{\boldsymbol{\sigma}}_C$ is composed of the vector of the extensive non-modeled variables $\bar{\boldsymbol{\sigma}}_{aC}$ and the vector of the intensive variables $\bar{\boldsymbol{\epsilon}}_{bC}$ (which is associated to $\bar{\boldsymbol{\sigma}}_b$); $\bar{\boldsymbol{\epsilon}}_G$ is composed of the vector of the intensive variables $\bar{\boldsymbol{\epsilon}}_{aG}$ (which is associated to $\bar{\boldsymbol{\sigma}}_a$) and the vector of the extensive modeled variables $\bar{\boldsymbol{\sigma}}_b$:

$$\begin{aligned} \bar{\boldsymbol{\sigma}}_C^T &= \left\{ \bar{\boldsymbol{\sigma}}_{aC}^T \bar{\boldsymbol{\epsilon}}_{bC}^T \right\} \\ \bar{\boldsymbol{\epsilon}}_G^T &= \left\{ \bar{\boldsymbol{\epsilon}}_{aG}^T \bar{\boldsymbol{\sigma}}_{bG}^T \right\} \end{aligned} \tag{8.112}$$

The physical constitutive matrix $\bar{\boldsymbol{H}}$ can be partitioned by dividing the cells into modeled and non-modeled quantities:

$$\bar{\boldsymbol{H}} = \begin{Bmatrix} \bar{\boldsymbol{H}}_{aa} & \bar{\boldsymbol{H}}_{ab} \\ \bar{\boldsymbol{H}}_{ba} & \bar{\boldsymbol{H}}_{bb} \end{Bmatrix} \tag{8.113}$$

where $\bar{\boldsymbol{H}}_{ab} = \bar{\boldsymbol{H}}_{ba}^T$.

In explicit form (see also Equation (8.39)) for each k layer:

$$\bar{\boldsymbol{H}}_{aa} = \begin{bmatrix} Q_{11} & Q_{12} & Q_{16} & 0 & 0 \\ Q_{12} & Q_{22} & Q_{26} & 0 & 0 \\ Q_{16} & Q_{26} & Q_{66} & 0 & 0 \\ 0 & 0 & 0 & -\varepsilon_{11} & -\varepsilon_{12} \\ 0 & 0 & 0 & -\varepsilon_{12} & -\varepsilon_{22} \end{bmatrix}, \quad \bar{\boldsymbol{H}}_{ab} = \begin{bmatrix} Q_{13} & 0 & 0 & -e_{31} \\ Q_{23} & 0 & 0 & -e_{32} \\ Q_{36} & 0 & 0 & -e_{36} \\ 0 & -e_{15} & -e_{14} & 0 \\ 0 & -e_{25} & -e_{24} & 0 \end{bmatrix}$$

$$\bar{\boldsymbol{H}}_{ba} = \begin{bmatrix} Q_{13} & Q_{23} & Q_{36} & 0 & 0 \\ 0 & 0 & 0 & -e_{15} & -e_{25} \\ 0 & 0 & 0 & -e_{14} & -e_{24} \\ -e_{31} & -e_{32} & -e_{36} & 0 & 0 \end{bmatrix}, \quad \bar{\boldsymbol{H}}_{bb} = \begin{bmatrix} Q_{33} & 0 & 0 & -e_{33} \\ 0 & Q_{55} & Q_{45} & 0 \\ 0 & Q_{45} & Q_{44} & 0 \\ -e_{33} & 0 & 0 & -\varepsilon_{33} \end{bmatrix} \tag{8.114}$$

The physical constitutive relations can be arranged according to the above partitioning (see also Equations (8.40) and (2.84)–(2.87)):

$$\bar{\boldsymbol{\sigma}}_{aC} = \bar{\boldsymbol{H}}_{aa}\bar{\boldsymbol{\epsilon}}_{aG} + \bar{\boldsymbol{H}}_{ab}\bar{\boldsymbol{\epsilon}}_{bG}, \quad \bar{\boldsymbol{\sigma}}_{bC} = \bar{\boldsymbol{H}}_{ba}\bar{\boldsymbol{\epsilon}}_{aG} + \bar{\boldsymbol{H}}_{bb}\bar{\boldsymbol{\epsilon}}_{bG} \tag{8.115}$$

From Equation (8.115), considering vectors $\bar{\boldsymbol{\sigma}}_C$ and $\bar{\boldsymbol{\epsilon}}_G$ in Equation (8.112), it is possible to write:

$$\bar{\boldsymbol{\sigma}}_{aC} = \tilde{\boldsymbol{C}}_{aa}\bar{\boldsymbol{\epsilon}}_{aG} + \tilde{\boldsymbol{C}}_{ab}\bar{\boldsymbol{\sigma}}_{bG}, \quad \bar{\boldsymbol{\epsilon}}_{bC} = \tilde{\boldsymbol{C}}_{ba}\bar{\boldsymbol{\epsilon}}_{aG} + \tilde{\boldsymbol{C}}_{bb}\bar{\boldsymbol{\sigma}}_{bG} \tag{8.116}$$

with:

$$\begin{aligned} \tilde{\boldsymbol{C}}_{aa} &= \bar{\boldsymbol{H}}_{aa} - \bar{\boldsymbol{H}}_{ab}(\bar{\boldsymbol{H}}_{bb})^{-1}\bar{\boldsymbol{H}}_{ba}, \quad \tilde{\boldsymbol{C}}_{ab} = \bar{\boldsymbol{H}}_{ab}(\bar{\boldsymbol{H}}_{bb})^{-1} \\ \tilde{\boldsymbol{C}}_{ba} &= -(\bar{\boldsymbol{H}}_{bb})^{-1}\bar{\boldsymbol{H}}_{ba}, \quad \tilde{\boldsymbol{C}}_{bb} = (\bar{\boldsymbol{H}}_{bb})^{-1} \end{aligned} \tag{8.117}$$

Matrix $\tilde{\boldsymbol{C}}$ in Equation (8.111) can be written as:

$$\tilde{\boldsymbol{C}} = \begin{Bmatrix} \tilde{\boldsymbol{C}}_{aa} & \tilde{\boldsymbol{C}}_{ab} \\ \tilde{\boldsymbol{C}}_{ba} & \tilde{\boldsymbol{C}}_{bb} \end{Bmatrix} \tag{8.118}$$

The explicit form of matrices $\tilde{\boldsymbol{C}}_{aa}$, $\tilde{\boldsymbol{C}}_{ba}$, $\tilde{\boldsymbol{C}}_{ab}$, and $\tilde{\boldsymbol{C}}_{bb}$ is given below for each k layer:

$$\tilde{C}_{aa11} = Q_{11} + \frac{Q_{33}e_{31}^2 - Q_{13}\,(2e_{31}e_{33} + Q_{13}\varepsilon_{33})}{e_{33}^2 + Q_{33}\varepsilon_{33}}$$

$$
\begin{aligned}
&\tilde{C}_{aa12} = \\
&\quad \frac{e_{33}\left(-Q_{13}e_{32} + Q_{12}e_{33}\right) + Q_{33}\left(e_{31}e_{32} + Q_{12}\varepsilon_{33}\right) - Q_{23}\left(e_{31}e_{33} + Q_{13}\varepsilon_{33}\right)}{e_{33}^2 + Q_{33}\varepsilon_{33}} \\
&\tilde{C}_{aa13} = Q_{16} + \frac{\left(Q_{33}e_{31} - Q_{13}e_{33}\right)e_{36} - Q_{36}\left(e_{31}e_{33} + Q_{13}\varepsilon_{33}\right)}{e_{33}^2 + Q_{33}\varepsilon_{33}} \\
&\tilde{C}_{aa14} = 0, \quad \tilde{C}_{aa15} = 0 \\
&\tilde{C}_{aa21} = \\
&\quad \frac{e_{33}\left(-Q_{13}e_{32} + Q_{12}e_{33}\right) + Q_{33}\left(e_{31}e_{32} + Q_{12}\varepsilon_{33}\right) - Q_{23}\left(e_{31}e_{33} + Q_{13}\varepsilon_{33}\right)}{e_{33}^2 + Q_{33}\varepsilon_{33}} \\
&\tilde{C}_{aa22} = Q_{22} + \frac{Q_{33}e_{32}^2 - Q_{23}\left(2e_{32}e_{33} + Q_{23}\varepsilon_{33}\right)}{e_{33}^2 + Q_{33}\varepsilon_{33}} \\
&\tilde{C}_{aa23} = Q_{26} + \frac{\left(Q_{33}e_{32} - Q_{23}e_{33}\right)e_{36} - Q_{36}\left(e_{32}e_{33} + Q_{23}\varepsilon_{33}\right)}{e_{33}^2 + Q_{33}\varepsilon_{33}} \\
&\tilde{C}_{aa24} = 0, \quad \tilde{C}_{aa25} = 0 \\
&\tilde{C}_{aa31} = Q_{16} + \frac{\left(Q_{33}e_{31} - Q_{13}e_{33}\right)e_{36} - Q_{36}\left(e_{31}e_{33} + Q_{13}\varepsilon_{33}\right)}{e_{33}^2 + Q_{33}\varepsilon_{33}} \\
&\tilde{C}_{aa32} = Q_{26} + \frac{\left(Q_{33}e_{32} - Q_{23}e_{33}\right)e_{36} - Q_{36}\left(e_{32}e_{33} + Q_{23}\varepsilon_{33}\right)}{e_{33}^2 + Q_{33}\varepsilon_{33}} \\
&\tilde{C}_{aa33} = Q_{66} + \frac{-2Q_{36}e_{33}e_{36} + Q_{33}e_{36}^2 - Q_{36}^2\varepsilon_{33}}{e_{33}^2 + Q_{33}\varepsilon_{33}} \\
&\tilde{C}_{aa34} = 0, \quad \tilde{C}_{aa35} = 0, \quad \tilde{C}_{aa41} = 0, \quad \tilde{C}_{aa42} = 0, \quad \tilde{C}_{aa43} = 0 \\
&\tilde{C}_{aa44} = \frac{Q_{55}e_{14}^2 - 2Q_{45}e_{14}e_{15} + Q_{44}e_{15}^2}{Q_{45}^2 - Q_{44}Q_{55}} - \varepsilon_{11} \\
&\tilde{C}_{aa45} = \frac{Q_{55}e_{14}e_{24} + Q_{44}e_{15}e_{25} - Q_{45}\left(e_{15}e_{24} + e_{14}e_{25}\right)}{Q_{45}^2 - Q_{44}Q_{55}} - \varepsilon_{12} \\
&\tilde{C}_{aa51} = 0, \quad \tilde{C}_{aa52} = 0, \quad \tilde{C}_{aa53} = 0 \\
&\tilde{C}_{aa54} = \frac{Q_{55}e_{14}e_{24} + Q_{44}e_{15}e_{25} - Q_{45}\left(e_{15}e_{24} + e_{14}e_{25}\right)}{Q_{45}^2 - Q_{44}Q_{55}} - \varepsilon_{12} \\
&\tilde{C}_{aa55} = \frac{Q_{55}e_{24}^2 - 2Q_{45}e_{24}e_{25} + Q_{44}e_{25}^2}{Q_{45}^2 - Q_{44}Q_{55}} - \varepsilon_{22}
\end{aligned}
\tag{8.119}
$$

$$\tilde{\boldsymbol{C}}_{ab} = \begin{bmatrix} \frac{e_{31}e_{33}+Q_{13}\varepsilon_{33}}{e_{33}^2+Q_{33}\varepsilon_{33}} & 0 & 0 & \frac{Q_{33}e_{31}-Q_{13}e_{33}}{e_{33}^2+Q_{33}\varepsilon_{33}} \\ \frac{e_{32}e_{33}+Q_{23}\varepsilon_{33}}{e_{33}^2+Q_{33}\varepsilon_{33}} & 0 & 0 & \frac{Q_{33}e_{32}-Q_{23}e_{33}}{e_{33}^2+Q_{33}\varepsilon_{33}} \\ \frac{e_{33}e_{36}+Q_{36}\varepsilon_{33}}{e_{33}^2+Q_{33}\varepsilon_{33}} & 0 & 0 & \frac{-Q_{36}e_{33}+Q_{33}e_{36}}{e_{33}^2+Q_{33}\varepsilon_{33}} \\ 0 & \frac{-Q_{45}e_{14}+Q_{44}e_{15}}{Q_{45}^2-Q_{44}Q_{55}} & \frac{Q_{55}e_{14}-Q_{45}e_{15}}{Q_{45}^2-Q_{44}Q_{55}} & 0 \\ 0 & \frac{-Q_{45}e_{24}+Q_{44}e_{25}}{Q_{45}^2-Q_{44}Q_{55}} & \frac{Q_{55}e_{24}-Q_{45}e_{25}}{Q_{45}^2-Q_{44}Q_{55}} & 0 \end{bmatrix} \tag{8.120}$$

$$\tilde{\boldsymbol{C}}_{ba} = \begin{bmatrix} -\frac{e_{31}e_{33}+Q_{13}\varepsilon_{33}}{e_{33}^2+Q_{33}\varepsilon_{33}} & -\frac{e_{32}e_{33}+Q_{23}\varepsilon_{33}}{e_{33}^2+Q_{33}\varepsilon_{33}} & -\frac{e_{33}e_{36}+Q_{36}\varepsilon_{33}}{e_{33}^2+Q_{33}\varepsilon_{33}} & 0 & 0 \\ 0 & 0 & 0 & \frac{Q_{45}e_{14}-Q_{44}e_{15}}{Q_{45}^2-Q_{44}Q_{55}} & \frac{Q_{45}e_{24}-Q_{44}e_{25}}{Q_{45}^2-Q_{44}Q_{55}} \\ 0 & 0 & 0 & \frac{-Q_{55}e_{14}+Q_{45}e_{15}}{Q_{45}^2-Q_{44}Q_{55}} & \frac{-Q_{55}e_{24}+Q_{45}e_{25}}{Q_{45}^2-Q_{44}Q_{55}} \\ \frac{-Q_{33}e_{31}+Q_{13}e_{33}}{e_{33}^2+Q_{33}\varepsilon_{33}} & \frac{-Q_{33}e_{32}+Q_{23}e_{33}}{e_{33}^2+Q_{33}\varepsilon_{33}} & \frac{Q_{36}e_{33}-Q_{33}e_{36}}{e_{33}^2+Q_{33}\varepsilon_{33}} & 0 & 0 \end{bmatrix} \tag{8.121}$$

$$\tilde{\boldsymbol{C}}_{bb} = \begin{bmatrix} \frac{1}{Q_{33}+e_{33}^2/\varepsilon_{33}} & 0 & 0 & -\frac{e_{33}}{e_{33}^2+Q_{33}\varepsilon_{33}} \\ 0 & \frac{1}{-Q_{45}^2/Q_{44}+Q_{55}} & \frac{Q_{45}}{Q_{45}^2-Q_{44}Q_{55}} & 0 \\ 0 & \frac{Q_{45}}{Q_{45}^2-Q_{44}Q_{55}} & \frac{1}{Q_{44}-Q_{45}^2/Q_{55}} & 0 \\ -\frac{e_{33}}{e_{33}^2+Q_{33}\varepsilon_{33}} & 0 & 0 & -\frac{1}{e_{33}^2/Q_{33}+\varepsilon_{33}} \end{bmatrix} \tag{8.122}$$

In analogy with PVD($\boldsymbol{u}$, Φ) in Equation (8.36), by substituting the constitutive and geometrical relations in the RMVT($\boldsymbol{u}$, Φ, $\boldsymbol{\sigma}_n$, $\boldsymbol{\mathcal{D}}_n$) variational statement in Equation (8.105), and referring to a multilayered structure by employing thickness functions for the kinematic description and shape functions for FE discretization, we obtain the following stiffness fundamental nucleus $\bar{\boldsymbol{K}}^{(e)k\tau sij}$ at the element level, which has the same form as in the previous sections. The FE approximation for the primary unknowns $\bar{\boldsymbol{U}}^T = \{u\ v\ w\ \Phi\ \mathcal{D}_z\ \sigma_{zz}\ \sigma_{xz}\ \sigma_{yz}\}$ and their virtual variation $\delta\bar{\boldsymbol{U}}$ is:

$$\bar{\boldsymbol{U}}_\tau^{(e)}(x, y) = N_i^{(e)}(\xi, \eta)\, \bar{\boldsymbol{Q}}_{\tau i}^{(e)} \tag{8.123}$$

$$\delta\bar{\boldsymbol{U}}_s^{(e)}(x, y) = N_j^{(e)}(\xi, \eta)\, \delta\bar{\boldsymbol{Q}}_{sj}^{(e)} \tag{8.124}$$

where the nodal values are $\bar{\boldsymbol{Q}}^T = \{\boldsymbol{Q}_u\ \boldsymbol{Q}_\Phi\ \boldsymbol{Q}_{\mathcal{D}_n}\ \boldsymbol{Q}_{\sigma_n}\}$. The explicit form of the fundamental stiffness nucleus at the element level is:

$$\bar{\boldsymbol{K}}^{(e)k\tau sij} = \begin{bmatrix} K_{11} & K_{12} & K_{13} & K_{14} & K_{15} & K_{16} & K_{17} & K_{18} \\ K_{21} & K_{22} & K_{23} & K_{24} & K_{25} & K_{26} & K_{27} & K_{28} \\ K_{31} & K_{32} & K_{33} & K_{34} & K_{35} & K_{36} & K_{37} & K_{38} \\ K_{41} & K_{42} & K_{43} & K_{44} & K_{45} & K_{46} & K_{47} & K_{48} \\ K_{51} & K_{52} & K_{53} & K_{54} & K_{55} & K_{56} & K_{57} & K_{58} \\ K_{61} & K_{62} & K_{63} & K_{64} & K_{65} & K_{66} & K_{67} & K_{68} \\ K_{71} & K_{72} & K_{73} & K_{74} & K_{75} & K_{76} & K_{77} & K_{78} \\ K_{81} & K_{82} & K_{83} & K_{84} & K_{85} & K_{86} & K_{87} & K_{88} \end{bmatrix}^{k\tau sij} \tag{8.125}$$

with:

$$\begin{aligned}
K_{11} &= \triangleleft F_s F_\tau \triangleright_A \tilde{C}_{aa11} \triangleleft N_{i,x} N_{j,x} \triangleright_\Omega + \triangleleft F_s F_\tau \triangleright_A \tilde{C}_{aa31} \triangleleft N_{i,y} N_{j,x} \triangleright_\Omega \\
&\quad + \triangleleft F_s F_\tau \triangleright_A \tilde{C}_{aa13} \triangleleft N_{i,x} N_{j,y} \triangleright_\Omega + \triangleleft F_s F_\tau \triangleright_A \tilde{C}_{aa33} \triangleleft N_{i,y} N_{j,y} \triangleright_\Omega \\
K_{21} &= \triangleleft F_s F_\tau \triangleright_A \tilde{C}_{aa31} \triangleleft N_{i,x} N_{j,x} \triangleright_\Omega + \triangleleft F_s F_\tau \triangleright_A \tilde{C}_{aa21} \triangleleft N_{i,y} N_{j,x} \triangleright_\Omega \\
&\quad + \triangleleft F_s F_\tau \triangleright_A \tilde{C}_{aa33} \triangleleft N_{i,x} N_{j,y} \triangleright_\Omega + \triangleleft F_s F_\tau \triangleright_A \tilde{C}_{aa23} \triangleleft N_{i,y} N_{j,y} \triangleright_\Omega \\
K_{31} &= 0, \quad K_{41} = 0 \\
K_{51} &= -\triangleleft F_s F_\tau \triangleright_A \triangleleft N_i N_{j,x} \triangleright_\Omega \tilde{C}_{ba11} - \triangleleft F_s F_\tau \triangleright_A \triangleleft N_i N_{j,y} \triangleright_\Omega \tilde{C}_{ba13} \\
K_{61} &= \triangleleft F_\tau F_{s,z} \triangleright_A \triangleleft N_i N_j \triangleright_\Omega, \quad K_{71} = 0 \\
K_{81} &= \triangleleft F_s F_\tau \triangleright_A \triangleleft N_i N_{j,x} \triangleright_\Omega \tilde{C}_{ba41} + \triangleleft F_s F_\tau \triangleright_A \triangleleft N_i N_{j,y} \triangleright_\Omega \tilde{C}_{ba43} \\
K_{12} &= \triangleleft F_s F_\tau \triangleright_A \tilde{C}_{aa13} \triangleleft N_{i,x} N_{j,x} \triangleright_\Omega + \triangleleft F_s F_\tau \triangleright_A \tilde{C}_{aa33} \triangleleft N_{i,y} N_{j,x} \triangleright_\Omega \\
&\quad + \triangleleft F_s F_\tau \triangleright_A \tilde{C}_{aa12} \triangleleft N_{i,x} N_{j,y} \triangleright_\Omega + \triangleleft F_s F_\tau \triangleright_A \tilde{C}_{aa32} \triangleleft N_{i,y} N_{j,y} \triangleright_\Omega \\
K_{22} &= \triangleleft F_s F_\tau \triangleright_A \tilde{C}_{aa33} \triangleleft N_{i,x} N_{j,x} \triangleright_\Omega + \triangleleft F_s F_\tau \triangleright_A \tilde{C}_{aa23} \triangleleft N_{i,y} N_{j,x} \triangleright_\Omega \\
&\quad + \triangleleft F_s F_\tau \triangleright_A \tilde{C}_{aa32} \triangleleft N_{i,x} N_{j,y} \triangleright_\Omega + \triangleleft F_s F_\tau \triangleright_A \tilde{C}_{aa22} \triangleleft N_{i,y} N_{j,y} \triangleright_\Omega
\end{aligned}$$

$$K_{32} = 0, \quad K_{42} = 0$$

$$K_{52} = -\triangleleft F_s F_\tau \triangleright_A \triangleleft N_i N_{j,x} \triangleright_\Omega \tilde{C}_{ba13} - \triangleleft F_s F_\tau \triangleright_A \triangleleft N_i N_{j,y} \triangleright_\Omega \tilde{C}_{ba12}$$

$$K_{62} = 0, \quad K_{72} = \triangleleft F_\tau F_{s,z} \triangleright_A \triangleleft N_i N_j \triangleright_\Omega$$

$$K_{82} = \triangleleft F_s F_\tau \triangleright_A \triangleleft N_i N_{j,x} \triangleright_\Omega \tilde{C}_{ba43} + \triangleleft F_s F_\tau \triangleright_A \triangleleft N_i N_{j,y} \triangleright_\Omega \tilde{C}_{ba42}$$

$$K_{13} = 0, \quad K_{23} = 0, \quad K_{33} = 0, \quad K_{43} = 0$$

$$K_{53} = \triangleleft F_{s,z} F_\tau \triangleright_A \triangleleft N_i N_j \triangleright_\Omega, \quad K_{63} = \triangleleft F_s F_\tau \triangleright_A \triangleleft N_i N_{j,x} \triangleright_\Omega$$

$$K_{73} = \triangleleft F_s F_\tau \triangleright_A \triangleleft N_i N_{j,y} \triangleright_\Omega, \quad K_{83} = 0, \quad K_{14} = 0, \quad K_{24} = 0, \quad K_{34} = 0$$

$$K_{44} = \triangleleft F_s F_\tau \triangleright_A \tilde{C}_{aa44} \triangleleft N_{i,x} N_{j,x} \triangleright_\Omega + \triangleleft F_s F_\tau \triangleright_A \tilde{C}_{aa54} \triangleleft N_{i,y} N_{j,x} \triangleright_\Omega$$
$$+ \triangleleft F_s F_\tau \triangleright_A \tilde{C}_{aa45} \triangleleft N_{i,x} N_{j,y} \triangleright_\Omega + \triangleleft F_s F_\tau \triangleright_A \tilde{C}_{aa55} \triangleleft N_{i,y} N_{j,y} \triangleright_\Omega$$

$$K_{54} = 0, \quad K_{64} = \triangleleft F_s F_\tau \triangleright_A \triangleleft N_i N_{j,x} \triangleright_\Omega \tilde{C}_{ba24} + \triangleleft F_s F_\tau \triangleright_A \triangleleft N_i N_{j,y} \triangleright_\Omega \tilde{C}_{ba25}$$

$$K_{74} = \triangleleft F_s F_\tau \triangleright_A \triangleleft N_i N_{j,x} \triangleright_\Omega \tilde{C}_{ba34} + \triangleleft F_s F_\tau \triangleright_A \triangleleft N_i N_{j,y} \triangleright_\Omega \tilde{C}_{ba35}$$

$$K_{84} = \triangleleft F_{s,z} F_\tau \triangleright_A \triangleleft N_i N_j \triangleright_\Omega$$

$$K_{15} = \triangleleft F_s F_\tau \triangleright_A \triangleleft N_j N_{i,x} \triangleright_\Omega \tilde{C}_{\mathrm{ab11}} + \triangleleft F_s F_\tau \triangleright_A \triangleleft N_j N_{i,y} \triangleright_\Omega \tilde{C}_{\mathrm{ab31}}$$

$$K_{25} = \triangleleft F_s F_\tau \triangleright_A \triangleleft N_j N_{i,x} \triangleright_\Omega \tilde{C}_{\mathrm{ab31}} + \triangleleft F_s F_\tau \triangleright_A \triangleleft N_j N_{i,y} \triangleright_\Omega \tilde{C}_{\mathrm{ab21}}$$

$$K_{35} = \triangleleft F_s F_{\tau,z} \triangleright_A \triangleleft N_i N_j \triangleright_\Omega, \quad K_{45} = 0$$

$$K_{55} = -\triangleleft F_s F_\tau \triangleright_A \triangleleft N_i N_j \triangleright_\Omega \tilde{C}_{\mathrm{bb11}}, \quad K_{65} = 0, \quad K_{75} = 0$$

$$K_{85} = \triangleleft F_s F_\tau \triangleright_A \triangleleft N_i N_j \triangleright_\Omega \tilde{C}_{\mathrm{bb41}}, \quad K_{16} = \triangleleft F_s F_{\tau,z} \triangleright_A \triangleleft N_i N_j \triangleright_\Omega, \quad K_{26} = 0$$

$$K_{36} = \triangleleft F_s F_\tau \triangleright_A \triangleleft N_j N_{i,x} \triangleright_\Omega$$

$$K_{46} = -\triangleleft F_s F_\tau \triangleright_A \triangleleft N_j N_{i,x} \triangleright_\Omega \tilde{C}_{\mathrm{ab42}} - \triangleleft F_s F_\tau \triangleright_A \triangleleft N_j N_{i,y} \triangleright_\Omega \tilde{C}_{\mathrm{ab52}}, \quad K_{56} = 0$$

$$K_{66} = -\triangleleft F_s F_\tau \triangleright_A \triangleleft N_i N_j \triangleright_\Omega \tilde{C}_{\mathrm{bb22}}, \quad K_{76} = -\triangleleft F_s F_\tau \triangleright_A \triangleleft N_i N_j \triangleright_\Omega \tilde{C}_{\mathrm{bb32}}$$

$$K_{86} = 0, \quad K_{17} = 0$$

$$K_{27} = \triangleleft F_s F_{\tau,z} \triangleright_A \triangleleft N_i N_j \triangleright_\Omega, \quad K_{37} = \triangleleft F_s F_\tau \triangleright_A \triangleleft N_j N_{i,y} \triangleright_\Omega$$

$$K_{47} = -\triangleleft F_s F_\tau \triangleright_A \triangleleft N_j N_{i,x} \triangleright_\Omega \tilde{C}_{\mathrm{ab43}} - \triangleleft F_s F_\tau \triangleright_A \triangleleft N_j N_{i,y} \triangleright_\Omega \tilde{C}_{\mathrm{ab53}}, \quad K_{57} = 0$$

$$K_{67} = -\triangleleft F_s F_\tau \triangleright_A \triangleleft N_i N_j \triangleright_\Omega \tilde{C}_{\mathrm{bb23}}, \quad K_{77} = -\triangleleft F_s F_\tau \triangleright_A \triangleleft N_i N_j \triangleright_\Omega \tilde{C}_{\mathrm{bb33}}$$

$$K_{87} = 0, \quad K_{18} = -\triangleleft F_s F_\tau \triangleright_A \triangleleft N_j N_{i,x} \triangleright_\Omega \tilde{C}_{\mathrm{ab14}} - \triangleleft F_s F_\tau \triangleright_A \triangleleft N_j N_{i,y} \triangleright_\Omega \tilde{C}_{\mathrm{ab34}}$$

$$K_{28} = -\triangleleft F_s F_\tau \triangleright_A \triangleleft N_j N_{i,x} \triangleright_\Omega \tilde{C}_{\mathrm{ab34}} - \triangleleft F_s F_\tau \triangleright_A \triangleleft N_j N_{i,y} \triangleright_\Omega \tilde{C}_{\mathrm{ab24}}$$

$$K_{38} = 0, \quad K_{48} = \triangleleft F_s F_{\tau,z} \triangleright_A \triangleleft N_i N_j \triangleright_\Omega, \quad K_{58} = \triangleleft F_s F_\tau \triangleright_A \triangleleft N_i N_j \triangleright_\Omega \tilde{C}_{\mathrm{bb14}}$$

$$K_{68} = 0, \quad K_{78} = 0, \quad K_{88} = -\triangleleft F_s F_\tau \triangleright_A \triangleleft N_i N_j \triangleright_\Omega \tilde{C}_{\mathrm{bb44}} \tag{8.126}$$

The in-plane and through-the-thickness integrals are defined as in the previous sections. The non-zero elements of the mass fundamental nucleus $\boldsymbol{M}^{(e)k\tau sij}$ (in analogy with the previous sections) are:

$$M_{11}^{k\tau sij} = M_{22}^{k\tau sij} = M_{33}^{k\tau sij} = \rho^k \triangleleft N_i N_j \triangleright_\Omega \triangleleft F_\tau F_s \triangleright_A \tag{8.127}$$

The obtained fundamental nuclei are used in the governing equations, which do not formally change with respect to the PVD($\boldsymbol{u}$,Φ) case (see Equation (8.47)).

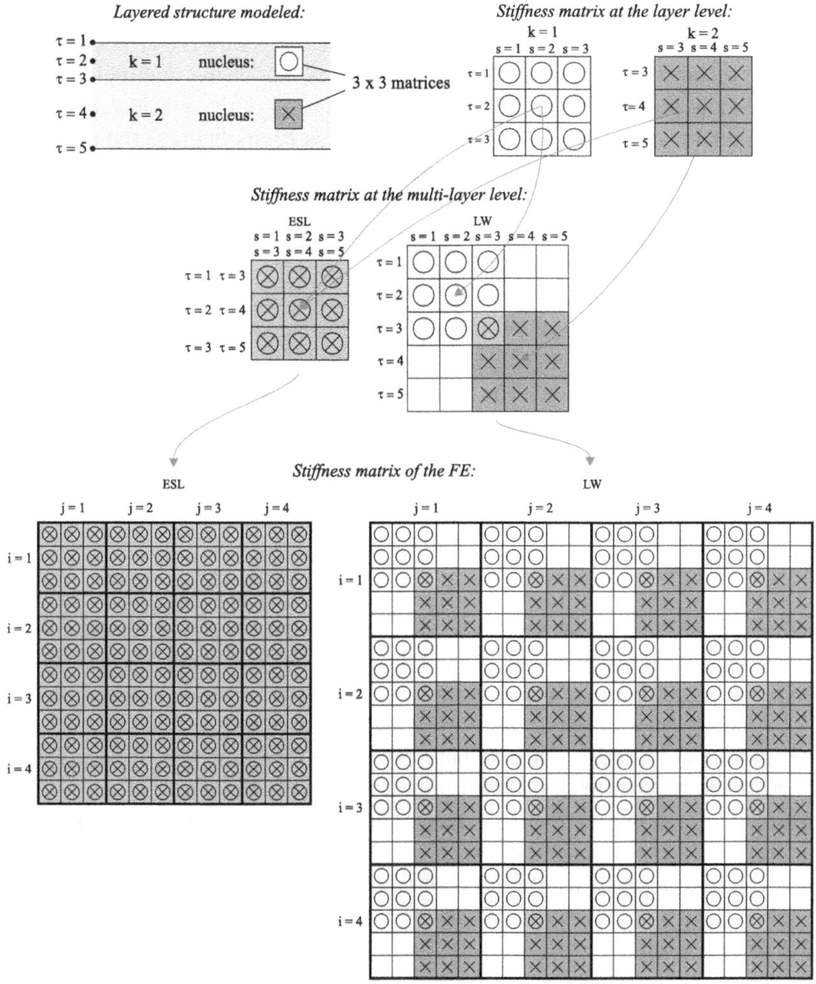

Figure 8.10 Assembly procedure for the FE stiffness matrix by using fundamental nucleus. Example of a two-layered plate: ED2($\boldsymbol{u}$) and LD2($\boldsymbol{u}$) for Q4 element.

8.7 FE assembly procedure and concluding remarks

The basic concepts related to the assembly of fundamental nuclei have already been given in Section 6.7, for the case of plate geometries for analytical closed-form solutions. When dealing with FE, shape functions are included in the procedure and two further indexes, i, j, are considered for the nodes. The assembly procedure followed to obtain the stiffness matrix of a generic FE in the CUF is then summarized in the scheme in Figure 8.10, where both ED2 and LD2 kinematic descriptions were chosen for the Q4 plate FE (see Section 6.8 for the acronyms). The procedure is the same as that of the multifield/mixed formulation, the only difference being the dimension of the considered fundamental nucleus. Once the FE stiffness matrix has been obtained, the standard assembly procedure leads to the stiffness matrix of the structure. The same procedure can be applied to obtain the mass matrix.

The system of acronyms for FE refined and advanced 2D models has already been described in Section 6.8 and Figure 6.12.

The electromechanical FE models (both refined and mixed models) degenerate into pure mechanical FE models in the same way as that described in Section 6.9 for the analytical closed-form plate solution. CLT and FSDT (for both pure mechanical and electromechanical problems) can be obtained as particular cases of an ESL model with linear expansion in the thickness direction whose details have already been discussed in Section 6.10.

References

Ballhause D, D'Ottavio M, Kröplin B, and Carrera E 2005 A unified formulation to assess multilayered theories for piezoelectric plates. *Comput. Struct.* **83**, 1217–1235.

Brischetto S 2009 Classical and mixed advanced models for sandwich plates embedding functionally graded cores. *J. Mech. Mater. Struct.* **4**, 13–33.

Brischetto S and Carrera E 2009 Refined 2D models for the analysis of functionally graded piezoelectric plates. *J. Intell. Mater. Syst. Struct.* **20**, 1783–1797.

Brischetto S and Carrera E 2010 Advanced mixed theories for bending analysis of functionally graded plates. *Comput. Struct.* **88**, 1474–1483.

Brischetto S, Carrera E, and Demasi L 2009a Improved bending analysis of sandwich plates using a zig-zag function. *Comp. Struct.* **89**, 408–415.

Brischetto S, Carrera E, and Demasi L 2009b Free vibration of sandwich plates and shells by using zig-zag function. *Shock Vib.* **16**, 495–503.

Brischetto S, Carrera E, and Demasi L 2009c Improved response of unsymmetrically laminated sandwich plates by using zig-zag functions. *J. Sandwich. Struct. Mater.* **11**, 257–267.

Carrera E 1995 A class of two-dimensional theories for anisotropic multilayered plates analysis. *Accad. Sci. Torino, Mem. Sci. Fis.* **19–20**, 1–39.

Carrera E 2002 Theories and finite elements for multilayered anisotropic, composite plates and shells. *Arch. Comput. Methods Eng.* **9**, 87–140.

Carrera E 2003 Historical review of zig-zag theories for multilayered plates and shells. *Appl. Mech. Rev.* **56**, 287–309.

Carrera E and Boscolo M 2007 Classical and mixed finite elements for static and dynamic analysis of piezoelectric plates. *Int. J. Numer. Methods. Eng.* **70**, 1135–1181.

Carrera E and Brischetto S 2007a Piezoelectric shell theories with "a priori" continuous transverse electro-mechanical variables. *J. Mech. Mater. Struct.* **2**, 377–398.

Carrera E and Brischetto S 2007b Reissner mixed theorem applied to static analysis of piezoelectric shells. *J. Intell. Mater. Syst. Struct.* **18**, 1083–1107.

Carrera E and Brischetto S 2008a Analysis of thickness locking in classical, refined and mixed multilayered plate theories. *Comp. Struct.* **82**, 549–562.

Carrera E and Brischetto S 2008b Analysis of thickness locking in classical, refined and mixed theories for layered shells. *Comp. Struct.* **85**, 83–90.

Carrera E and Brischetto S 2009a A survey with numerical assessment of classical and refined theories for the analysis of sandwich plates. *Appl. Mech. Rev.* **62**, 1–17.

Carrera E and Brischetto S 2009b A comparison of various kinematic models for sandwich shell panels with soft core. *J. Compos. Mater.* **43**, 2201–2221.

Carrera E, Boscolo M, and Robaldo A 2007 Hierarchic multilayered plate elements for coupled multifield problems of piezoelectric adaptive structures: formulation and numerical assessment. *Arch. Comput. Methods Eng.* **14**, 383–430.

Carrera E, Brischetto S, and Nali P 2008 Variational statements and computational models for multifield problems and multilayered structures. *Mech. Adv. Mater. Struct.* **15**, 182–198.

Demasi L 2008a ∞^3 hierarchy plate theories for thick and thin composite plates: the generalized unified formulation. *Comp. Struct.* **84**, 256–270.

Demasi L 2008b 2D, quasi 3D and 3D exact solutions for bending of thick and thin sandwich plates. *J. Sandwich. Struct. Mater.* **10**, 271–310.

Hsu T and Wang JT 1970 A theory of laminated cylindrical shells consisting of layers of orthotropic laminae. *AIAA J.* **8**, 2141–2146.

Hsu T and Wang JT 1971 Rotationally symmetric vibrations of orthotropic layered cylindrical shells. *J. Sound Vib.* **16**, 473–487.

Ikeda T 1996 *Fundamentals of Piezoelectricity*. Oxford University Press.

Librescu L and Schmidt R 1988 Refined theories of elastic anisotropic shells accounting for small strains and moderate rotations. *Int. J. Non-linear Mech.* **23**, 217–229.

Librescu L and Wu EM 1977 A higher-order theory of plate deformation. Part 2: laminated plates. *J. Appl. Mech.* **44**, 669–676.

Murakami H 1985 Laminated composite plate theory with improved in-plane responses. In *ASME Proceedings of Pressure Vessels & Piping Conference*.

Murakami H 1986 Laminated composite plate theory with improved in-plane responses. *J. Appl. Mech.* **53**, 661–666.

Reddy JN 2004 *Mechanics of Laminated Composite Plates and Shells; Theory and Analysis*. CRC Press.

Reissner E 1984 On a certain mixed variational theory and a proposed application. *Int. J. Numer. Methods Eng*. **20**, 1366–1368.

Robbins DH Jr and Reddy JN 1993 Modeling of thick composites using a layer-wise theory. *Int. J. Numer. Methods Eng*. **36**, 655–677.

Srinivas S 1973 A refined analysis of composite laminates. *J. Sound Vib*. **30**, 495–507.

Zienkiewicz OC, Taylor RL, and Zhu JZ 2005 *The Finite Element Method*. Elsevier Butterworth–Heinemann.

9

Numerical evaluation and assessment of classical and advanced theories using MUL2 software

MUL2 is the acronym of **MUL**tifield problems for **MUL**tilayered structures. It is academic software developed in-house for the thermo/electro/magneto/mechanical analysis of multilayered plates, shells, and beams. It has been implemented with the CUF, which has been extensively discussed in this book. Additional information about the MUL2 software can be found at http://www.mul2.com. In this chapter, the electromechanical MUL2 code for analytical closed-form solutions of plates and shells and the FE version of MUL2 for plate geometries are described with emphasis on the input and output files and the computing architecture. Simple examples are given for plate, shell, and beam geometries in order to point out the importance of refined models for the static and dynamic analysis of smart structures. Smart structures are multilayered configurations that embed piezoelectric and orthotropic layers. Since the use of classical theories could be inappropriate for these structures, both analytical closed-form and FE solutions are compared.

Plates and Shells for Smart Structures: Classical and Advanced Theories for Modeling and Analysis, First Edition. Erasmo Carrera, Salvatore Brischetto and Pietro Nali.

9.1 The MUL2 software for plates and shells: analytical closed-form solutions

The MUL2 software for the analytical solution of plates and shells considers only simply supported multilayered structures subjected to harmonic mechanical and electrical loads. The elastic coefficients Q_{16}, Q_{26}, Q_{36}, and Q_{45}, the piezoelectric coefficients e_{25}, e_{14}, and e_{36}, and the dielectric constant ε_{12} are set to zero in order to obtain a closed form of the governing equations (Reddy, 2004). Two main input files must be compiled in the code. The first file concerns embedded layers in multilayered structures (their thickness, the stacking sequence, and their elastic and piezoelectric properties) while the second one indicates the number, type, and magnitude of the considered electrical and mechanical loads, the geometry of the investigated plates and shells, the boundary conditions, the type of analysis, and the chosen 2D model. The output files give the amplitudes through the thickness direction of the displacements, stresses, strains, electric potential, electric field components, and electric displacements; these quantities can be obtained directly from the model if they are primary variables in the proposed 2D theory, otherwise they are obtained from opportune constitutive equations, after a dedicated post-processing.

The two main input files can also be compiled by means of opportune graphical interfaces, as described at http://www.mul2.com. The structures of these two files are given in Figures 9.1 and 9.2 for the material data and for the loading, the geometry, and the 2D theory data, respectively. In the first file, we indicate the total number of embedded layers in the multilayered piezoelectric plate or shell (N_l), and a number of blocks equal to N_l is obtained which describe the elastic and piezoelectric properties of each embedded layer. In the first line of each block, k indicates the layer, h_k is the thickness of the kth layer, and θ_k indicates the in-plane orthotropic angle of each k layer with respect to the global reference system (x,y,z) (Reddy 2004). The second line contains the three Young's moduli in the material reference system (1, 2, and 3 direction components), while the third and fourth lines contain the shear moduli and the Poisson ratio components, respectively. The fifth line indicates the mass density of the material and the sixth line contains the dielectric coefficients. Finally, the last line has the five piezoelectric coefficients that are different from zero in the case of the analytical closed-form solution (a schematic description of this file is given in Figure 9.1). The second file, which is briefly described in Figure 9.2, contains further information for correct functioning of the MUL2 software. The first line contains the number that indicates the chosen order of expansion in the thickness direction for the modeled mechanical and electrical variables (N from 1 to 4). The next line contains the index, which specifies whether the displacement components are modeled in ESL, LW, or ESL+ZZ form; all the other electromechanical variables are always modeled in LW form. The third

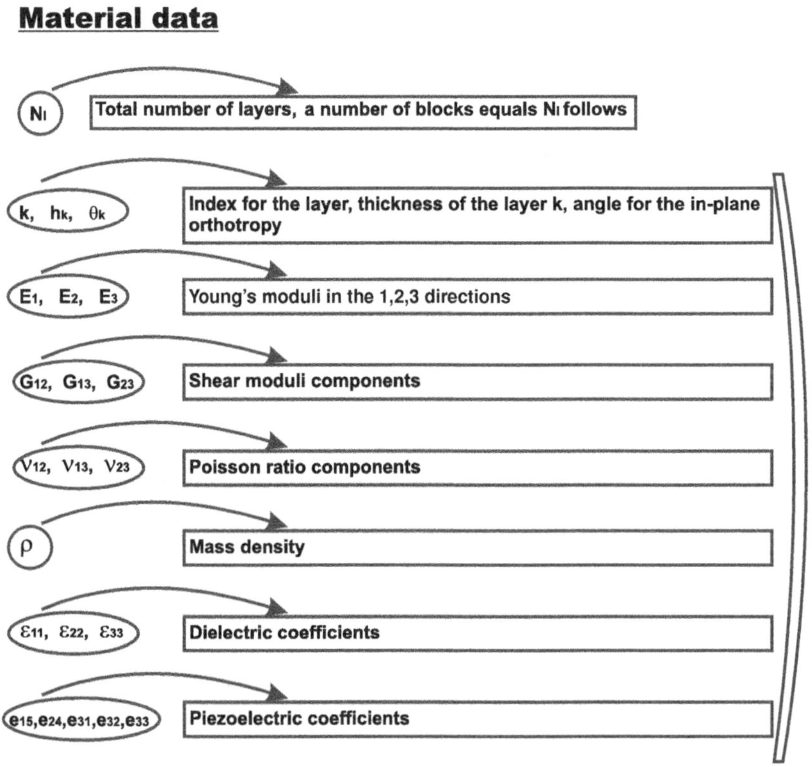

Figure 9.1 Analytical closed-form solution MUL2 software: description of the input file containing the material data of the embedded layers.

line permits one to choose the opportune variational statement, PVD (where the displacement and electric potential are primary variables), or one of the three possible extensions of RMVT (where the additional primary variables are the transverse stresses and/or the transverse normal electric displacement). In the fourth line, it is possible to specify whether the structure has a plate or shell geometry, and in the next line one chooses the type of analysis, which can be either static or dynamic (typical free-vibration problem). The MUL2 software also allows one to obtain classical theories, if an ESL theory with linear expansion in the thickness direction based on the PVD(u,Φ) variational statement is considered. The sixth line permits the CLT or FSDT analysis to be set by means of a typical penalty technique. The next line pertains to the geometry: a and b are the plate or shell in-plane dimensions, and $1/R_\alpha$ and $1/R_\beta$ are curvatures where R_α and R_β indicate the mean value of the radii of curvature in the α and β directions, respectively ($1/R_\alpha$ and $1/R_\beta$ are zero for

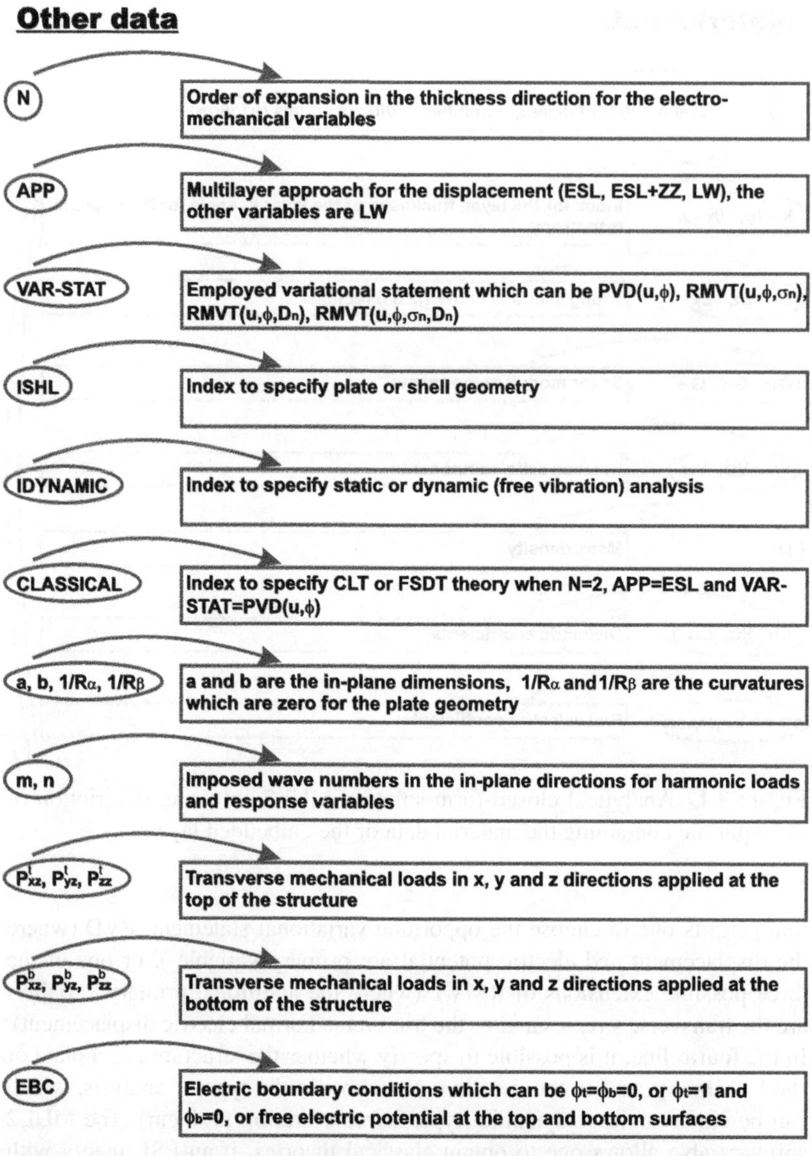

Figure 9.2 Analytical closed form solution MUL2 software: description of the input file containing the data for geometry, 2D approaches, and loading conditions.

a plate geometry). The eighth line indicates the wave numbers m and n in the in-plane directions for the harmonic form of the mechanical and electrical loads and for the response variables. The ninth and tenth lines contain the transverse mechanical loads in the x, y, and z directions which are applied at the top (t) and bottom (b) of the multilayered structure, respectively. The last line is for the electrical boundary and loading conditions, and it is here that one indicates whether the multilayered structure is an open-circuit configuration (free electric potential at the external surfaces) or a closed-circuit configuration (zero electric potential applied at the top and bottom surfaces). It is also possible to apply an electric potential at the top, while the bottom surface is set to zero.

Some examples of the output files are given in Figures 9.3 and 9.4. In the first figure, the three displacement components (u, v, and w) are given in the second, third, and fourth columns; the first column indicates the thickness coordinate z, which goes from $-h/2$ to $+h/2$, where h is the total thickness of the multilayered structure; u, v, and w are given in terms of maximum amplitudes in the plane. Figure 9.4 is an example of an output file for the three transverse stresses obtained from the governing equations, if a RMVT application is employed. In the case of a shell geometry, $\sigma_{\alpha z}$, $\sigma_{\beta z}$, and σ_{zz} are given in the second, third, and fourth columns, respectively, in terms of maximum amplitudes. The first column is for the thickness coordinate z.

Output file for displacement components

z	u	v	w
-0.50000000E+00	0.31159921E-06	0.00000000E+00	0.67308084E-06
-0.40000000E+00	0.29415142E-06	0.00000000E+00	0.67308084E-06
-0.30000000E+00	0.27670364E-06	0.00000000E+00	0.67308084E-06
-0.20000000E+00	0.25925585E-06	0.00000000E+00	0.67308084E-06
-0.10000000E+00	0.24180806E-06	0.00000000E+00	0.67308084E-06
0.00000000E+00	0.22436028E-06	0.00000000E+00	0.67308084E-06
0.10000000E+00	0.20691249E-06	0.00000000E+00	0.67308084E-06
0.20000000E+00	0.18946470E-06	0.00000000E+00	0.67308084E-06
0.30000000E+00	0.17201691E-06	0.00000000E+00	0.67308084E-06
0.40000000E+00	0.15456913E-06	0.00000000E+00	0.67308084E-06
0.50000000E+00	0.13712134E-06	0.00000000E+00	0.67308084E-06

Figure 9.3 Analytical closed-form solution MUL2 software: description of the output file containing the three displacement amplitudes in the thickness direction z.

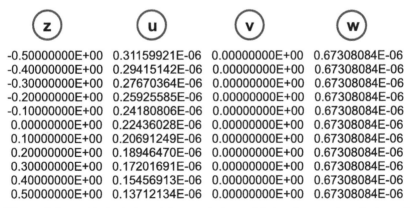

Output file for transverse stress components

z	$\sigma_{\alpha z}$	$\sigma_{\beta z}$	σ_{zz}
-0.50000E-01	0.27992E-03	0.27992E-03	0.00000E+00
-0.40000E-01	0.85946E+01	0.85946E+01	0.28002E-01
-0.30000E-01	0.15279E+02	0.15279E+02	0.10400E+00
-0.20000E-01	0.20053E+02	0.20053E+02	0.21600E+00
-0.10000E-01	0.22918E+02	0.22918E+02	0.35200E+00
0.00000E+00	0.23873E+02	0.23873E+02	0.50000E+00
0.10000E-01	0.22918E+02	0.22918E+02	0.64800E+00
0.20000E-01	0.20053E+02	0.20053E+02	0.78400E+00
0.30000E-01	0.15279E+02	0.15279E+02	0.89600E+00
0.40000E-01	0.85946E+01	0.85946E+01	0.97200E+00
0.50000E-01	0.27850E-03	0.27850E-03	0.10000E+01

Figure 9.4 Analytical closed-form solution MUL2 software: description of the output file containing the transverse stress amplitudes in the thickness direction z.

The computing architecture of the MUL2 software can be divided into eight main parts, which are summarized in Figure 9.5. The first part of the code reads the input data given in the two files shown in Figures 9.1 and 9.2, and in this way, MUL2 obtains all the information about the material properties of the embedded layers, the geometry of the structure and its stacking layer configuration, the loadings and boundary conditions, the chosen 2D theory, and the type of analysis, which can be either static or dynamic (free-vibration problem). In the second part of the code, the material coefficients (elastic, piezoelectric, and dielectric ones) are positioned in the matrices of the constitutive equations (see Equations (2.25)–(2.30)) (Ikeda, 1996). These matrices are transformed from the material reference system (1,2,3) to the problem reference system (x,y,z) by means of the rotation matrix which considers the orthotropic angle θ_k (Reddy, 2004). These matrices are ready for the PVD application; when one of the three possible extensions of the RMVT application is considered, the new coefficients are calculated (see $\hat{\boldsymbol{C}}$ in Equations (2.65) and (6.87)–(6.102), $\bar{\boldsymbol{C}}$ in Equations (2.79) and (6.139)–(6.154), and $\tilde{\boldsymbol{C}}$ in Equations (2.88) and (6.207)–(6.222)). The code calculates the in-plane dimensions and radii of curvature at each thickness coordinate z_k, the total thickness of the multilayered structure, and the reference system positioned at the middle surface of each layer from the geometrical data (in-plane dimensions, radii of curvature at the reference middle surface in the case of shells, and the thickness of each layer k). These first two parts constitute the pre-processing of the MUL2 software.

The MUL2 software contains all the fundamental nuclei in explicit algebraic form, as already shown in Equations (7.42)–(7.46) for PVD($\boldsymbol{u}$, Φ),

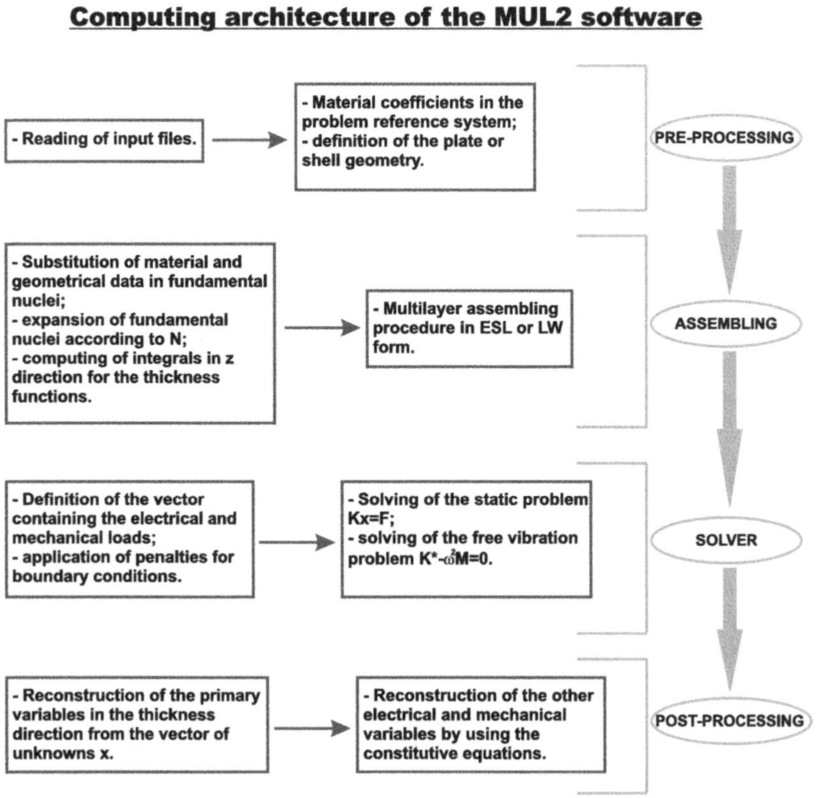

Figure 9.5 Computing architecture of the analytical MUL2 software.

Equations (7.75)–(7.83) for RMVT($\boldsymbol{u}$, Φ, $\boldsymbol{\sigma}_n$), Equations (7.111)–(7.119) for RMVT($\boldsymbol{u}$, Φ, $\boldsymbol{D}_n$), and Equations (7.156)–(7.171) for RMVT($\boldsymbol{u}$, Φ, $\boldsymbol{\sigma}_n$, $\boldsymbol{D}_n$) (Carrera *et al.*, 2008; 2010). These nuclei are only introduced for the shell geometry because they simply degenerate into those for cylindrical shell or plate geometries when one of the radii of curvature or both are infinite. The third part of the code substitutes the geometrical and material data in the opportune fundamental nuclei, depending on the chosen variational statement, then it expands the fundamental nuclei in the τ and s directions (see Figures 6.10 and 6.11) according to the chosen order of expansion in the thickness direction for the considered plate/shell theory. These expanded nuclei are given for each k layer embedded in the multilayered structure. Finally, the integrals in the z direction inside the fundamental nuclei (see Equations (6.41) for plates and (7.41) for shells) are numerically computed. The thickness functions F_τ and F_s, included in these integrals, can be seen in ESL (use of Taylor expansion) or LW (use of combinations of Legendre polynomials) form. The integrals in the z direction are computed by means of Gauss points and weights; six Gauss points are sufficient for an excellent approximation. These expanded fundamental nuclei are calculated for each k layer. The fourth part of the code permits the multilayer assembly procedure, as already shown in Figures 6.10 and 6.11, to be conducted. The obtained fundamental nuclei can be of type $\boldsymbol{K}_{uu}$, $\boldsymbol{K}_{u\sigma}$, $\boldsymbol{K}_{\sigma u}$, or $\boldsymbol{K}_{\sigma\sigma}$ (Carrera *et al.*, 2008; 2010). The first type is assembled as ESL in row and column directions, as in the case of the ESL theory, and as LW in row and column directions, as in the case of the LW theory. The second type is assembled as ESL in the row direction and LW in the column direction, as in the case of the ESL theory, and as LW in row and column directions, as in the case of the LW theory. The third type is assembled as LW in the row direction and ESL in the column direction, as in the case of the ESL theory, and as LW in row and column directions, as in the case of the LW theory. The last type is always assembled as LW in row and column directions for both the ESL and LW theories. The multilayer assembly procedure is accomplished by means of an opportune connectivity matrix, through the thickness direction, which permits one to understand when the matrices are simply summed or when the compatibility and/or equilibrium conditions at each layer interface must be enforced. These two blocks constitute the assembly procedure of the MUL2 software.

The algebraic governing equations, in closed form, have the structure that has already been described in Equations (6.30) and (7.30) for PVD($\boldsymbol{u}$, Φ), in Equations (6.58) and (7.58) for RMVT($\boldsymbol{u}$, Φ, $\boldsymbol{\sigma}_n$), in Equations (6.110) and (7.94) for RMVT($\boldsymbol{u}$, Φ, $\boldsymbol{D}_n$), and in Equations (6.162) and (7.130) for RMVT($\boldsymbol{u}$, Φ, $\boldsymbol{\sigma}_n$, $\boldsymbol{D}_n$). In the fifth part of the code, the mechanical loads at the top and bottom surfaces of the structure are introduced into the $\boldsymbol{p}^k_{us}$ vector, according to the direction and position of the transverse mechanical load. The electrical loads are given directly, in terms of electric potential, which is introduced into the Φ^k_τ vector; no variationally consistent electric loads $\boldsymbol{p}^k_{\Phi s}$ are considered. In the case of closed- and open-circuit configurations, the boundary conditions for the electric potential are directly imposed in the Φ^k_τ vector (Carrera and Brischetto, 2007a,b). In the sixth part of the code, the system of governing equations is solved; for a static problem, the code solves a general system of the type $\boldsymbol{K}\boldsymbol{x} = \boldsymbol{F}$, where $\boldsymbol{x}$ are the unknowns and $\boldsymbol{F}$ is the load vector. In the case of free-vibration analysis, a typical eigenvalue problem is solved for a general system of the type $\boldsymbol{K}^* - \omega^2\boldsymbol{M} = 0$; the eigenvalues are the frequencies and the relative eigenvector is computed for each value. This makes it possible to obtain the vibration modes of the structure in terms of primary variables. The matrix $\boldsymbol{K}^*$ is obtained after an opportune static condensation of the electromechanical matrices. The fifth and sixth parts constitute the solver of the MUL2 software.

The post-processing is divided into two parts: first the vector of unknowns $\boldsymbol{x}$ is employed to recover the values of the primary variables through the thickness direction. For example, the displacement components through the thickness direction (see Figure 9.3) in the case of an ESL approach are obtained using Equations (6.2) where the thickness functions are the Taylor polynomials of Equations (6.3); the unknowns in the vector $\boldsymbol{x}$ are u_τ, v_τ, and, w_τ and they permit one to obtain u, v, and w through an opportune substitution of the thickness coordinate in the thickness functions of Equations (6.3). In the case of a LW approach, Equations (6.9) are used, in which the thickness functions are a combination of Legendre polynomials, as given in Equations (6.11) and (6.12), the unknowns in the vector $\boldsymbol{x}$ are u_τ^k, v_τ^k, and, w_τ^k, and they permit u, v, and w to be obtained through an opportune substitution of the thickness coordinate in the thickness functions of Equations (6.11) and (6.12). The displacement is a primary variable in the PVD variational statement and in each extension of the RMVT approach. The transverse stresses shown in Figure 9.4 are given through the thickness, in such a form, when they are primary variables in the chosen governing equations (the RMVT($\boldsymbol{u}$, Φ, $\boldsymbol{\sigma}_n$) and RMVT($\boldsymbol{u}$, Φ, $\boldsymbol{\sigma}_n$, $\boldsymbol{D}_n$) cases). Transverse stresses are always recovered in LW form; as shown in Equations (6.16), the vector $\boldsymbol{x}$ contains the unknowns $\sigma_{xz\tau}^k$, $\sigma_{yz\tau}^k$, and $\sigma_{zz\tau}^k$ for the plate case and $\sigma_{\alpha z\tau}^k$, $\sigma_{\beta z\tau}^k$, and $\sigma_{zz\tau}^k$ for the shell case. The thickness functions, in terms of combinations of Legendre polynomials, are those in Equations (6.11) and (6.12) and the opportune value of z_k is substituted in these equations. In the case of RMVT($\boldsymbol{u}$,Φ,$\boldsymbol{\sigma}_n$,$\boldsymbol{\mathcal{D}}_n$) and RMVT($\boldsymbol{u}$,$\Phi$,$\boldsymbol{\mathcal{D}}_n$), the primary variable $\mathcal{D}_z^k$ is obtained by means of Equation (6.20), where the vector of unknowns $\boldsymbol{x}$ also contains $\mathcal{D}_{z\tau}^k$; this variable is always in LW form. Φ^k is a primary variable in each electromechanical problem and it is always considered in LW form by means of Equation (6.15), where the unknown Φ_τ^k is considered inside the vector $\boldsymbol{x}$. The second part of the post-processing permits one to recover the other quantities, which are not primary variables in the proposed governing equations through the thickness direction of the structure. In this case, we use the constitutive equations given in Equations (6.23)–(6.25) for the PVD($\boldsymbol{u}$, Φ) case, and in Equations (6.53)–(6.56) for RMVT($\boldsymbol{u}$, Φ, $\boldsymbol{\sigma}_n$), in Equations (6.105)–(6.108) for RMVT($\boldsymbol{u}$, Φ, $\boldsymbol{D}_n$), and in Equations (6.157)–(6.160) for RMVT($\boldsymbol{u}$, Φ, $\boldsymbol{\sigma}_n$, $\boldsymbol{D}_n$) cases. In order to use such constitutive equations, it is necessary to calculate some derivatives to obtain the mechanical strains and the electric field (see Equations (2.31), (2.32), (2.35), and (2.36) for shells, and Equations (2.38)–(2.41) for plates). The derivatives are calculated exactly, by means of the harmonic forms, for the analytical code. Another advantage of the closed form of the MUL2 software is that it does not need any post-processing in the plane because it only works on the amplitudes of the variables. If a value is needed in a different point in the xy-plane, it is sufficient to use the harmonic forms.

9.1.1 Classical plate/shell theories as particular cases in the MUL2 software

Classical plate/shell theories can be considered as particular cases of refined 2D theories on the basis of the CUF. Classical theories were obtained in Sections 6.10 and 7.10 as particular cases of the ESL theory with a linear order of expansion in the thickness direction (ED1) for plate and shell geometries, respectively. CLT and FSDT, employed for the analysis performed in this chapter, have been proposed in the framework of the MUL2 code and include the displacement kinematics, as hypothesized by Kirchhoff (Kirchhoff, 1850) and Reissner and Mindlin (Reissner 1945; Mindlin 1951), respectively, and the electric potential in LW form with linear expansion in the thickness direction.

MUL2 obtains CLT($\boldsymbol{u}$, Φ) and FSDT($\boldsymbol{u}$, Φ) from the ED1($\boldsymbol{u}$, Φ) theory. In the input files, the data are set for the ED1($\boldsymbol{u}$, Φ) model, which means a PVD($\boldsymbol{u}$, Φ) variational statement with linear displacement components in the thickness direction ($N = 2$ terms of expansion for each component) in ESL form and linear electric potential expanded in the z direction in LW form. When CLT or FSDT is set in CLASSICAL data (see Figure 9.2), the MUL2 software obtains these theories from the ED1($\boldsymbol{u}$, Φ) model, via typical penalty techniques, which will be discussed in the second part of this section.

By referring to Equation (6.231), it is possible to directly write the ED1($\boldsymbol{u}$, Φ) kinematic model for a spherical shell geometry (the MUL2 code analyzes cylindrical shells and plates as particular cases by setting one of the radii of curvature (R_α, R_β), or both, to infinity, in the fundamental nuclei):

$$
\begin{aligned}
u(\alpha, \beta, z) &= u_0(\alpha, \beta) + z u_1(\alpha, \beta) \\
v(\alpha, \beta, z) &= v_0(\alpha, \beta) + z v_1(\alpha, \beta) \\
w(\alpha, \beta, z) &= w_0(\alpha, \beta) + z w_1(\alpha, \beta) \\
\Phi^k(\alpha, \beta, z) &= F_t \Phi_t^k(\alpha, \beta) + F_b \Phi_b^k(\alpha, \beta)
\end{aligned}
\tag{9.1}
$$

The three displacement components in Equation (9.1) show a typical linear Taylor expansion ($N = 2$ terms for each component through the thickness) and the electric potential is linearly expanded for each k layer in LW form (where the subscripts t and b indicate the top and bottom of each layer k, respectively, and F_t and F_b are the thickness functions obtained as combinations of Legendre polynomials). The CLT($\boldsymbol{u}$, Φ) and FSDT($\boldsymbol{u}$, Φ) theories have the same expansion as the electric potential discussed in Equation (9.1), while the displacement kinematics are those already given in Equation (3.3) of Section 3.3.1 and in Equation (3.4) of Section 3.3.2 for CLT($\boldsymbol{u}$, Φ) and

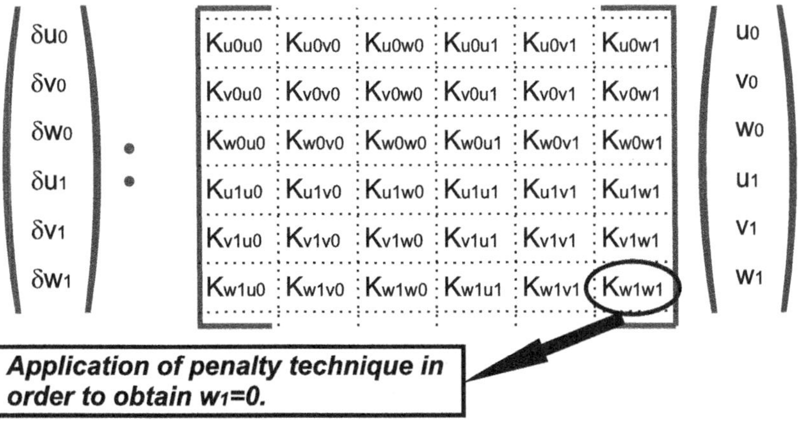

Figure 9.6 Penalty applied to the stiffness matrix of ED1($\boldsymbol{u}$, Φ) theory to impose zero w_1 component.

FSDT($\boldsymbol{u}$, Φ), respectively (see Kirchhoff 1850, Mindlin 1951, and Reissner 1945 for further details).

FSDT($\boldsymbol{u}$, Φ) is obtained from an ED1($\boldsymbol{u}$, Φ) model by simply imposing a constant transverse displacement through the thickness direction z, which means a zero w_1 component is obtained. In this way, the components u_1 and v_1 are the typical rotations of the FSDT model around the in-plane axes. Figure 9.6 shows the fundamental nucleus for the mechanical part that is expanded in the τ and s directions, according to the kinematic model of Equation (9.1). By means of a penalty application to the term K_{w1w1}, we obtain the model for FSDT($\boldsymbol{u}$, Φ), where $w(\alpha, \beta, z) = w_0(\alpha, \beta) =$ constant and the electric potential is linear in LW form.

CLT($\boldsymbol{u}$, Φ) is obtained from the FSDT($\boldsymbol{u}$, Φ) theory by simply considering an infinite shear rigidity which permits one to obtain the zero transverse shear strains γ_{yz} and γ_{xz}. In this way, on the basis of the FSDT($\boldsymbol{u}$, Φ) kinematic displacement model, the two terms u_1 and v_1 can be written as the partial derivatives of the middle transverse displacement $\partial w_0/\partial x$ and $\partial w_0/\partial y$ (for the plate case). The MUL2 software obtains such a result by acting on the constitutive equations and penalizing the elastic coefficients Q_{44} and Q_{55}; in the constitutive equations of the MUL2 software, such coefficients are multiplied by the shear correction factor χ, which can also be used to improve the FSDT results. If this shear correction factor is set to infinity, we obtain $\gamma_{yz} = \gamma_{xz} = 0$ by means of a typical penalty technique (see Figure 9.7 for further details). The coefficients Q_{16}, Q_{26}, Q_{36}, and Q_{45} in Figure 9.7 are set to zero in order to obtain analytical closed-form solutions (see Equation (2.16) for comparison purposes).

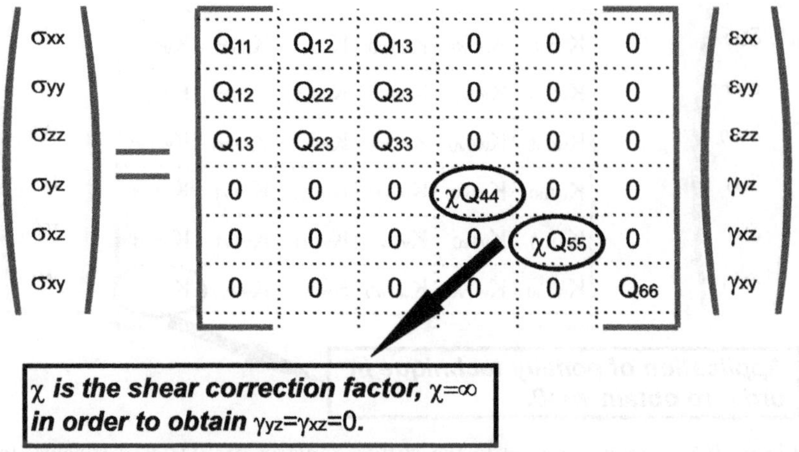

Figure 9.7 Penalty applied to the elastic coefficients matrix to impose zero γ_{yz} and γ_{xz} transverse shear strains.

The thickness locking (TL) mechanism, also known as the Poisson locking phenomenon, appears in each ESL theory with constant or linear expansion of the transverse displacement through the thickness (which means zero or constant transverse normal strain, respectively). This is caused by the use of simplified kinematic assumptions in the plate/shell analysis (Carrera and Brischetto 2008a,b). Two-dimensional plate/shell structures can be analyzed as particular cases of a three-dimensional (3D) continuum by eliminating, via a priori integration, the thickness coordinate z. Such an integration can be made according to two different methods: asymptotic expansion methods or axiomatic methods. The introduction of axiomatic and/or asymptotic approximations could introduce some undesirable mechanisms which are not found in the 3D solution. One of these is Poisson locking, which is related to the use of the plane strain/plane stress hypothesis in thin plate/shell theories. The analysis of thin plate/shell problems is in fact often associated to plane stress assumptions (thin surface problem), while a plane strain hypothesis usually refers to a beam theory. Discussions on plane strain, plane stress, and/or plane elastostatic problems can also be found in Carrera and Brischetto (2008a,b) and Sokolnikoff (1956). However, in most ESL theories, the assumptions on strain fields are conflicting. The *plane strain assumptions* are:

$$\gamma_{yz} = \gamma_{xz} = \epsilon_{zz} = 0 \tag{9.2}$$

and they are used in place of the more natural *plane stress assumptions*:

$$\sigma_{yz} = \sigma_{xz} = \sigma_{zz} = 0 \tag{9.3}$$

This contradiction introduces a "locking mechanism" that makes the plate/shell model no longer applicable in some cases. Thickness locking (also known as Poisson locking) is the name that has been given to this mechanism: TL does not permit that ESL analyses with constant or linear transverse displacement w through the thickness lead to 3D solutions in thin plate problems. One technique that can contrast TL consists of modifying the elastic stiffness coefficients by forcing the "contradictory" condition known as the *transverse normal stress zero condition*:

$$\sigma_{zz} = 0 \tag{9.4}$$

This method is used in the MUL2 software to contrast the Poisson locking that appears in CLT, FSDT, ED1 theories for pure mechanical problems, while Poisson locking is not corrected in the MUL2 code for electromechanical analysis. Poisson locking appears if, and only if, a plate theory shows a constant distribution of transverse normal strain ϵ_{zz}; in other words, to avoid TL, the plate/shell theories would require at least a parabolic distribution of the transverse displacement component w. Hooke's law presented in Chapter 2 is suitable for such theories. For these reasons, all the MUL2 theories that are different from CLT, FSDT, and ED1 models do not have Poisson locking phenomena and do not need any correction of the elastic coefficients. For further details, see the complete discussion reported in the books by Librescu (1975), Reddy (2004), Washizu (1968) as well as the discussion quoted in Carrera and Brischetto (2008a,b).

The modified stiffness coefficients (also known as *reduced stiffness coefficients*) can be obtained by imposing the condition $\sigma_{zz} = 0$ in the equations in Figure 9.7. These reduced elastic coefficients are indicated as $\tilde{Q}_{ij}$, and they are used in ESL theories, with constant or linear transverse displacement w in the thickness direction, in order to avoid the Poisson locking phenomena:

$$\sigma_{xx} = Q_{11}\epsilon_{xx} + Q_{12}\epsilon_{yy} + Q_{13}\epsilon_{zz} \tag{9.5}$$

$$\sigma_{yy} = Q_{12}\epsilon_{xx} + Q_{22}\epsilon_{yy} + Q_{23}\epsilon_{zz} \tag{9.6}$$

$$\sigma_{zz} = Q_{13}\epsilon_{xx} + Q_{23}\epsilon_{yy} + Q_{33}\epsilon_{zz} \tag{9.7}$$

$$\sigma_{yz} = Q_{44}\gamma_{yz} \tag{9.8}$$

$$\sigma_{xz} = Q_{55}\gamma_{xz} \tag{9.9}$$

$$\sigma_{xy} = Q_{66}\gamma_{xy} \tag{9.10}$$

A new form of the transverse normal strain ϵ_{zz}, coherent with the physical strain of the problem, is obtained by imposing the condition $\sigma_{zz} = 0$ in Equation (9.7). This form is used in the constitutive equations in place of the geometrical relation of ϵ_{zz}, which is incoherent with the physical strain and causes the Poisson locking phenomena. This substitution leads to the modified elastic coefficients $\tilde{Q}_{ij}$ which correct the Poisson locking phenomena:

$$\epsilon_{zz} = -\frac{Q_{13}}{Q_{33}}\epsilon_{xx} - \frac{Q_{23}}{Q_{33}}\epsilon_{yy} \tag{9.11}$$

By substituting Equation (9.11) in Equations (9.5)–(9.10), we obtain:

$$\sigma_{xx} = \left(Q_{11} - \frac{Q_{13}^2}{Q_{33}}\right)\epsilon_{xx} + \left(Q_{12} - \frac{Q_{13}Q_{23}}{Q_{33}}\right)\epsilon_{yy} \tag{9.12}$$

$$\sigma_{yy} = \left(Q_{12} - \frac{Q_{23}Q_{13}}{Q_{33}}\right)\epsilon_{xx} + \left(Q_{22} - \frac{Q_{23}^2}{Q_{33}}\right)\epsilon_{yy} \tag{9.13}$$

$$\sigma_{yz} = Q_{44}\gamma_{yz} \tag{9.14}$$

$$\sigma_{xz} = Q_{55}\gamma_{xz} \tag{9.15}$$

$$\sigma_{xy} = Q_{66}\gamma_{xy} \tag{9.16}$$

In this way, the new reduced elastic coefficients used to avoid the Poisson locking phenomena in the MUL2 software, in the case of pure mechanical problems, are:

$$\begin{aligned}
\tilde{Q}_{11} &= \left(Q_{11} - \frac{Q_{13}^2}{Q_{33}}\right)\\
\tilde{Q}_{12} &= \left(Q_{12} - \frac{Q_{13}Q_{23}}{Q_{33}}\right)\\
\tilde{Q}_{22} &= \left(Q_{22} - \frac{Q_{23}^2}{Q_{33}}\right)\\
\tilde{Q}_{44} &= Q_{44}\\
\tilde{Q}_{55} &= Q_{55}\\
\tilde{Q}_{66} &= Q_{66}
\end{aligned} \tag{9.17}$$

LW theories, with a linear expansion in the thickness direction for w, do not show TL because the transverse normal strain has a piecewise constant distribution along the thickness direction z.

9.2 The MUL2 software for plates: FE solutions

The MUL2 software has also been implemented in FE form in order to overcome the main limitations given by the analytical closed-form solution (e.g., simply supported multilayered structures subjected to harmonic mechanical and electrical loads, the elastic coefficients Q_{16}, Q_{26}, Q_{36}, and Q_{45}, the piezoelectric coefficients e_{25}, e_{14}, and e_{36}, and the dielectric constant ε_{12} are set to zero in order to obtain a closed form for governing equations). The computing architecture of the FE version is very similar to that already proposed for the analytical closed-form version. In the FE code, the fundamental nuclei are those that were dealt with in Chapter 8 where the assembly procedure considers the indexes τ and s for the order of expansion, i and j for nodes, and k for the multilayer level. In the analytical closed-form version, the fundamental nuclei are those of Chapter 7 for shells (plates are considered as particular cases), where the indexes i and j for nodes are not considered because they work on amplitudes. The FE code has two main input files: the first file concerns the embedded layers in the multilayered structures (their thickness, the stacking sequence, and their elastic and piezoelectric properties) and the second one indicates the number, type, and magnitude of the considered electrical and mechanical loads, the geometry of the investigated plates, the boundary conditions, the type of analysis, the chosen 2D model, and the information for the post-processing. The output files give the values through the thickness direction of the displacements, stresses, strains, electric potential, electric field components, and electric displacements; these quantities can be obtained directly from the model if they are primary variables in the proposed 2D theory, otherwise they are obtained from the opportune constitutive equations after a dedicated post-processing. They are not the amplitude values, as in the analytical closed-form version, but they are generic values in particular points in the xy-plane.

The two main input files can also be compiled by means of opportune graphical interfaces, as described at http://www.mul2.com. The structures of these two files are briefly explained here. The material data file is the same one that was explained in Figure 9.1 and in the previous section. Its structure is identical to that of the analytical code: the total number of embedded layers in the multilayered piezoelectric plate (N_l) is indicated, therefore a number of blocks equal to N_l are obtained which describe the elastic and piezoelectric properties of each embedded layer. In the first line of each block, k indicates the layer, h_k is the thickness of the kth layer, and θ_k indicates the in-plane orthotropic angle of each layer k with respect to the global reference system (x,y,z) (Reddy 2004). The second line contains the three Young's moduli in a material reference system (1, 2, and 3 direction components), the third and fourth lines contain the shear moduli and Poisson ratio components, respectively. The fifth line indicates the mass density of the material and the sixth line contains the dielectric coefficients. Finally, the last line has the five piezoelectric coefficients (a schematic description of this file has already been given in Figure 9.1). The second file contains further information for a correct functioning of the FE MUL2 software. Some differences can be noted with respect to the analytical code. This input file for the FE code is briefly described in Figure 9.8 and the main differences, with respect to the analytical code file (see Figure 9.2), are pointed out. The missing data, compared to the analytical case, are ISHL (see Figure 9.2) and $(1/R_\alpha, 1/R_\beta)$ since the shell geometry is not implemented, (m,n) and loadings at the top and bottom, because there are many possibilities of load applications. The first line contains the number that indicates the chosen order of expansion in the thickness direction for the modeled mechanical and electrical variables (N from 1 to 4). The next line shows the index that specifies whether the displacement components are modeled in ESL, LW, or ESL+ZZ form. All the other electromechanical variables are always modeled in LW form. The third line permits one to choose the opportune variational statement, PVD, where displacement and electric potential are primary variables, or one of the three possible extensions of RMVT, where the additional primary variables are the transverse stresses and/or the transverse normal electric displacement. In the fourth line, it is possible to choose the type of analysis, which can either be static or dynamic (typical free-vibration problem). The MUL2 software also allows one to obtain classical theories, if an ESL theory with linear expansion in the thickness direction based on a PVD($\boldsymbol{u},\Phi$) variational statement is considered. The fifth line permits one to set the CLT or FSDT analysis by means of a typical penalty technique (for details see Sections 9.1 and 9.1.1 for the analytical closed-form case). The next line concerns the geometry, where a and b are the in-plane dimensions of the plate, and it is possible to specify the type of boundary conditions for each plate side (simply supported, clamped, and so on). The seventh line gives indications on the mesh, in other words, the number of elements in the x and y directions and the number of nodes for each element (Q4, Q8, and Q9 types of elements). The eighth line gives information about the loading conditions, which are more detailed compared to the analytical closed-form version. It is necessary to indicate whether the load is harmonic (and the imposed wave numbers), concentrated, uniform, or distributed, together with its position, magnitude, and direction. The ninth line is for the electrical boundary and loading conditions; here it is necessary to indicate whether the multilayer structure is an open-circuit configuration (free electric potential at the external surfaces) or a closed-circuit configuration (zero electric potential applied at the top and bottom surfaces). It is also possible to apply an electric potential at the top, with the bottom surface set to zero. The tenth line permits one to choose between several types of integrations in the xy-plane in order to prevent the numerical locking phenomena (normal, reduced, or selective

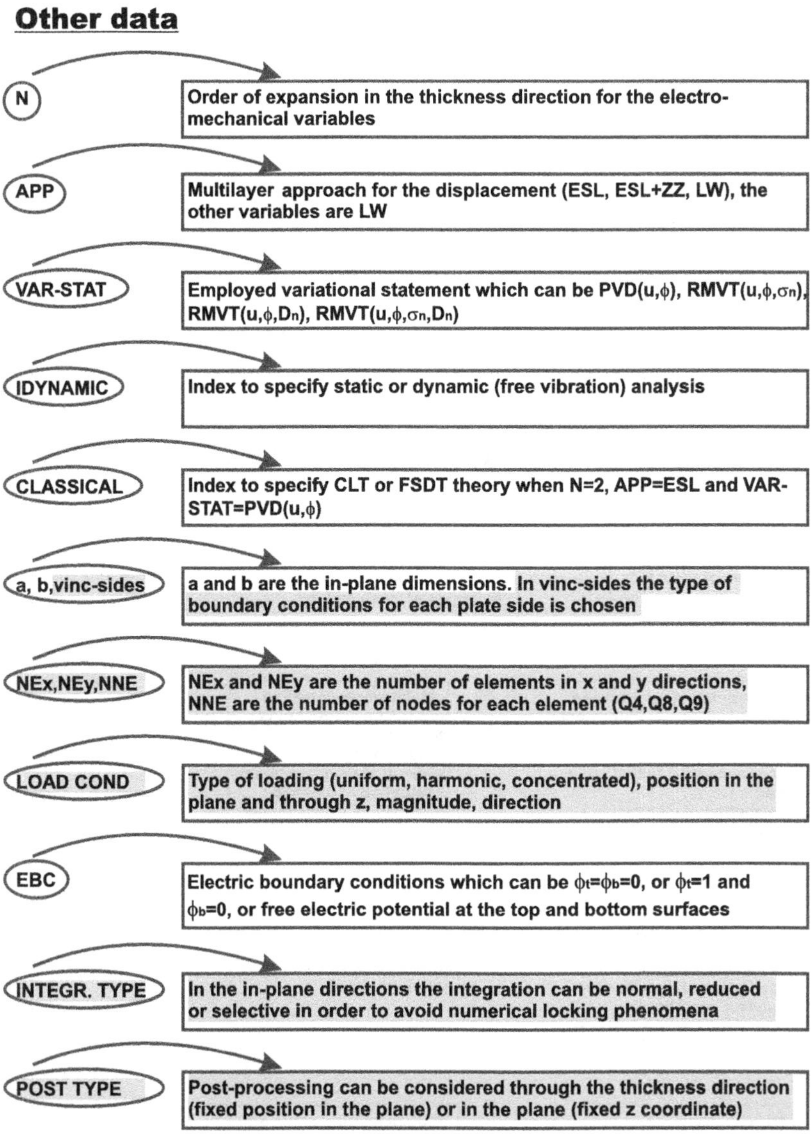

Figure 9.8 FE MUL2 software: description of the input file containing the data for geometry, 2D approaches, loading and boundary conditions, mesh, and post-processing.

integration). The last line gives information on the post-processing, which is more complete than in the analytical case, where only the amplitudes were plotted through the thickness direction; the position can be fixed in the *xy*-plane and the variables can be evaluated through the thickness direction, or the variables can be evaluated in the *xy*-plane for a fixed value of the *z* coordinate. When a point in the *xy*-plane is fixed, the output files have the same format as those already given for the amplitudes in the analytical case (see Figures 9.3 and 9.4).

The computing architecture of the FE version of the MUL2 software is quite similar to that already proposed for the analytical closed-form version. This architecture can be divided into eight main parts, which are summarized in Figure 9.9. The first part of the code reads the input data given in the

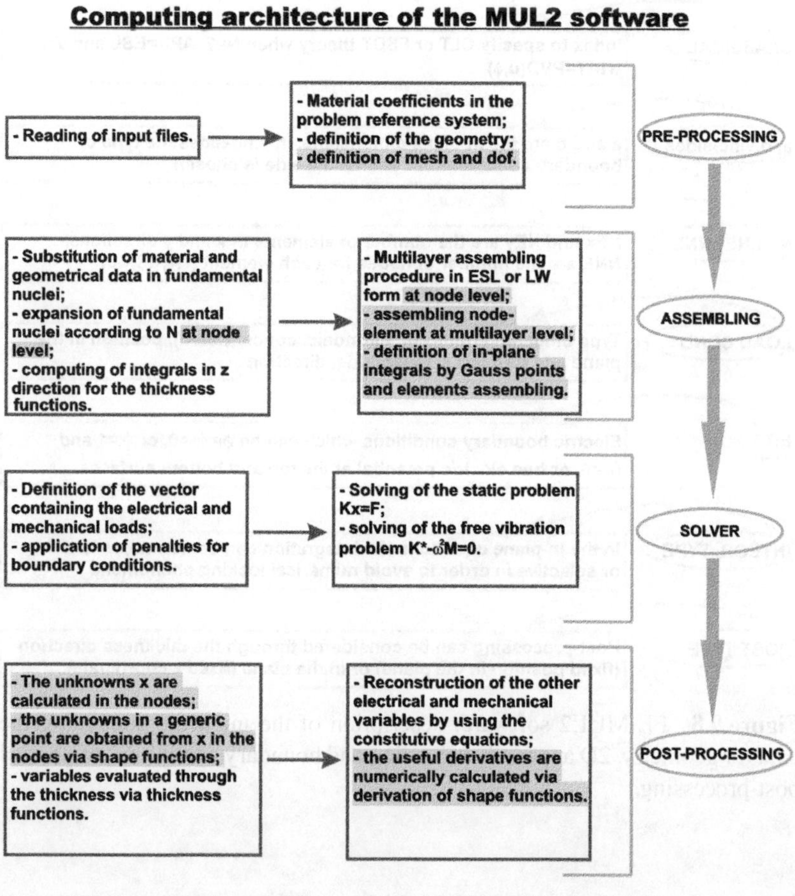

Figure 9.9 Computing architecture of the FE MUL2 software.

two files shown in Figures 9.1 and 9.8. In this way, MUL2 obtains all the information about the material properties of the embedded layers, the geometry of the structure and its stacking layer configuration, the loadings and boundary conditions, the chosen 2D theory, the type of analysis which can be either static or dynamic (free vibration problem), the mesh, and type of post-processing (Carrera 2002). In the second part of the code, the material coefficients (elastic, piezoelectric, and dielectric) are positioned in the matrices of the constitutive equations (for the FE approach, the constitutive equations are grouped in only one Equation (8.40) with the matrix of coefficients, as in Equation (8.39), and vectors containing all the electromechanical variables, as in Equations (8.37) and (8.38)) (Ikeda 1996). This matrix is transformed from the material reference system (1,2,3) to the problem reference system (x,y,z) by means of the rotation matrix which considers the orthotropic angle θ_k (Reddy 2004). This matrix is ready for the PVD application; when one of the three possible extensions of the RMVT application is considered, the new coefficients are calculated (see Sections 8.4, 8.5, and 8.6). From the geometrical data (in-plane dimensions and the thickness of each k layer) and the information concerning the mesh (number and type of elements employed), the code approximates the structure by means of the FE method. These first two parts constitute the pre-processing of the MUL2 software.

The FE MUL2 software contains all the fundamental nuclei, in explicit form, as one large equivalent fundamental nucleus containing all the information about the problem. For further details, see the variational statement and the fundamental nucleus for PVD($\boldsymbol{u}$, Φ) in Equations (8.47), (8.51), and (8.52), for RMVT($\boldsymbol{u}$, Φ, $\boldsymbol{\sigma}_n$) in Equations (8.47), (8.76), and (8.77), for RMVT($\boldsymbol{u}$, Φ, $\boldsymbol{D}_n$) in Equations (8.47), (8.101), and (8.102), and in Equations (8.47), (8.125), and (8.126) for RMVT($\boldsymbol{u}$, Φ, $\boldsymbol{\sigma}_n$, $\boldsymbol{D}_n$) (Carrera and Boscolo 2007; Carrera *et al.*, 2007; 2008). The third part of the code substitutes the geometrical and material data in the opportune fundamental nucleus, on the basis of the chosen variational statement, and then expands the fundamental nucleus in the τ and s directions at the node level (see Figure 8.10) according to the chosen order of expansion in the thickness direction for the considered plate theory. This expanded nucleus is given for each k layer embedded in the multilayered structure for a generic node; the integrals in the z direction inside the fundamental nuclei (see Sections 8.4, 8.5, and 8.6) are numerically computed. The thickness functions F_τ and F_s, included in these integrals, can be considered in ESL (use of Taylor polynomials) or LW (use of combinations of Legendre polynomials) form. The integrals in the z direction are computed by means of Gauss points and weights. These expanded fundamental nuclei are calculated for each k layer at the node level, and, remaining at the node level, the fourth part of the code conducts the multilayer assembly procedure which can be either ESL or LW. The multilayer assembly procedure is accomplished by means of an opportune connectivity matrix, through the thickness direction, which permits one to understand whether the matrices are simply summed or whether the compatibility and/or equilibrium conditions at each layer interface must be enforced. The other steps concern the nodes–element and the elements–structure assembly procedures; these two steps are new, if compared to the analytical MUL2 software application (Carrera 2002). In the FE version, numerical evaluation of the integrals of the shape functions in the xy-plane by means of opportune Gauss points and weights is fundamental. These two blocks constitute the assembly procedure of the MUL2 software.

The governing equations in FE form have the structure that has already been described in Equation (8.47) for PVD($\boldsymbol{u}$, Φ), RMVT($\boldsymbol{u}$, Φ, $\boldsymbol{\sigma}_n$), RMVT($\boldsymbol{u}$, Φ, $\boldsymbol{D}_n$), and RMVT($\boldsymbol{u}$, Φ, $\boldsymbol{\sigma}_n$, $\boldsymbol{D}_n$). Only the meanings and forms of matrix $\bar{\boldsymbol{K}}^{(e)k\tau sij}$, and vectors $\bar{\boldsymbol{Q}}^{(e)k}_{\tau i}$ and $\bar{\boldsymbol{F}}^{(e)k}_{sj}$, are different. In the FE version, the use of a single fundamental nucleus for each variational statement permits one to have the same compact form of the governing equation for each variational statement; the differences only concern the components that are included in each matrix and vector. In the fifth part of the code, the mechanical loads at the top and bottom surfaces of the structure and the electrical loads are introduced into the single vectors $\bar{\boldsymbol{F}}^{(e)k}_{sj}$ and $\bar{\boldsymbol{Q}}^{(e)k}_{\tau i}$ depending on their direction and position. Several forms of these loads can be considered, compared to the analytical case (harmonic, concentrated, distributed, and uniform loads in the xy-plane). In the case of closed- and open-circuit configurations, the boundary conditions for the electric potential are imposed directly in the Φ^k_τ vector (Carrera and Brischetto 2007a,b). In the FE code, penalty techniques are applied for both boundary loading conditions and boundary conditions on the plate sides. In the sixth part of the code, the system of governing equations is solved; for a static problem, the code solves a general system of the type$\boldsymbol{K}\boldsymbol{x} = \boldsymbol{F}$ where $\boldsymbol{x}$ are the unknowns at the nodes of the elements and $\boldsymbol{F}$ is the vector of the loads. In the case of free-vibration analysis, a typical eigenvalue problem is solved for a general system of the type $\boldsymbol{K}^* - \omega^2 \boldsymbol{M} = 0$; the eigenvalues are the frequencies, and the relative eigenvector is computed for each value. This permits one to obtain the vibration modes of the structure in terms of primary variables. The number of vibration modes obtained in a FE model depends on the total number of degrees of freedom, which means both the degrees of freedom of the employed 2D theory in the thickness direction and the degrees of freedom in the xy-plane, depending on the mesh size. The fifth and sixth parts constitute the solver of the MUL2 software.

The post-processing of the FE version of the MUL2 software is more complicated than that discussed for the analytical version in the previous section (in the analytical closed-form solution we only work on the amplitudes). However, the post-processing can be divided into two parts. First, the vector of unknowns $\boldsymbol{x}$ is employed to recover the values of the primary variables through the thickness direction (the thickness-direction evaluation is obtained by means of the thickness functions). In the FE version, the unknowns $\boldsymbol{x}$ are given in the nodes, therefore the code obtains the values at given generic points in the xy-plane by means of the shape functions. The evaluation of the variables through the thickness is given for a chosen point in the xy-plane. For example, the displacement components through the thickness direction are given in a file, as shown in Figure 9.3. The values for the analytical codes were the amplitudes, whereas in the FE version, the file is proposed for a given point in the xy-plane. In the case of the ESL approach, the evaluation in the thickness direction, for a given point in the xy-plane, is obtained using Equations (8.6) and (8.9), where the nodal values are multiplied by the thickness functions and by the shape functions. In the ESL approach, the thickness functions are Taylor polynomials, while the shape functions are the same for each multilayer approach. In the case of the LW procedure, Equations (8.16) and (8.17) are used, where the thickness functions are a combination of Legendre polynomials, and the shape functions do not change with respect to the ESL case. The displacement is a primary variable in the PVD variational statement and in each extension of the RMVT approach. The transverse stresses shown in Figure 9.4 are given through the thickness in such a form when they are primary variables in the chosen governing equations (the RMVT($\boldsymbol{u}$, Φ, $\boldsymbol{\sigma}_n$) and RMVT($\boldsymbol{u}$, Φ, $\boldsymbol{\sigma}_n$, $\boldsymbol{D}_n$) cases); transverse stresses are always recovered in LW form, as shown in Equations (8.26) and (8.28). The $\boldsymbol{x}$ vector contains the nodal unknowns, and, from these, we obtain the values through the thickness and in the xy-plane by means of the thickness functions and the shape functions, respectively. The electric potential is always a primary variable, whereas the transverse normal electric displacement is a primary variable in RMVT($\boldsymbol{u}$, Φ, $\boldsymbol{D}_n$) and in RMVT($\boldsymbol{u}$, Φ, $\boldsymbol{\sigma}_n$, $\boldsymbol{D}_n$). In these cases, both variables are calculated in LW form, as described in Equations (8.23) and (8.24) and in (8.32) and (8.33), respectively. The second part of the post-processing permits one to recover the other quantities through the thickness direction of the structure. These are not primary variables in the proposed governing equations. In this case, the constitutive equations are used as given in Equations (8.37)–(8.40) for the PVD($\boldsymbol{u}$, Φ) case, and in Equations (8.62)–(8.73) for RMVT($\boldsymbol{u}$, Φ, $\boldsymbol{\sigma}_n$), in Equations (8.87)–(8.98) for RMVT($\boldsymbol{u}$, Φ, $\boldsymbol{D}_n$), and in Equations (8.111)–(8.122) for RMVT($\boldsymbol{u}$, Φ, $\boldsymbol{\sigma}_n$, $\boldsymbol{D}_n$) cases. In order to use such constitutive equations, some derivatives must be calculated to obtain the mechanical strains and the electric field (see Equations (8.41)–(8.42), (8.58)–(8.61), (8.83)–(8.86), and (8.107)–(8.110) for PVD($\boldsymbol{u}$, Φ), RMVT($\boldsymbol{u}$, Φ, $\boldsymbol{\sigma}_n$), RMVT($\boldsymbol{u}$, Φ, $\boldsymbol{D}_n$), and RMVT($\boldsymbol{u}$, Φ, $\boldsymbol{\sigma}_n$, $\boldsymbol{D}_n$), respectively). The derivatives for the analytical code are calculated precisely by means of the harmonic forms. Another advantage of the analytical closed form of the MUL2 software is that it does not need a post-processing in the plane, as it only works on the amplitudes of the variables. The derivatives for the FE version are numerically calculated using the derivation of the shape functions. In the FE MUL2 software, a post-processing in the xy-plane is also considered, since it works on the nodal values.

The FE procedures for the degeneration of the refined theories into classical theories (CLT and FSDT) and the correction of the Poisson locking phenomena (in the case of pure mechanical problems) are the same as those already shown for the analytical code in Section 9.1.1.

9.3 Analytical closed-form solution for the electromechanical analysis of plates

This section discusses the main results of the electromechanical analysis of multilayered piezoelectric plates for an analytical closed-form solution obtained via the MUL2 code. Both actuator (applied electric voltage) and sensor (applied mechanical load) cases are investigated for the static analysis, while the free vibrations are evaluated, in the case of dynamic analysis, when the plate is in a closed-circuit configuration (which means zero imposed electric voltage at the top and bottom surfaces). In order to obtain an analytical closed-form solution, each plate is considered as simply supported with harmonic loads. The plates are square with in-plane dimensions $a = b = 0.04$ m, and total thickness $h_{tot} = 0.01$ m and $h_{tot} = 0.0008$ m for thickness ratios $a/h = 4$ and $a/h = 50$, respectively. Each plate can have two different stacking sequences:

- a three-layered configuration, where the internal layer is in isotropic aluminum alloy Al2024 ($h_2 = 0.8h_{tot}$) and the two external layers are in piezoelectric material PZT-4 ($h_1 = h_3 = 0.1h_{tot}$);
- a four-layered configuration, with two external layers in piezoelectric material PZT-4 ($h_1 = h_4 = 0.1h_{tot}$) and two internal layers in carbon fiber-reinforced material (Gr/Ep) ($h_2 = h_3 = 0.4h_{tot}$) with lamination sequence $0°/90°$.

The elastic and electrical properties of the embedded materials are given in Table 9.1. The actuator and sensor cases are described in Figure 9.10 where the bisinusoidal form of the electric voltage and mechanical load is clearly shown:

- bisinusoidal ($m = n = 1$) electric voltage with amplitude $\hat{\Phi} = 1$ V at the top with the bottom surface set to $\hat{\Phi} = 0$ V (actuator case);
- bisinusoidal transverse ($m = n = 1$) pressure with amplitude $\hat{p}_z = 1$ Pa applied at the top of the closed-circuit configuration plate with electric potential $\hat{\Phi} = 0$ V at the top and bottom (sensor case).

For the free-vibration analysis, the analytical code gives frequency values that correspond to the degrees of freedom, through the thickness direction of

Table 9.1 Electromechanical properties of the embedded materials in the proposed structures.

	PZT-4	Al2024	Gr/Ep
E_1 [GPa]	81.3	73.0	132.38
E_2 [GPa]	81.3	73.0	10.756
E_3 [GPa]	64.5	73.0	10.756
ρ [kg/m^3]	7600	2800	1578
$\nu_{12}[-]$	0.329	0.3	0.24
$\nu_{13}[-]$	0.432	0.3	0.24
$\nu_{23}[-]$	0.432	0.3	0.49
G_{23} [GPa]	25.6	28.0769	3.606
G_{13} [GPa]	25.6	28.0769	5.6537
G_{12} [GPa]	30.6	28.0769	5.6537
e_{15} [C/m^2]	12.72	—	—
e_{24} [C/m^2]	12.72	—	—
e_{31} [C/m^2]	−5.20	—	—
e_{32} [C/m^2]	−5.20	—	—
e_{33} [C/m^2]	15.08	—	—
ε_{11} [pc/V m]	13 060	101.8	30.9897
ε_{22} [pc/V m]	13 060	101.8	26.563
ε_{33} [pc/V m]	11 510	101.8	26.563

the 2D model employed, for imposed wave numbers $m = n = 1$, $m = n = 2$, and $m = n = 10$. The investigated plate is in a closed-circuit configuration with electric voltage $\hat{\Phi} = 0$ V at the top and bottom surfaces, as described in Figure 9.11. Only the cases of composite layers are illustrated in Figures 9.10 and 9.11 for the sake of brevity.

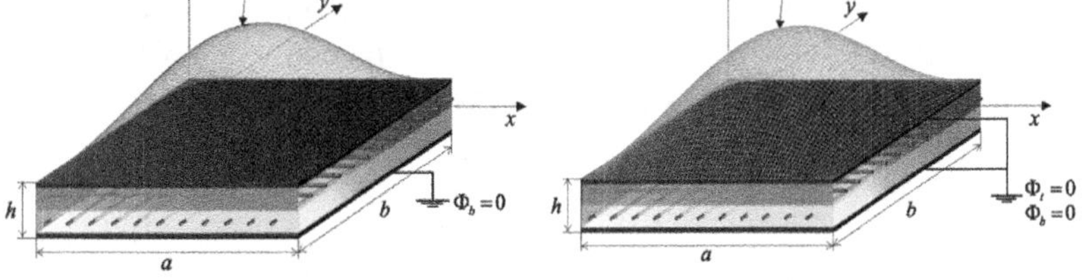

Figure 9.10 Multilayered piezoelectric plate: actuator configuration (left) and sensor configuration (right). Bisinusoidal distribution analyzed via analytical closed-form solution.

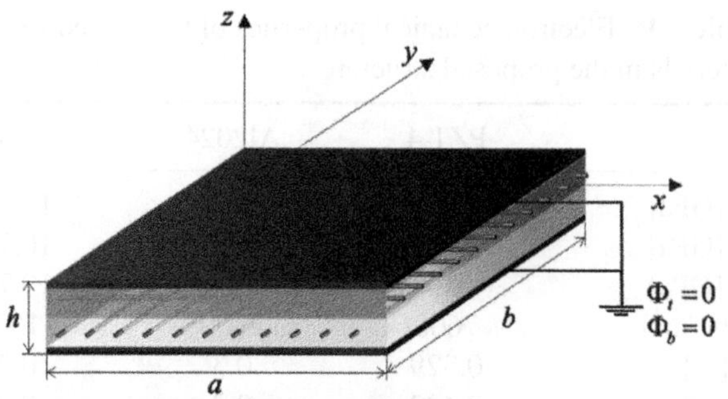

Figure 9.11 Closed-circuit configuration (electric potential Φ is zero at the external surfaces) for the free-vibration analysis of a plate.

The analytical solutions for the actuator case are given in Tables 9.2 and 9.3 for the isotropic and composite plates, respectively. Both thick and thin plates are compared by means of classical theories (CLT($\boldsymbol{u}$,Φ) and FSDT($\boldsymbol{u}$,Φ)) and an advanced mixed model. The advanced mixed model (here called MUL2 for the sake of brevity) is a LM4($\boldsymbol{u}$, Φ, $\boldsymbol{\sigma}_n$, $\boldsymbol{D}_n$) model, which means a LW theory where the four a priori modeled variables indicated in parentheses have a fourth-order expansion in the thickness plate direction. The electric potential is given in Table 9.2 at the interface between the piezoelectric and the isotropic layers; both values at the top of the isotropic layer and at the bottom of the piezoelectric one are given in order to evaluate the interlaminar

Table 9.2 Actuator case: thick and thin isotropic plate (bisinusoidal electric voltage $\hat{\Phi} = 1$ V applied at the top with $m = 1$ and $n = 1$). Analytical closed-form solution.

	$a/h = 4$			$a/h = 50$		
	MUL2	FSDT	CLT	MUL2	FSDT	CLT
$\Phi(2h/5)^t$ [V]	0.9926	0.9914	0.9917	0.9990	0.9988	0.9989
$\Phi(2h/5)^b$ [V]	0.9926	0.9914	0.9917	0.9990	0.9988	0.9989
$\sigma_{yy}(-h/2)$ [Pa]	15.531	−8.7290	−9.5384	−141.72	−57.176	−57.242
$\sigma_{zz}(2h/5)^t$ [Pa]	−1.0191	138.62	133.53	0.0046	226.75	226.34
$\sigma_{zz}(2h/5)^b$ [Pa]	−1.0191	5.0840	4.8369	0.0046	6.0227	6.0027
$\mathcal{D}_z(2h/5)^t$ [10^{-8} C/m^2]	−1.5782	−9.9583	−9.6000	−15.904	−16.570	−16.541
$\mathcal{D}_z(2h/5)^b$ [10^{-8} C/m^2]	−1.5782	−1.2607	−1.2607	−15.904	−15.871	−15.871

Table 9.3 Actuator case: thick and thin composite plate (bisinusoidal electric voltage $\hat{\Phi} = 1$ V applied at the top with $m = 1$ and $n = 1$). Analytical closed-form solution.

	$a/h = 4$			$a/h = 50$		
	MUL2	FSDT	CLT	MUL2	FSDT	CLT
$\Phi(0)^t$ [V]	0.4477	0.4461	0.4469	0.4996	0.4996	0.4996
$\Phi(0)^b$ [V]	0.4477	0.4461	0.4469	0.4996	0.4996	0.4996
$\sigma_{yy}(-h/2)$ [Pa]	27.783	−6.7320	−8.6435	−29.573	−12.836	−12.938
$\sigma_{zz}(0)^t$ [Pa]	−1.4650	0.4307	0.4074	0.0045	0.3872	0.3864
$\sigma_{zz}(0)^b$ [Pa]	−1.4650	0.3459	0.3409	0.0045	0.3804	0.3810
$\mathcal{D}_z(0)^t$ [10^{-8} C/m^2]	−0.3182	−0.3624	−0.3624	−4.1474	−4.1507	−4.1507
$\mathcal{D}_z(0)^b$ [10^{-8} C/m^2]	−0.3182	−0.2967	−0.2966	−4.1474	−4.1450	−4.1450

continuity of such a variable. However, interlaminar continuity is assured by each proposed theory because the electric potential is a primary variable in each considered model, and it is always considered in LW form. FSDT($\boldsymbol{u}$,Φ) and CLT($\boldsymbol{u}$,Φ) give a result for the thin plate which is very close to the quasi-3D description, while, for the thick case, the error given by classical theories is larger. The normal in-plane stress is evaluated at the bottom of the multilayered plate, and it is only calculated correctly by the MUL2 theory because the introduction of correct values of the derivatives of the main electromechanical variables is mandatory in the constitutive equations (employed to obtain this variable); the values given by CLT($\boldsymbol{u}$,Φ) and FSDT($\boldsymbol{u}$,Φ) for the thick plate are completely wrong. Both transverse normal stress and transverse normal electric displacement are given at the interface between the piezoelectric and the isotropic layers ($(2h/5)^t$ is the top of the isotropic layer and $(2h/5)^b$ is the bottom of the piezoelectric layer which is positioned at the top of the multilayered plate) in order to also evaluate their interlaminar continuity. The transverse normal stress is a primary variable in the MUL2 theory, in which it is obtained directly from the solution of the governing equations. CLT($\boldsymbol{u}$,Φ) and FSDT($\boldsymbol{u}$,Φ) obtain this stress via post-processing, using the constitutive equations in a contradictory manner because the mechanical part of the transverse stress should, by definition, be zero. For this reason, the correct values of σ_{zz} are only obtained using the MUL2 theory, which also guarantees their interlaminar continuity, CLT($\boldsymbol{u}$,Φ) and FSDT($\boldsymbol{u}$,Φ) do not give either correct values or interlaminar continuity of σ_{zz} (for both thick and thin plates). The transverse normal electric displacement is a primary variable in the governing equations of the MUL2 theory, and for this reason this model gives their correct values and ensures its interlaminar continuity. In CLT($\boldsymbol{u}$,Φ) and FSDT($\boldsymbol{u}$,Φ), the transverse normal electric displacement is obtained via post-processing (use of the constitutive equations), therefore its interlaminar continuity is not obtained, even though the values can be considered acceptable for the thin case. Table 9.3 proposes the same results, already discussed for the isotropic multilayered piezoelectric plate in Table 9.2, for the multilayered composite piezoelectric case. The values calculated at the interface are now considered at $(0)^t$ and $(0)^b$, which means at the interface between the two composite layers (top of the first fiber-reinforced layer and bottom of the second fiber-reinforced layer). All the considerations already made for the case in Table 9.2 are confirmed here for the composite plate. The interlaminar continuity of the variables is only ensured using 2D models, in which they are primary variables that are obtained directly from the governing equations. In the cases proposed in Table 9.3, CLT($\boldsymbol{u}$,Φ) and FSDT($\boldsymbol{u}$,Φ) appear to work better because the interface considered is between two layers made of the same material in which only the fiber orientation has been changed. For this reason, the anisotropy is smaller and even though the interlaminar discontinuity is confirmed through CLT($\boldsymbol{u}$,Φ) and FSDT($\boldsymbol{u}$,Φ) analysis, the values for transverse normal electric displacement and electric potential are calculated better than in the previous case (in particular for the thin plate). All the considerations made concerning Tables 9.2 and 9.3 are confirmed in Figure 9.12, where the electric potential through thickness z is evaluated for the thick isotropic piezoelectric multilayered plate and the transverse normal stress is shown through thickness z of the thin composite piezoelectric multilayered plate. A comparison is given for CLT($\boldsymbol{u}$,Φ), FSDT($\boldsymbol{u}$,Φ), and MUL2 theories; interlaminar continuity of the electric potential is obtained for each theory and the results are very close for each proposed 2D model (the small differences are due to the thick configuration).

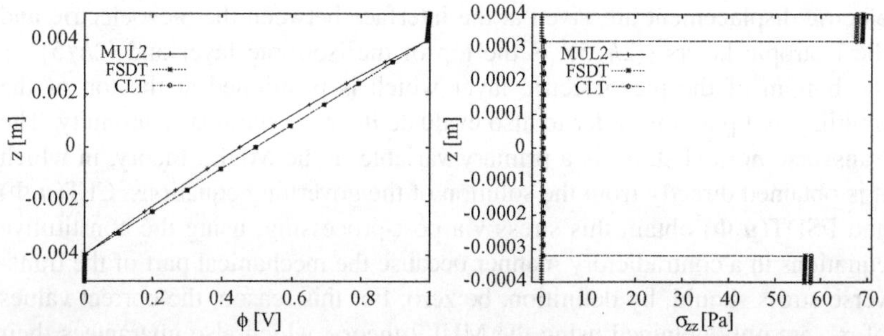

Figure 9.12 Actuator case: plate with bisinusoidal electric voltage $\hat{\Phi} = 1\,\text{V}$ applied at the top with $m = 1$ and $n = 1$. Analytical closed-form solution. Φ vs. z on the left (isotropic plate with $a/h = 4$) and σ_{zz} vs. z on the right (composite plate with $a/h = 50$).

Table 9.4 Sensor case: thick and thin isotropic plate in bending (bisinusoidal mechanical load $\hat{p}_z = 1$ Pa applied at the top with $m = 1$ and $n = 1$). Analytical closed-form solution.

	$a/h = 4$			$a/h = 50$		
	MUL2	FSDT	CLT	MUL2	FSDT	CLT
$w(h/2)\,[10^{-12}\,\text{m}]$	1.1272	0.9532	0.6609	1647.9	1294.4	1290.7
$\Phi(2h/5)^t\,[10^{-3}\,\text{V}]$	0.0386	0.0185	0.0165	0.6453	0.2073	0.2072
$\Phi(2h/5)^b\,[10^{-3}\,\text{V}]$	0.0386	0.0185	0.0165	0.6453	0.2073	0.2072
$\sigma_{yy}(h/2)$ [Pa]	4.3414	4.5114	4.5063	625.89	704.15	704.14
$\sigma_{zz}(2h/5)^t$ [Pa]	0.9673	2.1429	2.1754	0.9678	339.66	339.69
$\sigma_{zz}(2h/5)^b$ [Pa]	0.9673	1.3719	1.3735	0.9678	214.60	214.60
$\mathcal{D}_z(2h/5)^t$ $[10^{-13}\,\text{C/m}^2]$	−0.2120	430.82	202.30	−0.6383	33341	33112
$\mathcal{D}_z(2h/5)^b$ $[10^{-13}\,\text{C/m}^2]$	−0.2120	0.0000	0.0000	−0.6383	0.0000	0.0000

Interlaminar continuity of the transverse normal stress is only ensured for the MUL2 theory (CLT($\boldsymbol{u}$,Φ) and FSDT($\boldsymbol{u}$,Φ) give very large errors).

Tables 9.4 and 9.5 propose the same plates that have already been analyzed in Tables 9.2 and 9.3, but in a sensor configuration. The electric potential, the transverse normal stress, and the transverse normal electric displacement are

Table 9.5 Sensor case: thick and thin composite plate in bending (bisinusoidal mechanical load $\hat{p}_z = 1$ Pa applied at the top with $m = 1$ and $n = 1$). Analytical closed-form solution.

	$a/h = 4$			$a/h = 50$		
	MUL2	FSDT	CLT	MUL2	FSDT	CLT
$w(h/2)\,[10^{-12}\,\text{m}]$	3.1525	1.8488	0.9378	2354.0	1842.9	1831.5
$\Phi(0)^t\,[10^{-3}\,\text{V}]$	0.0611	0.0266	0.0211	0.9153	0.2943	0.2938
$\Phi(0)^b\,[10^{-3}\,\text{V}]$	0.0611	0.0266	0.0211	0.9153	0.2943	0.2938
$\sigma_{yy}(h/2)$ [Pa]	6.5642	5.9023	5.8912	786.52	920.54	920.53
$\sigma_{zz}(0)^t$ [Pa]	0.4984	0.0162	0.0163	0.5000	2.5403	2.5404
$\sigma_{zz}(0)^b$ [Pa]	0.4984	−0.0162	−0.0163	0.5000	−2.5403	−2.5404
$\mathcal{D}_z(0)^t$ $[10^{-13}\,\text{C/m}^2]$	0.5053	−0.1957	−0.1551	0.6116	−0.1672	−0.1669
$\mathcal{D}_z(0)^b$ $[10^{-13}\,\text{C/m}^2]$	0.5053	0.1957	0.1551	0.6116	0.1672	0.1669

evaluated at the same interfaces already seen for the actuator case, while the transverse displacement and the in-plane stress are calculated at the top of the plate (in-plane stress was considered at the bottom of the multilayer in the case of the actuator configuration). All the considerations already discussed for the actuator case are confirmed here for the sensor case, but some differences can be noted. FSDT($\boldsymbol{u}$,Φ) and CLT($\boldsymbol{u}$,Φ) work better for the sensor case than the actuator case, for the evaluation of the transverse displacement, while they work worse for the analysis of the electric potential. In fact, CLT($\boldsymbol{u}$,Φ) and FSDT($\boldsymbol{u}$,Φ) ensure interlaminar continuity of the electric voltage, but the values are far from the quasi-3D description given by the MUL2 theory; the values given by CLT($\boldsymbol{u}$,Φ) and FSDT($\boldsymbol{u}$,Φ) are not correct for the transverse displacement but they are reasonable and these considerations also lead to a better evaluation of the in-plane normal stress in the case of the sensor configuration. However, the main conclusion remains that, in order to obtain a quasi-3D description of the electromechanical variables in a sensor configuration plate, the use of the MUL2 theory is mandatory. This conclusion is confirmed from the results shown in Figure 9.13, in which the transverse displacement is evaluated through the thickness direction in the case of a thick plate and the electric potential is shown through the thickness direction for a thin plate. The transverse displacement cannot be considered constant for the thick plate as suggested in CLT($\boldsymbol{u}$,Φ) and FSDT($\boldsymbol{u}$,Φ), and for this reason the correct evaluation is given by MUL2, which considers it LW with a fourth-order expansion in the z direction. The quasi-3D description of the electric potential, through the thickness direction, is very difficult to obtain in the sensor configuration case, as clearly shown by the comparison between MUL2 and classical theories.

The free-vibration analysis in a closed-circuit configuration is proposed in Table 9.6. The first three vibration modes, through the thickness, are proposed

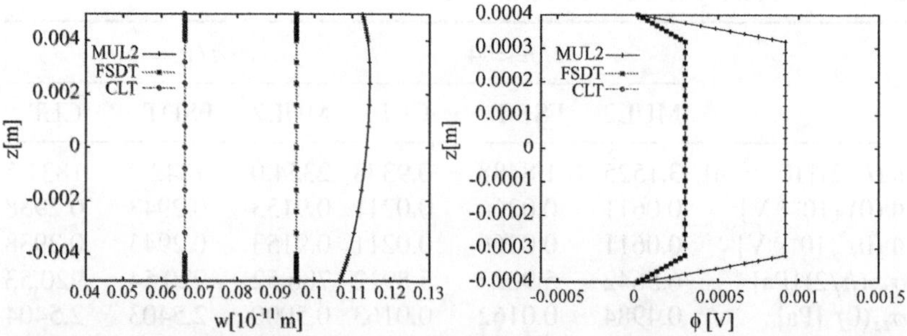

Figure 9.13 Sensor case: plate with bisinusoidal mechanical load $\hat{p}_z = 1$ Pa applied at the top with $m = 1$ and $n = 1$. Analytical closed-form solution. w vs. z on the left (isotropic plate with $a/h = 4$) and Φ vs. z on the right (composite plate with $a/h = 50$).

Table 9.6 Closed-circuit plate configuration, first three modes for different wave number values. Circular frequency $\omega^* = \omega/1000 = 2\pi f/1000$. Thick isotropic plate ($a/h = 4$) and thin composite plate ($a/h = 50$).

	$a/h = 4$			$a/h = 50$		
	MUL2	FSDT	CLT	MUL2	FSDT	CLT
			$m = 1, n = 1$			
First mode	146.48	161.53	187.83	13.803	15.604	15.652
Second mode	305.63	306.22	306.22	386.83	387.59	387.71
Third mode	523.81	592.21	592.21	497.23	535.53	535.62
			$m = 2, n = 2$			
First mode	429.85	463.53	612.44	54.326	61.737	62.494
Second mode	607.67	612.44	641.90	770.48	774.66	775.60
Third mode	936.43	961.67	1184.4	991.86	1070.5	1071.2
			$m = 10, n = 10$			
First mode	2494.6	2770.1	3062.2	969.40	1201.0	1476.3
Second mode	2500.1	2996.7	5160.9	3484.0	3804.4	3909.3
Third mode	2657.0	3062.2	5918.0	4298.8	5184.7	5356.2

for each 2D model. For imposed wave numbers $m = n = 1$, $m = n = 2$, and $m = n = 10$, the thick plate is the isotropic one ($a/h = 4$) and the thin case ($a/h = 50$) is the composite piezoelectric plate. Three effects can be noted in free-vibration analysis: the vibration mode related to the degrees of freedom through the thickness of the considered 2D theory; the imposed wave numbers in the plane, which permit one to investigate the higher order modes; and the plate configuration (thickness ratio and stacking sequence). In Table 9.6, it is clear how classical theories only work well for thin plates, in-plane modes, and $m = n = 1$; for other modes that are different from the in-plane ones, and for high values of wave numbers, the use of the MUL2 theory is fundamental. When wave numbers increase, the MUL2 theory is mandatory for a correct dynamic description of the multilayered plate; the error given by classical theories (CLT($\boldsymbol{u}$,Φ) and FSDT($\boldsymbol{u}$,Φ)) increases with the thickness, the wave numbers, and for frequency values that are different from the in-plane ones.

9.4 Analytical closed-form solution for the electromechanical analysis of shells

This section discusses the main results of the electromechanical analysis of multilayered piezoelectric shells, as already conducted for the relative plate cases, in order to note the possible curvature effects. The case of an analytical closed-form solution is obtained via the MUL2 code. Both actuator (applied electric voltage) and sensor (applied mechanical load) cases are investigated for the static analysis while, in the case of dynamic analysis, the free vibrations are evaluated when the shell is in a closed-circuit configuration (which means zero imposed electric voltage at the top and bottom surfaces). In order to obtain an analytical closed-form solution, each shell is considered as simply supported with harmonic loads; the shell geometry considers a radius of curvature $R_\alpha = 10\,\mathrm{m}$ in the α direction and infinite radius of curvature R_β in the β direction. The two in-plane dimensions are $a = \pi/3R_\alpha$ in the α direction and $b = 1\,\mathrm{m}$ in the β direction. The total thickness is $h_{tot} = 2.5\,\mathrm{m}$ and $h_{tot} = 0.2\,\mathrm{m}$ for thickness ratios $R_\alpha/h = 4$ and $R_\alpha/h = 50$, respectively. Each shell can have two different stacking sequences, as already seen in the previous section for the plate geometry. The elastic and electrical properties of the embedded materials are given in Table 9.1. The actuator and sensor cases are described in Figure 9.14, where the sinusoidal form of electric voltage and mechanical load is clearly shown:

- sinusoidal ($m = 1$ and $n = 0$, cylindrical bending) electric voltage with amplitude $\hat{\Phi} = 1\,\mathrm{V}$ at the top; bottom surface set to $\hat{\Phi} = 0\,\mathrm{V}$ (actuator case);
- sinusoidal transverse ($m = 1$ and $n = 0$, cylindrical bending) pressure with amplitude $\hat{p}_z = 1\,\mathrm{Pa}$ applied at the top of the closed-circuit configuration shell with electric potential $\hat{\Phi} = 0\,\mathrm{V}$ at the top and bottom (sensor case).

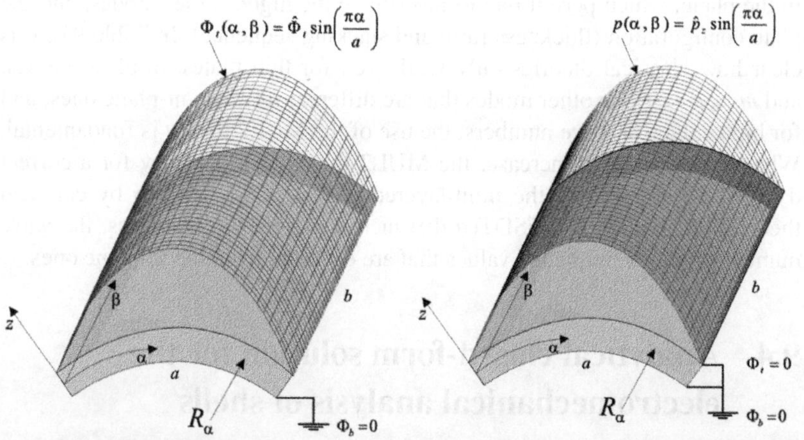

Figure 9.14 Multilayered piezoelectric shells: actuator configuration (left) and sensor configuration (right). Cylindrical distribution analyzed via analytical closed-form solution.

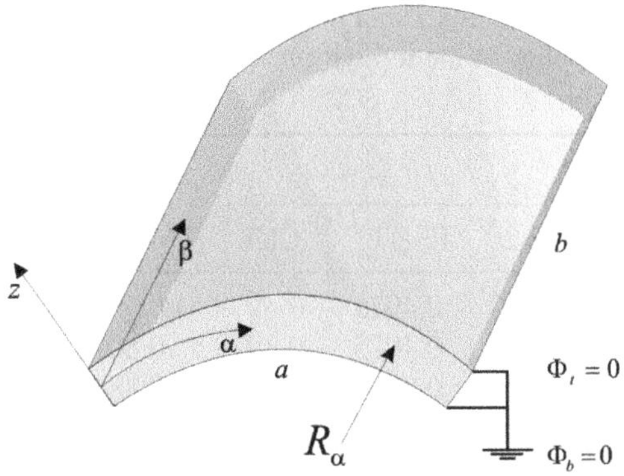

Figure 9.15 Closed circuit configuration (electric potential Φ zero at the external surfaces) for the free vibration analysis of shell.

For the free-vibration analysis, the analytical code gives frequency values that correspond to the degrees of freedom through the thickness direction of the employed 2D model, for imposed wave numbers $(m = 1, n = 0)$, $(m = 2, n = 0)$, and $(m = 10, n = 0)$. The investigated shell is in a closed-circuit configuration with electric voltage $\hat{\Phi} = 0\,\text{V}$ at the top and bottom surfaces, as described in Figure 9.15. Layer interfaces are not drawn in Figures 9.14 and 9.15 for graphical reasons.

The analytical results for the actuator case are given in Tables 9.7 and 9.8 for isotropic and composite shells, respectively. Both thick and thin shells are compared by means of classical theories (CLT($\boldsymbol{u}$,Φ) and FSDT($\boldsymbol{u}$,Φ)) and

Table 9.7 Actuator case: thick and thin isotropic shell (sinusoidal electric voltage $\hat{\Phi} = 1\,\text{V}$ applied at the top with $m = 1$ and $n = 0$). Analytical closed-form solution.

	$R_\alpha/h = 4$			$R_\alpha/h = 50$		
	MUL2	FSDT	CLT	MUL2	FSDT	CLT
$\Phi(2h/5)^t$ [V]	0.9968	0.9962	0.9963	0.9991	0.9989	0.9989
$\Phi(2h/5)^b$ [V]	0.9968	0.9962	0.9963	0.9991	0.9989	0.9989
$\sigma_{\beta\beta}(-h/2)$ [Pa]	0.0229	0.2913	−0.0345	−0.5238	0.1483	−0.2476
$\sigma_{zz}(2h/5)^t$ [Pa]	−0.0034	0.5423	0.2267	0.0010	1.2544	0.8759
$\sigma_{zz}(2h/5)^b$ [Pa]	−0.0034	0.1786	0.0033	0.0010	0.2365	0.0224
$\mathcal{D}_z(2h/5)^t$ [10^{-10} C/m^2]	−0.5135	−1.9538	−1.6901	−6.3052	−6.6808	−6.4114
$\mathcal{D}_z(2h/5)^b$ [10^{-10} C/m^2]	−0.5135	−0.5066	−0.5065	−6.3052	−6.3484	−6.3484

Table 9.8 Actuator case: thick and thin composite shell (sinusoidal electric voltage $\hat{\Phi} = 1$ V applied at the top with $m = 1$ and $n = 0$). Analytical closed-form solution.

	$R_\alpha/h = 4$			$R_\alpha/h = 50$		
	MUL2	FSDT	CLT	MUL2	FSDT	CLT
$\Phi(0)^t$ [V]	0.4980	0.4975	0.4979	0.5018	0.5018	0.5018
$\Phi(0)^b$ [V]	0.4980	0.4975	0.4979	0.5018	0.5018	0.5018
$\sigma_{\beta\beta}(-h/2)$ [Pa]	0.1005	0.9829	−0.0262	−0.3860	1.0517	−0.1842
$\sigma_{zz}(0)^t$ [Pa]	−0.0018	0.0914	−0.0004	0.0001	0.1041	−0.0091
$\sigma_{zz}(0)^b$ [Pa]	−0.0018	0.0660	−0.0003	0.0001	0.0752	−0.0066
$\mathcal{D}_z(0)^t$ [10^{-10} C/m^2]	−0.1304	−0.1326	−0.1326	−1.6591	−1.6531	−1.6531
$\mathcal{D}_z(0)^b$ [10^{-10} C/m^2]	−0.1304	−0.1322	−0.1322	−1.6591	−1.6653	−1.6653

an advanced mixed model. The proposed advanced mixed model is called the MUL2 theory for the sake of brevity, but it is, in fact, the well-known LM4($\boldsymbol{u}$, Φ, $\boldsymbol{\sigma}_n$, $\boldsymbol{D}_n$) theory, where the primary variables in parentheses are LW modeled with fourth-order expansion in the thickness direction. The introduction of the curvature for a shell geometry and the loading conditions in cylindrical bending in Tables 9.7 and 9.8 do not introduce any further effects, compared to the plate cases already discussed in Tables 9.2 and 9.3. The variables are investigated for the same thickness coordinates and interface positions already discussed for the relative plate case. Classical theories show remarkable difficulties in obtaining the correct values of the in-plane stresses for both stacking sequences and thickness ratios. The electric potential is well described by each 2D theory proposed, even though the use of the MUL2 theory is suitable for thick shells. Transverse normal stresses and transverse normal electric displacement are only primary variables in the MUL2 theory, and this permits a quasi-3D description of such variables for each thickness ratio and stacking sequence proposed. Classical theories obtain erroneous values for transverse normal stresses (they evaluate them in a contradictory way) and transverse normal electric displacements, and they do not satisfy their interlaminar continuity, because they are obtained from the constitutive equations via post-processing. However, the transverse normal electric displacement via CLT($\boldsymbol{u}$,Φ) and FSDT($\boldsymbol{u}$,Φ) theories, in the case of thin shells, does not appear to be very far from the correct values, even though interlaminar continuity is not ensured. Some of these results are summarized in Figure 9.16 where the in-plane normal stress is shown through the thickness direction for the case of a thick isotropic shell, and the electric potential is evaluated through the z

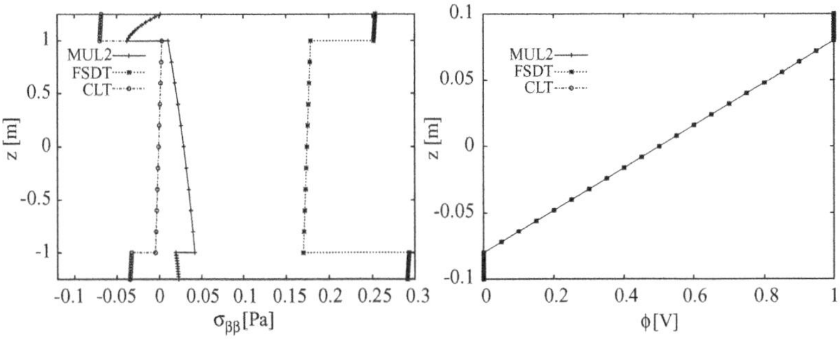

Figure 9.16 Actuator case: shell with sinusoidal electric voltage $\hat{\Phi} = 1\,\mathrm{V}$ applied at the top with $m = 1$ and $n = 0$. Analytical closed-form solution. $\sigma_{\beta\beta}$ vs. z on the left (isotropic shell with $R_\alpha/h = 4$) and Φ vs. z on the right (composite shell with $R_\alpha/h = 50$).

direction for the case of a thin composite shell. The stress is described well by the MUL2 theory, as it is able to correctly calculate the variables and their derivatives employed in the constitutive equations (post-processing). It is important to remember that interlaminar continuity is not requested for in-plane stresses. For actuator cases, the electric potential is described well by each theory, in particular for the case in Figure 9.16 which considers a thin shell.

The relative sensor configurations for shell geometries described in this chapter are investigated in Tables 9.9 and 9.10 for isotropic multilayered piezoelectric and composite multilayered piezoelectric shells, respectively. Compared to the actuator case, classical theories appear to work better for the analysis of transverse displacements, even though the quasi-3D description is only given by the MUL2 theory. The difficulties in recovering the quasi-3D description of the electric potential by means of classical theories (CLT($\boldsymbol{u}$,Φ) and FSDT($\boldsymbol{u}$,Φ)) are greater than for the relative actuator case; CLT($\boldsymbol{u}$,Φ) and FSDT($\boldsymbol{u}$,Φ) obtain the interlaminar continuity of the electric potential, but the values are not as correct as the values given by the MUL2 theory. For transverse normal stresses and transverse normal electric displacement, it is clear how the use of the MUL2 theory is mandatory in order to achieve the quasi-3D description and interlaminar continuity. The transverse normal electric displacement through the z direction for the thick isotropic multilayered piezoelectric shell is correctly described in Figure 9.17 by the MUL2 theory, which also ensures interlaminar continuity. The MUL2 theory also gives a quasi-3D description of the transverse normal stress along the thickness direction that is continuous and which satisfies the loading boundary conditions for the sensor configuration ($\sigma_{zz} = 1\,\mathrm{Pa}$ at the top and $\sigma_{zz} = 0\,\mathrm{Pa}$ at the bottom).

The free vibration analysis, in a closed-circuit configuration, is proposed in Table 9.11 for the shell geometry. The first three vibration modes, through the thickness, are proposed for each 2D model for imposed wave numbers

Table 9.9 Sensor case: thick and thin isotropic shell in bending (sinusoidal mechanical load $\hat{p}_z = 1$ Pa applied at the top with $m = 1$ and $n = 0$). Analytical closed-form solution.

	$R_\alpha/h = 4$			$R_\alpha/h = 50$		
	MUL2	FSDT	CLT	MUL2	FSDT	CLT
$w(h/2)$ $[10^{-10}$ m]	16.277	13.663	11.361	25 309	19 869	19 843
$\Phi(2h/5)^t$ $[10^{-1}$ V]	0.2919	0.1090	0.1040	3.9949	1.2799	1.2790
$\Phi(2h/5)^b$ $[10^{-1}$ V]	0.2919	0.1090	0.1040	3.9949	1.2799	1.2790
$\sigma_{\beta\beta}(h/2)$ [Pa]	−4.9730	−17.793	−16.170	−18593	−14071	−14052
$\sigma_{zz}(2h/5)^t$ [Pa]	0.6900	−3.7152	−2.1432	−2.8898	−13797	−13778
$\sigma_{zz}(2h/5)^b$ [Pa]	0.6900	−1.7334	−0.8602	−2.8898	−7763.1	−7752.4
$\mathcal{D}_z(2h/5)^t$ $[10^{-12}$ C/m^2]	0.2156	714.69	583.34	3.6718	1 032 200	1 030 800
$\mathcal{D}_z(2h/5)^b$ $[10^{-12}$ C/m^2]	0.2156	0.1472	0.1328	3.6718	1.4763	1.4752

Table 9.10 Sensor case: thick and thin composite shell in bending (sinusoidal mechanical load $\hat{p}_z = 1$ Pa applied at the top with $m = 1$ and $n = 0$). Analytical closed-form solution.

	$R_\alpha/h = 4$			$R_\alpha/h = 50$		
	MUL2	FSDT	CLT	MUL2	FSDT	CLT
$w(h/2)$ $[10^{-10}$ m]	31.597	21.152	14.026	31 812	25 088	25 008
$\Phi(0)^t[10^{-1}$ V]	0.4158	0.1567	0.1377	0.5039	1.6294	1.6280
$\Phi(0)^b[10^{-1}$ V]	0.4158	0.1567	0.1377	0.5039	1.6294	1.6280
$\sigma_{\beta\beta}(h/2)$ [Pa]	−12.298	−2.1443	2.8287	−23086	−17418	−17356
$\sigma_{zz}(0)^t$ [Pa]	−0.0179	−1.1538	−0.6966	−7.1456	−1742.4	−1736.8
$\sigma_{zz}(0)^b$ [Pa]	−0.0179	−0.8340	−0.5036	−7.1456	−1259.6	−1255.5
$\mathcal{D}_z(0)^t$ $[10^{-13}$ C/m^2]	−0.6599	−0.6007	−0.6863	−445.67	−126.59	−126.60
$\mathcal{D}_z(0)^b$ $[10^{-13}$ C/m^2]	−0.6599	−0.1965	−0.3482	−445.67	−127.24	−127.25

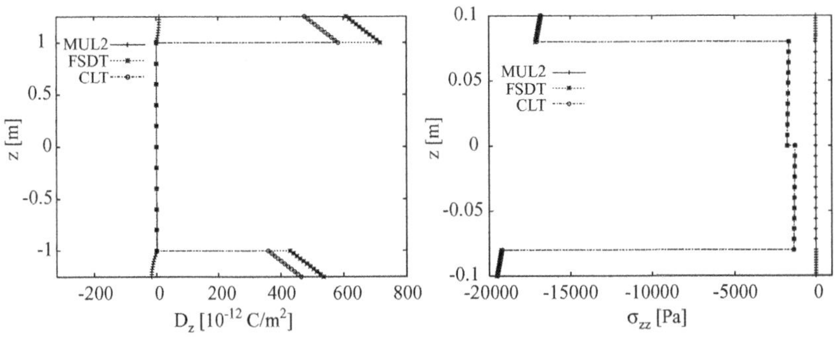

Figure 9.17 Sensor case: shell with sinusoidal mechanical load $\hat{p}_z = 1$ Pa applied at the top with $m = 1$ and $n = 0$. Analytical solution in closed form. $\mathcal{D}_z$ vs. z on the left (isotropic shell with $R_\alpha/h = 4$) and σ_{zz} vs. z on the right (composite shell with $R_\alpha/h = 50$).

Table 9.11 Closed-circuit shell configuration, first three modes for different wave number values. Circular frequency $\omega = 2\pi f$. Thick isotropic shell ($R_\alpha/h = 4$) and thin composite shell ($R_\alpha/h = 50$).

	$R_\alpha/h = 4$			$R_\alpha/h = 50$		
	MUL2	FSDT	CLT	MUL2	FSDT	CLT
			$m = 1, n = 0$			
First mode	250.95	279.31	304.46	22.697	25.559	25.600
Second mode	828.00	828.73	829.35	586.75	586.75	586.75
Third mode	1461.8	1641.2	1648.5	1679.8	1772.2	1772.6
			$m = 2, n = 0$			
First mode	929.24	1013.7	1275.0	102.39	115.66	116.40
Second mode	1647.7	1653.5	1658.7	1173.5	1173.5	1173.5
Third mode	2727.5	3126.3	3176.0	3218.8	3405.5	3408.8
			$m = 10, n = 0$			
First mode	6705.3	7467.2	8293.5	2153.5	2598.7	2954.5
Second mode	7073.0	7931.8	13 199	5867.3	5867.5	5867.5
Third mode	7799.2	8763.0	16 355	14 216	16 418	16 858

($m = 1, n = 0$), ($m = 2, n = 0$), and ($m = 10, n = 0$); the thick shell is the isotropic one ($R_\alpha/h = 4$) and the thin shell ($R_\alpha/h = 50$) is the composite piezoelectric shell. Three effects can be noted in the free vibration analysis: the vibration mode related to the degrees of freedom of the 2D theory considered through the thickness; the imposed wave numbers in the plane, which permit one to investigate the higher order modes; and the shell configuration (thickness ratio and stacking sequence). In Table 9.11, it is clear how classical theories only give acceptable results for thin shells, in-plane modes, and $m = 1$; for other modes that are different from in-plane ones, the use of the MUL2 theory is fundamental. When wave numbers increase, the MUL2 theory is mandatory for a correct dynamic description of the multilayered shell; the error given by classical theories (CLT($\boldsymbol{u}$,Φ) and FSDT($\boldsymbol{u}$,Φ)) increases with the thickness, wave numbers, and for frequency values that are different from the in-plane ones. The introduction of curvature does not modify the main conclusions already given for the relative plate case.

9.5 FE solution for the electromechanical analysis of beams

This section gives the main results of the electromechanical analysis of multilayered piezoelectric beams. A FE analysis via the MUL2 code permits one to investigate boundary and loading conditions that are different from those seen for the analytical solution (which were simply supported structures and harmonic mechanical and electric loads). Both actuator (applied electric voltage) and sensor (applied mechanical load) cases have been investigated for the static analysis, while in the case of dynamic analysis, the free vibrations have been evaluated when the beam is in an open-circuit configuration (which means free electric voltage at the top and bottom surfaces). The investigated beam has an in-plane dimensional ratio a/b equal to 4 and the same total thickness values h_{tot} already considered for the plate case; the clamped side is the shortest one. Each beam can have two different stacking sequences, as already seen in the previous section for the plate geometry. The elastic and electrical properties of the embedded materials are given in Table 9.1. The actuator and sensor cases are described in Figure 9.18, where the uniform electric voltage and mechanical load are clearly observable:

- uniform distribution of electric voltage with value $\Phi = 1\,\text{V}$ at the top; bottom surface set to $\Phi = 0\,\text{V}$ (actuator case);
- uniform distribution of pressure with value $p_z = 1\,\text{Pa}$ applied at the top of the closed-circuit configuration beam with electric potential $\Phi = 0\,\text{V}$ at the top and bottom (sensor case).

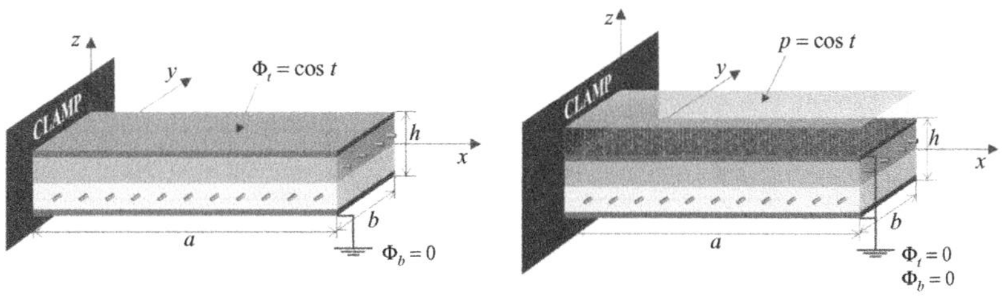

Figure 9.18 Multilayered cantilever beams: actuator configuration (left) and sensor configuration (right). FE solution.

The FE free-vibration analysis of the proposed beams gives a number of frequencies that are equal to the total degrees of freedom of the employed 2D theory. This number depends on both the degrees of freedom through the thickness for the 2D approximation and the degrees of freedom in relation to the mesh in the in-plane direction (number of elements and nodes for each element). For this reason, the first six modes are given in the tables without specifying the wave numbers which are typical of the analytical form solution. The investigated cantilever beam is an open-circuit configuration with free electric voltage at the top and bottom surfaces, as described in Figure 9.19.

The beam in an actuator configuration is analyzed in Tables 9.12 and 9.13 for an isotropic multilayered piezoelectric beam and for a composite multilayered piezoelectric beam, respectively. A comparison has already been made between classical and advanced 2D theories in the previous sections concerning closed-form solutions, while for the FE analysis, we propose a complete study of the mesh using a LM4($\boldsymbol{u}$, Φ, $\boldsymbol{\sigma}_n$, $\boldsymbol{D}_n$) model as the MUL2 theory where the primary variables in parentheses are a priori LW modeled with a fourth-order expansion in the thickness direction. The electromechanical variables given through the thickness are all considered at the mid-point of the tip edge of the beam (the amplitude was considered for the case of the analytical solutions).

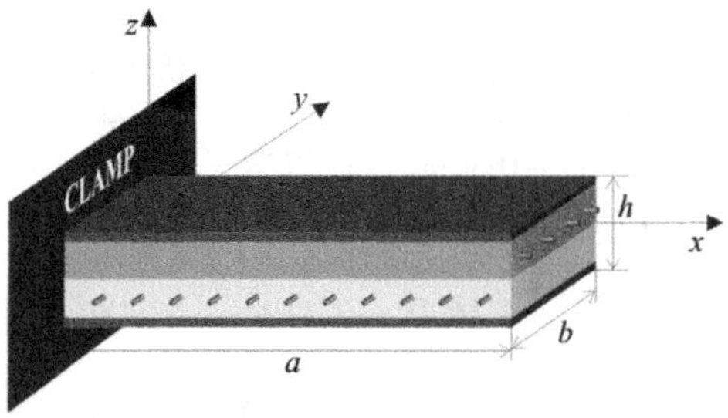

Figure 9.19 Open-circuit configuration (free electric potential Φ at the external surfaces) for the free-vibration analysis of beam.

Table 9.12 Actuator case: thick and thin isotropic beam (uniform electric voltage $\Phi = 1$ V). FE solution provided at the mid-point of the tip edge of the beam.

	$a/h = 4$			$a/h = 50$		
	MUL2 7×3	MUL2 15×3	MUL2 21×3	MUL2 7×3	MUL2 15×3	MUL2 21×3
$w(-h/2)$ $[10^{-13}$ m]	2.2584	2.9628	3.1733	2.0928	2.1207	2.1534
$\Phi(2h/5)^t$ [V]	0.9992	0.9992	0.9992	0.9991	0.9991	0.9991
$\Phi(2h/5)^b$ [V]	0.9992	0.9992	0.9992	0.9991	0.9991	0.9991
$\sigma_{yy}(-h/2)$ [Pa]	−6.0397	−4.6370	−4.0853	−139.28	−138.35	−137.30
$\sigma_{zz}(2h/5)^t$ [Pa]	0.4864	−0.8960	−1.9236	0.0049	−0.2578	−0.5583
$\sigma_{zz}(2h/5)^b$ [Pa]	0.4864	−0.8960	−1.9236	0.0049	−0.2578	−0.5583
$\mathcal{D}_z(2h/5)^t$ $[10^{-8}$ C/m^2]	−1.2704	−1.2706	−1.2707	−15.878	−15.878	−15.878
$\mathcal{D}_z(2h/5)^b$ $[10^{-8}$ C/m^2]	−1.2704	−1.2706	−1.2707	−15.878	−15.878	−15.878

Table 9.13 Actuator case: thick and thin composite beam (uniform electric voltage $\Phi = 1$ V). FE solution provided at the mid-point of the tip edge of the beam.

	$a/h = 4$			$a/h = 50$		
	MUL2 7×3	MUL2 15×3	MUL2 21×3	MUL2 7×3	MUL2 15×3	MUL2 21×3
$w(-h/2)$ $[10^{-13}$ m]	−8.3831	−8.3414	−8.3482	−1926.4	−2010.1	−2028.7
$\Phi(0)^t$ [V]	0.5000	0.5000	0.5000	0.5000	0.5000	0.5000
$\Phi(0)^b$ [V]	0.5000	0.5000	0.5000	0.5000	0.5000	0.5000
$\sigma_{yy}(-h/2)$ [Pa]	−0.0808	0.2155	0.3217	−21.948	−21.199	−20.595
$\sigma_{zz}(0)^t$ [Pa]	−0.1924	−0.3490	−0.3734	0.0151	−0.2979	−0.6234
$\sigma_{zz}(0)^b$ [Pa]	−0.1924	−0.3490	−0.3734	0.0151	−0.2979	−0.6234
$\mathcal{D}_z(0)^t$ $[10^{-9}$ C/m^2]	−3.3190	−3.3190	−3.3190	−41.486	−41.486	−41.486
$\mathcal{D}_z(0)^b$ $[10^{-9}$ C/m^2]	−3.3190	−3.3190	−3.3190	−41.486	−41.486	−41.486

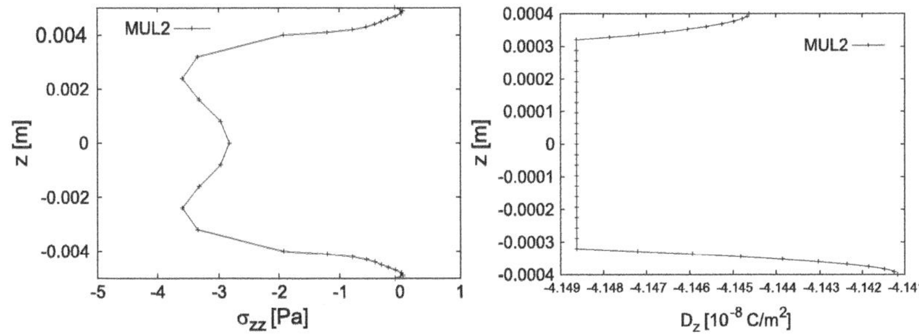

Figure 9.20 Actuator case: beam with uniform electric voltage $\Phi = 1$ V applied at the top. FE solution provided at the mid-point of the tip edge of the beam. σ_{zz} vs. z on the left (isotropic beam with $a/h = 4$) and $\mathcal{D}_z$ vs. z on the right (composite beam with $a/h = 50$). Mesh 21×3.

For each considered stacking sequence, thickness ratio, and electromechanical variables, it is clear how a 21×3 mesh is sufficient to obtain a convergence value; the beam has a dimension in the plane which is predominant with respect to the other in-plane dimensions and for this reason there are a larger number of elements in the largest dimension. The proposed advanced mixed model ensures interlaminar continuity of each electromechanical variable. This fact can be confirmed in Figure 9.20 where the MUL2 theory (using the 21×3 convergence mesh) gives a quasi-3D description of the transverse normal stress through the thickness (thick isotropic piezoelectric multilayered beam) and of the transverse normal electric displacement through the thickness (thin composite piezoelectric multilayered beam). MUL2 ensures interlaminar continuity of these two variables and the boundary loading conditions for the transverse normal stress ($\sigma_{zz} = 0$ at the top and bottom of the beam in the actuator configuration).

The sensor configuration of the beam is analyzed in Tables 9.14 and 9.15 for the isotropic multilayered piezoelectric beam and for the composite multilayered piezoelectric beam, respectively. A convergence study has been proposed using a LM4($\boldsymbol{u}$, Φ, $\boldsymbol{\sigma}_n$, $\boldsymbol{D}_n$) model as the MUL2 theory. All the variables given through the thickness are calculated at the mid-point of the tip edge of the beam. The same 21×3 mesh used for the actuator case is sufficient to recover a quasi-3D description of all the mechanical and electrical variables; 7×3 and 15×3 meshes are also given for comparison purposes. The MUL2 theory (using the 21×3 convergence mesh) in Figure 9.21 gives a quasi-3D description of the transverse displacement through the thickness (thick isotropic piezoelectric multilayered beam) and the transverse normal electric displacement through the thickness (thin composite piezoelectric multilayered beam). MUL2 ensures interlaminar continuity of each electromechanical variable and their quasi-3D description through the thickness direction z.

Table 9.14 Sensor case: thick and thin isotropic beam in bending (uniform mechanical load $p_z = 1$ Pa applied at the top). FE solution provided the mid-point of the tip edge of the beam.

	$a/h = 4$			$a/h = 50$		
	MUL2 7×3	MUL2 15×3	MUL2 21×3	MUL2 7×3	MUL2 15×3	MUL2 21×3
$w(h/2)$ $[10^{-12}$ m]	41.669	42.848	43.083	70 899	73 795	74 383
$\Phi(2h/5)^t$ $[10^{-3}$ V]	−0.0118	−0.0160	−0.0166	0.4604	0.1927	0.1201
$\Phi(2h/5)^b$ $[10^{-3}$ V]	−0.0118	−0.0160	−0.0166	0.4604	0.1927	0.1201
$\sigma_{yy}(h/2)$ [Pa]	−0.1397	0.0303	0.0291	−50.707	52.140	56.941
$\sigma_{zz}(2h/5)^t$ [Pa]	0.9822	1.0175	1.0394	1.4247	0.9722	0.9862
$\sigma_{zz}(2h/5)^b$ [Pa]	0.9822	1.0175	1.0394	1.4247	0.9722	0.9862
$\mathcal{D}_z(2h/5)^t$ $[10^{-13}$ C/m^2]	−0.6963	1.2824	1.8192	−27.738	−47.931	−52.117
$\mathcal{D}_z(2h/5)^b$ $[10^{-13}$ C/m^2]	−0.6963	1.2824	1.8192	−27.738	−47.931	−52.117

Table 9.15 Sensor case: thick and thin composite beam in bending (uniform mechanical load $p_z = 1$ Pa applied at the top). FE solution provided at the mid-point of the tip edge of the beam.

	$a/h = 4$			$a/h = 50$		
	MUL2 7×3	MUL2 15×3	MUL2 21×3	MUL2 7×3	MUL2 15×3	MUL2 21×3
$w(h/2)$ $[10^{-12}$ m]	67.004	68.719	69.077	89 630	92 750	93 373
$\Phi(0)^t$ $[10^{-3}$ V]	−0.0094	−0.0134	−0.0115	0.4849	0.1341	0.0698
$\Phi(0)^b$ $[10^{-3}$ V]	−0.0094	−0.0134	−0.0115	0.4849	0.1341	0.0698
$\sigma_{yy}(h/2)$ [Pa]	0.1229	0.4405	0.4156	25.780	25.904	25.430
$\sigma_{zz}(0)^t$ [Pa]	0.3427	0.2609	0.2595	0.4026	0.4524	0.4398
$\sigma_{zz}(0)^b$ [Pa]	0.3427	0.2609	0.2595	0.4026	0.4524	0.4398
$\mathcal{D}_z(0)^t$ $[10^{-13}$ C/m^2]	0.8361	0.7350	0.7208	−116.67	−10.247	2.8072
$\mathcal{D}_z(0)^b$ $[10^{-13}$ C/m^2]	0.8361	0.7350	0.7208	−116.67	−10.247	2.8072

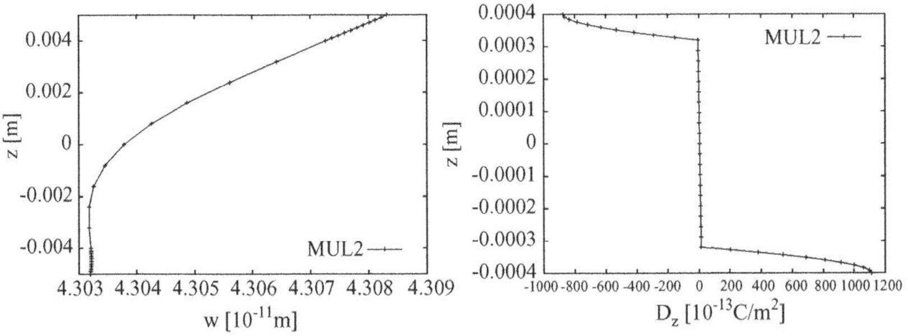

Figure 9.21 Sensor case: beam with uniform mechanical load $p_z = 1$ Pa applied at the top. FE solution provided at the mid-point of the tip edge of the beam. w vs. z on the left (isotropic beam with $a/h = 4$) and $\mathcal{D}_z$ vs. z on the right (composite beam with $a/h = 50$). Mesh 21×3.

Table 9.16 gives the first six vibration modes for the cantilever beam in an open-circuit configuration. The thick beam ($a/h = 4$) is an isotropic multilayered piezoelectric one, while the thin beam ($a/h = 50$) is a composite multilayered piezoelectric structure. Convergence depends on the considered vibration mode; for low values of frequency, the 21×3 mesh is sufficient, but

Table 9.16 Open-circuit beam configuration, first six modes for FE analysis. Circular frequency $\omega = 2\pi f$. Thick isotropic beam ($a/h = 4$) and thin composite beam ($a/h = 50$).

	$a/h = 4$			$a/h = 50$		
	MUL2 7×3	MUL2 15×3	MUL2 21×3	MUL2 7×3	MUL2 15×3	MUL2 21×3
First mode	29 961.59	29 113.34	28 915.68	2 544.353	2 533.182	2 531.100
Second mode	30 455.28	29 839.71	29 820.49	16 194.42	15 972.12	15 877.41
Third mode	94 138.20	93 903.63	93 868.45	16 658.37	16 096.10	16 081.36
Fourth mode	154 638.2	148 628.9	145 936.7	33 849.93	33 298.46	33 209.36
Fifth mode	156 433.1	150 042.3	149 518.2	51 118.17	45 433.86	44 708.65
Sixth mode	182 947.8	178 356.2	181 469.0	52 123.36	50 619.43	50 414.25

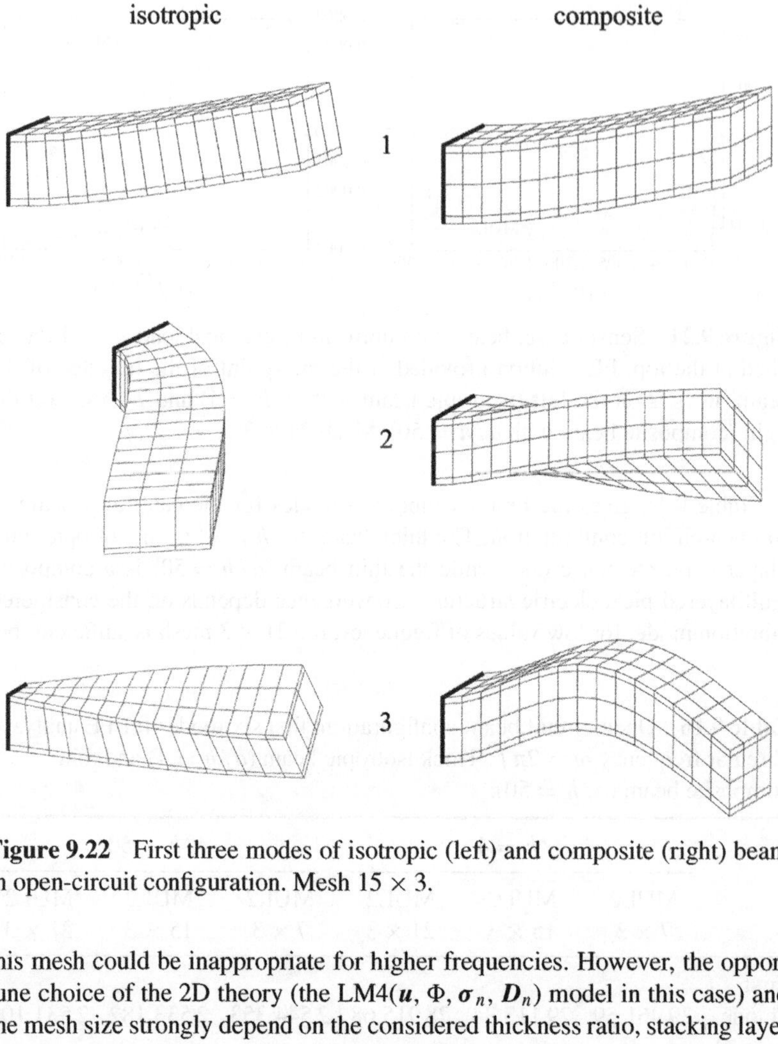

Figure 9.22 First three modes of isotropic (left) and composite (right) beam in open-circuit configuration. Mesh 15 × 3.

this mesh could be inappropriate for higher frequencies. However, the opportune choice of the 2D theory (the LM4($\boldsymbol{u}$, Φ, $\boldsymbol{\sigma}_n$, $\boldsymbol{D}_n$) model in this case) and the mesh size strongly depend on the considered thickness ratio, stacking layer sequence and vibration modes (low- or high-frequency values). Figure 9.22 proposes the first three vibration modes of the cantilever beam with isotropic or composite internal layers. The MUL2 model is able to capture the 3D behavior exactly in terms of displacements.

9.6 FE solution for the electromechanical analysis of plates

The last analysis proposed in this chapter considers a FE analysis of the same simply supported plates already seen in Section 9.3 for the case of an analytical

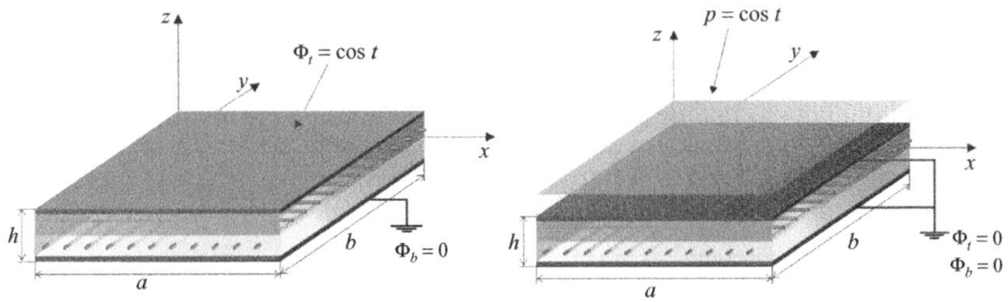

Figure 9.23 Multilayered piezoelectric plates: actuator configuration (left) and sensor configuration (right). Uniform in-plane distribution analyzed via FE solution.

closed-form solution when bisinusoidal electrical and mechanical loads were considered. The stacking layer sequences, the geometry, the boundary conditions, and the embedded materials are the same as those already discussed in Section 9.3. Both actuator (applied electric voltage) and sensor (applied mechanical load) cases are investigated for the static analysis, while in the case of dynamic analysis, the free vibrations are evaluated when the plate is in an open-circuit configuration (which means free electric voltage at the top and bottom surfaces). The actuator and sensor cases are described in Figure 9.23, where the uniform electric voltage and mechanical load are clearly shown:

- uniform distribution of electric voltage with the value $\Phi = 1$ V at the top, and the bottom surface set to $\Phi = 0$ V (actuator case);
- uniform distribution of pressure with the value $p_z = 1$ Pa applied at the top of the closed-circuit configuration plate with electric potential $\Phi = 0$ V at the top and bottom (sensor case).

The proposed FE free-vibration analysis of the plates is quite similar to the one that was proposed for a beam geometry in the previous section. The FE model gives a number of frequencies that are equal to the total degrees of freedom of the employed 2D theory. This number depends on both the degrees of freedom through the thickness for the 2D approximation and the degrees of freedom in relation to the mesh in the in-plane direction (number of elements and nodes for each element). For this reason the first six modes have been given in the proposed tables without specifying the wave numbers which are typical of an analytical closed-form solution. The plates investigated in an open-circuit configuration with free electric voltage at the top and bottom surfaces have already been described in Figure 9.11 for the closed-circuit case.

Tables 9.17 and 9.18 propose the static analysis of the actuator configuration of the isotropic multilayered piezoelectric plate and composite multilayered piezoelectric structure, respectively. The MUL2 theory is a LM4($\boldsymbol{u}$, Φ, $\boldsymbol{\sigma}_n$, $\boldsymbol{D}_n$) model which calculates the electromechanical variables through the thickness

Table 9.17 Actuator case: thick and thin isotropic plate (uniform electric voltage $\Phi = 1$ V). FE solution provided in the middle of the plate.

	$a/h = 4$			$a/h = 50$		
	MUL2 7×7	MUL2 15×15	MUL2 21×21	MUL2 7×7	MUL2 15×15	MUL2 21×21
$w(-h/2)$ $[10^{-12}$ m]	−8.2748	−8.3591	−8.3618	−8.1258	−8.3203	−8.3765
$\Phi(2h/5)^t$ [V]	1.0126	0.9990	0.9991	1.0269	0.9992	0.9991
$\Phi(2h/5)^b$ [V]	1.0126	0.9990	0.9991	1.0269	0.9992	0.9991
$\sigma_{yy}(-h/2)$ [Pa]	0.6428	−8.1936	−12.585	−145.87	−147.52	−147.94
$\sigma_{zz}(2h/5)^t$ [Pa]	−11.639	3.5889	−0.5818	−0.0487	0.0585	−0.0106
$\sigma_{zz}(2h/5)^b$ [Pa]	−11.639	3.5889	−0.5818	−0.0487	0.0585	−0.0106
$\mathcal{D}_z(2h/5)^t$ $[10^{-8}$ C/m^2]	−1.4524	−1.2725	−1.2739	−16.370	−15.882	−15.877
$\mathcal{D}_z(2h/5)^b$ $[10^{-8}$ C/m^2]	−1.4524	−1.2725	−1.2739	−16.370	−15.882	−15.877

Table 9.18 Actuator case: thick and thin composite plate (uniform electric voltage $\Phi = 1$ V). FE solution provided in the middle of the plate.

	$a/h = 4$			$a/h = 50$		
	MUL2 7×7	MUL2 15×15	MUL2 21×21	MUL2 7×7	MUL2 15×15	MUL2 21×21
$w(-h/2)$ $[10^{-12}$ m]	−12.923	−12.772	−12.772	−12.839	−13.218	−13.311
$\Phi(0)^t$ [V]	0.4983	0.4986	0.4986	0.5113	0.5000	0.5000
$\Phi(0)^b$ [V]	0.4983	0.4986	0.4986	0.5113	0.5000	0.5000
$\sigma_{yy}(-h/2)$ [Pa]	39.762	7.8803	7.8804	−29.761	−35.216	−35.576
$\sigma_{zz}(0)^t$ [Pa]	−0.0298	1.9433	0.7708	−0.0264	−0.0444	−0.0077
$\sigma_{zz}(0)^b$ [Pa]	−0.0298	1.9433	0.7708	−0.0264	−0.0444	−0.0077
$\mathcal{D}_z(0)^t$ $[10^{-8}$ C/m^2]	−0.3287	−0.3320	−0.3319	−4.2624	−4.1490	−4.1485
$\mathcal{D}_z(0)^b$ $[10^{-8}$ C/m^2]	−0.3287	−0.3320	−0.3319	−4.2624	−4.1490	−4.1485

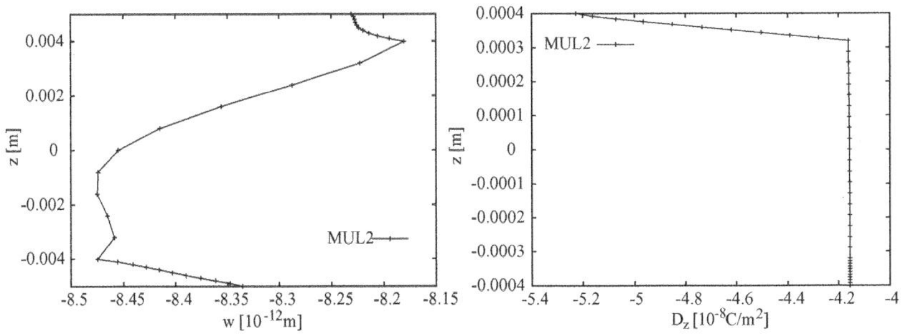

Figure 9.24 Actuator case: plate with uniform electric voltage $\Phi = 1$ V applied at the top. FE solution provided in the middle of the plate. w vs. z on the left (isotropic plate with $a/h = 4$) and $\mathcal{D}_z$ vs. z on the right (composite plate with $a/h = 50$). Mesh 21×21.

in the middle of the plate. The proposed meshes are square, because of the square geometry of the plate (7×7, 15×15, and 21×21); the 21×21 mesh applied in the LM4($\boldsymbol{u}$, Φ, $\boldsymbol{\sigma}_n$, $\boldsymbol{D}_n$) theory gives a quasi-3D description of each variable, regardless of the stacking layer sequence, embedded materials, and thickness ratio. The interlaminar continuity of the electric potential, transverse normal stress, and transverse normal electric displacement are ensured by the proposed model because they are primary variables. Figure 9.24 proposes the transverse displacement through the thickness direction z for the thick isotropic plate and the transverse normal electric displacement along z for the thin composite plate. The MUL2 theory is employed using a 21×21 mesh. The zigzag form of the displacement is clearly displayed because a LW theory has been employed and interlaminar continuity of the transverse normal electric displacement has been obtained because it was a priori modeled in the proposed variational statement.

The relative sensor cases, in which a uniform mechanical load has been applied to the top of an isotropic multilayered piezoelectric plate and a composite multilayered piezoelectric plate, are illustrated in Tables 9.19 and 9.20, respectively. The electromechanical variables have been evaluated in the middle of the plate using a LM4($\boldsymbol{u}$, Φ, $\boldsymbol{\sigma}_n$, $\boldsymbol{D}_n$) model as MUL2 theory; the values are closer to the quasi-3D solution when the mesh increases (until 21×21). No further considerations are obtained compared to the relative actuator case. Interlaminar continuity is ensured for each electromechanical variable. Figure 9.25 shows the transverse shear stress, through the thickness direction z, for the thick isotropic multilayered piezoelectric plate and the electric potential along the z direction for the thin composite multilayered piezoelectric structure. The employed theory is a LM4($\boldsymbol{u}$, Φ, $\boldsymbol{\sigma}_n$, $\boldsymbol{D}_n$) model with a 21×21 mesh; these features ensure a quasi-3D description of such variables through the thickness

Table 9.19 Sensor case: thick and thin isotropic plate in bending (uniform mechanical load $p_z = 1$ Pa applied at the top). FE solution provided in the middle of the plate.

	$a/h = 4$			$a/h = 50$		
	MUL2 7×7	MUL2 15×15	MUL2 21×21	MUL2 7×7	MUL2 15×15	MUL2 21×21
$w(h/2)$ $[10^{-12}$ m]	1.5876	1.6105	1.6136	2312.7	2357.1	2363.2
$\Phi(2h/5)^t$ $[10^{-3}$ V]	0.0545	0.0554	0.0553	0.8437	0.8527	0.8540
$\Phi(2h/5)^b$ $[10^{-3}$ V]	0.0545	0.0554	0.0553	0.8437	0.8527	0.8540
$\sigma_{yy}(h/2)$ [Pa]	5.7633	5.6740	5.6389	833.33	829.25	828.46
$\sigma_{zz}(2h/5)^t$ [Pa]	0.9953	0.9712	0.9701	0.9924	0.9704	0.9702
$\sigma_{zz}(2h/5)^b$ [Pa]	0.9953	0.9712	0.9701	0.9924	0.9704	0.9702
$\mathcal{D}_z(2h/5)^t$ $[10^{-13}$ C/m^2]	−0.3947	−0.5013	−0.5115	−0.5820	−0.5385	−0.5377
$\mathcal{D}_z(2h/5)^b$ $[10^{-13}$ C/m^2]	−0.3947	−0.5013	−0.5115	−0.5820	−0.5385	−0.5377

Table 9.20 Sensor case: thick and thin composite plate in bending (uniform mechanical load $p_z = 1$ Pa applied at the top). FE solution provided in the middle of the plate.

	$a/h = 4$			$a/h = 50$		
	MUL2 7×7	MUL2 15×15	MUL2 21×21	MUL2 7×7	MUL2 15×15	MUL2 21×21
$w(h/2)$ $[10^{-12}$ m]	4.4415	4.4838	4.4894	3409.3	3479.4	3488.9
$\Phi(0)^t$ $[10^{-3}$ V]	0.0861	0.0873	0.0875	1.2430	1.2585	1.2606
$\Phi(0)^b$ $[10^{-3}$ V]	0.0861	0.0873	0.0875	1.2430	1.2585	1.2606
$\sigma_{yy}(h/2)$ [Pa]	8.7359	8.4014	8.2450	1079.7	1075.0	1073.8
$\sigma_{zz}(0)^t$ [Pa]	0.5222	0.5092	0.5058	0.5142	0.5001	0.5000
$\sigma_{zz}(0)^b$ [Pa]	0.5222	0.5092	0.5058	0.5142	0.5001	0.5000
$\mathcal{D}_z(0)^t$ $[10^{-13}$ C/m^2]	0.5803	0.5752	0.5752	0.5512	0.5475	0.5473
$\mathcal{D}_z(0)^b$ $[10^{-13}$ C/m^2]	0.5803	0.5752	0.5752	0.5512	0.5475	0.5473

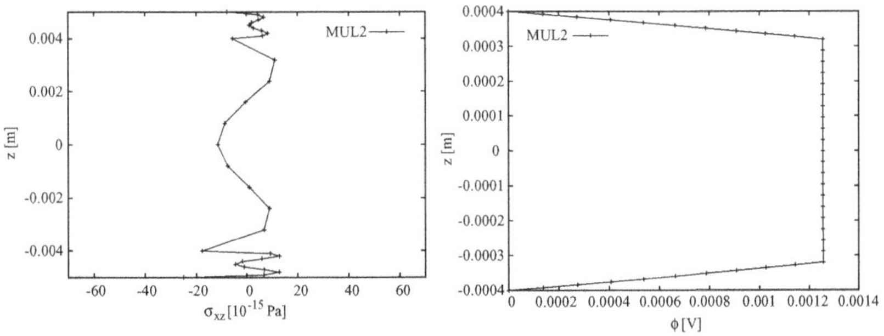

Figure 9.25 Sensor case: plate with uniform mechanical load $p_z = 1$ Pa applied at the top. FE solution provided in the middle of the plate. σ_{xz} vs. z on the left (isotropic plate with $a/h = 4$) and Φ vs. z on the right (composite plate with $a/h = 50$). Mesh 21×21.

and their interlaminar continuity. Both the transverse shear stress and electric potential satisfy the boundary loading conditions which are, for the proposed sensor case, $\sigma_{xz} = 0$ and $\Phi = 0$ at the top and bottom surfaces of the plate.

Table 9.21 gives the first six vibration modes for the simply supported plate in an open-circuit configuration. The thick plate ($a/h = 4$) is an isotropic multilayered piezoelectric one, while the thin plate ($a/h = 50$) is a composite multilayered piezoelectric structure. The convergence depends on the considered vibration mode; the 19×19 mesh is sufficient for low values of frequency but this mesh could be inappropriate for higher frequencies. However, the opportune choice of the 2D theory (the LM4($\boldsymbol{u}$, Φ, $\boldsymbol{\sigma}_n$, $\boldsymbol{D}_n$) model as MUL2

Table 9.21 Open-circuit plate configuration, first six modes for FE analysis. Circular frequency $\omega = 2\pi f$. Thick isotropic plate ($a/h = 4$) and thin composite plate ($a/h = 50$).

	$a/h = 4$			$a/h = 50$		
	MUL2 3×3	MUL2 7×7	MUL2 19×19	MUL2 3×3	MUL2 7×7	MUL2 19×19
First mode	163 431	152 013	149 866	16 382.3	13 655.7	13 596.8
Second mode	226 517	218 362	216 787	66 010.3	36 302.6	35 848.8
Third mode	226 517	218 362	216 787	66 010.3	36 302.6	35 849.0
Fourth mode	377 208	328 127	283 111	97 071.0	56 557.9	55 176.7
Fifth mode	416 535	338 610	311 258	160 684	79 561.4	76 178.2
Sixth mode	416 536	351 890	347 236	160 684	79 561.8	76 178.7

theory in this case) and the mesh size depend to a great extent on the considered thickness ratio, stacking layer sequence, and vibration modes (low- or high-frequency values).

References

Carrera E 2002 Theories and finite elements for multilayered anisotropic, composite plates and shells. *Arch. Comput. Methods Eng.* **9**, 87–140.

Carrera E and Boscolo M 2007 Classical and mixed finite elements for static and dynamic analysis of piezoelectric plates. *Int. J. Numer Methods Eng.* **70**, 1135–1181.

Carrera E and Brischetto S 2007a Piezoelectric shell theories with "a priori" continuous transverse electromechanical variables. *J. Mech. Mater. Struct.* **2**, 377–398.

Carrera E and Brischetto S 2007b Reissner mixed theorem applied to static analysis of piezoelectric shells. *J. Int. Mater. Syst. Struct.* **18**, 1083–1107.

Carrera E and Brischetto S 2008a Analysis of thickness locking in classical, refined and mixed multilayered plate theories. *Comp. Struct.* **82**, 549–562.

Carrera E and Brischetto S 2008b Analysis of thickness locking in classical, refined and mixed theories for layered shells. *Comp. Struct.* **85**, 83–90.

Carrera E, Boscolo M, and Robaldo A 2007 Hierarchic multilayered plate elements for coupled multifield problems of piezoelectric adaptive structures: formulation and numerical assessment. *Arch. Comput. Methods Eng.* **14**, 383–430.

Carrera E, Brischetto S, and Nali P 2008 Variational statements and computational models for multifield problems and multilayered structures. *Mech. Adv. Mater. Struct.* **15**, 182–198.

Carrera E, Brischetto S, and Cinefra M 2010 Variable kinematics and advanced variational statements for free vibrations analysis of piezoelectric plates and shells. *Comput. Model. Eng. Sci.* **65**, 259–341.

Ikeda T 1996 *Fundamentals of Piezoelectricity*. Oxford University Press.

Kirchhoff G 1850 Über das Gleichgewicht und die Bewegung einer elastischen Scheibe. *J. Reine Angew. Math.* **40**, 51–88.

Librescu L 1975 *Elasto Static and Kinetics of Anisotropic and Heterogeneous Shell-type Structures*. Nordhoff International.

Mindlin RD 1951 Influence of rotatory inertia and shear on flexural motions of isotropic, elastic plates. *J. Appl. Mech.* **18**, 31–38.

Reddy JN 2004 *Mechanics of Laminated Composite Plates and Shells: Theory and Analysis*. CRC Press.

Reissner E 1945 The effect of transverse shear deformation on the bending of elastic plates. *J. Appl. Mech.* **12**, 69–77.

Sokolnikoff IS 1956 *Mathematical Theory of Elasticity*. McGraw-Hill.

Washizu K 1968 *Variational Methods in Elasticity and Plasticity*. Pergamon Press.

Index

Plates and Shells for Smart Structures: Classical and Advanced Theories for Modeling and Analysis, First Edition.
Erasmo Carrera, Salvatore Brischetto and Pietro Nali.